Advances in Biological Research into Shrimps, Crabs and Lobsters

Advances in Biological Research into Shrimps, Crabs and Lobsters

Guest Editors

Lutz Auerswald
Xianliang Meng
Yafei Duan

MDPI

Basel • Beijing • Wuhan • Barcelona • Belgrade • Novi Sad • Cluj • Manchester

Guest Editors

Lutz Auerswald
Department of Animal Sciences
Stellenbosch University
Stellenbosch
South Africa

Xianliang Meng
Yellow Sea Fisheries Research Institute
Chinese Academy of Fishery Sciences
Qingdao
China

Yafei Duan
South China Sea Fisheries Research Institute
Chinese Academy of Fishery Sciences
Guangzhou
China

Editorial Office
MDPI AG
Grosspeteranlage 5
4052 Basel, Switzerland

This is a reprint of the Special Issue, published open access by the journal *Biology* (ISSN 2079-7737), freely accessible at: https://www.mdpi.com/journal/biology/special_issues/4T9X0P039J.

For citation purposes, cite each article independently as indicated on the article page online and as indicated below:

Lastname, A.A.; Lastname, B.B. Article Title. *Journal Name* **Year**, *Volume Number*, Page Range.

ISBN 978-3-7258-5225-3 (Hbk)
ISBN 978-3-7258-5226-0 (PDF)
https://doi.org/10.3390/books978-3-7258-5226-0

Contents

Preface

This Special Issue, "Advances in Biological Research into Shrimps, Crabs and Lobsters", presents a curated collection of seventeen research articles that highlight the dynamism and breadth of modern crustacean science. Crustaceans are a group of profound ecological and economic importance, and understanding their complex biology is fundamental to their conservation and the sustainable management of global fisheries and aquaculture in an era of environmental change.

The scope of this issue is intentionally broad, designed to bridge the gap between fundamental and applied research. The articles within explore a vast range of topics, from molecular genetics, phylogenetics, and the neuroendocrine control of life processes to the physiological responses of crustaceans to environmental stressors like hypercapnia, ammonia, and salinity fluctuations. The collection also addresses practical challenges in fisheries and aquaculture, including the development of DNA markers for disease tolerance, the optimization of cryopreservation techniques for germplasm banking, and the assessment of fishery dynamics and stock health.

The primary motivation for this work is to consolidate recent advancements and provide a valuable resource for the scientific community. By examining organisms from shrimp and crabs to lobsters and crayfish, these studies collectively advance our knowledge and offer new avenues for research. This Special Issue is addressed to a wide audience, including researchers, academics, students, fishery managers, and conservation practitioners. It is our hope that this synthesis will not only inform but also inspire further investigation into the intricate and vital world of crustaceans.

Lutz Auerswald, Xianliang Meng, and Yafei Duan
Guest Editors

biology

MDPI

Editorial

From Genes to Fisheries: A Synthesis of Current Research in Crustacean Biology and Management

Yafei Duan [1,2,3], Lutz Auerswald [4] and Xianliang Meng [1,2,*]

1 State Key Laboratory of Mariculture Biobreeding and Sustainable Goods, Yellow Sea Fisheries Research Institute, Chinese Academy of Fishery Sciences, Qingdao 266071, China; duanyafei89@163.com
2 Laboratory for Marine Fisheries Science and Food Production Processes, Qingdao Marine Science and Technology Center, Qingdao 266237, China
3 Key Laboratory of South China Sea Fishery Resources Exploitation & Utilization, Ministry of Agriculture and Rural Affairs, South China Sea Fisheries Research Institute, Chinese Academy of Fishery Sciences, Guangzhou 510300, China
4 Department of Animal Sciences, Stellenbosch University, Stellenbosch 7600, South Africa; lutz.auerswald@gmail.com
* Correspondence: xlmeng@ysfri.ac.cn

Crustaceans represent an extraordinarily diverse group of arthropods, occupying a vast array of aquatic and terrestrial habitats worldwide [1]. Their ecological significance is profound, influencing food web dynamics and nutrient cycling, and serving as crucial indicators of ecosystem health [2,3]. Economically, many species are pillars of global fisheries and the rapidly expanding aquaculture sector [4]. A deep understanding of their complex biology—from their genetic architecture and physiological responses to environmental stresses and anthropogenic impacts—is essential for their effective conservation and sustainable management, particularly in the context of global environmental change.

This collection of 17 research articles presents recent advancements across the broad spectrum of crustacean biology. These contributions, from a multitude of researchers, highlight the dynamic nature of the field and the innovative approaches being employed to address both fundamental biological questions and pressing applied challenges. The studies explore critical areas, including molecular genetics and evolution, physiological adaptations to environmental stressors, innovations in aquaculture techniques, neuroendocrine control of vital life processes, fisheries science, and sensory biology.

Significant strides have been made in understanding the genetic underpinnings of crustacean life. Modern molecular tools have provided unprecedented insights into phylogenetic relationships, the genetic diversity within and between populations, and the impacts of genetic factors such as inbreeding on viability and performance [5–9]. These genetic studies are not only crucial for evolutionary biology but also form the bedrock for developing genetically improved strains for aquaculture, focusing on traits such as disease resistance and stress tolerance.

A substantial body of research is dedicated to elucidating how crustaceans respond and adapt to a changing environment. Investigations into the physiological and molecular responses to stressors such as fluctuating salinity, ammonia exposure, and hypercapnia have revealed the complex mechanisms involved in maintaining homeostasis [10–16]. Transcriptomic, proteomic, and metabolomic analyses have identified the key genes, pathways, and molecular markers associated with stress adaptation, offering a deeper understanding of crustacean resilience and vulnerability.

The aquaculture of crustaceans plays a crucial role in global food production, and research in this area is focused on enhancing sustainability and efficiency. Studies have explored the efficacy of novel dietary supplements, including seaweeds and probiotics, to improve growth, health, and disease resistance [17–19]. Concurrently, advancements in reproductive technologies, such as sperm cryopreservation, are critical for establishing germplasm banks, facilitating selective breeding programs, and overcoming challenges related to asynchronous maturation [20,21].

The intricate neuroendocrine systems that regulate crucial life-cycle events in crustaceans, such as molting and reproduction, continue to be an active area of investigation [22–25]. Identifying neuropeptides and their receptors and understanding their roles in these physiological processes can lead to new strategies for manipulating growth and reproduction in cultured species.

Furthermore, research into the dynamics of wild crustacean populations and their sensory biology provides essential data for informed fisheries management and conservation strategies [26]. Studies assessing stock status, life history traits, and the impacts of fishing pressure are vital for ensuring the long-term viability of commercially important fisheries. Investigations into sensory systems, such as statocysts, contribute to our understanding of how crustaceans perceive and interact with their environment.

To provide a comprehensive overview and a quick reference for readers to navigate the diverse topics covered in this Special Issue, Table 1 summarizes the 17 published studies, categorizing them by their primary research focus and highlighting their key findings and contributions.

Table 1. Summary of 17 research articles on crustacean biology, highlighting species studied, primary research focus, key findings/contribution, and authors.

Species Studied	Primary Research Focus	Key Finding/Contribution	Author(s)
Fenneropenaeus chinensis (shrimp)	Elimination effects of inbreeding on genotype frequency in larval stages.	Showed significant inbreeding depression in larval survival, with selection favoring heterozygotes and increasing population heterozygosity post-elimination.	Fu et al. [5]
Fenneropenaeus merguiensis (shrimp)	Nutritional value, antioxidant potential, and genetic diversity from three Chinese regions.	Identified a population with superior nutritional quality (ash, protein, EAAs) and relatively higher genetic diversity, highlighting its potential for aquaculture.	Li et al. [7]
Litopenaeus vannamei (shrimp)	Nutritional composition, physiological indicators, and genetic diversity across different aquaculture populations.	Found consistent nutritional content but significant differences in genetic diversity in different aquaculture populations.	Li et al. [6]
Litopenaeus vannamei (shrimp)	Genome-wide identification and expression of neuropeptides; expression patterns after CHH gene RNAi.	Identified 125 neuropeptide-encoding genes, noted gene family expansions (ACP, CHH, and PDH), and found most neuropeptide genes downregulated after CHH/VIH silencing.	Zhang et al. [23]
Litopenaeus vannamei (shrimp)	Development of DNA markers for Acute Hepatopancreatic Necrosis Disease (AHPND) tolerance using GWAS.	Identified four candidate SNPs and 17 InDels associated with AHPND tolerance using DArT sequencing.	Whankaew et al. [27]
Metapenaeus monoceros (shrimp)	Life history traits and fishery dynamics along the Saudi Arabian Red Sea coast.	Estimated growth parameters, indicated overfishing, and suggested current fishing pressure reduces spawning stock biomass to 23% of unexploited levels.	Gireesh et al. [28]
Penaeus monodon (shrimp)	Metabolic response to acute ammonia nitrogen stress.	Revealed severe tissue damage, conversion of ammonia to urea for detoxification, increased SOD activity, and inhibited caspase activity, indicating complex metabolic and stress responses.	Ding et al. [10]

Table 1. *Cont.*

Species Studied	Primary Research Focus	Key Finding/Contribution	Author(s)
Penaeus monodon (shrimp)	Regulatory role of microRNAs under moderately low salinity stress.	Identified differentially expressed miRNAs in hepatopancreas after low salinity exposure, targeting genes in metabolism, immune response, and stress adaptation pathways.	Shi et al. [13]
Penaeus monodon (shrimp)	Effects of dietary *Gracilaria lichenoides* and *Bacillus amyloliquefaciens* on growth, antioxidant capacity, and intestinal health.	G. lichenoides supplementation significantly improved growth performance; additives enhanced lipase activity and antioxidant gene expression and modulated intestinal microbiota.	Tian et al. [17]
Penaeus monodon (shrimp)	Optimization of sperm cryopreservation protocol.	Optimized protocol using 10% DMSO with natural seawater, programmable freezing, and 37 °C thawing achieved 85% sperm viability after 15 days with minor acrosomal damage.	Kong et al. [20]
Procambarus clarkii (crayfish)	Transcriptomic responses to different salinity stress.	Revealed metabolic pathways as primary responses, differential expression of immune/antioxidant genes, and specific signaling pathways (Foxo, Wnt, Hippo, and Notch) activated under high-salinity.	Luo et al. [11]
Jasus lalandii (lobster)	Influence of hypercapnia on physiology of ovigerous females and embryonic development.	Berried females efficiently respond to acute and chronic hypercapnia by increasing hemolymph bicarbonate; embryo growth and development are not impacted by chronic hypercapnia.	Ritter et al. [12]
Nephrops norvegicus (lobster)	Statocyst ultrastructure.	Described the fine morphology of the statocyst sensory setae using SEM and TEM, contributing to understanding sound perception.	Solé [29]
genus Panulirus (lobsters)	Shifts in catch composition in the artisanal spiny lobster fishery.	Revealed considerable shifts in catch composition of Spiny lobsters over time compared to 1970s data, highlighting changes in species dominance in Kenya.	Kulmiye [26]
Portunus trituberculatus (crab)	Role of corazonin in stimulating ecdysteroid synthesis during molting.	Demonstrated that PtCrz peptide stimulates ecdysteroid levels and related gene expression in vitro and in vivo, potentially via PtCrzR and affecting PtETH expression.	Xie et al. [22]
Portunus trituberculatus (crab)	Comparative transcriptome analysis of response to *Vibrio parahaemolyticus* and low-salinity stress.	Identified differentially expressed genes involved in ion transport, immunoregulation, apoptosis, and Hippo signaling pathway under dual stress conditions.	Sun et al. [14]
genus Tuerkayana (crabs)	Molecular phylogeny and evolution based on whole mitochondrial genome sequences.	Provided mitochondrial evidence for *Tuerkayana*, clarified Gecarcinidae genera, and identified selective pressure in the *nad6* gene.	Wang et al. [30]

In conclusion, this Special Issue provides a valuable and timely overview of the current state of crustacean biology. The breadth of topics and the depth of investigation highlight the vibrancy of this field. The collective findings underscore the necessity of continued research, fostering integrative and interdisciplinary approaches. Such endeavors are crucial for advancing our fundamental knowledge of these fascinating organisms and developing scientifically informed strategies to ensure the sustainable management of crustacean populations and their ecosystems in the face of ongoing and future challenges.

Conflicts of Interest: The authors declare no conflicts of interest.

References

1. Giribet, G.; Edgecombe, G.D. Reevaluating the arthropod tree of life. *Annu. Rev. Entomol.* **2012**, *57*, 167–186. [CrossRef]
2. Robinson, J.M.; Barnes, A.D.; Fickling, N.; Costin, S.; Sun, X.; Breed, M.F. Food webs in food webs: The micro–macro interplay of multilayered networks. *Trends Ecol. Evol.* **2024**, *39*, 913–922. [CrossRef]
3. Flecker, A.S.; Twining, C.W.; Schmitz, O.J.; Cooke, S.J.; Hammerschlag, N. Aquatic predators influence micronutrients: Important but understudied. *Trends Ecol. Evol.* **2019**, *34*, 882–883. [CrossRef] [PubMed]
4. Bostock, J.; McAndrew, B.; Richards, R.; Jauncey, K.; Telfer, T.; Lorenzen, K.; Little, D.; Ross, L.; Handisyde, N.; Gatward, I.; et al. Aquaculture: Global status and trends. *Philos. Trans. R. Soc. B Biol. Sci.* **2010**, *365*, 2897–2912. [CrossRef]
5. Fu, Q.; Zhou, J.; Luan, S.; Dai, P.; Lyu, D.; Chen, B.; Luo, K.; Kong, J.; Meng, X. Analysis of elimination effects of inbreeding on genotype frequency in larval stages of Chinese Shrimp. *Biology* **2024**, *13*, 268. [CrossRef]
6. Li, Y.; Cao, S.; Jiang, S.; Huang, J.; Yang, Q.; Jiang, S.; Yang, L.; Zhou, F. Comparative study of nutritional composition, physiological indicators, and genetic diversity in *Litopenaeus vannamei* from different aquaculture populations. *Biology* **2024**, *13*, 722. [CrossRef] [PubMed]
7. Li, Y.; Chen, J.; Cao, S.; Jiang, Z.; Jiang, S.; Yang, Q.; Yang, L.; Huang, J.; Shi, J.; Ma, Z.; et al. A comprehensive assessment of the nutritional value, antioxidant potential, and genetic diversity of *Fenneropenaeus merguiensis* from three different regions in China. *Biology* **2024**, *13*, 1002. [CrossRef] [PubMed]
8. Richter, S.; Møller, O.; Wirkner, C. Advances in crustacean phylogenetics. *Arthropod. Syst. Phylogeny* **2009**, *67*, 275–286. [CrossRef]
9. Peres, P.A.; Mantelatto, F.L. Macrogenetics approach reveals spatial trends and drivers of mitochondrial genetic diversity at different biological organization levels in tropical Western Atlantic decapods. *Ecol. Evol.* **2025**, *15*, e71372. [CrossRef]
10. Ding, Y.; Jiang, S.; Jiang, S.; Li, Y.; Yang, Q.; Yang, L.; Huang, J.; Shi, J.; Li, P.; Diao, H.; et al. Metabolic response of black tiger shrimp (*Penaeus monodon*) to acute ammonia nitrogen stress. *Biology* **2025**, *14*, 501. [CrossRef]
11. Luo, L.; Yang, L.S.; Huang, J.H.; Jiang, S.G.; Zhou, F.L.; Li, Y.D.; Jiang, S.; Yang, Q.B. Effects of different salinity stress on the transcriptomic responses of freshwater crayfish (*Procambarus clarkii*, Girard, 1852). *Biology* **2024**, *13*, 530. [CrossRef]
12. Ritter, A.; Bridges, C.R.; Auerswald, L. Investigation of the influence of hypercapnia on the physiology of ovigerous West Coast rock lobsters, *Jasus lalandii*, and their embryonic development. *Biology* **2025**, *14*, 132. [CrossRef] [PubMed]
13. Shi, J.; Jiang, S.; Ding, Y.; Diao, H.; Li, W.; Li, Y.; Huang, J.; Yang, L.; Yang, Q.; Zhou, F. Insights into the regulatory role of microRNAs in *Penaeus monodon* under moderately low salinity stress. *Biology* **2025**, *14*, 440. [CrossRef] [PubMed]
14. Sun, D.; Lv, J.; Li, Y.; Wu, J.; Liu, P.; Gao, B. Comparative transcriptome analysis of the response to *Vibrio parahaemolyticus* and low-salinity stress in the swimming crab *Portunus trituberculatus*. *Biology* **2023**, *12*, 1518. [CrossRef] [PubMed]
15. Wang, D.; Liu, X.; Zhang, J.; Gao, B.; Liu, P.; Li, J.; Meng, X. Identification of neuropeptides using long-read RNA-Seq in the swimming crab *Portunus trituberculatus*, and their expression profile under acute ammonia stress. *Front. Physiol.* **2022**, *13*, 910585. [CrossRef]
16. Lu, Y.; Liu, Y.; Cao, J.; Zhang, Y.; Zheng, Y.; Wang, F. Waterborne ammonia toxicity damages crustacean hemocytes via lysosome-dependent autophagy: A case study of swimming crabs *Portunus trituberculatus*. *Environ. Res.* **2025**, *272*, 120985. [CrossRef]
17. Tian, J.; Wang, Y.; Huang, J.; Yan, H.; Duan, Y.; Wang, J.; Zhou, C.; Huang, Z. Effects of dietary *Gracilaria lichenoides* and *Bacillus amyloliquefaciens* on growth performance, antioxidant capacity, and intestinal health of *Penaeus monodon*. *Biology* **2024**, *13*, 252. [CrossRef]
18. Li, W.; Li, E.; Wang, S.; Liu, J.; Wang, M.; Wang, X.; Qin, C.; Qin, J.; Chen, L. Comparative effects of four feed attractants on growth, appetite, digestion and absorption in juvenile Chinese mitten crab, *Eriocheir sinensis*. *Aquaculture* **2025**, *594*, 741441. [CrossRef]
19. Shen, Q.; Li, S.; Zhang, S.; Xu, J.; Chen, H.; Luo, Q.; Yang, R.; Chen, J. The impact of *Neoporphyra haitanensis* dietary supplement on astaxanthin esters and fatty acids accumulation associated with immune promotion in Pacific white shrimp, *Litopenaeus vannamei*. *Aquaculture* **2024**, *593*, 741347. [CrossRef]
20. Kong, D.; Jiang, S.; Shi, J.; Yang, Q.; Huang, J.; Li, Y.; Ding, Y.; Wang, J.; Qi, X.; Liu, T.; et al. Cryopreservation protocol optimization for *Penaeus monodon* sperm: Reagent screening and parameter refinement. *Biology* **2025**, *14*, 408. [CrossRef]
21. Rakbanjong, N.; Okutsu, T.; Chotigeat, W.; Songnui, A.; Wonglapsuwan, M. Cryopreservation of germ cells of banana shrimp, *Fenneropenaeus merguiensis* and black tiger shrimp, *Penaeus monodon*. *Mar. Biotechnol.* **2021**, *23*, 590–601. [CrossRef] [PubMed]
22. Xie, X.; Zhang, J.; Tu, S.; Zhou, Q.; Zhu, D. Corazonin stimulates ecdysteroid synthesis during the molting process of the swimming crab, *Portunus trituberculatus*. *Biology* **2024**, *13*, 630. [CrossRef] [PubMed]

23. Zhang, L.; Sun, L.; Song, G.; Wang, B.; Cui, Y.; Liu, F.; Li, Y.; Wang, Z. Genome-wide identification and expression of neuropeptides and their expression patterns after RNAi of CHH genes in Pacific white shrimp *Litopenaeus vannamei*. *Biology* **2024**, *13*, 1038. [CrossRef] [PubMed]
24. Wang, M.; Ye, H.; Miao, L.; Li, X. Role of short neuropeptide F in regulating eyestalk neuroendocrine systems in the mud crab *Scylla paramamosain*. *Aquaculture* **2022**, *560*, 738493. [CrossRef]
25. Wang, D.; Liu, X.; Shang, Y.; Yu, X.; Gao, B.; Lv, J.; Li, J.; Liu, P.; Li, J.; Meng, X. Ammonia stress disturbs moult signaling in juvenile swimming crab *Portunus trituberculatus*. *Biology* **2023**, *12*, 409. [CrossRef]
26. Kulmiye, A.J. Evidence of considerable shifts in catch composition in the artisanal spiny lobster fishery in Kenya. *Biology* **2023**, *12*, 1477. [CrossRef]
27. Whankaew, S.; Suksri, P.; Sinprasertporn, A.; Thawonsuwan, J.; Sathapondecha, P. Development of DNA markers for acute hepatopancreatic necrosis disease tolerance in *Litopenaeus vannamei* through a genome-wide association study. *Biology* **2024**, *13*, 731. [CrossRef]
28. Gireesh, S.; Tıra¸sın, E.M.; Muthu Palani, G.B.; Charles, S.K.; Kadengal, S.T.; Santucci, R.G.; Garla, R.C.; Okba, Z.; Adam, A.M.S.; Dimech, M. Life history traits and fishery dynamics of speckled shrimp, *Metapenaeus monoceros* (Fabricius, 1798), along the Saudi Arabian Red Sea coast. *Biology* **2025**, *14*, 406. [CrossRef]
29. Solé, M. Statocyst ultrastructure in the Norwegian lobster (*Nephrops norvegicus*). *Biology* **2024**, *13*, 325. [CrossRef]
30. Wang, Z.; Zheng, Y.; Zhao, X.; Xu, X.; Xu, Z.; Cui, C. Molecular phylogeny and evolution of the *Tuerkayana* (Decapoda: Brachyura: Gecarcinidae) genus based on whole mitochondrial genome sequences. *Biology* **2023**, *12*, 974. [CrossRef]

biology

MDPI

Article

Molecular Phylogeny and Evolution of the *Tuerkayana* (Decapoda: Brachyura: Gecarcinidae) Genus Based on Whole Mitochondrial Genome Sequences

Zhengfei Wang [1,*,†], Yuqing Zheng [1,†], Xinyue Zhao [1], Xinyi Xu [1,2], Zhiwen Xu [1] and Chong Cui [1,3]

1 Jiangsu Key Laboratory for Bioresources of Saline Soils, Jiangsu Synthetic Innovation Center for Coastal Bio-Agriculture, Jiangsu Provincial Key Laboratory of Coastal Wetland Bioresources and Environmental Protection, School of Wetlands, Yancheng Teachers University, Yancheng 224001, China; 19895621808@163.com (Y.Z.); 15251075879@163.com (X.Z.); 18082165776@163.com (X.X.); 15722573182@163.com (Z.X.); ccyyznlxsa@163.com (C.C.)

2 Ministry of Education Key Laboratory of Contemporary Anthropology, Department of Anthropology and Human Genetics, School of Life Sciences, Fudan University, Shanghai 200000, China

3 College of Life Sciences, Henan Normal University, Xinxiang 453007, China

* Correspondence: wangzf01@yctu.edu.cn; Tel./Fax: +86-515-88233587

† These authors contributed equally to this work.

Simple Summary: The current classification system of the Brachyura based on complete mitochondrial genomes lacks representative species for many genera and even families. This leads to a confusing and incomplete taxonomy within the Brachyura suborder. The target species in this study belong to Gecarcinidae: *Tuerkayana*, which is an intriguing genus proposed in recent years but lacks complete mitochondrial genome phylogenetic evidence. In our research, we sequenced the complete mitochondrial genomes of four species and compared with the existing dataset of 202 mitochondrial genomes of crabs. Our analyses provide mitochondrial evidence for *Tuerkayana* and shedding light on the division of genera within Gecarcinidae. Furthermore, using the dataset of 206 crab mitochondrial genomes to examine selective pressure in individual codons, and the selective pressure in the *nad6* gene, suggesting its potential role in the evolutionary history of Gecarcinidae.

Abstract: *Tuerkayana* is of particular interest because it has been separated, in recent years, from *Cardisoma* and *Discoplax* but studies of its taxonomic status, especially from a whole mitochondrial genome perspective, have been lacking. In this study, the mitogenomes of four species (*Tuerkayana magnum*, *Tuerkayana rotundum*, *Tuerkayana hirtipes*, and *Tuerkayana celeste*) of *Tuerkayana* are sequenced and contrasted with other species in Brachyura for the first time. The phylogenetic tree of Brachyura, which includes 206 crab species (189 species of Brachyuran and 17 Anomura species) with a complete mitogenome, was constructed to evaluate the phylogenetic position of *Tuerkayana* and Gecarcinidae within Brachyuran, and explore the monophyly of Gecarcinidae. Furthermore, two single gene trees based on *cox1* and *16SrRNA* separately within interspecies of Gecarcinidae were reconstructed, providing molecular evidence for *Tuerkayana* and further clarifying the division of genera in Gecarcinidae. Based on the mitogenome dataset of 206 crabs, the branch-site model was utilized to explore selective pressure in individual codons with CodeML. The strong selective pressure shown in *nad6* indicates that it may have played a significant role in the evolution of Gecarcinidae.

Keywords: Decapoda; mitogenome; phylogeny; selective pressure; *Tuerkayana*

1. Introduction

Due to its rapid evolutionary rate, maternal inheritance, small size, conserved gene content, and relatively easy acquisition, the mitochondrial genome (mitogenome) plays a key role in evolutionary studies [1]. The complete mitochondrial sequence contains more biological information than single gene molecular markers and can effectively avoid the interference of nuclear pseudogenes [2]. The mitogenomes within metazoan are generally 14–20 kb in length, comprising two ribosomal RNA genes (*12SrRNA* and *16SrRNA*), 22 transfer RNA genes (tRNAs), 13 protein-coding genes (PCGs), and 1 control region (CR, or D-loop region). Complete mitogenome sequences are usually utilized in phylogenetic research, comparative genomics, population genetics, and molecular evolution [3]. The important information provided by mitogenome data regarding sequences and genomic rearrangements, which are generally complex and diverse, partly reflects the evolution of animals [4].

At present, Gecarcinidae is considered to have 7 genera (*Cardisoma*, *Discoplax*, *Gecarcinus*, *Gecarcoidea*, *Johngarthia*, *Epigrapsus*, and *Tuerkayana*) and 26 species. The main object of our study, *Tuerkayana*, is a newly established genus in recent years [5], which includes five species separated from *Discoplax* (*Tuerkayana celeste* and *Tuerkayana magnum*), *Cardisoma* (*Tuerkayana rotundum*, *Tuerkayana hirtipes*), and *Tuerkayana latens*. However, the taxonomic status of the genus, especially from a whole mitochondrial genome perspective, has been lacking. Mitogenome information in Gecarcinidae is available for only four sequenced species belonging to genera *Cardisoma* (*Cardisoma armatum* and *Cardisoma carnifex*) and *Gecarcoidea* (*Gecarcoidea lalandii* and *Gecarcoidea natalis*), and other genera still lack representative species. This situation, with the phylogeny not yet resolved and requiring further investigation, has caused difficulties in confirming the monophyly of *Tuerkayana*.

In addition, beyond that, we have taken note of the adaptability that may be manifested by mitochondrial selection pressure. Gecarcinidae crabs, a semi-terrestrial family in the Brachyura group, are generally distributed in coastal burrows, rock crevices, and coastal thickets [6,7]. Adult crabs of Gecarcinidae are generally terrestrial, but they seasonally spawn in seawater, and their larva grows in water prior to maturity [8]. For example, the Christmas Island red crab (*Gecarcoidea natalis*) migrates yearly to the coast for breeding [9]. On account of the terrestrial characteristics of Gecarcinidae, many studies have been published regarding its morphological structure, reproduction, tolerance level, feeding habit, and other aspects [9–13]. Evidence of periodic transitions from marine to terrestrial environments suggests land crabs could be the result of single adaptive radiation from a marine ancestor which invaded terrestrial habitats [14]. Land crabs may therefore be hypoxic-adapted [15]. Oxygen plays a central role in the mitochondrial respiratory chain during the OXPHOS process, acting as the terminal electron acceptor of the electron transport chain (ETC) and enabling the synthesis of adenosine triphosphate (ATP) [16]. Accompanying inflation and a posterior narrowing of the carapace, the reduction of gill filaments, and the occurrence of branchiostegal lungs [10,17], hypoxia may have imposed important selective constraints on the evolution of the mitogenome in terrestrial crabs.

T. magnum, *T. rotundum*, *T. hirtipes*, and *T. celeste* are distributed in the Indo–West Pacific region and are highly terrestrial, usually settling in the vegetation above shorelines and shallow burrows near freshwater. In this study, the complete mitogenome sequences of four species in *Tuerkayana* were determined and analyzed for the first time. To ensure the conclusions we reached were reliable, we used a variety of existing complete mitogenomes in this study and compared their molecular features with *Tuerkayana*. The phylogenetic tree of Brachyuran was reconstructed, based on the sequences of 13 PCGs from 206 species, to analyze the selection pressure of 13 PCGs. The reconstructed phylogenetic tree of Brachyuran enables the division of genera in the family Gecarcinidae to be better understood. The tree was constructed to provide evidence for the establishment of *Tuerkayana* at the mitochondrial level. In addition, based on the results of the selection pressure analysis, the influence of the terrestrial adaptation of Gecarcinidae in the mitochondrial genes was investigated.

2. Materials and Methods

2.1. Samples and the Extraction of DNA

Two samples of each of the four species (*T. magnum*, *T. rotundum*, *T. hirtipes*, and *T. celeste*) were purchased from Indonesia. The samples entered China in accordance with the Animal Quarantine Law of the People's Republic of China and the Animal Quarantine Control Measures. One of the samples was stored as a copy in the freezer at −80 °C. The other was dissected for muscle samples, which were flash-frozen under liquid nitrogen and stored in the freezer at −80 °C in Yancheng Teachers University for extracting the total genomic DNA as per the manufacturer's instructions of the Aidlab Genomic DNA Extraction Kit (Aidlab Biotech, Beijing, China); 1% agarose gel electrophoresis (0.2 g agarose, 20 mL 1 × TAE, and 2 μL EB) was used to evaluate the quality of the extracted DNA samples, which were stored at −20 °C for further polymerase chain reaction (PCR) if they were of sufficiently high quality.

2.2. PCR Amplification and Sequencing

The conserved sequences were amplified by PCR using the universal primer *cox1* and *16SrRNA* (Table S2) [4,18]. PCR was undertaken with a 25 μL mixed system (12.5 μL of 2 × F8 PCR MasterMix; 0.5 μL of forward primer and reverse primer, respectively; 1.5 μL DNA template; and 10 μL ddH_2O) on a DNA amplification apparatus (ABI9700). The thermocycling profile was operated in 94 °C for 5 min, followed by 34 cycles, 95 °C for 30 s, 53–59 °C for 30 s, 72 °C for 30 s, followed by a 10 min extension at 72 °C. Subsequently, the quantity and the quality were assessed with 1% agarose gel electrophoresis.

The complete mitogenomes were sequenced using next-generation sequencing. *T. magnum* was sequenced with Shanghai Origingene Bio-pharm Technology Co. Ltd., Shanghai, China (Illumina HisSeq 4000), and the other three species (*T. celeste*, *T. rotundum*, and *T. hirtipes*) were sequenced with majorbio Bio-Pharm Technology Co. Ltd., Shanghai, China (Illumina HisSeq 6000). To ensure consistency between species names and sequences, the assembled mitochondrial genes were identified by checking the *cox1* barcode sequence on NCBI with NCBI's Basic Local Alignment Search Tool (BLAST) search function for nucleotide sequences (BLASTn; available at: https://www.ncbi.nlm.nih.gov/; accessed on 7 November 2021) [19]. BLASTn searches, using putative orthologous sequences from other crab species obtained from NCBI, was used to query the non-redundant nucleotide (nr/nt) sequence database in order to discover annotated putative orthologue genes. Clean data were assembled from scratch without sequencing adapters using NOVOPlasty [20]. To evaluate the single-base accuracy of the assembled mitogenome (assembled mitochondrial genome) with two confirmed sequences, it was contrasted with Sanger sequencing. The complete mitogenome was uploaded to GenBank (accession number ON990061, OQ948153-OQ948155, available at NCBI).

2.3. Sequence Analysis and Gene Annotation

The PCGs were identified using NCBI and MITOS Web Server [20] (http://mitos2.bioinf.unileipzig.de/index.py; accessed on 20 March 2022) to analyze the mitogenome code of invertebrates. Codon usage was assessed with MEGA 7.0 [21] based on PCGs with incomplete stop codons removed. The secondary structures of tRNAs were determined with MITOS Web Server and tRNAscan-SE Web Server (http://lowelab.ucsc.edu/tRNAscan-SE/; accessed on 23 March 2022) [22]. Moreover, MEGA was applied to confirm compositional skews and calculated through the following formulas: AT-skew = (A − T)/(A + T) and GC-skew = (G − C)/(G + C) [23], and calculated genetic distance analysis (p-distance) [24].

The gene maps of the mitogenomes of four species were generated and visualized with Organellar Genome DRAW (https://chlorobox.mpimp-golm.mpg.de/OGDraw.html; accessed on 25 March 2022) [25]. In addition, the RNAfold WebServer (http://rna.tbi.univie.ac.at/; accessed on 2 May 2022) was used to predict secondary structures of rRNA and CR, and the Tandem Repeats Finder server (http://tandem.bu.edu/trf/trf.html; accessed on 10 May 2022) was used to identify tandem repeats of the control region.

2.4. Phylogenetic Analysis and Gene Rearrangement

In conformity with the phylogenetic trees conducted with the 13 PCGs of all crabs downloaded from GenBank, the taxonomic status of *Tuerkayana* was confirmed. Except for the newly sequenced mitogenomes, the complete mitogenomes of 189 Brachyuran species belonged to 39 families and 17 Anomura species as the outgroup (Table S1) were included. MEGA7.0 [26] was used to align the 13 PCGs of 206 mitogenomes with the default nucleotide and amino acid sequences set in MUSCLE 3.8 [27]. The aligned PCGs were then concatenated into datasets. Before reconstruction, we used DAMBE [28] to detect the nucleotide substitution saturation rate of the aligned PCGs. The phylogenetic tree was reconstructed by the concatenated set of the nucleotide sequence and amino acid sequences, respectively. MtArt + I + G, which was selected by ProtTest, was determined as the best model for analyzing nucleotide sequences [29]. Using Bayesian Inference (BI) and Maximum Likelihood (ML) methods, four phylogenetic trees were constructed with MrBayes v3.2.6 [30] and IQ-TREE [31], respectively. ML bootstrap values (BS) $\geq$ 75% and Bayesian posterior probabilities (BPP) $\geq$ 0.95 were considered significant. In the analysis of BI, we made two simultaneous 10,000,000 generation runs to encourage swapping among the Markov chain Monte Carlo (MCMC) chains and sampled trees every 1000 generations. The convergence of sampled parameters and potential autocorrelation (the effective sampling size of all parameters > 200) were investigated by Tracer 1.6 (http://tree.bio.ed.ac.uk/software/tracer; accessed on 5 July 2022), and the average standard deviation of split frequencies was inspected between runs (<0.01). The first 25% of trees were in the burn-in stage. These were removed, and the BPPs could be obtained from the 50% majority rule consensus of the postburn-in trees sampled at stationarity. Subsequently, the final phylogenetic trees were visualized and edited using FigTreev1.4.2.

Two single gene phylogenetic trees (Table S1) were constructed based on the *cox1* and *16SrRNA* from 11 Gecarcinidae species (including all *Discoplax*, *Tuerkayana*, and *Cardisoma* species) and a Menippidae species (*Pseudocarcinus gigas*), respectively. Similar to the above phylogenetic trees of Brachyura, the datasets of nucleotide sequences were used to construct the trees using BI and ML.

Regarding gene rearrangement, MITOS [20] and NCBI [32] were used to re-annotate all the different mitogenomes to ensure the accuracy, consistency, and comparability of the studied species. If differences in the mitochondrial genome are found, they are repaired artificially. In order to examine the underlying evolutionary mechanisms, the CREx Web Server was used. The Common Interval Rearrangement Explorer (CREx) [33] was used to flexibly calculate complex rearrangement mechanisms by means of mathematical models. Under normal circumstances, a single transformation was frequently utilized to determine the most economical mechanism from a variety of possibilities.

2.5. Selective Pressure Detection

The alignments and consensus trees were used to perform the analysis of selective pressure. The analyses of selective pressure compared the nonsynonymous/synonymous substitution ratios (ω = dN/dS) [34] based on the phylogeny with the codon-based maximum likelihood (CodeML) method in the PAML4.7 package [35]. A Values of $\omega < 1$, =1, and >1 correspond to purifying selection, neutral evolution, and positive selection, respectively. The improved branch-site model A (test 2) was performed for every gene in each foreground lineage. Additionally, all the positively selected sites in branch-site models were identified by using Bayes Empirical Bayes (BEB) analysis with posterior probabilities of $\geq$0.80. Likelihood ratio tests (LRTs) were used to evaluate the significance of differences between the two nested models following a Chi-square distribution by calculating twice the log-likelihood (2ΔL) of the difference. The degrees of freedom refer to the difference in the number of free parameters between models.

3. Results

3.1. Mitogenome Organization and Base Composition

The current study sequenced and fully annotated the mitogenome of four species (*T. magnum, T. rotundum, T. hirtipes, T. celeste*) in *Tuerkayana*. The mitogenomes in *Tuerkayana* are 15,556 (*T. magnum, T. celeste*), 15,559 (*T. hirtipes*), 15,562 (*T. rotundum*) base pairs (bp) long. Each species has a standard set of 37 genes, including 13 PCGs, 2 rRNAs, 22 tRNAs, and a control region (CR). The distribution was the typical of Decapoda, with 23 genes distributed on the heavy (+) strand, and 14 genes distributed on the light (−) strand (Figure 1 and Table S3). An amount of 14 overlapping regions existed among the mitochondrial genes of species in *Tuerkayana*, and these ranged from 1 bp to 25 bp. The number of overlapping bases is the same except for the difference in *T. magnum* between *cox2* and *trnK*. The longest overlapping region existed between *trnL1* and *rrnL* (Table S3). In addition, a total of 803 (*T. hirtipes*) and −828 (*T. celeste*) bp non-coding regions were present in the mitogenomes of *Tuerkayana*, of which 181 (*T. hirtipes*) −208 (*T. celeste*) bp were distributed in 15 intergenic spacers. The other longest non-coding regions, the CRs, were identically situated between *rrnS* and *trnI* with high A + T content (76.01–79.56%).

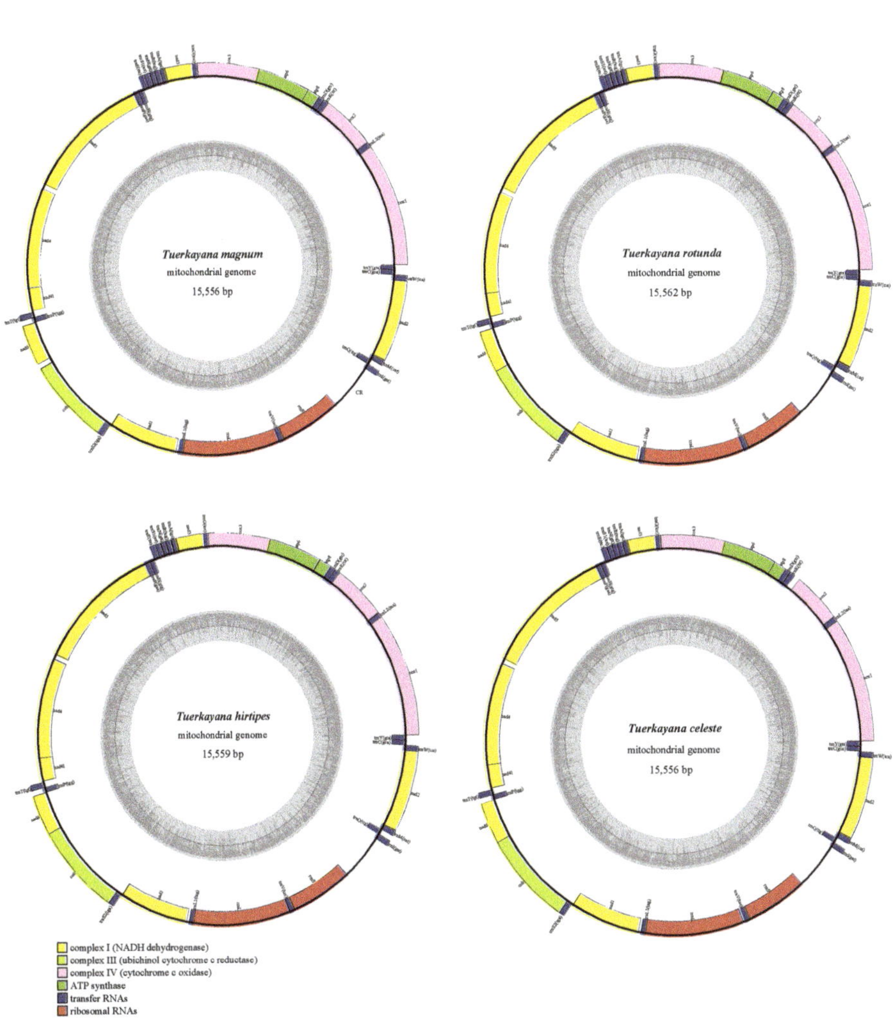

Figure 1. Mitochondrial genome maps of *Tuerkayana*. Protein-coding genes are color-coded (*cox*: lavender; *nad*: yellow; *atp*: green; *cob*: kelly); rRNA genes are in red; tRNA genes are in blue. Abbreviations of protein-coding genes are: *atp6* and *atp8* for ATP synthase subunits 6 and 8, *cox1-3* for *cytochrome oxidase subunits 1-3*, *cob* for *cytochrome b*, *nad1-6* and *nad4l* for *NADH dehydrogenase subunits 1-6* and 4 L, *rrnL* and *rrnS* for large and small rRNA subunits, CR for control region.

The nucleotide bias, which occurred in high A and T representation, was the general characteristic of metazoan mitogenomes [36]. These characteristics resulted in a subsequent bias in homologous encoded amino acids. The AT contents of mitogenome in *Tuerkayana* are high (70.23–71.48%), as is typical of species in Brachyuran (Table S3). In Figure S1, the genera of the Gecarcinidae family are represented separately in a three-dimensional Cartesian coordinate system (*X*-axis represents AT-skew, *Y*-axis represents GC-skew, and *Z*-axis represents AT content). The GC-skews were negative, indicating that Cs were more abundant than Gs. AT-skew and GC-skew resulted from the nucleotides being distributed differentially between the two DNA strands, which caused further DNA strand asymmetry [37].

3.2. Protein-Coding Genes and Codon Usage

PCGs (protein-coding genes) in *Tuerkayana* occupied approximately 71.77–71.82% (11,165 bp, 111,72 bp), including one *cytochrome b* (*cob*), two *ATP synthase* (*atp6* and *atp8*), seven *NADH dehydrogenases* (*nad1-6* and *nad4l*), and three *cytochrome c oxidases* (*cox1-3*). The sizes of the 13 PCGs ranged from 159 bp (*atp8*) to 1719 bp (*nad5*) (Table S3). Similar to a typical mitogenome in Brachyura, 9 PCGs were encoded in the heavy strand (*cox1-3, cob, atp6, atp8, nad2-3, and nad6*), and 4 PCGs were encoded on the light strand (*nad1, nad4-5,* and *nad4l*) in every mtDNA of *Tuerkayana*. The start codon of the 13 PCGs used the typical start codon ATN, embracing ATG, ATA, and ATT. In addition, the stop codon of PCGs also used the representative stop codon TNN, including TAA, TAG, and the incomplete stop codon: T (*cox2* in *T. magnum* and *cob*). The peculiar incomplete stop codon (T) was possibly completed as TAA through post-transcriptional polyadenylation and would not affect the normal synthesis of amino acids [38].

RSCU (Relative Synonymous Codon Usage) is a reference value to evaluate the frequency of codons encoding the same amino acid. When the RSCU results were higher than 1, this suggested that the codon appears many times and with a high frequency. On the contrary, a value of less than one indicates that the frequency of this codon is low and the number of occurrences is small. As shown in Figure 2, the RSCU of single amino acids of Gecarcinidae were compared, and they were all significantly different, especially Leucine (Leu) and Serine (Ser) in terms of the two patterns in the first codons (Leu: CUN, UUA/UUG; Ser: AGN, UCN). The RSCU ratio of each species varies greatly, as UUA, UCU, CCU, and GCU are used relatively frequently, and the frequency of CUG, CCG, GCG, and UGC was low. According to the results, the RSCU values varied widely, indicating that there is a great bias in the usage frequency of codons. The RSCU of NNU and NNA codons were larger than one, indicating that the codons with the third site of the A and T base were used more frequently. This bias in codon usage is consistent with a strong AT bias in the third site of the codon of the protein-coding gene. In addition, we counted the single amino acid usage count. A frequency of 20 amino acids showed a strong bias. The frequencies of leucine (Leu), isoleucine (Ile), phenylalanine (Phe), and serine (Ser) were high. These amino acids were all composed of T or TA. The frequencies of arginine (Arg), aspartate (Asp), and cysteine (Cys) determined by CG-rich base codons were relatively low. This phenomenon is consistent with the mitochondrial genome showing strong AT bias (Figure 3).

3.3. Transfer RNAs, Ribosomal RNAs, and CR

The mitogenomes of *Tuerkayana* all included 22 tRNAs, ranging from 62 bp (*trnC* in *T. magnum* and *T. celeste*) to 73 bp (*trnV*) in size. Eight tRNA genes were distributed on the light strand [*trnP, trnQ, trnV, trnC, trnY, trnH, trnL1* (CUN), and *trnF*], and the other 14 tRNAs were distributed on the other strand (Figure 1 and Table S3), these distributions were consistent with other Gecarcinidae crabs. As shown in Figures S2 and S3, the majority of tRNAs exhibited typical cloverleaf structures. However, the *trnS* (AGN) lacked the entire dihydrouridine (DHU) arm, which was simplified as a loop. The absence of the DHU arm in the secondary structure of *trnS* (AGN) is common in the mitogenome of metazoans [39,40]. Previous studies have verified that the lack of the D-arm does not impact the function of tRNA in metazoans. Additionally, many mismatches (G-U, A-C, U-U, C-U) have been found in tRNAs, most of which are G-U pairs. This often occurs in tRNAs of other Crustacea species [41], and these mismatches have been amended in the tRNA modification [42].

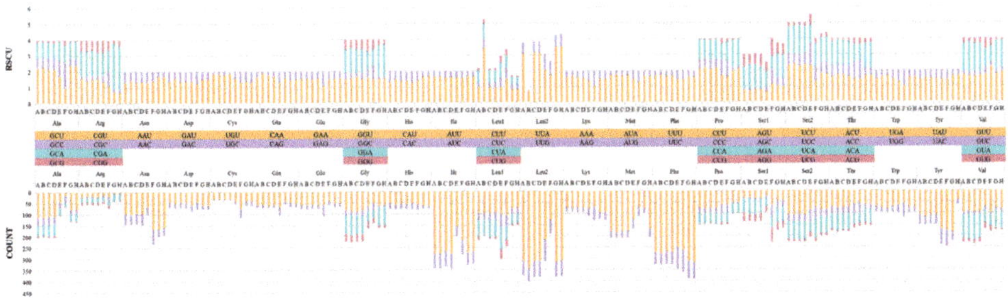

Figure 2. The relative synonymous codon usage (RSCU) in the mitogenomes of Tuerkayana. A: Tuerkayana magnum; B: Tuerkayana rotundum; C: Tuerkayana hirtipes; D: Tuerkayana celeste; E: Cardisoma armatum; F: Cardisoma carnifex; G: Gecarcoidea lalandii; H: Gecarcoidea natalis.

Mitogenomes of all species in Gecarcinidae showed that the small coding subunit (*12SrRNA*) and large coding subunit (*16SrRNA*) were separated by *trnV*. In addition, all were located on the light strand, which is also a characteristic shared by most species of Brachyura. Due to the specific function of rRNAs, their sequences are often conserved, such that the secondary structure and three-dimensional composition are not significantly altered (Figure S4).

The CR typically was heavily A + T biased. Its length was 620 (*T. celeste*) −622 (*T. hirtipes*) bp with an A+T content of 76.01% (*T. magnum*) −78.58% (*T. rotundum*). The AT-skew and GC-skew in CR were −0.072–0.074 and −0.306–0.315, respectively. On account of in-depth research undertaken in recent years, CR is no longer considered a pure non-coding. However, its specific function remains to be studied. There were many conserved motifs, such as the poly T-stretch, TA(A)n-like stretch, G(A)nT motif, TATA motif, and hairpin loop structures (Figure S5), and these motifs have been identified as initiation sites for replication and transcription [43]. Notably, Tandem Repeats Finder was used to search for tandem repeats, but none were found, which is unusual in crustaceans.

3.4. Phylogenetic Relationships

The paucity of existing mitochondrial genome data constrained the scope of this study. Complete data were only available for the *Tuerkayana* we sequenced. For other species belonging to the same family, information was limited to *cox1* and *16SrRNA* sequences available in the database. In order to facilitate a more comprehensive examination of the taxonomic relationships involving *Tuerkayana*, we reconstructed two single-gene phylogenetic trees (Figure S6 and Table S1) of *cox1* and *16SrRNA* sequences from 11 different species belonging to the Gecarcinidae family (including species from *Discoplax, Tuerkayana,* and *Cardisoma*) and a Menippidae species (*Pseudocarcinus gigas*). In the two single gene phylogenetic trees, three genera *Tuerkayana, Discoplax,* and *Cardisoma* were closely related and appeared to be monophyletic, forming the relationship of (*Cardisoma* (*Discoplax* + *Tuerkayana*)).

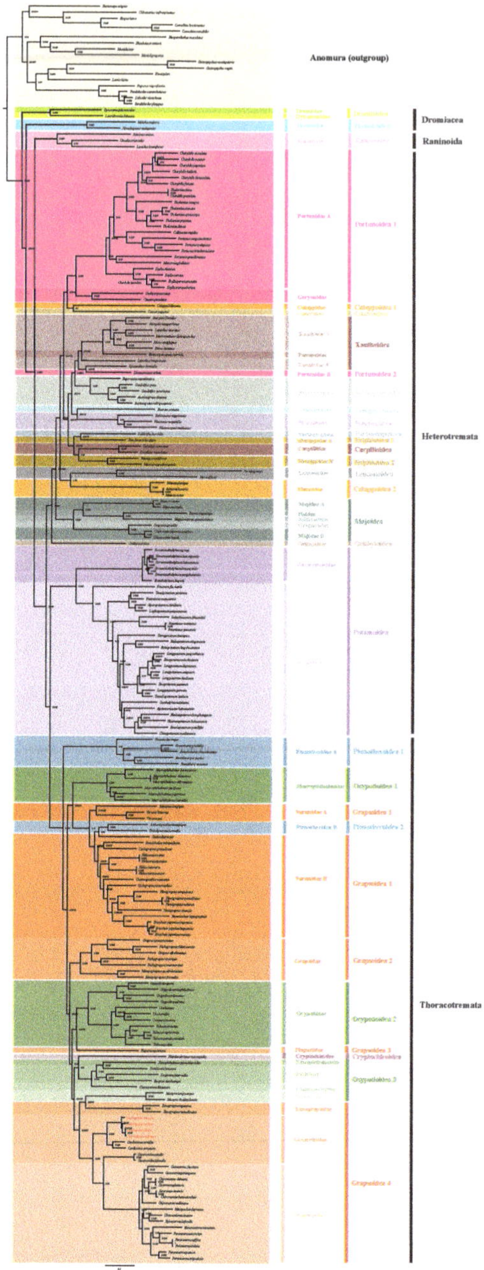

Figure 3. The phylogenetic tree was inferred from the nucleotide sequences of 13 mitogenome PCGs using BI and ML methods, including 189 species of Brachyuran belonging to 39 families and 17 Anomura species as the outgroup. Numbers on branches indicate posterior probabilities (BI) and bootstrap values (ML). The dashed lines on the right represent the families and superfamilies of these species, respectively.

In addition, two types of phylogenetic trees (ML and BI tree) were reconstructed with complete mitogenomes in order to further investigate the phylogenetic position of *Tuerkayana* and Gecarcinidae within Brachyuran. This phylogenetic tree included 189 Brachyuran crabs belonging to 39 families and 17 Anomura species as the outgroup (Table S1). The phylogenetic tree was estimated using the dataset of nucleotide sequences and amino acid sequences of 13 PCGs. In front of the building, the substitution saturation of PCGs was measured by DAMBE (Figure S7), all *iss* were smaller than *iss.c*, indicating that PCGs were not saturated and contained accurate phylogenetic information. Through BI and ML methods, most of the same topological structures were produced from nucleotide sequences and amino acid sequences, respectively. Based on the analysis of support values, the BI topologies were used to present both support values, including the bootstrap values for the ML tree (BS) and the posterior probabilities for Bayesian analysis (BPP), to show the results of nucleotide sequences (N-tree) and amino acid sequences (AA-tree) separately (Figure 2 and Figure S8).

The phylogenetic relationships of *Tuerkayana* can be inferred from its representation in the N-tree as (*T. rotundum* (*T. hirtipes* (*T. celeste* + *T. magnum*))) and in the AA-tree as ((*T. rotundum* + *T. hirtipes*) (*T. celeste* + *T. magnum*)). Both of these topologies support the monophyly of *Tuerkayana* and imply a close evolutionary affinity between *T. celeste* and *T. magnum*.

Notably, the N-tree and AA-tree display divergent phylogenetic information. The first and fundamental question is "Is the G family monophyletic?". The answer to this question depends mainly on the position of the genus *Gecarcoidea* in the phylogenetic trees (Figure S6). According to the N-tree, the *Gecarcoidea* and Sesarmidae exhibit a sister group relationship with high BPP and low BS, indicating that Gecarcinidae is partially paraphyletic with respect to Sesarmidae, presenting a ((*Tuerkayana* + *Cardisoma*) (*Gecarcoidea* + Sesarmidae)) topology. Thus, Gecarcinidae appears to be polyphyletic. However, the AA-tree presents a different scenario, as it shows that Gecarcinidae is monophyletic, featuring a (*Gecarcoidea* (*Tuerkayana* + *Cardisoma*)) topology with robust support values for both BS and BPP. To further investigate this issue, a genetic distance analysis (p-distance) was conducted (Figure S9 and Table S4). The results indicate that the p-distance between *Gecarcoidea* and (*Tuerkayana* + *Cardisoma*) is 0.1485–0.1659 (average 0.1587), whereas the p-distance between *Gecarcoidea* and Sesarmidae is 0.1512–0.1711 (average 0.1602). The p-distance between Gecarcinidae and Sesarmidae was 0.1773. Based solely on genetic distance values, the p-distance between *Gecarcoidea* and (*Tuerkayana* + *Cardisoma*) was slightly smaller than between Gecarcinidae and Sesarmidae, indicating a closer relationship between *Gecarcoidea* and the genera *Tuerkayana* and *Cardisoma*, which supports the current classification status of Gecarcinidae. Nevertheless, further studies combining morphology and biogeography are necessary to validate this conclusion. Additionally, a notable divergence can be observed in the classification of Gecarcinidae within the N-tree and AA-tree (Figure S6). In the N-tree, Gecarcinidae is closely associated with Sesarmidae and Xenograpsidae, which are typically positioned in the Grapsoidea 4. In different forms, the AA-tree exhibits a more intricate configuration, with Gecarcinidae classified as (Gecarcinidae (Camptandriidae (Dotillidae (Dotillidae + Xenophthalmidae)) + Sesarmidae)) rather than associated solely with Sesarmidae. The utilization of the genetic distance analysis reveals that the distance between Gecarcinidae or Sesarmidae and Ocypodoidea 4 is greater than between Gecarcinidae and Sesarmidae, providing evidence to support the closer relationship between Gecarcinidae and Sesarmidae and thus justifying the N-tree topology. Additionally, comparisons made with previous studies lent further credence to this result [44,45]. Third, despite the lack of substantial differences in the overall status of the superfamily, the distribution of individual families within it has resulted in significant variations. Following a thorough comparison with the phylogenetic tree presented in previous studies, we posit that the N-tree represents a more precise depiction of the evolutionary relationships between families. Consequently, we elected to concentrate on the N-tree.

Having been added to the dataset, 33 of the 39 families included in this tree formed a monophyletic clade, and from the other families arose branches. This finding is not consistent with previous research [46,47]. Therefore, it is necessary to increase the mitogenome of species for the research of phylogeny. In addition, *Lissocarcinus arkati* (Portunidae species) occurred unexpectedly in Xanthidae. Through the analysis and comparison of 13 PCGs, *L. arkati* was determined as closer to *Eriphia* species, and we speculated that *L. arkati* should belong to *Eriphia* or that there may have been an identification error or a contamination in the original study. Furthermore, at the superfamily level, Ocypodoidea and Grapsoidea were polyphyletic, with Grapsoidea and Ocypodoidea mutually intersecting. Many studies have suggested that Grapsoidea and Ocypodoidea could be polyphyletic and combined into a large taxon [48–50]. This conjecture is consistent with our phylogenetic analysis, and the reconstruction of the phylogenetic tree in this study used more location data than in previous studies, enhancing the persuasiveness of this conjecture. At the same time, some taxonomists have proposed that the morphological classification of Ocypodoidea and Grapsoidea is indeed inconsistent with molecular phylogeny, but the creation of these superfamilies was primarily for practical purposes and convenience rather than phylogenetic accuracy [51,52].

3.5. Mitogenome Gene Rearrangement

The mitogenome order of species in Gecarcinidae is presented in Figure S10. This figure shows that all the species in the Gecarcinidae have the same gene order pattern as the ancestors of Brachyura. This is consistent with a previous study regarding the evolution of Brachyura (Crustacea: Decapoda) based on mitochondrial arrangement and gene order rearrangements [4].

The mitogenome arrangements of a Grapsoidea lineage (Sesarmidae and Xenograpsidae) have been researched (Figure S10) and were determined to be located in the same branch as "Grapsoidea 4" (Figure S6). We studied the sequence evolution of genes from three aspects (Figure 4), namely the (A) evolutionary process from the ancestor of Brachyura to Sesarmidae, (B) evolutionary process from the ancestor of Brachyura to *Xenograpsus ngatama*, and (C) evolutionary process from the ancestor of Brachyura to *Xenograpsus testudinatus*. The mitogenome gene arrangements of all Gecarcinidae species were the same. It is clear, therefore, that the mitogenome arrangements of Sesarmidae species are identical and differ from the ancestral arrangement of the Brachyura only by the inversion of *trnI-trnQ*, which has been described in previous studies of Sesarmidae species [53,54].

As a semi-terrestrial family of Grapsoidea, the species of Xenograpsidae usually inhabit the harsh environment of the Epipelagic thermal spring ecosystem [55]. Xenograpsidae contained one genus (*Xenograpsus*), and three known species are included in this category, two (*X. ngatama* and *X. testudinatus*) of which have published complete mitogenomes [56]. In this study, CREx was used to preliminarily predict the putative evolutionary rearrangement mechanisms in *X. ngatama* and *X. testudinatus*. The results of this analysis are summarized in Figure 4B. Two tandem duplication-random loss (TDRL) events were inferred from the Brachyura ancestor to *X. ngatama*. The first TDRL event occurred between *trnH* and *trnS2*. Then, *trnH-trnF-nad5-nad4-nad4l*, *trnP*, *trnT*, and *nad6-cob-trnS2* were randomly absent. The second TDRL event went from *trnA* to *trnY*. Then, six redundant fragments were deleted (*trnA-trnR*, *trnI*, *trnW*, *trnN-trnS1-trnE-trnT-nad6-cob-trnS2-trnH-trnF-nad5-nad4-nad4l-trnP-nad1-trnL1-rrnL-trnV-rrnS-CR*, *trnQ-trnM-nad2*, *trnC-trnY*). Moreover, from Brachyura ancestors to *X. testudinatus*, a complex genetic rearrangement was determined, which is illustrated in Figure 4C. This genetic rearrangement is more complicated than that of the Brachyura ancestor to *X. ngatama*, which started with a small-scale TDRL with *trnN* and *trnR* transposed and accompanied by *trnY* reversed. Subsequently, the TDRL event went from *trnH* to *trnS2*. Then four redundant fragments were lost (*trnH-trnF-nad5-nad4-nad4l*, *trnP*, *trnT*, *nad6-cob-trnS2*), with *trnA* reversed. After two consecutive TDRLs and gene reversals, two large-range TDRLs were experienced again, which changed the position of *trnA*, *trnC*, *trnY* and *trnS1*, *trnI*, *trnW*, respectively. The three families are closely related but have completely different rearrangement mechanisms.

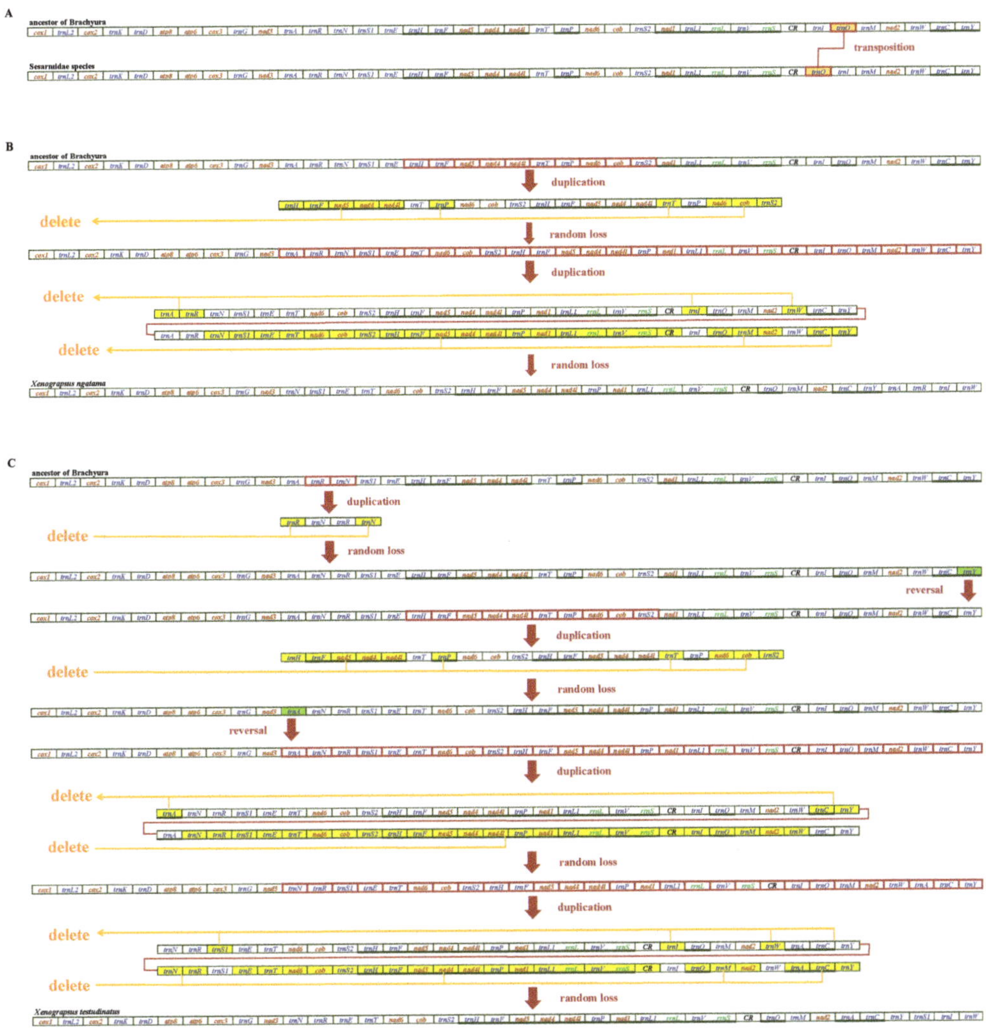

Figure 4. (**A**) Evolutionary procession from the ancestors of Brachyura to Sesarmidae; (**B**) Evolutionary procession from the ancestors of Brachyura to *Xenograpsus ngatama*. (**C**) Evolutionary procession from the ancestors of Brachyura to *Xenograpsus testudinatus*.

3.6. Selective Pressure in Gecarcinidae

To explore the adaptive evolution of Brachyura in the mitochondrial genes and assess the evolutionary patterns of PCGs, the branch-site model was used to detect selective pressure in individual codons for 206 crabs (Table S1) with CodeML, and the dN/dS(ω) were calculated. The codon *nad6* was found under positive selection, where LRTs of the branch-site model A were statistically significant (Table 1). Except for *nad6*, the ω values of the other 12 PCGs were far lower than one, indicating that the other 12 PCGs had a low evolutionary rate, and *nad6* was subjected to strong positive selection ($p << 0.05$, $\omega 2 >> 1$).

Table 1. Selective pressure analyses (branch-site model) of mitochondrial protein-encoding genes (13 genes) in all crabs datasets.

Gene	Model	lnL	2lnL	*p* Level	Parameters	Positive Selected Sites (Posterior Probabilities)
atp8	Ge					
	ma	−14,486.44134			ω0 = 0.087 ω1 = 1.0 ω2 = 1.0	19 I 0.852;
	ma0	−14,486.44134	0	1	ω0 = 0.087 ω1 = 1.0 ω2 = 1.0	
cox2	Ge					
	ma	−39,595.89288			ω0 = 0.018 ω1 = 1.0 ω2 = 1.628	
	ma0	−39,595.89284	-7.4×10^{-5}	1	ω0 = 0.018 ω1 = 1.0 ω2 = 1.0	
cox3	Ge					
	ma	−44,309.9257			ω0 = 0.016 ω1 = 1.0 ω2 = 1.605	
	ma0	−44,309.92564	−0.00012	1	ω0 = 0.016 ω1 = 1.0 ω2 = 1.0	
cob	Ge					
	ma	−69,109.49107			ω0 = 0.017 ω1 = 1.0 ω2 = 1.0	
	ma0	−69,109.49113	0.00012	0.991259787	ω0 = 0.017 ω1 = 1.0 ω2 = 1.0	
cox1	Ge					
	ma	−77,798.40391			ω0 = 0.008 ω1 = 1.0 ω2 = 2.421	
	ma0	−77,798.40376	−0.000292	1	ω0 = 0.008 ω1 = 1.0 ω2 = 1.0	
atp6	Ge					
	ma	−42,775.44489			ω0 = 0.02 ω1 = 1.0 ω2 = 11.004	
	ma0	−42,775.44974	0.009696	0.921560467	ω0 = 0.02 ω1 = 1.0 ω2 = 1.0	
nad4l	Ge					
	ma	−19,516.44828			ω0 = 0.023 ω1 = 1.0 ω2 = 1.0	
	ma0	−19,514.55719	−3.782182	1	ω0 = 0.023 ω1 = 1.0 ω2 = 1.0	
nad1	Ge					
	ma	−55,412.74594			ω0 = 0.017 ω1 = 1.0 ω2 = 1.0	
	ma0	−55,412.74594	0	1	ω0 = 0.017 ω1 = 1.0 ω2 = 1.0	
nad3	Ge					
	ma	−24,126.5674			ω0 = 0.026 ω1 = 1.0 ω2 = 1.0	
	ma0	−24,126.5674	2×10^{-6}	0.998871621	ω0 = 0.026 ω1 = 1.0 ω2 = 1.0	
nad2	Ge					
	ma	−90,455.6106			ω0 = 0.042 ω1 = 1.0 ω2 = 1.0	
	ma0	−90,455.6106	0	1	ω0 = 0.042 ω1 = 1.0 ω2 = 1.0	
nad5	Ge					
	ma	−123,998.0812			ω0 = 0.034 ω1 = 1.0 ω2 = 1.0	
	ma0	−123,998.0812	0	1	ω0 = 0.034 ω1 = 1.0 ω2 = 1.0	
nad4	Ge					
	ma	−91,823.48875			ω0 = 0.03 ω1 = 1.0 ω2 = 1.0	386 I 0.728;
	ma0	−91,823.48875	0	1	ω0 = 0.03 ω1 = 1.0 ω2 = 1.0	
nad6	Ge					
	ma	−45,717.39912			ω0 = 0.043 ω1 = 1.0 ω2 = 137.937	28 L 0.686; 36 V 0.556; 88 I 0.997;
	ma0	−45,721.31165	7.825062	0.005152668	ω0 = 0.043 ω1 = 1.0 ω2 = 1.0	

4. Discussion

In this study, the complete mitogenome of four species in *Tuerkayana* was determined and described. Based on the phylogenetic tree constructed from the most complete dataset of mitogenomes in Brachyura to date, the classification within Brachyura was further investigated, stabilizing the phylogenetic positions of *Tuerkayana* and Gecarcinidae and furthered our understanding. Our four primary conclusions are given below.

First, *Tuerkayana* is an independent genus. Based on morphological analysis, four species previously assigned to *Discoplax* (*T. celeste* and *T. magnum*) and *Cardisoma* (*T. rotundum* and *T. hirtipes*) were divided into the new genus, *Tuerkayana*. The specific manifestation is proepistome dome-shaped wide but low, pleonal somite six broad and short, telson short, bluntly tipped, no suborbital ridge, and no stridulatory apparatus [5]. Based on the results of the two single gene phylogenetic trees, it is evident that the three genera (*Tuerkayana*, *Discoplax*, and *Cardisoma*) are closely related and show monophyly of each branch. This consequence is consistent with the results of morphological studies, which support *Tuerkayana* as an independent genus.

Secondly, the mitogenomes in Gecarcinidae are structurally stable, indicating a common origin or common derivation from adaptation to a similar terrestrial environment. In Gecarcinidae, the mitogenomes have 37 genes. Their size varied from 15,545 (*G. natalis*) to 15,597 bp (*C. carnifex*), with a maximum difference of only 52bp [5,57,58]. These lengths are smaller in metazoans, and the genes are compact. The range of A + T content of species in Gecarcinidae is from 68.86% to 75.22%, with an abundance of A and T. As a highly mutated region in the mitogenome, CR is the most prone to gene mutation and length change. While in Gecarcinidae, the CR region is highly conserved, and the size, position (*rrnS*-CR-*trnI*), and stem–loop structures of CR did not demonstrate significant variation. Research of the complete mitogenome demonstrates that the gene order in vertebrates is almost fixed, and gene rearrangement rarely occurs [59]. In contrast, in invertebrates, the probabilities of gene rearrangement is significantly higher, and a variety of gene rearrangement models are derived [60]. However, the genetic order of Gecarcinidae is the same as that of the presumed ancestor of Brachyura. As the most derived family within Brachyura, it exhibited no genetic rearrangement, which is relatively rare among invertebrates. All of the above evidence suggests that the mitogenome of Gecarcinidae is structurally stable. We conjecture that this is on account of common origin or derived from adaptations to a similar terrestrial environment.

Third, *Gecarcoidea* belongs to the monophyletic Gecarcinidae. However, the monophyly of Gecarcinidae remains uncertain due to the inconsistent positions of *Gecarcoidea* on the N-tree and AA-tree. Although a preliminary analysis based on genetic distance indicates a closer relationship between *Gecarcoidea* and other species in Gecarcinidae, this inference lacks conclusive evidence. To resolve the taxonomic position of Gecarcinidae, additional evidence from morphology and geographic distribution must be considered. Morphologically, species of Gecarcinidae can be distinguished from other Grapsoidea by the unique characteristics of zoea larvae, antennal and telson morphology, and setation of the second maxilliped endopod [51]. Geographically, members of Gecarcinidae are restricted to tropical island regions (such as Indonesia, the Spratly Islands, the Philippines) and typically inhabit damp crevices, migrating to the sea only during the breeding season. In contrast, Sesarmidae have a much broader distribution, encompassing nearly all coastal regions in tropical and subtropical areas. Given these factors, we tentatively support the monophyly of Gecarcinidae while recognizing that the evidence remains inconclusive. Furthermore, the presence of polyphyly in the N-tree raises the possibility of homology and a close relationship between Gecarcinidae and Sesarmidae.

Fourth, *nad6* may play a significant role in the adaptation within Gecarcinidae. According to its terrestrial characteristics, Gecarcinidae would inevitably face selection pressure to adapt to the habitat, so the branch-site has been conducted here. It is remarkable that *nad6* was subjected to strong positive selection ($p << 0.05$, $\omega 2 >> 1$); *nad6* is an essential part of the NADH dehydrogenase (complex I) whose alteration can have a significant impact on organisms [61]. NADH dehydrogenase is an important enzyme that catalyzes the oxidation of NADH into NAD^+. NAD^+ maintains redox homeostasis, energy metabolism, DNA repair, gene expression, adaptive stress responses, metabolism, mitochondrial homeostasis, and cellular bioenergetics [62–64]. The dynamic NAD^+ rewires biological processes with post-synthesis modification of fundamental biomolecules to enable cells the adaption to environmental changes [65]. Given that Gecarcinidae repeatedly transitioned from marine to terrestrial environments and suffered from hypoxia stress, the larger numbers of nonsynonymous substitutions, which accumulated in *nad6*, indicate its critical effect on the adaptive process. The adaptive evolution of *nad6* has been previously reported in the study of mammals (*Equus caballus*) [66] subjected to a low temperature and hypoxia environment. These results indicate that *nad6* was significantly determined by the regulation of hypoxia. In consequence, unusual selection pressures acting at the molecular level for organisms that were subjected to hypoxia stress would be disclosed by comparative analyses of complete mitogenomes in Gecarcinidae. This study lays an important foundation for exploring the process of terrestrial Gecarcinidae and provides new insight into possible molecular adaptation mechanisms in crabs under hypoxia. However, the analyses of molecular evolution that this study provided could be strengthened. Only 1 out of 13 genes was found to indicate a signal of selection that could be random or a species-specific pattern. Future analyses of the biochemical or protein structure are, therefore, needed.

5. Conclusions

The complete mitogenome sequences of four species in *Tuerkayana* were determined and analyzed for the first time. The most comprehensive phylogenetic tree of Brachyura involving 206 crabs was constructed. The research evaluated the phylogenetic position of *Tuerkayana* and Gecarcinidae in Brachyuran, further supported the establishment of *Tuerkayana* and the division of the genus in Gecarcinidae at the mitochondrial level, and investigated the monophyly of Gecarcinidae. The mitochondrial genome structure of Gecarcinidae is stable. The strong selective pressure shown in *nad6* suggested that it may play a crucial part in the evolution of Gecarcinidae.

Supplementary Materials: The following supporting information can be downloaded at: https://www.mdpi.com/article/10.3390/biology12070974/s1, Figure S1: The three-dimensional Cartesian coordinate system within Gecarcinidae. X-axis represents AT-skew, Y-axis represents GC-skew, and Z-axis represents AT content. *Tuerkayana* are represented by circles, *Cardisoma* by triangles, and *Gecarcoidea* by squares; Figure S2: Predicted secondary cloverleaf structure for the tRNAs of *Tuerkayana magnum*; Figure S3: Predicted secondary cloverleaf structure for the tRNAs of *Tuerkayana rotunda*, *Tuerkayana hirtipes*, and *Tuerkayana celeste*; Figure S4: Predicted secondary structure for *16SrRNA* and *12SrRNA* in *Tuerkayana*; Figure S5: Predicted secondary structure for CR (control region) in *Tuerkayana*; Figure S6: Nucleotide substitution saturation analysis of PCGs. X-axis represents F84 distance and Y-axis represents s and v; Figure S7: (A,B): Two single gene phylogenetic trees were inferred from *cox1* and *16SrRNA* using BI and ML methods, including 11 Gecarcinidae species (including all *Discoplax* species, *Tuerkayana* species and *Cardisoma* species have been sequenced) and a Menippidae species (*Pseudocarcinus gigas*). Numbers on branches indicate posterior probability (BI) and bootstrap values (ML). (C,D): The configurational difference of trees in nucleotide and amino acid sequences by PCGs, especially the taxonomic status of Gecarcinidae; Figure S8: The phylogenetic tree was inferred from the amino acid sequences of 13 mitogenome PCGs using BI and ML methods; Figure S9: The genetic distance analysis (p-distance) between species in Gecarcinidae and Sesarmidae; Figure S10: The mitogenome genetic sequence of species in Gecarcinidae, Sesarmidae, Xenograpsidae, and the ancestor of Brachyura. All genes are transcribed from left to right. CR represents control region. *rrnL* and *rrnS* are the large and small ribosomal RNA subunits, respectively; Table S1: The GenBank accession numbers of 189 Brachyura species and 17 Anomura species in this study; Table S2: Sequences of universal primers; Table S3: Summary of mitochondrial genomes of *Tuerkayana magnum*, *Tuerkayana rotunda*, *Tuerkayana hirtipe*, *Tuerkayana celeste*; Table S4: Estimates of Evolutionary Divergence.

Author Contributions: Conceptualization, Z.W. and Y.Z.; Data curation, Y.Z., X.Z., X.X. and C.C.; Formal analysis, Y.Z. and X.Z.; Funding acquisition, Z.W.; Investigation, Y.Z.; Methodology, Z.W.; Project administration, Z.W.; Software, Y.Z., X.Z., X.X. and Z.X.; Supervision, Z.W.; Validation, Y.Z., X.Z., X.X., Z.X. and C.C.; Visualization, Y.Z., X.Z., Z.X. and C.C.; Writing—original draft, Z.W. and Y.Z.; Writing—review & editing, Z.W. and Y.Z. All authors have read and agreed to the published version of the manuscript.

Funding: This study was funded by the National Natural Science Foundation of China (grant number 31702014), Doctoral Scientific Research Foundation of Yancheng Teachers University to Z.F.W, and Open Foundation of Jiangsu Key Laboratory for Bioresources of Saline Soils (grant number JKLBS2019006).

Institutional Review Board Statement: Not applicable.

Informed Consent Statement: Not applicable.

Data Availability Statement: The data that support this study are available at NCBI and are provided in Table S1.

Acknowledgments: We thank the editor and anonymous reviewers for their comments that helped improve this work.

Conflicts of Interest: The authors declare there are no competing interest.

References

1. Broughton, R.E.; Milam, J.E.; Roe, B.A. The Complete Sequence of the Zebrafish (*Danio rerio*) Mitochondrial Genome and Evolutionary Patterns in Vertebrate Mitochondrial DNA. *Genome Res.* **2001**, *11*, 1958–1967. [CrossRef] [PubMed]
2. Ingman, M.; Kaessmann, H.; Pääbo, S.; Gyllensten, U. Mitochondrial genome variation and the origin of modern humans. *Nature* **2000**, *408*, 708–713. [CrossRef] [PubMed]
3. Xu, X.; Wang, Q.; Wu, Q.; Xu, J.; Wang, J.; Wang, Z. The entire mitochondrial genome of *Macrophthalmus abbreviatus* reveals insights into the phylogeny and gene rearrangements of Brachyura. *Biochem. Genet.* **2021**, *59*, 617–636. [CrossRef] [PubMed]
4. Wang, Q.; Wang, J.; Wu, Q.; Xu, X.; Wang, P.; Wang, Z. Insights into the evolution of Brachyura (Crustacea: Decapoda) from mitochondrial sequences and gene order rearrangements. *Int. J. Biol. Macromol.* **2021**, *170*, 717–727. [CrossRef]
5. Guinot, D.; Ng, N.K.; Moreno, P.A.R. Review of grapsoid families for the establishment of a new family for Leptograpsodes Montgomery, 1931, and a new genus of Gecarcinidae H. Milne Edwards, 1837 (Crustacea, Decapoda, Brachyura, Grapsoidea MacLeay, 1838). *Zoosystema* **2018**, *40*, 547–604. [CrossRef]
6. Jiménez, C.; Rubio, A.O.; Cárdenas, S.A.; Arnaud, G. Ecological aspects of the land crab *Gecarcinus planatus* (Decapoda: Gecarcinidae) in Socorro Island, Mexico. *Biol. Conserv.* **1994**, *69*, 9–13. [CrossRef]
7. Green, P.T. Burrow Dynamics of the Red Land Crab *Gecarcoidea natalis* (Brachyura, Gecarcinidae) in Rain Forest on Christmas Island (Indian Ocean). *J. Crustac. Biol.* **2004**, *24*, 340–349. [CrossRef]
8. Hartnoll, R.G.; Clark, P.F. A mass recruitment event in the land crab *Gecarcinus ruricola* (Linnaeus, 1758) (Brachyura: Grapsoidea: Gecarcinidae), and a description of the megalop. *Zool. J. Linn. Soc.* **2010**, *146*, 149–164. [CrossRef]
9. Adamczewska, A.M.; Morris, S. Ecology and behavior of *Gecarcoidea natalis*, the Christmas Island red crab, during the annual breeding migration. *Biol. Bull.* **2001**, *200*, 305–320. [CrossRef]
10. Farrelly, C.A.; Greenaway, P. Land crabs with smooth lungs: Grapsidae, Gecarcinidae, and Sundathelphusidae ultrastructure and vasculature. *J. Morphol.* **1993**, *215*, 245–260. [CrossRef]
11. Cuesta, J.A.; Anger, K. Larval morphology and salinity tolerance of a land crab from West Africa, *Cardisoma armatum* (Brachyura: Grapsoidea: Gecarcinidae). *J. Crustac. Biol.* **2005**, *25*, 640–654. [CrossRef]
12. Allardyce, B.J.; Linton, S.M. Functional morphology of the gastric mills of carnivorous, omnivorous, and herbivorous land crabs. *J. Morphol.* **2010**, *271*, 61–72. [CrossRef]
13. Mendes, R.A.S.; Silva, J.R.F.; Neto, J.S.; Hazin, F.H.V. Reproductive biology of the land crab *Cardisoma guanhumi* (Decapoda: Gecarcinidae) in north-eastern Brazil. *J. Mar. Biol. Assoc. UK* **2013**, *93*, 761–768. [CrossRef]
14. Schubart, C.D.; Diesel, R.; Hedges, S.B. Rapid evolution to terrestrial life in Jamaican crabs. *Nature* **1998**, *393*, 363–365. [CrossRef]
15. Watson-Zink, V.M. Making the grade: Physiological adaptations to terrestrial environments in decapod crabs. *Arthropod Struct. Dev.* **2021**, *64*, 101089. [CrossRef] [PubMed]
16. López-Barneo, J.; Pardal, R.; Ortega-Sáenz, P. Cellular mechanism of oxygen sensing. *Annu. Rev. Physiol.* **2001**, *63*, 259–287. [CrossRef]
17. Farrelly, C.A.; Greenaway, P. Morphology and ultrastructure of the gills of terrestrial crabs (Crustacea, Gecarcinidae and Grapsidae): Adaptations for air-breathing. *Zoomorphology* **1992**, *112*, 39–49. [CrossRef]
18. Simon, C.; Buckley, T.R.; Frati, F.; Stewart, J.B.; Beckenbach, A.T. Incorporating Molecular Evolution into Phylogenetic Analysis, and a New Compilation of Conserved Polymerase Chain Reaction Primers for Animal Mitochondrial DNA. *Annu. Rev. Ecol. Evol. Syst.* **2006**, *37*, 545–579. [CrossRef]
19. Altschul, S.F.; Madden, T.L.; Schäffer, A.A.; Zhang, J.; Zhang, Z.; Miller, W.; Lipman, D.J. Gapped BLAST and PSI-BLAST: A new generation of protein database search programs. *Nucleic Acids Res.* **1997**, *25*, 3389–3402. [CrossRef]
20. Bernt, M.; Donath, A.; Jühling, F.; Externbrink, F.; Florentz, C.; Fritzsch, G.; Pütz, J.; Middendorf, M.; Stadler, P.F. MITOS: Improved de novo metazoan mitochondrial genome annotation. *Mol. Phylogenet. Evol.* **2013**, *69*, 313–319. [CrossRef]
21. Sudhir, K.; Stecher, G.; Li, M.; Knyaz, C.; Tamura, K. MEGA X: Molecular Evolutionary Genetics Analysis across computing platforms. *Mol. Biol. Evol.* **2018**, *35*, 1547–1549.
22. Lowe, T.M.; Chan, P.P. tRNAscan-SE On-line: Integrating search and context for analysis of transfer RNA genes. *Nucleic Acids Res.* **2016**, *44*, W54–W57. [CrossRef] [PubMed]
23. Peng, Y.; Zhou, L.; Zhou, X.; Yang, W.; Zhang, J.; Zhang, X.; Wang, Y.; Gui, J. Unusual AT-skew of *Sinorhodeus microlepis* mitogenome provides new insights into mitogenome features and phylogenetic implications of bitterling fishes. *Int. J. Biol. Macromol.* **2019**, *129*, 339–350.
24. Tamura, K.; Peterson, D.; Peterson, N.; Stecher, G.; Nei, M.; Kumar, S. MEGA5: Molecular Evolutionary Genetics Analysis Using Maximum Likelihood, Evolutionary Distance, and Maximum Parsimony Methods. *Mol. Biol. Evol.* **2011**, *28*, 2731–2739. [CrossRef] [PubMed]
25. Lohse, M.; Drechsel, O.; Bock, R. OrganellarGenomeDRAW (OGDRAW): A tool for the easy generation of high-quality custom graphical maps of plastid and mitochondrial genomes. *Curr. Genet.* **2007**, *52*, 267–274. [CrossRef] [PubMed]
26. Sudhir, K.; Glen, S.; Koichiro, T. MEGA7: Molecular Evolutionary Genetics Analysis Version 7.0 for Bigger Datasets. *Mol. Biol. Evol.* **2016**, *33*, 1870–1874.
27. Edgar, R.C. MUSCLE: Multiple sequence alignment with high accuracy and high throughput. *Nucleic Acids Res.* **2004**, *32*, 1792–1797. [CrossRef]

28. Xia, X. DAMBE6: New Tools for Microbial Genomics, Phylogenetics, and Molecular Evolution. *J. Hered.* **2017**, *108*, 431–437. [CrossRef]
29. Darriba, D.; Taboada, G.L.; Doallo, R.; Posada, D. ProtTest 3: Fast selection of best-fit models of protein evolution. *Bioinformatics* **2011**, *27*, 1164–1165. [CrossRef]
30. Ronquist, F.; Teslenko, M.; Mark, P.; Ayres, D.L.; Darling, A.; Höhna, S.; Larget, B.; Liu, L.; Suchard, M.A.; Huelsenbeck, J.P. MrBayes 3.2: Efficient Bayesian Phylogenetic Inference and Model Choice Across a Large Model Space. *Syst. Biol.* **2012**, *61*, 539–542. [CrossRef]
31. Trifinopoulos, J.; Nguyen, L.; Haeseler, A.; Minh, B.Q. W-IQ-TREE: A fast online phylogenetic tool for maximum likelihood analysis. *Nucleic Acids Res.* **2016**, *44*, W232–W235. [CrossRef] [PubMed]
32. Wang, W. The Molecular Detection of *Corynespora Cassiicola* on Cucumber by PCR Assay Using DNAman Software and NCBI. In Proceedings of the International Conference on Computer and Computing Technologies in Agriculture, Beijing, China, 27–30 September 2015; Springer: Cham, Switzerland, 2015; pp. 248–258.
33. Bernt, M.; Merkle, D.; Ramsch, K.; Fritzsch, G.; Perseke, M.; Bernhard, D.; Schlegel, M.; Stadler, P.F.; Middendorf, M. CREx: Inferring genomic rearrangements based on common intervals. *Bioinformatics* **2007**, *23*, 2957–2958. [CrossRef]
34. Ohta, T. The nearly neutral theory of molecular evolution. *Annu. Rev. Ecol. Syst.* **1992**, *23*, 263–286. [CrossRef]
35. Yang, Z.; Nielsen, R. Estimating synonymous and nonsynonymous substitution rates under realistic evolutionary models. *Mol. Biol. Evol.* **2000**, *17*, 32–43. [CrossRef] [PubMed]
36. Cha, S.Y.; Yoon, H.J.; Lee, E.M.; Yoon, M.H.; Hwang, J.S.; Jin, B.R.; Han, Y.S.; Kim, I. The complete nucleotide sequence and gene organization of the mitochondrial genome of the bumblebee, *Bombus ignitus* (Hymenoptera: Apidae). *Gene* **2007**, *392*, 206–220. [CrossRef]
37. Reyes, A.; Gissi, C.; Pesole, G.; Saccone, C. Asymmetrical directional mutation pressure in the mitochondrial genome of mammal. *Mol. Biol. Evol.* **2018**, *15*, 957–966. [CrossRef] [PubMed]
38. Ojala, D.; Montoya, J.; Attardi, G. tRNA punctuation model of RNA processing in human mitochondria. *Nature* **1981**, *290*, 470–474. [CrossRef]
39. Lavrov, D.V.; Brown, W.M.; Boore, J.L. A novel type of RNA editing occurs in the mitochondrial tRNAs of the centipede *Lithobius forficatus*. *Proc. Natl. Acad. Sci. USA* **2000**, *97*, 13738–13742. [CrossRef]
40. Kilpert, F.; Podsiadlowski, L. The complete mitochondrial genome of the common sea slater, *Ligia oceanica* (Crustacea, Isopoda) bears a novel gene order and unusual control region features. *BMC Genom.* **2006**, *7*, 241. [CrossRef]
41. Wang, Q.; Wang, Z.; Tang, D.; Xu, X.; Tao, Y.; Ji, C.; Wang, Z. Characterization and comparison of the mitochondrial genomes from two Alpheidae species and insights into the phylogeny of Caridea. *Genomics* **2020**, *112*, 65–70. [CrossRef]
42. Yuan, M.L.; Zhang, Q.L.; Guo, Z.L.; Wang, J.; Shen, Y.Y. Comparative mitogenomic analysis of the superfamily Pentatomoidea (Insecta: Hemiptera: Heteroptera) and phylogenetic implications. *BMC Genom.* **2015**, *16*, 460. [CrossRef] [PubMed]
43. Taanman, J. The mitochondrial genome: Structure, transcription, translation and replication. *Biochim. Biophys. Acta (BBA)-Bioenerg.* **1999**, *1410*, 103–123. [CrossRef]
44. Sun, S.E.; Jiang, W.; Yuan, Z.; Sha, Z. Mitogenomes Provide Insights Into the Evolution of Thoracotremata (Brachyura: Eubrachyura). *Front. Mar. Sci.* **2022**, *9*, 848203. [CrossRef]
45. Zhang, Y.; Gong, L.; Lu, X.; Miao, Z.; Jiang, L.; Liu, B.; Liu, L.; Li, P.; Zhang, X.; Lü, Z. Comparative mitochondrial genome analysis of Varunidae and its phylogenetic implications. *Acta Oceanol. Sin.* **2022**, *41*, 119–131. [CrossRef]
46. Zhang, Y.; Gong, L.; Lu, X.; Jiang, L.; Liu, B.; Liu, L.; Lü, Z.; Li, P. Gene rearrangements in the mitochondrial genome of *Chiromantes eulimene* (Brachyura: Sesarmidae) and phylogenetic implications for Brachyura. *Int. J. Biol. Macromol.* **2020**, *162*, 704–714. [CrossRef] [PubMed]
47. Zhang, Y.; Gao, Y.; Gong, L.; Lu, X.; Jiang, L.; Liu, B.; Liu, L.; Lü, Z.; Li, P. Mitochondrial Genome of *Episesarma lafondii* (Brachyura: Sesarmidae) and Comparison with Other Sesarmid Crabs. *J. Ocean. Univ. China* **2021**, *20*, 1545–1556. [CrossRef]
48. Basso, A.; Babbucci, M.; Pauletto, M.; Riginella, E.; Patarnello, T.; Negrisolo, E. The highly rearranged mitochondrial genomes of the crabs *Maja crispata* and *Maja squinado* (Majidae) and gene order evolution in Brachyura. *Sci. Rep.* **2017**, *7*, 4096. [CrossRef]
49. Tsang, L.M.; Ahyong, S.T.; Shih, H.; Ng, P.K. Further polyphyly of pinnotheroid crabs: The molecular phylogenetic position of the polychaete-associated Aphanodactylidae. *Invertebr. Syst.* **2018**, *32*, 92–99. [CrossRef]
50. Kobayashi, G.; Itoh, H.; Fukuda, H.; Kojima, S. The complete mitochondrial genome of the sand bubbler crab *Scopimera globosa* and its phylogenetic position. *Genomics* **2021**, *113*, 831–839. [CrossRef]
51. Ng, P.K.; Guinot, D.; Davie, P.J. Systema Brachyurorum: Part I. An annotated checklist of extant brachyuran crabs of the world. *Raffles Bull. Zool.* **2008**, *17*, 1–286.
52. Davie, P.J.; Guinot, D.; Ng, P.K. Systematics and classification of Brachyura. In *Treatise on Zoology-Anatomy, Taxonomy, Biology*; Brill Academic Publisher: Leiden, The Netherlands, 2015; Volume 9, pp. 1049–1130.
53. Xin, Z.Z.; Liu, Y.; Zhang, D.Z.; Chai, X.Y.; Wang, Z.F.; Zhang, H.B.; Zhou, C.L.; Tang, B.P.; Liu, Q.N. Complete mitochondrial genome of *Clistocoeloma sinensis* (Brachyura: Grapsoidea): Gene rearrangements and higher-level phylogeny of the Brachyura. *Sci. Rep.* **2017**, *7*, 4128. [CrossRef] [PubMed]
54. Wang, Z.; Wang, Z.; Shi, X.; Wu, Q.; Tao, Y.; Guo, H.; Ji, C.; Bai, Y. Complete mitochondrial genome of *Parasesarma affine* (Brachyura: Sesarmidae): Gene rearrangements in Sesarmidae and phylogenetic analysis of the Brachyura. *Int. J. Biol. Macromol.* **2018**, *118 Pt A*, 31–40. [CrossRef]

55. Ng, N.K.; Davie, P.J.F.; Schubart, C.D.; Ng, P.K.L. Xenograpsidae, a new family of grapsoid crabs (Crustacea: Brachyura) associated with shallow water hydrothermal vents. *Raffles Bull. Zool.* **2007**, *16*, 233–256.
56. Ki, J.S.; Dahms, H.U.; Hwang, J.S.; Lee, J.S. The complete mitogenome of the hydrothermal vent crab *Xenograpsus testudinatus* (Decapoda, Brachyura) and comparison with brachyuran crabs. *Comp. Biochem. Physiol. Part D Genom. Proteom.* **2009**, *4*, 290–299. [CrossRef]
57. Tan, M.H.; Gan, H.M.; Lee, Y.P.; Linton, S.; Grandjean, F.; Bartholomei-Santos, M.L.; Miller, A.D.; Austin, C.M. ORDER within the chaos: Insights into phylogenetic relationships within the Anomura (Crustacea: Decapoda) from mitochondrial sequences and gene order rearrangements. *Mol. Phylogenet. Evol.* **2018**, *127*, 320–331. [CrossRef]
58. Gan, H.M.; Linton, S.M.; Austin, C.M. Two reads to rule them all: Nanopore long read-guided assembly of the iconic Christmas Island red crab, Gecarcoidea natalis (Pocock, 1888), mitochondrial genome and the challenges of AT-rich mitogenomes. *Mar. Genom.* **2019**, *45*, 64–71. [CrossRef]
59. Lombardo, G.; Migliore, N.R.; Colombo, G.; Capodiferro, M.R.; Formenti, G.; Caprioli, M.; Moroni, E.; Caporali, L.; Lancioni, H.; Secomandi, S.; et al. The Mitogenome Relationships and Phylogeography of Barn Swallows (*Hirundo rustica*). *Mol. Biol. Evol.* **2022**, *39*, msac113. [CrossRef]
60. Tang, P.; Zhu, J.C.; Zheng, B.Y.; Wei, S.J.; Sharkey, M.J.; Chen, X.X.; Vogler, A. Mitochondrial phylogenomics of the Hymenoptera. *Mol. Phylogenet. Evol.* **2019**, *131*, 8–18. [CrossRef] [PubMed]
61. Bai, Y.; Attardi, G. The mtDNA-encoded *ND6* subunit of mitochondrial NADH dehydrogenase is essential for the assembly of the membrane arm and the respiratory function of the enzyme. *EMBO J.* **1998**, *17*, 4848–4858. [CrossRef] [PubMed]
62. Kushnareva, Y.; Murphy, A.N.; Andreyev, A. Complex I-mediated reactive oxygen species generation: Modulation by cytochrome c and $NAD(P)^+$ oxidation–reduction state. *Biochem. J.* **2002**, *368*, 545–553. [CrossRef]
63. Yang, H.; Yang, T.; Baur, J.A.; Perez, E.; Matsui, T.; Carmona, J.J.; Lamming, D.W.; Souza-Pinto, N.C.; Bohr, V.A.; Rosenzweig, A.; et al. Nutrient-Sensitive Mitochondrial NAD^+ Levels Dictate Cell Survival. *Cell* **2007**, *130*, 1095–1107. [CrossRef] [PubMed]
64. Nakahata, Y.; Kaluzova, M.; Grimaldi, B.; Sahar, S.; Hirayama, J.; Chen, D.; Guarente, L.P.; Sassone-Corsi, P. The NAD^+-dependent deacetylase SIRT1 modulates CLOCK-mediated chromatin remodeling and circadian control. *Cell* **2008**, *134*, 329–340. [CrossRef] [PubMed]
65. Xie, N.; Zhang, L.; Gao, W.; Huang, C.; Huber, P.E.; Zhou, X.; Li, C.; Shen, G.; Zou, B. NAD^+ metabolism: Pathophysiologic mechanisms and therapeutic potential. *Signal Transduct. Target. Ther.* **2020**, *5*, 227. [CrossRef] [PubMed]
66. Ning, T.; Xiao, H.; Li, J.; Hua, S.; Zhang, Y.P. Adaptive evolution of the mitochondrial *ND6* gene in the domestic horse. *Genet. Mol. Res.* **2010**, *1*, 144–150. [CrossRef]

biology

MDPI

Article

Evidence of Considerable Shifts in Catch Composition in the Artisanal Spiny Lobster Fishery in Kenya

Abdirahman J. Kulmiye †

Department of Zoology, University of Nairobi, P.O. Box 30197, Nairobi 00100, Kenya; akulmiye@kaadinstitute.edu.so
† Current affiliation: KAAD Institute, MRK-ZA2-040, Mubarak, Garowe, Somalia.

Simple Summary: In Kenya, the artisanal lobster fishery is important socioeconomically for supporting local fishing communities and generating revenue for the government, yet detailed knowledge of many aspects of the fishery is lacking. In this study, the population structure and catch composition of spiny lobsters caught by divers were investigated, and the findings were compared to lobster survey data from the 1970s to identify potential changes in the artisanal landings that may have occurred over time.

Abstract: The artisanal lobster fishery in Kenya is small in world terms but important locally both in terms of supporting local fishing communities and generating revenue for the government. Despite its socioeconomic importance, detailed knowledge of many aspects of the fishery is lacking. The study reported herein aimed to investigate and provide information on the population structure and catch composition of spiny lobsters caught by artisanal fishers off six major landing sites along the coastline. A total of 2711 lobsters representing five palinurid species were collected during the study period (November 2000–March 2001). Among the five species, *Panulirus longipes* dominated the catches in Msambweni (75%) and Shimoni (58%), *P. homarus* in Mambrui (70%) and Kipini (72%), *P. ornatus* in Lamu (49%), and *P. penicillatus* in Kilifi (39%). *P. versicolor* was the rarest species observed in the catches across the six sites. The overall catch consisted of 33% *P. ornatus*, 32% *P. homarus*, 28% *P. longipes*, 6% *P. penicillatus* and 2% *P. versicolor*. Sitewise, Lamu contributed 31% of the overall catch, Kipini 23%, Shimoni 20%, Mambrui 13%, Msambweni 7%, and Kilifi 6%. A comparison of the results of this study and lobster abundance data from 1970s surveys revealed considerable shifts in the catch composition of artisanal landings over time. Future work on this fishery should concentrate on the lobster populations in the decades-old marine protected areas to obtain unfished reference data to assess the fishery and establish the underlying cause(s) of the observed shifts in catch composition.

Keywords: crustacea; population structure; distribution; exploitation; Kenya

1. Introduction

Kenya has both an artisanal and an offshore lobster fishery, of which the artisanal fishery is the most important in terms of production, employment creation, and revenue generation. The offshore production is based on deep-water lobster species (*Metanephrops andamanicus*, *Puerulus angulatus*, Scyllarides) caught as bycatch during semi-industrial prawn trawling off the north coast. The artisanal fishery, on the other hand, exploits five species of shallow-water tropical spiny lobsters belonging to the genus Panulirus, namely, *Panulirus ornatus*, *P. longipes*, *P. homarus*, *P. versicolor* and *P. penicillatus* [1].

Artisanal lobster fishing is undertaken along the entire coastline, with the major landing sites being Shimoni, Msambweni, Kilifi, Mambrui, Kipini, and Lamu. Spiny lobsters are harvested year-round, but the peak fishing season occurs during the northeast monsoon (November–April) when weather conditions are favorable. Even though artificial shelters have been shown to attract spiny lobsters at Gazi Bay on the south coast [2], breath-hold diving remains the most popular harvest method. Fishers capture lobsters in shallow waters (<10 m) during the daytime, rarely venturing beyond three nautical miles offshore due to the depth limits of breath-hold diving. Most divers access fishing grounds with dug-out canoes, outriggers, Arab dhows, and other small, non-mechanized crafts [3]. Landed lobsters are either sold to local tourist hotels or exported live to overseas markets, primarily China.

Although the artisanal spiny lobster production is small in world terms, averaging only 100 metric tons annually, the fishery is socioeconomically important in supporting many fishers and their families and earning the government hard currency through the export of lobsters. Lobster fishing is popular among fisher folks because of the high price lobsters fetch compared to finfish and other crustaceans. Unfortunately, the ever-increasing demand for seafood has led to a tremendous increase in fishing effort in recent years, resulting in overfishing [4]. Declines in catch rates and lobster sizes have been reported for some species [5–7].

Despite the socioeconomic importance of the resource and the apparent overfishing, the fishery is open access with a few applicable management controls, such as a minimum legal weight of retention (250 g) for all species, prohibitions on retaining berried females, and the use of scuba gear, spears, dynamite, and tangle nets. Four small marine protected areas (MPAs) are also in place where no fishing is permitted. However, these management controls are insufficient to ensure the long-term conservation and sustainable use of lobster resources, even if they are strictly enforced, since they are based mainly on limited scientific information. More comprehensive research is therefore needed to obtain the biological and population parameters in order to formulate stringent, science-based management regulations. Previous studies on Kenyan spiny lobsters have either dealt with a single species or populations in a small geographic area [3,4,6,8,9]. Although both Mutegyera [1] and Mueni et al. [7] focused on the entire coastline, their findings varied substantially, especially regarding the relative abundance of lobster species. This study aimed to investigate and provide information on the species composition, relative abundance, spatial distribution, and population structure of spiny lobsters harvested along the Kenya coast.

2. Materials and Methods

2.1. Study Areas

The Kenyan coastline is approximately 640 km long, running in a south-westerly direction from the Kenya–Somalia border in the north to the Kenya–Tanzania border in the south (Figure 1). It is fringed by a continuous, well–developed reef that lies between 0.5 and 2 km offshore, except where major rivers (Tana and Sabaki) discharge into the Indian Ocean. The coral reefs, along with seagrasses and mangroves, provide an ideal habitat for spiny lobsters. Generally, the coastline is divided into the north and south coasts, with Mombasa Island serving as the reference point.

For this study, six sampling sites were selected based on their being major lobster landing centers: Shimoni and Msambweni on the south coast, Kilifi, Mambrui, Kipini, and Lamu on the north coast.

2.2. Sample Collection

This study was conducted between November 2000 and March 2001, coinciding with the lobster fishing season. A six-day trip was made to each of the selected six landing sites, where five divers were randomly selected from the respective local fishers and hired to assist with sample collection. In order for their catches to represent the typical landings of a given site, the divers were instructed to fish as they usually would and bring all their lobster catches into a makeshift laboratory at the landing site for analysis. Once brought to the laboratory, lobster samples were identified to the species level, sexed based on external sexual dimorphism, and measured to the nearest 0.1 mm (carapace length, CL) before being returned to the fishers.

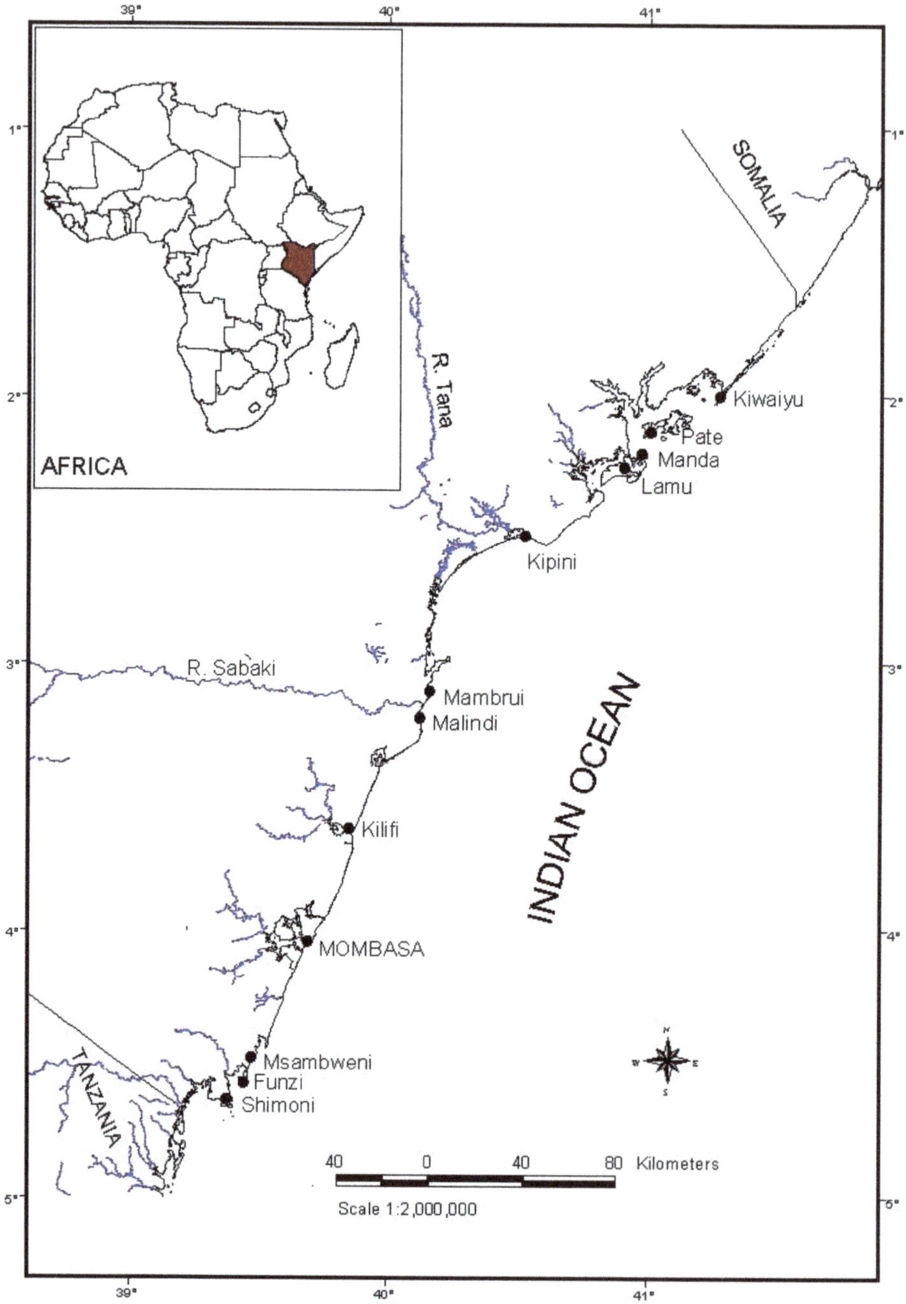

Figure 1. Map of the Kenyan coast showing sampling sites and other lobster landing centers.

2.3. Data Analysis

Lobster CL data were tested for normality (Kolmogorov–Smirnov test) and homogeneity of variance (Levene test) prior to formal analysis. For lobster size comparisons, CL data were grouped by landing site and compared for differences using the Kruskal–Wallis test. When the result was statistically significant ($p < 0.05$), a post hoc test was used to determine which comparisons had significant differences. The sex ratio was determined for each sampling site, and deviation from the expected ratio of 1:1 was tested using the Chi-square (χ^2) test. *Panulirus penicillatus* and *P. versicolor* were omitted from size and sex composition analyses due to their low proportions or absence from some landing sites. Species composition was calculated based on the number of each species within the catch. Relative abundance was estimated by dividing the number of lobsters in each species by the total number of lobsters in all species. All statistical analyses were performed using SPSS for Windows (version 22.0, IBM, Armonk, NY, USA), and p-values lower than 0.05 were considered statistically significant.

2.4. Limitations of This Study

The results provided in this study were derived from data collected approximately 22 years ago. Due to the cumulative effects of fishing on exploited stocks, these results may not reflect the current landings of the artisanal lobster fishery in Kenya. However, the findings are important since they can serve as a valuable reference for future assessments of the spiny lobsters in the region.

3. Results

3.1. Species Distribution

In this study, five species of spiny lobsters belonging to the genus Panulirus were encountered in the landed catches at the six sampling sites (Table 1). These were *Panulirus ornatus, P. longipes, P. homarus, P. penicillatus,* and *P. versicolor. Panulirus ornatus, P. longipes* and *P. homarus* were the most widely distributed of the five species, occurring in all six sampling sites. *Panulirus versicolor* was found at four sites, while *P. penicillatus* was encountered at only three sites. Among the six sites, Shimoni, Kilifi, and Lamu had the highest number of spiny lobster species, with all five species present. These were followed by Msambweni and Kipini, with four species each. Mambrui had the least number of species, hosting only three.

Table 1. Distribution of spiny lobster species along the Kenya coast based on samples caught off the six sampling sites during the study period (November 2000–March 2001) *.

Species	Shimoni	Msambweni	Kilifi	Mambrui	Kipini	Lamu
Panulirus ornatus	x	x	x	x	x	x
Panulirus homarus	x	x	x	x	x	x
Panulirus longipes	x	x	x	x	x	x
Panulirus versicolor	x	x			x	x
Panulirus penicillatus	x		x			x

* Symbol (x) indicates the presence of a species in a particular site.

3.2. Catch Composition

A total of 2711 lobsters were collected during the study period, of which 887 belonged to *P. ornatus*, 866 to *P. homarus*, 747 to *P. longipes*, 167 to *P. penicillatus* and 44 to *P. versicolor*. Figure 2 shows the composition of lobster species by landing site. *P. ornatus* was the most abundant species in Lamu, accounting for 49.3% of the sampled catch. It also constituted a substantial proportion of the catches landed at the other five sites. *P. homarus* was most prevalent in the estuarine sites of Mambrui (70%) and Kipini (72%) but least common on the south coast sites of Msambweni and Shimoni, contributing just 1% of the catch. *P. penicillatus* appeared to be the most harvested species in Kilifi. In Msambweni and Shimoni, *P. longipes* accounted for 75% and 58% of the catch, respectively, making it the most important species on the south coast. *P. versicolor* was the rarest species observed in the catches across the sites. Pooling all the catches together (Figure 3), *P. ornatus* had the highest relative abundance (33%), followed by *P. homarus* (32%), *P. longipes* (28%), and *P. penicillatus* (6%), while *P. versicolor* had the lowest relative abundance of just 2%. Sitewise, Lamu contributed most of the pooled catch (31%), followed by Kipini (23%) and Shimoni (20%). The contribution of the other three sites ranged from 6% to 13% (Figure 4). The northern sites of Lamu, Kipini, and Mambrui together accounted for 67% of the lobster landings.

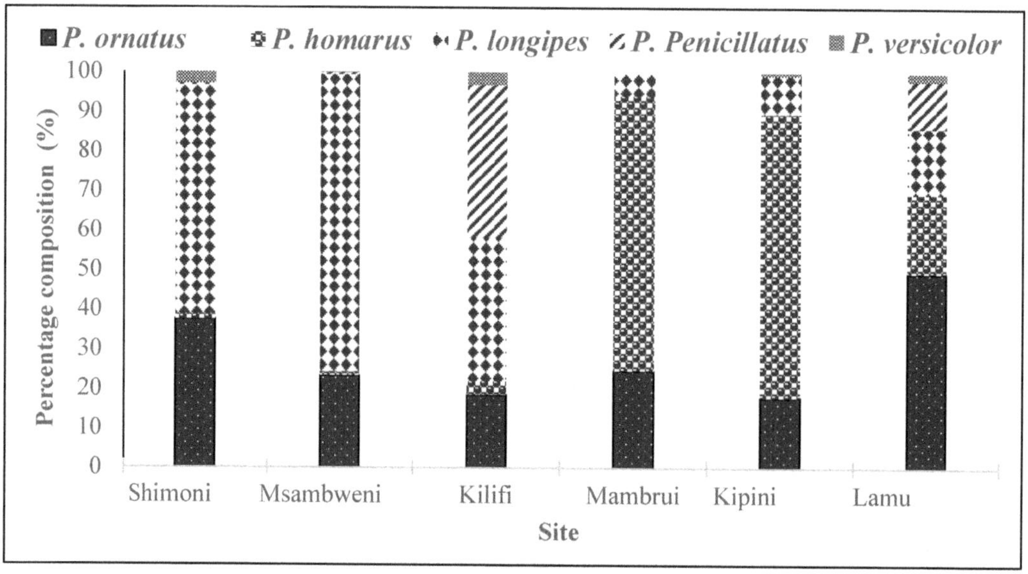

Figure 2. Species composition of spiny lobsters caught off the six sampling sites during the study period (November 2000–March 2001).

3.3. Size and Sex Composition

The descriptive statistics of the carapace length of males and females of the three most abundant species (i.e., *P. ornatus*, *P. homarus* and *P. longipes*) collected from the six sites are presented in Table 2. The overall mean and standard deviation of carapace length were 82 mm ± 22 mm SD for *P. ornatus*, 69 mm ± 12 mm SD for *P. homarus* and 70 mm ± 11 mm SD for *P. longipes*. In terms of the mean size of lobsters by landing sites, lobster samples from Kipini had the largest mean size, while those from Mambrui had the smallest mean size. Carapace length frequency distributions constructed for males and females exhibited a unimodal distribution for all three species over the range of individuals measured (Figures 5–7), except for *P. ornatus* males and females from Shimoni and Kilifi and *P. longipes* females from Kilifi, which displayed a bimodal distribution. Analysis of variance (Kruskal–Wallis) test showed significant differences in the lobster size distributions among the landing sites for all three species (Kruskal–Wallis test, $p < 0.05$). Subsequent Dunn's post hoc tests also revealed significant differences in most paired site comparisons (Table 3).

The sex ratio of *P. ornatus*, *P. homarus* and *P. longipes* determined by site did not depart from the expected male-to-female ratio of 1:1 for all sites (Table 4), except Shimoni and Kipini, where males of *P. ornatus* and *P. longipes* significantly outnumbered the females in the population. In terms of pooled data, *P. ornatus* samples were male-biased, with the sex ratio significantly deviating from a 1:1 ratio. In contrast, male-to-female ratios of *P. homarus* and *P. longipes* in the pooled catch were near parity.

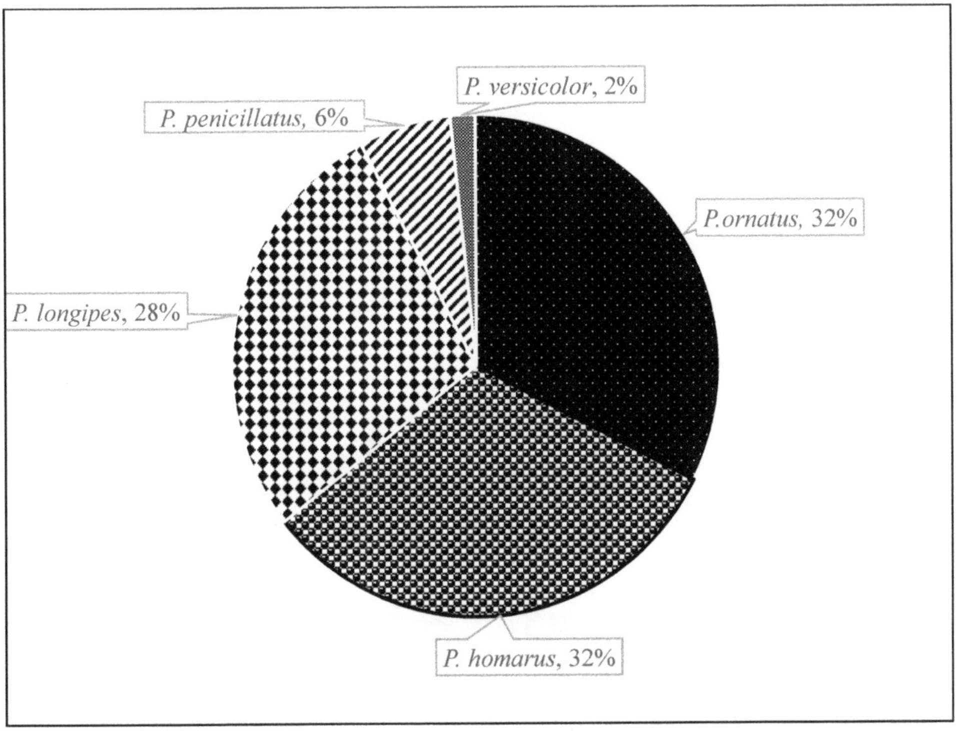

Figure 3. Catch composition of spiny lobsters in the pooled data from all sampling sites.

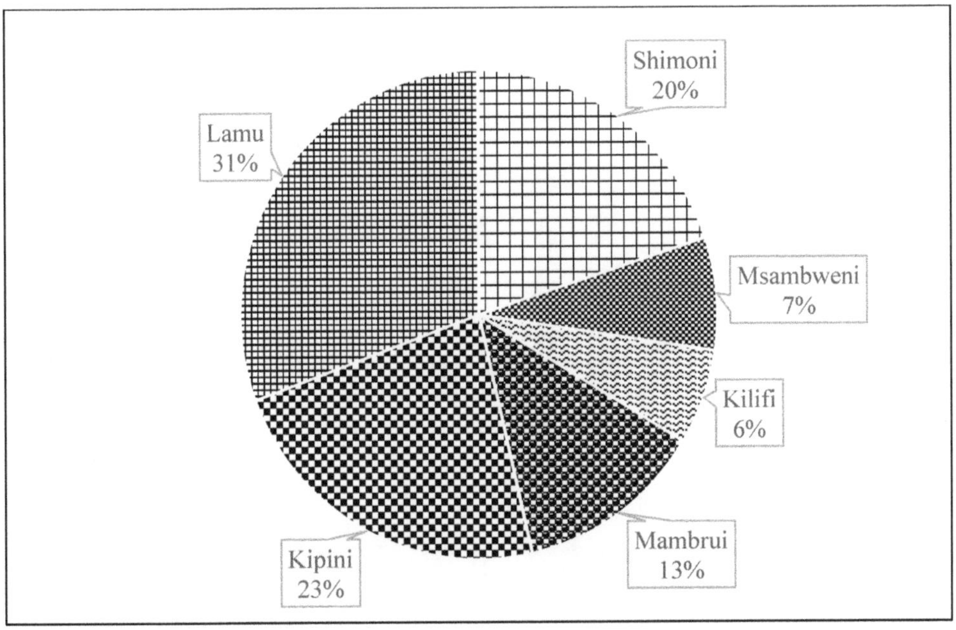

Figure 4. Catch contributions of the respective sampling sites to the pooled spiny lobster data.

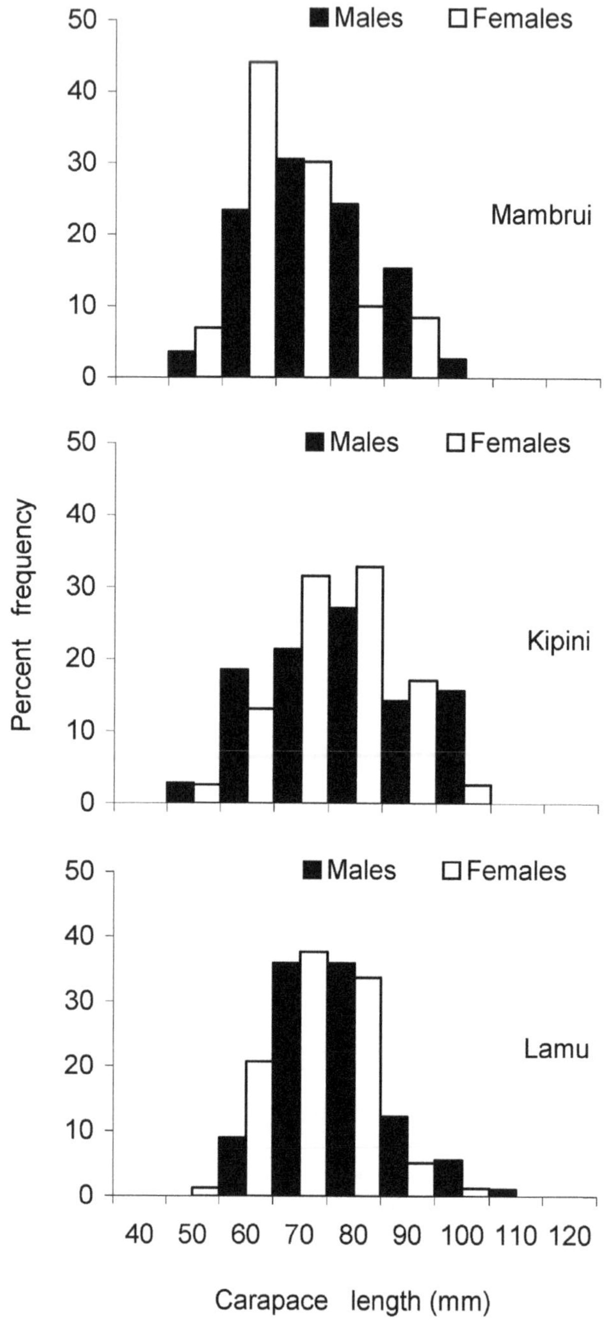

Figure 5. Size frequency distributions of *Panulirus homarus* from three sampling sites during the study period (November 2000–March 2001).

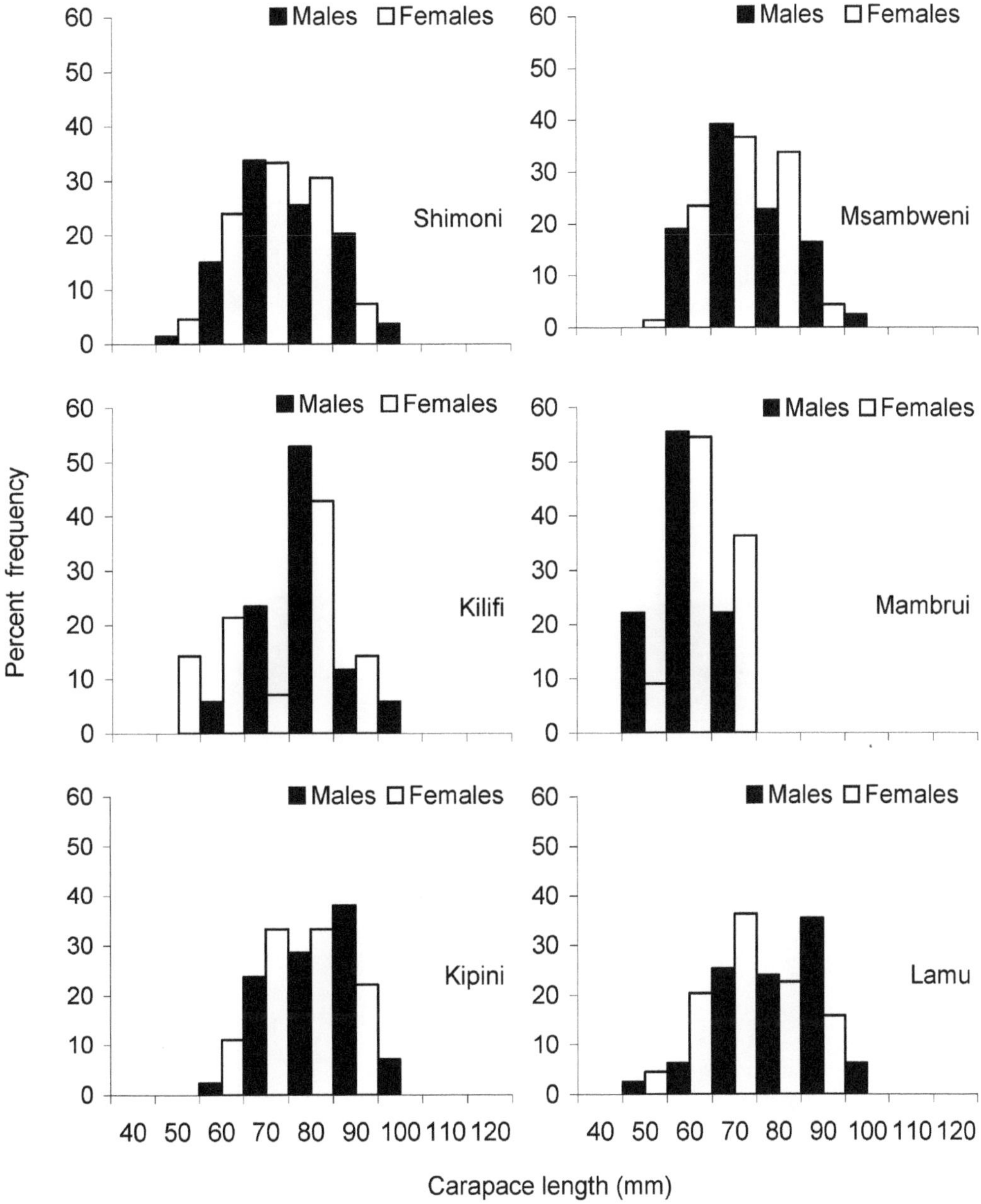

Figure 6. Size frequency distributions of *Panulirus longipes* from six sampling sites during the study period (November 2000–March 2001).

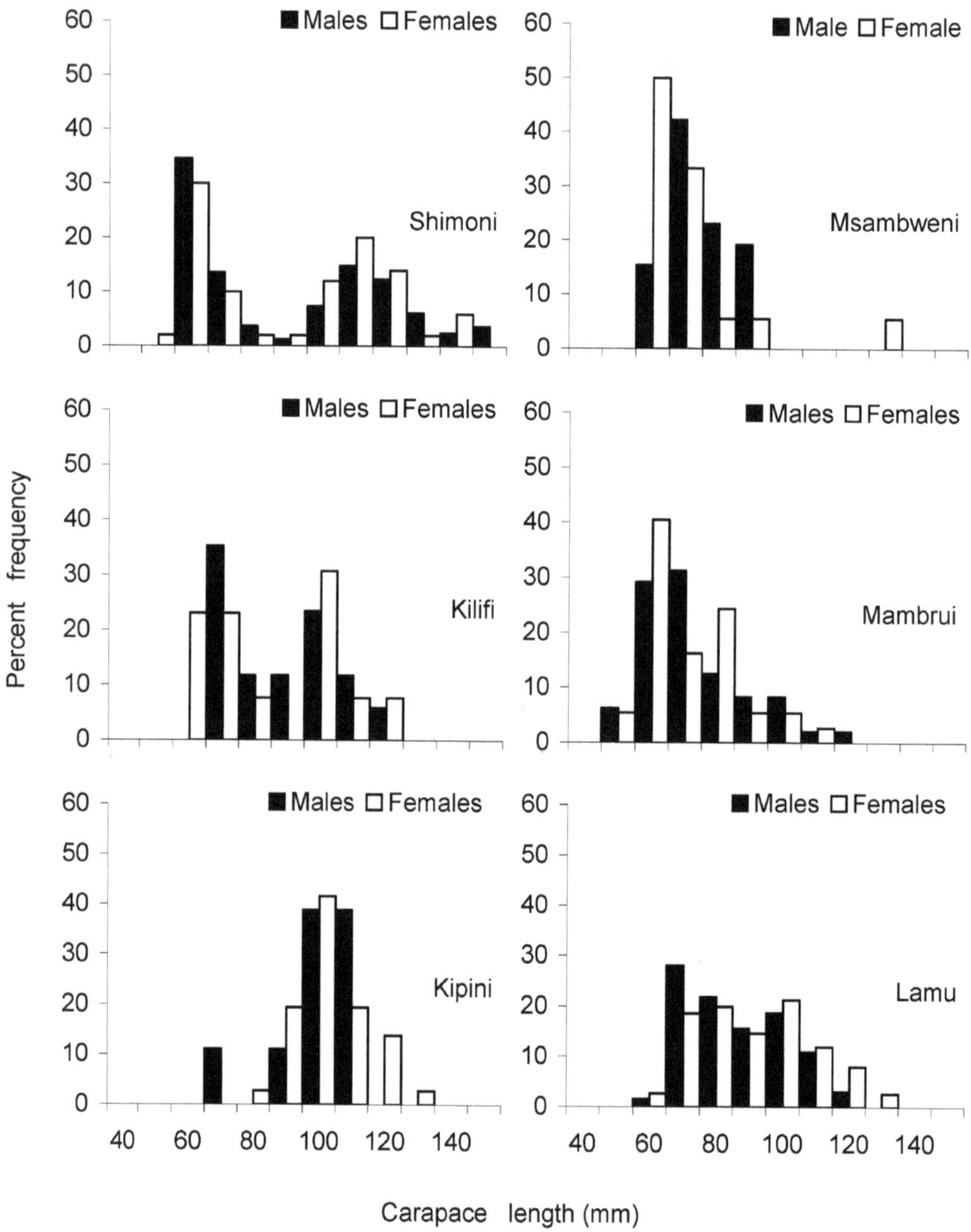

Figure 7. Size frequency distributions of *Panulirus ornatus* from six sampling sites during the study period (November 2000–March 2001).

Table 2. Descriptive statistics of the carapace length of male and female spiny lobsters caught off the six sampling sites during the study period (November 2000–March 2001).

Species		Shimoni		Msambweni		Kilifi		Mambrui		Kipini		Lamu		All
		M	F	M	F	M	F	M	F	M	F	M	F	
Panulirus ornatus	N	124	79	27	20	18	13	48	37	50	61	215	195	887
	Mean (CL) ± SD	86 ± 31	87 ± 28	69 ± 11	66 ± 17	84 ± 18	79 ± 21	69 ± 16	67 ± 15	95 ± 11	94 ± 10	82 ± 15	87 ± 17	82 ± 22
	Min	55	50	53	53	61	53	42	45	70	79	60	54	42
	Max	155	140	90	125	120	112	116	106	108	127	114	127	155
Panulirus homarus	N	4	3	1	1	4	2	111	129	213	232	89	77	866
	Mean (CL) ± SD	72 ± 9	70 ± 11			71 ± 7	64 ± 7	69 ± 12	62 ± 10	74 ± 14	70 ± 11	73 ± 10	68 ± 9	69 ± 12
	Min	61	59	68	57	65	60	45	44	49	49	67	50	44
	Max	82	80			80	69	97	86	100	99	103	93	103
Panulirus longipes	N	143	170	79	73	33	29	9	11	42	18	75	65	747
	Mean (CL) ± SD	72 ± 11	67 ± 9	70 ± 10	67 ± 8	73 ± 9	67 ± 12	57 ± 7	59 ± 6	77 ± 9	72 ± 10	76 ± 11	68 ± 11	70 ± 11
	Min	45	48	54	50	55	48	45	50	55	51	45	50	45
	Max	97	98	97	89	93	81	68	67	95	87	95	87	97

Table 3. Kruskal–Wallis and Dunn's post hoc test results for the carapace length of spiny lobsters caught off the six sampling sites during the study period (November 2000–March 2001). Statistically significant ($p < 0.05$) values in bold.

P. ornatus	Kruskal–Wallis Test			*P. longipes*	Kruskal–Wallis Test			*P. homarus*	Kruskal–Wallis Test		
	H	**Degree of Freedom**	***p* Value**		**H**	**Degree of Freedom**	***p* Value**		**H**	**Degree of Freedom**	***p* Value**
Site	89.791	5	**<0.001**	Site	60.357	5	**<0.001**		35.937	4	**<0.001**
Dunn's post hoc test				Dunn's post hoc test				Dunn's post hoc test			
Peer-to-peer comparison	Test static	Standard error	*p* value	Peer-to-peer comparison	Test static	Standard error	*p* value	Peer-to-peer comparison	Test static	Standard error	*p* value
Mambrui vs. Msambweni	−2.409	25.918	0.926	Mambrui vs. Msambweni	−177.724	42.934	**<0.001**	Mambrui vs. Kilifi	57.844	67.464	0.391
Mambrui vs. Kalifi	93.692	29.636	**0.002**	Mambrui vs. Shimoni	−192.952	41.929	**<0.001**	Mambrui vs. Lamu	74.679	16.477	**<0.001**
Mambrui vs. Shimoni	102.079	19.437	**<0.001**	Mambrui vs. Kilifi	227.731	51.687	**<0.001**	Mambrui vs. Kipini	91.055	17.132	**<0.001**
Mambrui vs. Lamu	116.511	19.216	**<0.001**	Mambrui vs. Lamu	258.016	43.450	**<0.001**	Mambrui vs. Shimoni	94.903	62.586	0.129
Mambrui vs. Kipini	195.918	24.286	**<0.001**	Mambrui vs. Kipini	306.742	46.531	**<0.001**	Kilifi vs. Lamu	−16.835	67.830	0.804
Msambweni vs. Kilifi	91.283	33.043	**0.006**	Msambweni vs. Shimoni	−15.227	18.805	0.418	Kilifi vs. Kipini	−33.211	67.992	0.625
Msambweni vs. Shimoni	−99.670	24.317	**<0.001**	Msambweni vs. Kilifi	50.006	35.596	0.160	Kilifi vs. Shimoni	−37.060	90.810	0.683
Msambweni vs. Lamu	114.103	24.140	**<0.001**	Msambweni vs. Lamu	80.292	21.988	**<0.001**	Lamu vs. Kipini	16.376	18.520	0.377
Msambweni vs. Kipini	193.509	28.342	**<0.001**	Msambweni vs. Kipini	129.017	27.581	**<0.001**	Lamu vs. Shimoni	20.224	62.980	0.748
Kilifi vs. Shimoni	−8.387	28.246	0.767	Shimoni vs. Kilifi	34.779	34.378	0.312				
Kilifi vs. Lamu	−22.819	28.095	0.417	Shimoni vs. Lamu	65.064	19.956	**0.001**				
Kilifi vs. Kipini	−102.226	31.778	**0.001**	Shimoni vs. Kipini	113.790	25.990	**<0.001**				
Shimoni vs. Lamu	14.433	16.994	0.396	Kilifi vs. Lamu	−30.285	36.217	0.403				
Shimoni vs. Kipini	93.839	22.568	**<0.001**	Kilifi vs. Kipini	−79.011	39.862	**0.047**				
Lamu vs. Kipini	79.407	22.378	**<0.001**	Lamu vs. Kipini	48.726	28.378	0.086				

Table 4. Sex ratios of three spiny lobster species from the six sampling sites during the study period (November 2000–March 2001). Statistically significant ($p < 0.05$) values in bold.

Species		Shimoni	Msambweni	Kilifi	Mambrui	Kipini	Lamu	All
Panulirus ornatus	N	203	47	31	85	111	410	887
	Male	124	27	18	48	50	215	482
	Female	79	20	13	37	61	195	405
	Sex ratio (M:F)	1:0.64	1:0.74	1:0.72	1:0.77	1:1.22	1:0.91	1:084
	χ^2	9.98	1.04	0.81	1.42	1.09	0.98	6.68
	p-value	**0.01**	0.31	0.37	0.23	0.30	0.32	**0.01**
Panulirus homarus	N	7	2	6	240	445	166	866
	Male	4	1	4	111	213	89	422
	Female	3	1	2	129	232	77	444
	Sex ratio (M:F)	1: 0.75	1:1	1:0.50	1: 1.16	1:1.08	1:0.87	1:1.11
	χ^2	0.14	0.0	0.67	1.35	0.81	0.87	0.56
	p-value	0.71	1.0	0.41	0.21	0.37	0.35	0.46
Panulirus longipes	N	313	152	62	20	60	140	747
	Male	143	79	33	11	42	75	381
	Female	170	73	29	9	18	65	366
	Sex ratio (M:F)	1:1.19	1:0.92	1:0.79	1:1.20	1:0.42	1:0.87	1:0.96
	χ^2	2.33	0.24	0.26	0.20	9.60	0.71	0.58
	p-value	0.13	0.63	0.61	0.66	**0.01**	0.40	0.30

4. Discussion

4.1. Species Distribution

In this study, the five species of spiny lobsters previously recorded in Kenya [1,7,8,10] were encountered in the catches, albeit with varying spatial distributions along the coastline. Of the five species, three were found at all six sites, one at four sites, and the remaining at three. Spiny lobster distributions are influenced by various physical factors, including but not limited to seawater quality, wave action, tidal range, and substrate characteristics [11–14]. While the ecological preferences of the five palinurid lobsters were not investigated in this study, it appears that the nearshore biotopes (coral reefs, mangroves, seagrass beds, and estuaries) are more habitable for *P. ornatus*, *P. homarus*, *P. longipes*, and *P. versicolor* than for *P. penicillatus*, given the former four species' broader distributions. This is entirely in agreement with the views of George [13], who classifies the former four palinurids as continental and coral species that thrive in coastal areas where the climate of adjacent land has a major influence and/or where there are extensive coral formations. In contrast, *P. penicillatus* is found in optimal conditions around oceanic islands away from the influence of terrestrial runoff and rarely occurs in estuarine areas like Mambrui and Kipini, an observation made in this study.

4.2. Catch Composition

The catch composition of the artisanal lobster fishery analysed in this study differed significantly from that of the 1970s. According to Muteryera [1], who conducted lobster assessment surveys between 1975 and 1978, *P. ornatus* often accounted for 80% and, in some cases, 100% of the landings, while *P. homarus* and *P. longipes* each contributed 5% to the catch during that period. The contribution of the remaining two species (*P. penicillatus* and *P. versicolor*) was negligible, rarely exceeding 1% of the total catch. As evidenced by subsequent lobster studies undertaken in 2000 [8], 2000–2001 [present study] and 2013–2014 [7], however, the catch composition appears to have shifted considerably over time in favor of *P. homarus* and *P. longipes*. Both Fielding and Everett [8] and Mueni et al. [7] reported a significant drop in the relative abundance of *P. ornatus* (45%, 55%) and an inverse considerable increase in the proportions of *P. longipes* (20%, 26%) and *P. homarus* (20%, 13%), respectively, compared to the findings of 1970s lobster surveys [1,10]. Mueni et al. [7] did not survey the estuarine sites of Mambrui and Kipini, where *P. homarus* is the most abundant species, which may have underestimated this species' proportions in the overall catch. In the present study, *P. ornatus* dominated the catch at only one of the six sampling sites, indicating that its historically dominant status as the most harvested species has been reversed along much of the coastline (Table 2). Moreover, its contribution to the overall catch was just 33%, slightly more than that of *P. homarus* (32%) and *P. longipes* (28%), both of which have now become as important in the catch as *P. ornatus*.

Variations in spiny lobster catch composition and abundance across landing sites are determined by various factors, including depth, substratum type, and harvesting method [15]. However, the observed change in the catch composition cannot be explained by these factors alone, given that neither the fishing method nor the lobster fishing grounds have changed over the last seventy years. The increased relative abundance of *P. longipes* and *P. homarus* observed both in this research and in the 2000 and 2013/2014 studies suggest that these two species are more resilient to high levels of fishing-induced mortality than *P. ornatus*, probably due to a faster growth rate, higher reproductive output, more successful recruitment, and juvenile survival [8]. In such a scenario, *P. longipes* and *P. homarus* would ultimately outcompete *P. ornatus* under prolonged, intense fishing pressure, provided all other variables remained constant. Although the effects of fishing may partly explain the observed shifts in the catch composition in the artisanal lobster fishery, changes in lobster habitats due to climate-induced coral die-off cannot be ruled out as one of the likely causes of the decreased abundance of *P. ornatus* in recent years. Repeated bleaching and subsequent mortality of shallow-water hard coral may have altered the epi-benthic cover, particularly in terms of shelter and food availability, affecting the recruitment and survival of *P. ornatus* [8].

4.3. Size Composition

The size composition of lobsters sampled at the six sites varied, with the Mambrui samples containing the smallest specimens among the sites. Variations in the size composition of lobster catches across different geographic locations can be attributed to various factors. These factors include fishing pressure [16–18], density-dependent factors [19,20], food availability [21], migration [22], and the combined effects of differences in exploitation levels, oceanographic conditions, and ecological conditions [23]. Although the influence of such factors on the Kenyan spiny lobster populations is unknown and cannot be ruled out, there appears to be a relationship between lobster size composition and the level of fishing intensity at the different sampling sites. Lobsters from less intensely fished areas were considerably larger in size than those from heavily fished grounds. Fishing intensity is higher in areas where both fishing grounds and markets are readily accessible to fishers than in locations where both are remote enough to make lobster fishing less profitable for fishers and dealers. Among the sites, Mambrui and Kipini exemplify the two extremes of the fishing intensity spectrum.

Mambrui is very close to the popular tourist town of Malindi, with a consistent and substantial demand for lobsters throughout the year. Moreover, the absence of mangroves along this stretch of coastline makes it much easier for fishers to reach the lobster fishing grounds quickly. In contrast, the Kipini lobster fishing grounds lie several kilometers offshore and are further made less accessible by the deltaic nature of the area. The latter site is also located halfway between Malindi and Lamu towns, away from lobster markets, with no reliable road link to either one until recently.

Because of the artisanal nature of the fishery, the more quantifiable data on fishing effort (i.e., number of lobster boats, fishers, permits, etc.) are not collected systematically across landing sites. The Fisheries Department does not register fishers based on the type of fishing they engage in or their fishing method. Moreover, many fishers do not readily register with the Fisheries Department for fear of paying the required licence fees and other levies.

Irrational exploitation of spiny lobsters in the form of capture and retention of immature and gravid lobsters has been widespread ever since the commercial harvesting of lobsters started in the mid-1950s. Brusher [10] found that 50% of the fishery comprised undersized lobsters. Mutegyera [1] also noted that 89% of all *P. ornatus* females landed at Kizingitini (Lamu) were below the mean size of the berried females (98.9 mm CL), indicating that a significant percentage of the lobsters were harvested before they had a chance to reproduce. Okechi [24] estimated the mean sizes of male and female *P. ornatus* at first capture at 77.5 mm CL and 67.5 mm CL, respectively, suggesting the capture and retention of immature lobsters. More recently, Mueni et al. [7] reported that *P. ornatus*, *P. homarus* and *P. versicolor* were harvested before they reached the minimum legal weight of retention (250 g) based on the length at first capture they estimated for the three species. In the present study, more than half of all *P. ornatus* collected were smaller than the size (84 mm), at which 50% of females reached size at first maturity [7], confirming the continued harvest of undersized lobsters. During a 2001 survey [5], the most frequent complaints of local fishers along the coastline were the scarcity of lobsters in coastal waters and the declining sizes of lobsters caught.

5. Conclusions

When compared to the data from lobster surveys conducted in the 1970s, the results of this study show a significant change in the catch composition of Kenya's artisanal lobster fishery over time. In particular, the results show that the traditionally dominant species, *P. ornatus*, has lost its position as the most harvested species along the coastline and now (at the time of this study) ranks marginally higher in terms of abundance than *P. homarus* and *P. longipes*. An interesting question that comes to mind is why only *P. ornatus* has been negatively affected and not the other four species with which it cooccurs in nearshore habitats under the same environmental conditions. Unlike the affected *P. ornatus*, for example, *P. homarus* and *P. longipes* have thrived and become as important in the catch as *P. ornatus* over time, with a combined contribution of 60%, up from 10% in the 1970s. Although several factors are known to influence catch composition and abundance of lobsters, the available data are insufficient to draw any conclusion on the specific factor(s) responsible for the observed shifts in lobster catch composition. Additional research is therefore required to determine the underlying causes of this anomaly, which appears to have varying impacts on the lobster species. In the short term, research on the fishery should focus on the decades-old marine protected areas in order to obtain unfished reference data to empirically assess the status of the fishery. In the long term, however, continuous monitoring of the fishery is required to identify trends and potential issues, such as overfishing or population shifts, for timely regulatory interventions. Alternatively, researchers can use length-based stock assessment models such as length-based spawning potential ratio (LBSPR), length-based indicators (LBI) or length-based Bayesian biomass approach (LBB) to assess the status of the lobster stocks and establish reference points for their management. Length-based models are a viable alternative to conventional catch-based methods for assessing fisheries that lack substantial datasets (i.e., time series of total catch, absolute or relative abundance, fishing effort, life-history parameters, etc.), given their robustness to produce less biased and reliable estimates as well as the simplicity and cost-effectiveness of length data collection. Moreover, the application of length-based models in lobster management can support efforts to assess stock health, evaluate management effectiveness, and explore different length-based management scenarios.

Funding: This study was funded through a scholarship awarded to the author by the Kenya Belgium Flemish Interuniversity Cooperation (VLIR-IUC-UONBI) Programme.

Institutional Review Board Statement: Not applicable.

Informed Consent Statement: Not applicable.

Data Availability Statement: The data presented in this study are available from the author upon request.

Acknowledgments: I want to thank the various divers from the six landing sites who assisted in lobster sample collection.

Conflicts of Interest: The author declares no conflict of interest.

References

1. Mutegyera, W.B. Some observations on the Kenyan lobster fishery. *East Afr. Agric. For. J.* **1978**, *43*, 401–407. [CrossRef]
2. Okechi, J.K.; Polovina, D.E. An evaluation of artificial shelters in the artisanal spiny lobster fishery in Gazi Bay, Kenya. *S. Afr. J. Mar. Sci.* **1995**, *16*, 373–376. [CrossRef]
3. Kulmiye, A.J.; Mavuti, K.M.; Groeneveld, J.C. Size at onset of maturity of *Panulirus homarus homarus* at Mambrui, Kenya. *Afr. J. Mar. Sci.* **2006**, *28*, 51–55. [CrossRef]
4. Kulmiye, A.J. Reproductive seasonality and fecundity of the spiny lobster *Panulirus* homarus homarus (Linnaeus, 1758) (Decapoda: Achelata: Palinuridae) in Kenya: Implications for sustainable management. *J. Crustac. Biol.* **2022**, *42*, ruac028. [CrossRef]
5. Marshall, N.; Milledge, S.A.H.; Alfonso, P.S. (Eds.) *Trade Review: Stormy Seas for Marine Invertebrates—Trade in Sea Cucumbers, Seashells and Lobsters in Kenya, Tanzania, and Mozambique*; Traffic Eastern/Southern Africa: Nairobi, Kenya, 2001.
6. Maina, G.W.; Samoilys, M.A. *An Assessment of the State of the Lobster Fishery in Kiunga National Marine Reserve (KNMR), Kenya*; CORDIO East Africa/UNDP SGP Report; CORDIO East Africa: Mombasa, Kenya, 2010.
7. Mueni, E.M.; Manyala, J.O.; Kimani, E.N.; Maina, G.W.; Fulanda, B.M.; Olendo, M.; Nyongesa, P.; Mzingirwa, F.; Athman, A. *Stock Assessment of Kenya Lobster Fishery*; KMFRI-KCDP Research Report; KMFRI: Mombasa, Kenya, 2016.
8. Fielding, P.J.; Everett, B.I. The spiny lobster fishery in the Kiunga Marine Reserve, northern Kenya. *S. Afr. Assoc. Mar. Biol. Res.* **2000**, *194*, 1–43.
9. Kulmiye, A.J.; Mavuti, K.M. Growth and moulting of captive *Panulirus homarus homarus* in Kenya, western Indian Ocean. *N. Z. J. Mar. Freshw. Res.* **2005**, *39*, 539–549. [CrossRef]
10. Brusher, H. Report to the Government of Kenya on the evaluation of prawn (Penaeidae) and spiny lobster (Palinuridae) marine resources. In *United Nations Development Programme KEN/68/9, No. TA 3006*; Food and Agriculture Organisation: Rome, Italy, 1971.
11. De Bruin, G.H.P. The ecology of spiny lobsters *Panulirus* spp. of Ceylon waters. *Bull. Fish. Res. Stn. Ceylon* **1969**, *20*, 171–189.
12. Berry, P.F. The spiny lobster (Palinuridae) of the coast of Southern Africa: Distribution and ecological notes. *South Afr. Assoc. Mar. Biol. Res. Investig. Rep.* **1971**, *27*, 1–22.
13. George, R.W. Coral reefs and rock lobster ecology in the Indo-Pacific region. In Proceedings of the Second International Coral Reef Symposium; University of Queensland: Brisbane, Australia, 1974; Volume 1, pp. 321–325.
14. Pitcher, R. *Spiny Lobster*; Wright, A., Hill, L., Eds.; Nearshore Marine Resources of the South Pacific; Institute of Fisheries Agency, Honiara and International Centre for Ocean Development: Halifax, Canada, 1993; pp. 539–608.
15. Ardill, J.D.; Sanders, M.J. (Eds.) Review of the Marine Fishery sub-sector for Kenya. In *Priorities for Fisheries Management and Development in the Southwest Indian Ocean*, Proceedings of a Seminar Held 3–5 September 1991, Albion, Mauritius; FAO Report No. 457; Food and Agriculture Organisation of the United Nations: Rome, Italy, 1991; pp. 53–59.
16. Skud, B.E. The effect of fishing on size composition and sex ratio of offshore lobster stocks. *Skr. Ser. Havundersögelser* **1969**, *15*, 295–309.
17. Lyons, W.G.; Barber, D.G.; Foster, S.M.; Kennedy, F.S.; Milano, G.R. *The Spiny Lobster Panulirus Argus in the Middle and Upper Florida Keys: Population Structure, Seasonal Dynamics, and Reproduction*; Florida Marine Research Publication: Petersburg, FL, USA, 1981; pp. 1–45.
18. Al-Abdulsalaam, T.Z. Effects of Exploitation on Life History Characteristics of Spiny Lobster Populations (*Panulirus homarus*) off Oman. Master's Thesis, Oregon State University, Corvallis, OR, USA, 1989; pp. 1–59.
19. Addison, T.J. Density mortality and the relationship between size composition and fishing effort in lobster populations. *Can. J. Fish. Aquat. Sci.* **1986**, *43*, 2360–2367. [CrossRef]
20. Polovina, D.E. Density-dependence in spiny lobster, *Panulirus marginatus*, in the North-western Hawaiian Islands. *Can. J. Fish. Aquat. Sci.* **1989**, *46*, 660–665. [CrossRef]
21. Pollock, D.E.; Cockcroft, A.C.; Goosen, P.C. A note on reduced rock lobster growth rates and related environmental anomalies in the Southern Beguile, 1988–1995. *S. Afr. J. Mar. Sci.* **1997**, *18*, 287–293. [CrossRef]
22. Skewes, T.D.; Pitcher, C.R.; Trendall, J.T. Changes in the size structure, sex ratio and moulting activity of a population of ornate rock lobster *Panulirus ornatus* caused by an annual maturation moult and migration. *Bull. Mar. Sci.* **1994**, *54*, 38–48.
23. Mohan, R. Size structure and reproductive variation of the spiny lobster *Panulirus homarus* over a relatively small geographic range along Dhofar coast in the Sultanate of Oman. *Mar. Freshw. Res.* **1997**, *48*, 1085–1091. [CrossRef]
24. Okechi, J.K. *The Fisheries Growth, Mortality, Exploitation, and Recruitment Rates of the Spiny Lobster Panulirus ornatus (Fabricius 1798) from Lamu Archipelago Kenya*; A Paper Presented at the 1st Pan African Fisheries Congress and Exhibition 30 July–4 August 1995; UNEP: Nairobi, Kenya, 1995.

biology MDPI

Article

Comparative Transcriptome Analysis of the Response to *Vibrio parahaemolyticus* and Low-Salinity Stress in the Swimming Crab *Portunus trituberculatus*

Dongfang Sun [1,†], Jianjian Lv [1,2,†], Yukun Li [1], Jie Wu [1], Ping Liu [1,2] and Baoquan Gao [1,2,*]

1 National Key Laboratory of Mariculture Biobreeding and Sustainable Goods, Yellow Sea Fisheries Research Institute, Chinese Academy of Fishery Sciences, Qingdao 266071, China; sundf@ysfri.ac.cn (D.S.); lvjj@ysfri.ac.cn (J.L.); liyukun969@163.com (Y.L.); 17806287246@163.com (J.W.); liuping@ysfri.ac.cn (P.L.)

2 Laboratory for Marine Fisheries Science and Food Production Processes, Laoshan Laboratory, Qingdao 266237, China

* Correspondence: gaobq@ysfri.ac.cn

† These authors contributed equally to this work.

Simple Summary: In this study, we investigated the effects of pathogen infection and low-salinity stress on organismal immunity during *Portunus trituberculatus* culture by utilizing transcriptome technology. We depicted the dynamic expression process of the transcriptome under pathogenic infection and low-salinity stress. In addition, several immune genes were screened that were jointly affected by immune infection and low-salinity stress. These results contribute to our understanding of the impact of environmental factors on the immunoregulatory mechanisms of *P. trituberculatus*.

Abstract: *Vibrio parahaemolyticus* is one of the main pathogenic bacteria of *Portunus trituberculatus* and causes mass mortality of *P. trituberculatus* in aquaculture. In addition, low-salinity stimulation makes *P. trituberculatus* more susceptible to *V. parahaemolyticus* infections. In order to elucidate the molecular mechanism of resistance to *V. parahaemolyticus* in *P. trituberculatus*, comparative transcriptomic analysis of blood cells stimulated by low salinity and *V. parahaemolyticus* was carried out in this study. Transcriptome sequencing of low-salinity stress and pathogen infection at different time points was completed using Illumina sequencing technology. A total of 5827, 6432, 5362 and 1784 differentially expressed genes (DEGs) involved in pathways related to ion transport and immunoregulation were found under low-salinity stress at 12, 24, 48 and 72 h compared with the control at 0 h. In contrast, 4854, 4814, 5535 and 6051 DEGs, which were significantly enriched in Toll and IMD signaling pathways, were found at 12, 24, 48 and 72 h compared with the control at 0 h under *V. parahaemolyticus* infection. Among them, 952 DEGs were shared in the two treatment groups, which were mainly involved in apoptosis and Hippo signaling pathway. Cluster analysis screened 103 genes that were differentially expressed in two factors that were negatively correlated, including immunoglobulin, leukocyte receptor cluster family, scavenger receptor, macroglobulin and other innate-immune-related genes. These results provide data support for the analysis of the mechanisms of immunity to *V. parahaemolyticus* under low-salinity stress in *P. trituberculatus* and help to elucidate the molecular mechanisms by which environmental factors affect immunity.

Keywords: *Portunus trituberculatus*; *Vibrio parahaemolyticus*; low salinity

1. Introduction

The swimming crab *Portunus trituberculatus* belongs to Crustacea, Decapoda, Portunidae, and is widely distributed in the coastal waters of China, Korea and Japan. The swimming crab possesses the characteristics of fast growth and strong adaptability. In China, the annual aquaculture production of swimming crab, which has become another important economic crustacean following shrimps and blue crabs, reaches 120,000 tons.

However, with the increasing scale of aquaculture, intensive aquaculture results in the frequent occurrence of diseases [1,2], among which *Vibrio parahaemolyticus* is a major pathogen causing diseases in *P. trituberculatus* [3,4]. This pathogen causes hepatopancreatic tissue damage and slow growth in swimming crabs, thus generating huge economic losses in the aquaculture industry. However, the outbreak of disease is commonly associated with alterations in the environment. Environmental stresses induce decreased immunity and increased sensitivity to pathogenic infections in crustaceans [5]. For instance, an increase or decrease in salinity can lead to a decrease in the immunity of tiger shrimp [6]. In addition, changes in temperature, dissolved oxygen, ammonia nitrogen and biological contaminants also affect the immune response [5]. Immunological studies on *V. parahaemolyticus* in crustaceans have been reported; e.g., immune genes showed corresponding expression patterns in response to infection [7,8]. However, fewer reports are available on the mechanism of immune response in swimming crabs under environmental stress.

Innate immunity is an important way in which crustaceans resist pathogen invasion, and immune-related genes perform vital roles in this process. Transcriptome sequencing technology is widely applied to mine genes related to traits such as immunity, body color and sex in aquatic animals [9–14], and its utilization has been reported in species such as fish, shrimp and shellfish to explore the expression patterns of genes related to growth, development and sex [9,12,13]. More studies have reported utilizing comparative transcriptomics to mine crustacean immune-related genes, mainly focusing on economic crustaceans, including *Litopenaeus vannamei, Eriocheir sinensis* and *Scylla paramamosain* [15–21], which provided an essential foundation for analyzing the immune mechanisms of crustaceans.

In this study, blood cells at 0, 12, 24, 48 and 72 h post low-salinity stress and *V. parahaemolyticus* treatment were analyzed for the comparative analysis of dynamic transcriptome sequencing. The results are of great significance in analyzing the immune mechanism of *P. trituberculatus*, as well as contributing to the understanding of the molecular mechanisms of the effects of environmental stress on the immune system.

2. Materials and Methods

2.1. Animals

Swimming crabs (80 days, 25.6 ± 3.2 g) were collected from a full-sib family line from a local aquaculture farm in Weifang, Shandong Province, China. Prior to commencing the formal experiment, the swimming crabs underwent a one-week acclimation period. During this period, the water temperature was consistently maintained at 24 °C, oxygen was continually introduced into the aquatic environment, and miscellaneous fish were periodically fed according to the previous study [22]. The assay was divided into a low-salinity group (Ls, salinity 11‰) and a challenge group (Vp, *V. parahaemolyticus* concentration of 10^5 CFU/mL) with 60 swimming crabs in each group. In the challenge group, 100 μL of *V. parahaemolyticus* was injected into the basement membrane of the first joint at the base of the swimming foot. The *V. parahaemolyticus* used in this study was obtained from the Research Laboratory of Disease Control and Molecular Pathology of Cultured Organisms, Yellow Sea Fisheries Research Institute. At the same time, we obtained authorization from the Yellow Sea Fisheries Research Institute to use the pathogen for infection experiments. The feeding conditions during the experiment were consistent with those during the temporary feeding period. The blood cells were taken from three crabs randomly selected from each group at 0 h, 12 h, 24 h, 48 h and 72 h, and 0 h served as the control group. Swimming crabs were treated with hypothermia to induce cold shock, and sampling of blood cells was performed according to the previous procedure [22]. Three samples were taken at each time point, and each sample was generated from a mixture of blood cells from five crabs.

2.2. Library Preparation and RNA-Seq

The RNA was isolated from the blood cells by utilizing the TRIzol method according to the manufacturer's instructions. RNA integrity was assessed using the RNA Nano 6000 Assay Kit of the Bioanalyzer 2100 system (Agilent Technologies, Santa Clara, CA,

USA). A total amount of 1 μg RNA per sample was used as input material for the RNA sample preparations. Briefly, mRNA was purified from total RNA using poly-T oligo-attached magnetic beads. Fragmentation was carried out using divalent cations under elevated temperature in First Strand Synthesis Reaction Buffer(5×). First-strand cDNA was synthesized using random hexamer primer and M-MuLV Reverse Transcriptase (RNase H-). Second-strand cDNA synthesis was subsequently performed using DNA Polymerase I and RNase H. Remaining overhangs were converted into blunt ends via exonuclease/polymerase activities. After adenylation of 3′ ends of DNA fragments, adaptors with a hairpin loop structure were ligated to prepare for hybridization. In order to select cDNA fragments preferentially 370~420 bp in length, the library fragments were purified with the AMPure XP system (Beckman Coulter, Beverly, CA, USA). Then, PCR was performed with Phusion High-Fidelity DNA polymerase, Universal PCR primers and Index (X) Primer. Finally, PCR products were purified (AMPure XP system) and library quality was assessed on the Agilent Bioanalyzer 2100 system. The library preparations were sequenced on an Illumina Novaseq platform, and 150 bp paired-end reads were generated.

2.3. Data Analysis

The raw data were processed for quality control using Fastp software (v0.23.1) to obtain cleandata. All the downstream analyses were based on the clean data with high quality. The reference genome (GCA_020740555.1) was indexed using Hisat2 (v2.0.5), and clean reads were aligned to the reference genome by utilizing Hisat2 with default parameters. FeatureCounts (v1.5.0) was used to count the reads numbers mapped to each gene. And then FPKM (fragments per kilobase of exon model per million mapped fragments) of each gene was calculated based on the length of the gene and reads count mapped to the gene. Differentially expressed genes (DEGs) analysis of two groups was performed using the DESeq2 (v1.20.0) with default parameters. DESeq2 provides statistical routines for determining differential expression in digital gene expression data using a model based on the negative binomial distribution. The resulting p-values were adjusted using Benjamini and Hochberg's approach for controlling the false discovery rate. Only genes with p-value < 0.05 calibrated by false discovery rate (FDR) and absolute value of log2FoldChange > 0 were considered as DEGs.

2.4. GO and KEGG Enrichment Analysis of DEGs

GO functional enrichment analysis of differential gene sets was performed using clusterProfile software (v4.0), with padj < 0.05 as the threshold for being significantly enriched. KEGG is a database resource for understanding high-level functions and utilities of the biological system, such as the cell, the organism and the ecosystem, from molecular-level information, especially large-scale molecular datasets generated by genome sequencing and other high-through put experimental technologies (http://www.genome.jp/kegg/, accessed on 1 April 2023). We used clusterProfiler R package to test the statistical enrichment of differential expression genes in KEGG pathways.

2.5. Quantitative Real-Time PCR Validation

The RNA was isolated from the blood cells by utilizing the Trizol method according to the manufacturer's instruction. Blood cells from five swimming crabs were mixed into one sample, and three biological replicates were used for each time point. cDNA was reverse-transcribed from 1 μg of RNA using the HiScript II Q RT SuperMix for RT-qPCR (+gDNA wiper) kit (Vazyme, Nanjing, China) according to the instructions. RT-qPCR was performed following a 5-fold dilution of the cDNA. The volume of the qPCR reaction system was 20 μL, containing 10 μL of 2 × SYBR Master Mix (Takara, Osaka, Japan), 0.4 μL of primer (Table S1) and 0.2 μL of ROX Reference Dye II, 2 μL of cDNA template and RNase-free water. qPCR was carried out in a 7500 Fast system (ThermoFisher, Waltham, MA, USA) according to previous research [22]. The β-actin gene was used as an endogenous control, and relative expression levels were determined utilizing the $2^{-\Delta\Delta Ct}$ method [23]. The RT-qPCR data were presented as the mean ± S.D.

3. Result

3.1. Information Obtained from the Sequencing Data

In this study, in order to study the interaction between environment and immunity, Illumina Hiseq sequencing technology was used to sequence the transcriptome of blood cell samples from 0 (control), 12, 24, 48 and 72 h under low salinity and pathogen challenge. The mortality rates at 12, 24, 48 and 72 h of pathogenic infection were 18.2%, 9.4%, 3.22% and 1.02%, respectively. The sequencing results show that a total of 174.85G of raw data was obtained, with an average coverage depth of 7.83. The QC results show that the average values of Q20 and Q30 were above 97% and 92%, respectively, with an average mapping rate of 86.46% (Table S2). Ten differentially expressed genes were randomly selected for data validation, and the results show that the RT-qPCR results are basically consistent with the FPKM trend, indicating the accuracy of the sequencing results (Figure S1). All raw data were uploaded to the NCBI (National Center for Biotechnology Information, accession number: PRJNA1039533 and PRJNA1039325).

3.2. Dynamic Expression of the Transcriptome under Low Salinity and Pathogenic Challenge

A total of 5827, 6432, 5362 and 1784 DEGs were found under low-salinity stress at 12, 24, 48 and 72 h compared with the control at 0 h. The number of DEGs tended to increase and then decrease as the duration of stress increased (Figure 1A). In contrast, 4854, 4814, 5535 and 6051 DEGs were found at 12, 24, 48 and 72 h compared with the control at 0 h. The number of DEGs increased gradually with the duration of pathogen stress (Figure 1B). Under low-salinity stress, DEGs were mainly enriched in biological processes related to energy metabolism and ion transport, such as the ATP metabolic process and ion transmembrane transport. These differential genes were mainly involved in apoptosis and proteasome pathways (Figure 2A–D). Surprisingly, they were also enriched for the immune-related pathways Toll and Imd signaling pathways (Figures 3A–D and S2). Differential genes under pathogen stress were mainly related to peptidase synthesis and activity, while differential genes were significantly enriched in immune-related signaling pathways, such as Toll and IMD signaling pathways (Figures 2E–H and 3E–H).

3.3. Expression Pattern Clustering of DEGs

A cluster analysis was performed on the DEGs in the infection and low-salinity groups, and the results reveal that each of the two groups of DEGs was clustered into six categories (Figure 4). In the infection group, subcluster 3 contained the largest number of DEGs, with an initial trend of downregulated expression (Figure 4A). Subcluster 6 was the largest cluster of DEGs in the low-salinity group, with an overall expression trend of upregulation (Figure 4B). According to the results of cluster analysis, 2212 genes showed an overall upregulation trend post infection (subclusters_4, 5 and 6), and 2076 genes showed an overall downregulation trend post infection (subclustesr_1, 2 and 3); there were 2377 genes that showed an overall upward regulation trend after salinity stress (subclusters_5 and 6), and 2236 genes that showed an overall downregulation trend (subclusters_1, 2, 3 and 4).

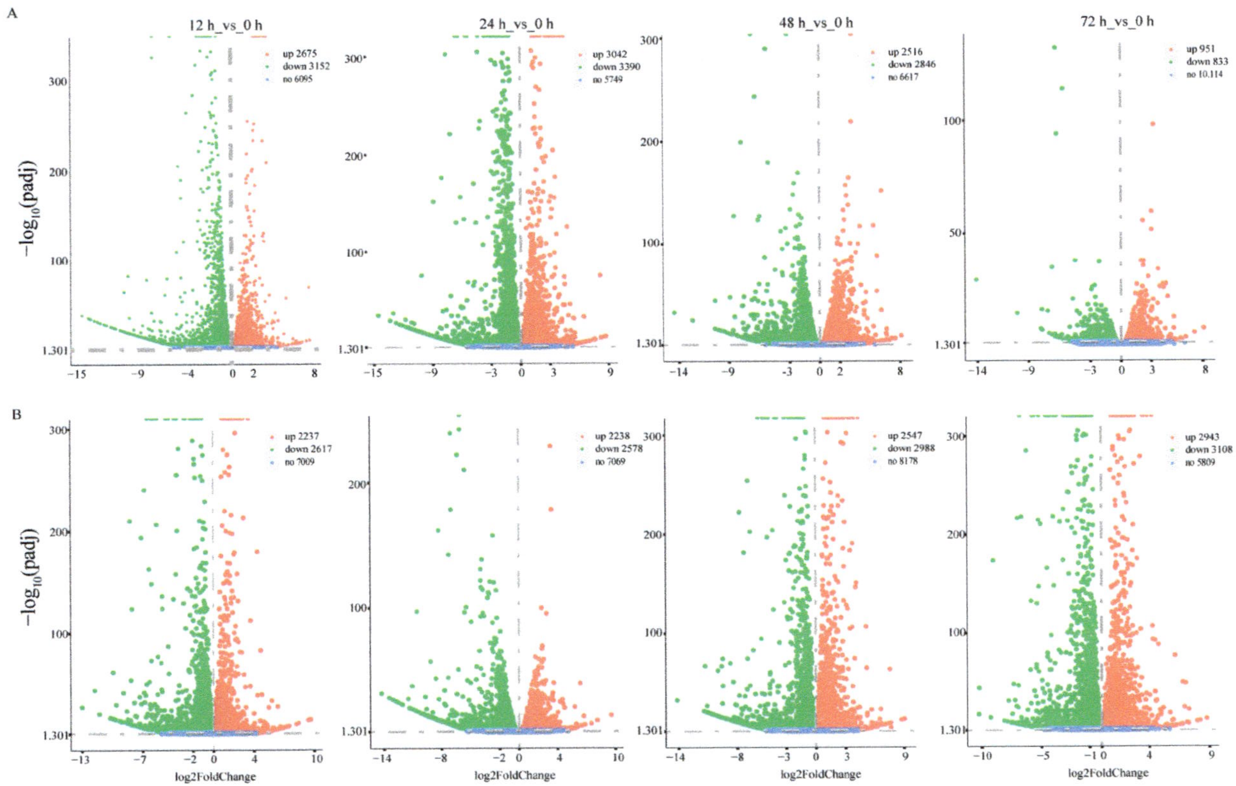

Figure 1. The number of DEGs at different time points post *V. parahaemolyticus* (**A**) and low-salinity (**B**) stress. "up" indicates the number of DEGs that were upregulated, "down" indicates the number of DEGs that were downregulated, and "no" indicates the number of genes that were not significantly different.

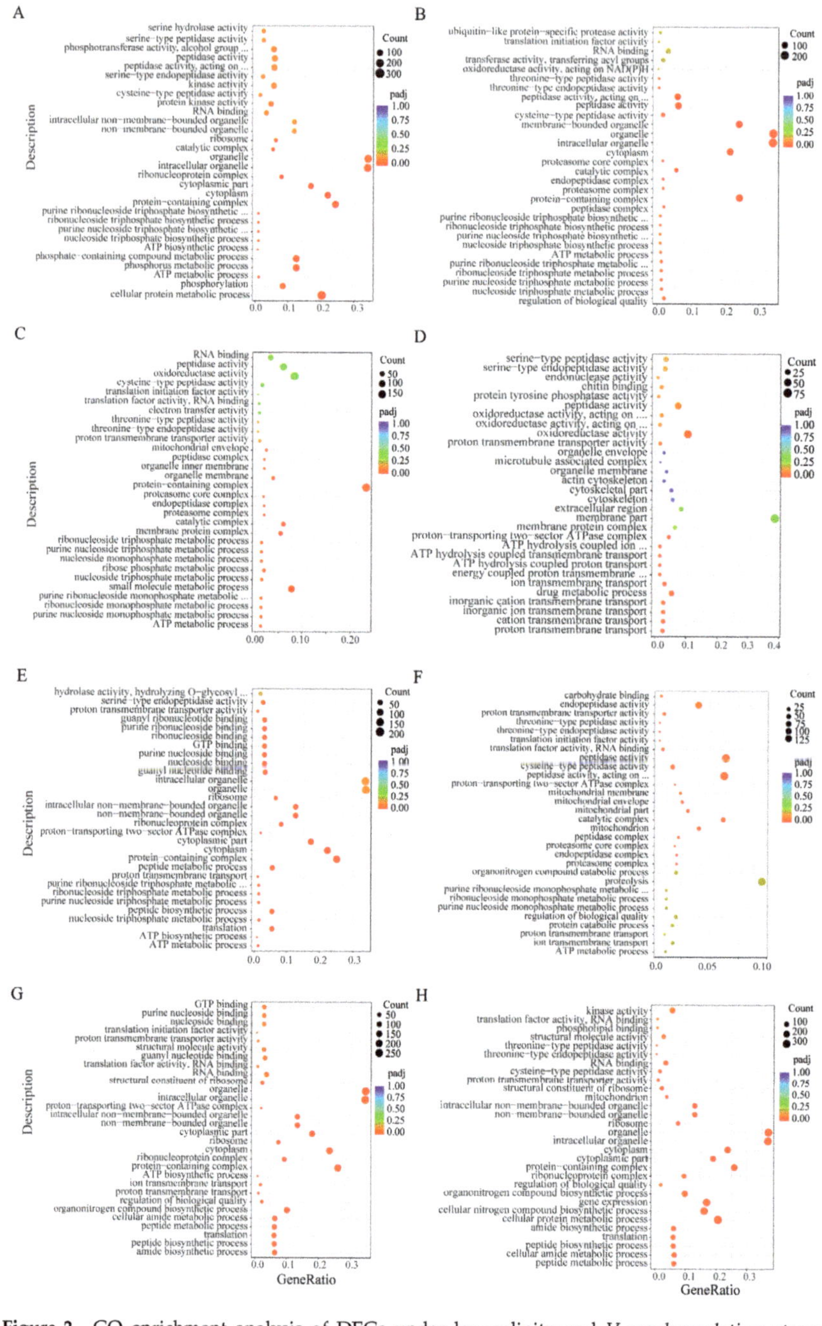

Figure 2. GO enrichment analysis of DEGs under low-salinity and *V. parahaemolyticus* stress. (**A–D**) GO enrichment analysis of DEGs at 12, 24, 48 and 72 h under low-salinity stress compared with 0 h. (**E–H**) GO enrichment analysis of DEGs at 12, 24, 48 and 72 h under *V. parahaemolyticus* stress compared with 0 h.

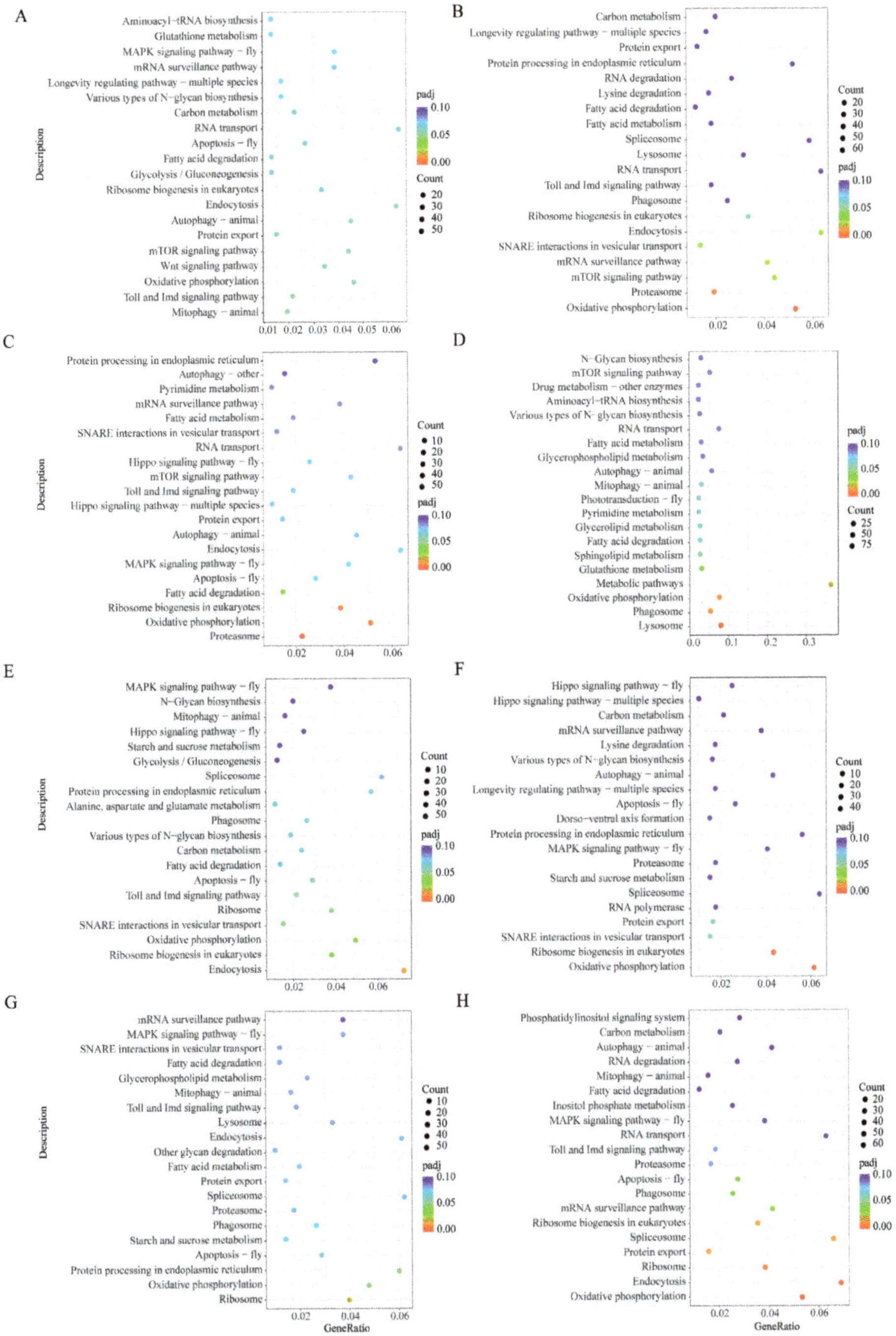

Figure 3. KEGG annotation of DEGs under low salinity and *V. parahaemolyticus*. **(A–D)** KEGG annotation of DEGs at 12, 24, 48 and 72 h under low-salinity stress compared with 0 h. **(E–H)** KEGG annotation of DEGs at 12, 24, 48 and 72 h under *V. parahaemolyticus* stress compared with 0 h.

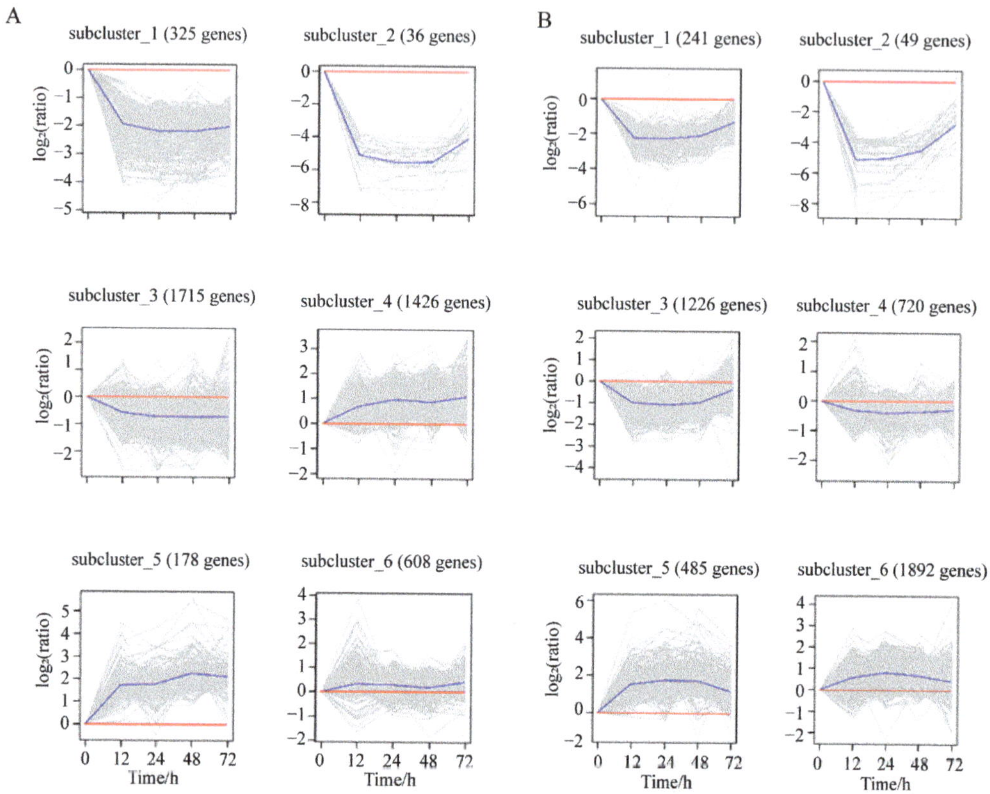

Figure 4. Cluster analysis of differential genes. (**A**) DEG clustering analysis of *V. parahemolyticus* challenge group. (**B**) DEG clustering analysis of low-salinity group. Each subplot shows the expression levels of genes in a subcluster, where the *X*-axis represents the stress treatment time and the *Y*-axis represents the relative expression levels of DEGs. Each gray line is the relative expression of a specific gene at different stress times, while the blue line represents the average relative expression of all genes in the subcluster, and the red line represents the baseline.

3.4. Screening of Reverse DEGs after Exposure to the Two Kinds of Stress

A total of 1281 genes were found to be consistently differentially expressed throughout the low-salinity treatment (Figure 5A). Surprisingly, these genes were mainly enriched in immune-related signaling pathways (Figure S3A). In contrast, a total of 2998 genes were consistently differentially expressed under pathogen stress (Figure 5B), and they were mainly enriched in immune-related signaling pathways (Figure S3B). Looking at the intersection of the DEGs under the treatment of two stress factors, we screened 952 shared genes (Figure 5C). These genes were mainly involved in biological processes related to osmotic pressure regulation, such as proton transmembrane transport (Figure 5D). Furthermore, they were involved in signaling pathways related to immunity and osmotic pressure regulation, such as apoptosis (Figure 5E).

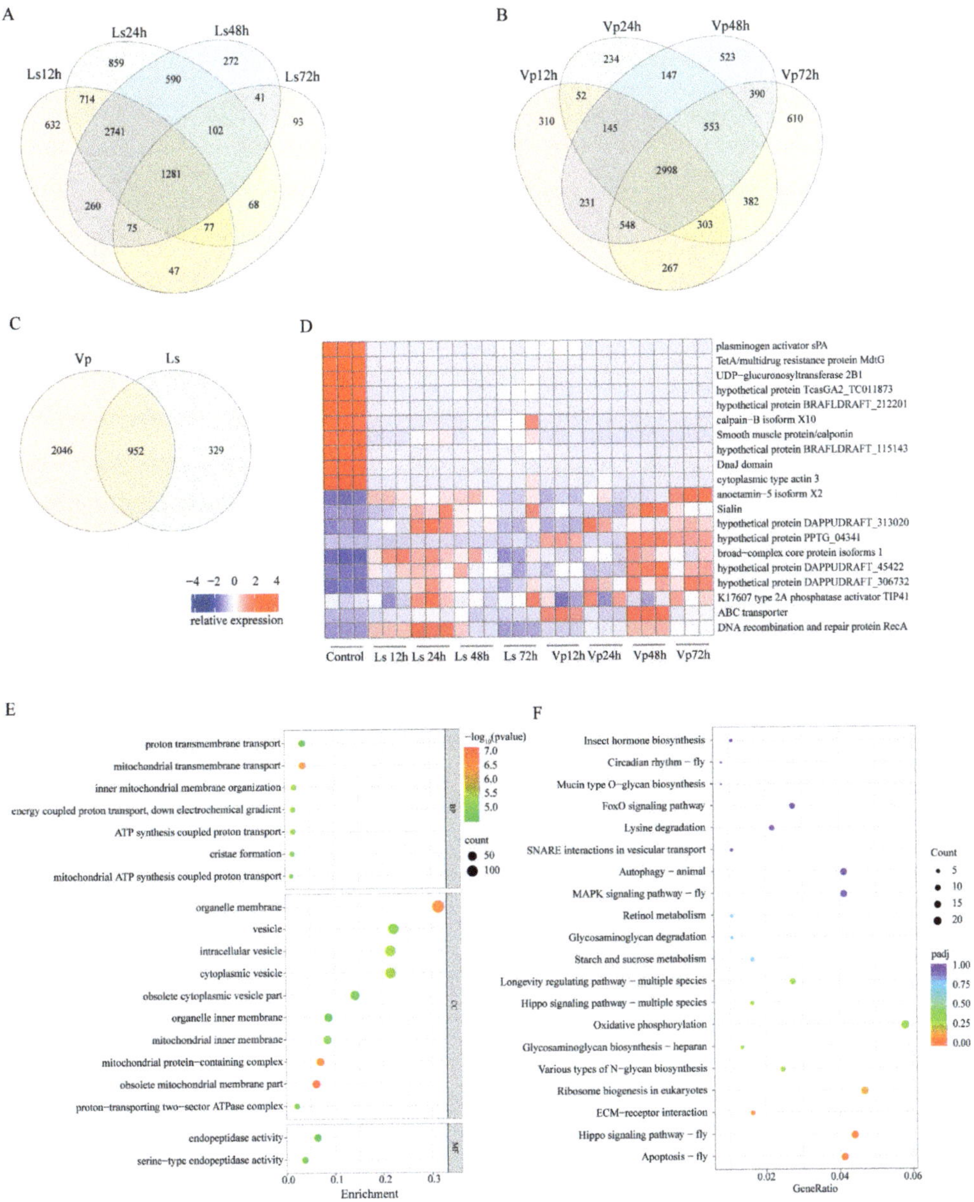

Figure 5. Co-analysis of DEGs under low salinity and pathogenic infections. (**A**) Number of shared DEGs at four time points under low-salinity stress. (**B**) Number of shared DEGs at four time points under pathogen infection. (**C**) Number of shared differential genes under low-salinity and pathogen stress. (**D**) Expression pattern of top 10 DEGs under low-salinity and pathogen stress. (**E**) GO enrichment analysis of shared DEGs. (**F**) KEGG enrichment analysis of shared DEGs.

Based on the results of cluster analysis, DEGs with opposite trends in expression patterns after pathogen infection and low-salinity stress were screened out. Among them, 44 DEGs were upregulated post pathogen infection and downregulated following low-salinity stress (vp_up and ls_down). This was mainly due to the significant upregulation of expression by pathogen infection after 72 h, including for immunoglobulin genes, leukocyte receptor cluster family genes, scavenger receptors, macroglobulins and other genes related to innate immunity. In addition, 59 DEGs were downregulated following pathogen infection and upregulated following low-salinity stress (vp_down and ls_up). This was mainly due to the significant upregulation of expression under low-salinity stress, including chloride ion channels, calcium channels and other genes. The results of KEGG enrichment analysis show that the vp_up and ls_down group was mainly enriched in the Wnt signaling pathway, and the vp_down and ls_up group was not enriched in a significant pathway (Table S3).

4. Discussion

Immune protection in crustaceans is mainly accomplished by blood cells, which play an important role in the innate immune response of crustaceans. Blood cells are distributed throughout the body via blood circulation and play a role in the host's immune response in terms of recognition, phagocytosis, melanogenesis, cytotoxicity and cellular communication as well as immune defenses such as phagocytosis, nodules and cysts [24]. *V. parahaemolyticus* is a major pathogen in the culture of *P. trituberculatus*, which can cause severe pathological damage to tissues, slow growth and difficulty in shelling [25]. Salinity is an environmental factor that cannot be ignored in aquaculture, especially in outdoor pond culture mode. During the rainy season, the reduced salinity of farmed seawater puts *P. trituberculatus* in a low-salinity stress environment, which increases its susceptibility to *V. parahaemolyticus* infection. In this study, the transcriptome dynamics analysis of blood cells posts *V. parahaemolyticus* and salinity stress was carried out. A large number of post-stress DEGs were identified, and correlation analysis of post-stress differential genes between the two factors helped us to analyze the mechanisms of immunity and stress resistance in this species.

P. trituberculatus is a hyperosmotic-regulated species [26]. In the early stage of low-salinity stress, a large number of differential genes were involved in osmotic pressure regulation in the swimming crab, involving a variety of ionic regulation and capacity metabolic pathways. At the later stage of stress, the osmotic pressure in the swimming crab maintained a relative equilibrium with the environmental osmotic pressure, and the number of differential genes decreased. Meanwhile, we found that the differential genes were significantly enriched in signaling pathways related to immunity, which is consistent with studies in other crustaceans and suggests that environmental factors may induce immune responses [27,28]. *V. parahaemolyticus* is a major pathogen of crustaceans. In this study, a large number of individuals died 12 h post pathogen injection, followed by a gradual decrease in mortality. As the duration of pathogen challenge increased, more genes responded to the infection stimulus and participated in the immune response. Many crustaceans have shown that the Toll, IMD and Jak-STAT signaling pathways are involved in the immune regulation of *V. parahaemolyticus* infection [21,29,30]. In this study, the differential genes were significantly enriched in the Toll and IMD signaling pathways, in line with the findings in other crustaceans.

In this study, the expression of differential genes during salinity and pathogen treatments provided us with the dynamic expression patterns of *P. trituberculatus* in response to different stresses and clarified the transcriptional regulatory mechanisms of *P. trituberculatus* in response to salinity and pathogen stresses. The DEGs following pathogenic infection were associated with immunity, such as chitin and lectin [31,32], suggesting that they participate in the resistance of *P. trituberculatus* to invasion by parahaemolyticus disease. In contrast to the gill transcriptome results [33], fewer specific pathways were enriched for DEGs associated with salinity acclimation in hemocytes after low-salinity stress. Therefore, it was hypothesized that hemocytes do not possess a major function in salinity adaptation and that these differential genes may be aberrantly expressed only in low-salinity stimulation.

Notably, the number of DEGs shared by the two groups was 952, and they were mainly enriched in apoptosis and the Hippo signaling pathway. Apoptosis plays a necessary role in the removal of unwanted or abnormal cells in multicellular organisms. The addition of AP to feed reduces ROS accumulation and apoptosis in shrimp, thereby protecting the shrimp from *V. albicans infection* [34]. The Hippo signaling pathway is an evolutionarily conserved pathway that controls organ size by regulating cell proliferation, apoptosis and stem cell self-renewal [35]. These pathways have been reported to be associated with innate immunity of crustaceans [36,37], suggesting that these pathways were involved in the resistance to *V. parahaemolyticus* and were affected by low-salinity stress in *P. trituberculatus*.

Under low-salt stress, crustaceans enhance their immunity by increasing lipid metabolism, glucose metabolism, detoxification and osmotic sensitivity regulation, suggesting that lower immune responses under low-salinity stress indicate a state of immunosuppression and increased disease susceptibility [5]. In the crustacean culture, a large number of deaths were induced after heavy rainfall. *P. trituberculatus* under low-salinity stress was more sensitive to *V. parahaemolyticus* infection, and it was hypothesized that low-salinity stress may lead to reduced immunity by affecting immunity or altering the expression patterns of relevant genes. In this study, we screened a total of 103 DEGs, including 44 DEGs that were upregulated after pathogenic infection and downregulated by low-salinity stress and 59 DEGs that were downregulated after pathogenic infection and upregulated after low-salinity stress. These genes included immunoglobulin genes, leukocyte receptor cluster families, scavenger receptors, macroglobulins and other genes related to innate immunity and stress resistance [38–45], which may play important roles in the reduction in immunity in *P. trituberculatus* under low-salinity stress.

5. Conclusions

In this study, we analyzed the dynamic expression patterns of differential genes during low-salinity and pathogen stresses. The differential genes after pathogen stress were mainly enriched in apoptosis, ribosome biogenesis in eukaryotes, Hippo signaling and other pathways. The number of DEGs shared by the two groups was 952, suggesting that a large number of genes may function in both disease resistance and stress adaptation. Our results provide data support for analyzing the mechanism of immunity to *V. parahaemolyticus* in *P. trituberculatus* and help to elucidate the molecular mechanisms by which salinity affects immunity.

Supplementary Materials: The following supporting information can be downloaded at: https://www.mdpi.com/article/10.3390/biology12121518/s1. Figure S1: qPCR validation of transcriptome data. The RT-qPCR data were presented as the mean ± S.D. Figure S2: qPCR validation of genes in the Toll and IMD signaling pathways. TLR1, Toll-like receptors 1; TLR2, Toll-like receptors 2; SR, scavenger receptor; Mapkkk4, mitogen-activated protein kinase kinase kinase 4; Mapkkk7, mitogen-activated protein kinase kinase kinase 7; Mapkk6, mitogen-activated protein kinase kinase 6. Different lowercase letters indicate significant differences ($p < 0.05$) in RNA-seq; different uppercase letters indicate significant differences ($p < 0.05$) in qPCR. Figure S3: KEGG annotation of DEGs under low salinity and *V. parahaemolyticus*. (A) KEGG annotation of shared DEGs under low-salinity stress. (B) KEGG annotation of shared DEGs under pathogen stress. Table S1: primer sequences for qPCR. Table S2: statistics of the transcriptome sequencing from blood samples of *P. trituberculatus*. Table S3: KEGG of DEGs with opposite trends in expression patterns post pathogen infection and low-salinity stress.

Author Contributions: D.S.: Investigation, Writing—review and editing. J.L.: Formal analysis, Investigation. Y.L.: Writing—review and editing. J.W.: Formal analysis, Investigation. P.L.: Conceptualization, Investigation. B.G.: Conceptualization, Methodology, Funding acquisition, Project administration. All authors have read and agreed to the published version of the manuscript.

Funding: This work was supported by grants from the National Key R and D Program of China (2022YFD2400104), the National Natural Science Foundation of China (42076116, 41976106), the earmarked fund for CARS-48 and the Central Public-interest Scientific Institution Basal Research Fund, CAFS (NO. 2023TD50 and 2023XT0204).

Institutional Review Board Statement: Not applicable.

Informed Consent Statement: Not applicable.

Data Availability Statement: Data are contained within the article and supplementary materials.

Conflicts of Interest: The authors declare that they have no known competing financial interests or personal relationships that could have appeared to influence the work reported in this paper.

References

1. Flegel, T.W. Historic emergence, impact and current status of shrimp pathogens in Asia. *J. Invertebr. Pathol.* **2012**, *110*, 166–173. [CrossRef] [PubMed]
2. Senapin, S.; Phiwsaiya, K.; Gangnonngiw, W.; Flegel, T.W. False rumours of disease outbreaks caused by infectious myonecrosis virus (IMNV) in the whiteleg shrimp in Asia. *J. Negat. Results Biomed.* **2011**, *10*, 10. [CrossRef] [PubMed]
3. Lee, C.T.; Chen, I.T.; Yang, Y.T.; Ko, T.P.; Huang, Y.T.; Huang, J.Y.; Huang, M.F.; Lin, S.J.; Chen, C.Y.; Lin, S.S.; et al. The opportunistic marine pathogen *Vibrio parahaemolyticus* becomes virulent by acquiring a plasmid that expresses a deadly toxin. *Proc. Natl. Acad. Sci. USA* **2015**, *112*, 10798–10803. [CrossRef] [PubMed]
4. Theethakaew, C.; Nakamura, S.; Motooka, D.; Matsuda, S.; Kodama, T.; Chonsin, K.; Suthienkul, O.; Iida, T. Plasmid dynamics in *Vibrio parahaemolyticus* strains related to shrimp Acute Hepatopancreatic Necrosis Syndrome (AHPNS). *Infect. Genet. Evol.* **2017**, *51*, 211–218. [CrossRef] [PubMed]
5. Wang, Z.; Zhang, Y.; Yao, D.; Zhao, Y.; Tran, N.T.; Li, S.; Ma, H.; Aweya, J.J. Metabolic reprogramming in crustaceans: A vital immune and environmental response strategy. *Rev. Aquac.* **2021**, *14*, 1094–1119. [CrossRef]
6. Wang, F.I.; Chen, J.C. Effect of salinity on the immune response of tiger shrimp *Penaeus monodon* and its susceptibility to *Photobacterium damselae* subsp. *damselae*. *Fish Shellfish Immunol.* **2006**, *20*, 671–681. [CrossRef]
7. Zhou, Z.K.; Gu, W.B.; Cong, W.; Zhou, Y.L.; Tu, D.D.; Liu, Z.P.; Zhu, Q.H.; Shu, M.A. Seven transcripts from the chitinase gene family of the mud crab *Scylla paramamosain*: Their expression profiles during development and moulting and under environmental stresses. *Aquac. Res.* **2018**, *49*, 3296–3308. [CrossRef]
8. Zhang, J.; Sun, Y.; Li, F.; Huang, B.; Xiang, J. Molecular characterization and expression analysis of chitinase (Fcchi-3) from Chinese shrimp, *Fenneropenaeus chinensis*. *Mol. Biol. Rep.* **2010**, *37*, 1913–1921. [CrossRef]
9. Lin, X.; Zhou, D.; Zhang, X.; Li, G.; Zhang, Y.; Huang, C.; Zhang, Z.; Tian, C. A first insight into the gonad transcriptome of Hong Kong Catfish (*Clarias fuscus*). *Animals* **2021**, *11*, 1131. [CrossRef]
10. Zhang, Y.; Liu, J.; Peng, L.; Ren, L.; Zhang, H.; Zou, L.; Liu, W.; Xiao, Y. Comparative transcriptome analysis of molecular mechanism underlying gray-to-red body color formation in red crucian carp (*Carassius auratus*, red var.). *Fish Physiol. Biochem.* **2017**, *43*, 1387–1398. [CrossRef]
11. Chen, X.; Chen, J.; Shen, Y.; Bi, Y.; Hou, W.; Pan, G.; Wu, X. Transcriptional responses to low-salinity stress in the gills of adult female *Portunus trituberculatus*. *Comp. Biochem. Physiol. Part D* **2019**, *29*, 86–94. [CrossRef]
12. Wei, J.; Zhang, X.; Yu, Y.; Huang, H.; Li, F.; Xiang, J. Comparative transcriptomic characterization of the early development in Pacific white shrimp *Litopenaeus vannamei*. *PLoS ONE* **2014**, *9*, e106201. [CrossRef] [PubMed]
13. Zuo, S.; Jiang, K.; Li, D.; Yan, X.; Nie, H. Transcriptomic analysis of Manila clam *Ruditapes philippinarum* under lipopolysaccharide challenge provides molecular insights into immune response. *Fish Shellfish. Immunol.* **2020**, *106*, 110–119. [CrossRef]
14. Cao, M.; Zhang, M.; Yang, N.; Fu, Q.; Su, B.; Zhang, X.; Li, Q.; Yan, X.; Thongda, W.; Li, C. Full length transcriptome profiling reveals novel immune-related genes in black rockfish (*Sebastes schlegelii*). *Fish Shellfish Immunol.* **2020**, *106*, 1078–1086. [CrossRef] [PubMed]
15. Interaminense, J.A.; Vogeley, J.L.; Gouveia, C.K.; Portela, R.S.; Oliveira, J.P.; Silva, S.; Coimbra, M.R.M.; Peixoto, S.M.; Soares, R.B.; Buarque, D.S.; et al. Effects of dietary Bacillus subtilis and Shewanella algae in expression profile of immune-related genes from hemolymph of Litopenaeus vannamei challenged with *Vibrio parahaemolyticus*. *Fish Shellfish Immunol.* **2019**, *86*, 253–259. [CrossRef] [PubMed]
16. Panigrahi, A.; Sundaram, M.; Saranya, C.; Swain, S.; Dash, R.R.; Dayal, J.S. Carbohydrate sources deferentially influence growth performances, microbial dynamics and immunomodulation in Pacific white shrimp (*Litopenaeus vannamei*) under biofloc system. *Fish Shellfish Immunol.* **2019**, *86*, 1207–1216. [CrossRef]
17. Pan, L.; Zhang, X.; Yang, L.; Pan, S. Effects of Vibro harveyi and Staphyloccocus aureus infection on hemocyanin synthesis and innate immune responses in white shrimp *Litopenaeus vannamei*. *Fish Shellfish Immunol.* **2019**, *93*, 659–668. [CrossRef]
18. Pang, H.; Wang, G.; Zhou, S.; Wang, J.; Zhao, J.; Hoare, R.; Monaghan, S.J.; Wang, Z.; Sun, C. Survival and immune response of white shrimp *Litopenaeus vannamei* following single and concurrent infections with WSSV and *Vibrio parahaemolyticus*. *Fish Shellfish Immunol.* **2019**, *92*, 712–718. [CrossRef]
19. Zheng, X.; Zhang, L.; Jiang, W.; Abasubong, K.P.; Zhang, C.; Zhang, D.; Li, X.; Jiang, G.; Chi, C.; Liu, W. Effects of dietary icariin supplementation on the ovary development-related transcriptome of Chinese mitten crab (*Eriocheir sinensis*). *Comp. Biochem. Physiol. Part D* **2021**, *37*, 100756. [CrossRef]
20. Yang, Z.G.; Zhou, J.Y.; Wei, B.H.; Cheng, Y.X.; Zhang, L.; Zhen, X.M. Comparative transcriptome analysis reveals osmotic-regulated genes in the gill of Chinese mitten crab (*Eriocheir sinensis*). *PLoS ONE* **2019**, *14*, e0210469. [CrossRef]

21. Kong, T.; Lin, S.; Ren, X.; Li, S.; Gong, Y. Transcriptome and metabolome integration analysis of mud crab *Scylla paramamosain* challenged to *Vibrio parahaemolyticus* infection. *Fish Shellfish Immunol.* **2020**, *103*, 430–437. [CrossRef] [PubMed]
22. Song, L.; Lv, J.J.; Wang, L.; Sun, D.F.; Gao, B.Q.; Liu, P. Characterization of a chitinase-1 gene (PtCht-1) from a marine crab *Portunus trituberculatus* and its response to immune stress. *Gene* **2020**, *741*, 144523. [CrossRef] [PubMed]
23. Schmittgen, T.D.; Livak, K.J. Analyzing real-time PCR data by the comparative C(T) method. *Nat. Protoc.* **2008**, *3*, 1101–1108. [CrossRef] [PubMed]
24. Cerenius, L.; Jiravanichpaisal, P.; Liu, H.P.; Söderhill, I. Crustacean immunity. *Adv. Exp. Med. Biol.* **2010**, *708*, 239–259.
25. Hao, J.W.; Gao, B.Q.; Wang, C.; Meng, X.L.; Wan, X.Y.; Li, X.P.; Liu, P.; Zhang, Q.L. Natural infection of *Portunus trituberculatus* with acute hepatopancreas necrosis disease causing by *Vibrio parahaemolyticus* (VpAHPND). *J. Fish. China* **2019**, *43*, 1647–1660.
26. Ma, J.W.; Lv, J.J.; Liu, P.; Gao, B.Q.; Li, J. Effects of abrupt salinity stress on serum osmolarity and ion concentration of "Huangxuan No.1" *Portunus trituberculatus*. *Prog. Fish. Sci.* **2016**, *37*, 58–62.
27. Chen, T.; Li, Z.; Liu, J.; Liang, C.; Yuan, L. Transcriptome of hepatopancreas in kuruma shrimp *Marsupenaeus japonicus* under low-salinity stress. *J. Oceanol. Limnol.* **2021**, *40*, 745–765. [CrossRef]
28. Tang, D.; Wu, Y.; Huang, S.; Wu, L.; Luo, Y.; Wang, Z. Transcriptome reveals the mechanism of immunity in the low salinity stress of the Chinese shrimp (*Fenneropenaeus chinensis*). *Thalass. Int. J. Mar. Sci.* **2022**, *38*, 977–987. [CrossRef]
29. Chen, D.; Guo, L.; Yi, C.; Wang, S.; Ru, Y.; Wang, H. Hepatopancreatic transcriptome analysis and humoral immune factor assays in red claw crayfish (*Cherax quadricarinatus*) provide insight into innate immunomodulation under *Vibrio parahaemolyticus* infection. *Ecotoxicol. Environ. Saf.* **2021**, *217*, 112266. [CrossRef]
30. Ren, X.; Liu, P.; Li, J. Comparative transcriptomic analysis of *Marsupenaeus japonicus* hepatopancreas in response to *Vibrio parahaemolyticus* and white spot syndrome virus. *Fish Shellfish Immunol.* **2019**, *87*, 755–764. [CrossRef]
31. Cheng, A.C.; Shiu, Y.L.; Chiu, S.T.; Ballantyne, R.C.; Liu, H. Effects of chitin from Daphnia similis and its derivative, chitosan on the immune response and disease resistance of white shrimp, *Litopenaeus vannamei*. *Fish Shellfish Immunol.* **2021**, *119*, 329–338. [CrossRef] [PubMed]
32. Zhang, Y.; Ni, M.; Zhang, P.; Bai, Y.; Zhou, B.; Zheng, J.; Cui, Z. Identification and functional characterization of C-type lectins and crustins provide new insights into the immune response of *Portunus trituberculatus*. *Fish Shellfish Immunol.* **2022**, *129*, 170–181. [CrossRef]
33. Lv, J.J.; Liu, P.; Wang, Y.; Gao, B.Q.; Chen, P.; Li, J. Transcriptome analysis of *Portunus trituberculatus* in response to salinity stress provides insights into the molecular basis of osmoregulation. *PLoS ONE* **2013**, *8*, e82155. [CrossRef] [PubMed]
34. Yin, X.; Zhuang, X.; Liao, M.; Cui, Q.; Yan, C.; Huang, J.; Jiang, Z.; Huang, L.; Luo, W.; Liu, Y.; et al. Andrographis paniculata improves growth and non-specific immunity of shrimp *Litopenaeus vannamei*, and protects it from *Vibrio alginolyticus* by reducing oxidative stress and apoptosis. *Dev. Comp. Immunol.* **2022**, *139*, 104542. [CrossRef] [PubMed]
35. Zhao, B.; Tumaneng, K.; Guan, K.L. The Hippo pathway in organ size control, tissue regeneration and stem cell self-renewal. *Nat. Cell Biol.* **2011**, *13*, 877–883. [CrossRef] [PubMed]
36. Wang, L.; Zhi, B.; Wu, W.; Zhang, X. Requirement for shrimp caspase in apoptosis against virus infection. *Dev. Comp. Immunol.* **2008**, *32*, 706–715. [CrossRef]
37. Zuo, H.; Chang, C.; Yun, G.; Lin, J.; Jin, C.; Wei, W. Regulation of shrimp PjCaspase promoter activity by WSSV VP38 and VP41B. *Fish Shellfish Immunol.* **2011**, *30*, 1188–1191. [CrossRef]
38. Nakamura, R.; Pedrosa-Gerasmio, I.R.; Alenton, R.R.R.; Nozaki, R.; Kondo, H.; Hirono, I. Anti-PirA-like toxin immunoglobulin (IgY) in feeds passively immunizes shrimp against acute hepatopancreatic necrosis disease. *J. Fish Dis.* **2019**, *42*, 1125–1132. [CrossRef]
39. Rodríguez-Nunez, I.; Wcisel, D.J.; Litman, G.W.; Yoder, J.A. Multigene families of immunoglobulin domain-containing innate immune receptors in zebrafish: Deciphering the differences. *Dev. Comp. Immunol.* **2014**, *46*, 24–34. [CrossRef]
40. Piewbang, C.; Tattiyapong, P.; Techangamsuwan, S.; Surachetpong, W. Tilapia lake virus immunoglobulin G (TiLV IgG) antibody: Immunohistochemistry application reveals cellular tropism of TiLV infection. *Fish Shellfish Immunol.* **2021**, *116*, 115–123. [CrossRef]
41. Wu, L.; Gao, A.; Li, L.; Chen, J.; Li, J.; Ye, J. A single-cell transcriptome profiling of anterior kidney leukocytes from Nile tilapia (*Oreochromis niloticus*). *Front. Immunol.* **2021**, *12*, 783196. [CrossRef]
42. Yoder, J.A.; Mueller, M.G.; Wei, S.; Corliss, B.C.; Prather, D.M.; Willis, T.R.; Litman, T.; Djeu, J.Y.; Litman, G.W. Immune-type receptor genes in zebrafish share genetic and functional properties with genes encoded by the mammalian leukocyte receptor cluster. *Proc. Natl. Acad. Sci. USA* **2001**, *98*, 6771–6776. [CrossRef] [PubMed]
43. Yang, M.C.; Yang, H.T.; Li, J.; Sun, J.J.; Bi, W.J.; Niu, G.J.; Zhang, Q.; Shi, X.Z.; Zhao, X.F.; Wang, J.X. Scavenger receptor C promotes bacterial clearance in kuruma shrimp *Marsupenaeus japonicus* by enhancing hemocyte phagocytosis and AMP expression. *Fish Shellfish Immunol.* **2017**, *67*, 254–262. [CrossRef] [PubMed]
44. Tang, M.; Li, X.; Yang, L.; Wang, Q.; Li, W. A class B scavenger receptor mediates antimicrobial peptide secretion and phagocytosis in Chinese mitten crab (*Eriocheir sinensis*). *Dev. Comp. Immunol.* **2020**, *103*, 103496. [CrossRef] [PubMed]
45. Chaikeeratisak, V.; Somboonwiwat, K.; Tassanakajon, A. Shrimp alpha-2-macroglobulin prevents the bacterial escape by inhibiting fibrinolysis of blood clots. *PLoS ONE* **2012**, *7*, e47384. [CrossRef]

biology

MDPI

Article

Effects of Dietary *Gracilaria lichenoides* and *Bacillus amyloliquefaciens* on Growth Performance, Antioxidant Capacity, and Intestinal Health of *Penaeus monodon*

Jialin Tian [1,2], Yun Wang [2,3,*], Jianhua Huang [2], Hailiang Yan [1,2], Yafei Duan [2], Jun Wang [2,3], Chuangpeng Zhou [2,3] and Zhong Huang [4]

1 College of Fisheries and Life Science, Shanghai Ocean University, Shanghai 201306, China; jltian2023@163.com (J.T.); hailiangyan813@163.com (H.Y.)
2 Key Laboratory of Aquatic Product Processing, Ministry of Agriculture and Rural Affairs, South China Sea Fisheries Research Institute, Chinese Academy of Fishery Sciences, Guangzhou 510300, China; hjh210440@sina.com.cn (J.H.); duanyafei89@163.com (Y.D.); junnywang@163.com (J.W.); chpzhou@163.com (C.Z.)
3 Key Laboratory of Efficient Utilization and Processing of Marine Fishery Resources of Hainan Province, Hainan Engineering Research Center of Deep-Sea Aquaculture and Processing, Sanya Tropical Fisheries Research Institute, Sanya 572018, China
4 Shenzhen Base of South China Sea Fisheries Research Institute, Chinese Academy of Fishery Sciences, Shenzhen 518121, China; huangzhongnhs@163.com
* Correspondence: wangyun@scsfri.ac.cn; Tel./Fax: +86-020-89221832

Simple Summary: Seaweeds and probiotics are commonly utilized as additives in the diet of aquatic animals. However, limited information exists regarding the effects of *Gracilaria lichenoides* and *Bacillus amyloliquefaciens*, either alone or in combination, on *Penaeus monodon*. In our study, we examined the impact of dietary supplementation with *G. lichenoides* and *B. amyloliquefaciens*, both individually and in combination, on the growth performance, antioxidant capacity, and intestinal function of *P. monodon*. Our findings indicate a beneficial influence of dietary *G. lichenoides* and *B. amyloliquefaciens* on *P. monodon*. These findings offer a theoretical foundation for the judicious utilization of *G. lichenoides* and *B. amyloliquefaciens* in the cultivation of black tiger shrimp, as well as for future investigations into their metabolic mechanisms.

Abstract: This research sought to assess the effects of dietary supplements with *Gracilaria lichenoides* and *Bacillus amyloliquefaciens*, either individually or combined, on the growth performance, antioxidant capacity, and intestinal function of *Penaeus monodon*. A total of 840 shrimps were randomly assigned to 28 tanks with an average initial weight of (1.04 ± 0.03) g (30 shrimp per tank) with 7 different treatment groups and 4 replicates per treatment. The control treatment (C) consisted of a basal diet; in contrast, the experimental groups were complement with varying levels of *G. lichenoides* (3% or 8%), either alone (S3 and S8) or in combination with *B.amyloliquefaciens* at different concentrations (3% *G. lichenoides* and 10^9 CFU/g—S3B9; 8% *G. lichenoides* and 10^{11} CFU/g *B. amyloliquefaciens*—S8B11; 10^9 CFU/g *B. amyloliquefaciens*—S9; 10^{11} CFU/g *B. amyloliquefaciens*—B11). The results indicated that the maximum values of final body weight (FBW) (10.49 ± 0.90) g, weight gain rate (WGR) (908.94 ± 33.58) g, and specific growth rate (SGR) (4.20 ± 0.06) g were perceived in the 3% *G. lichenoide* diet treatment, and compared with the control group, the difference was significant ($p < 0.05$). The whole-body lipid content of shrimp in the B9 group was significantly higher than that in the B11 group ($p < 0.05$), but no significant difference was observed when compared with shrimp fed other diets ($p > 0.05$). The ash content of shrimp in the B9 group was found to be significantly higher than that in the S3B9 group ($p < 0.05$). Furthermore, the lipase activity in the stomach and intestines of the experimental groups exhibited a statistically significantly increase compared to the control ($p < 0.05$). In comparison to the control group, the hepatopancreas of the S3 group exhibited a significant increase in the activities of glutathione peroxidase (GSH-Px), superoxide dismutase (SOD), and antioxidant genes [*SOD*, catalase (*CAT*), *GSH-Px*, thioredoxin (*Trx*), *Hippo*, and NF-E2-related factor 2 (*Nrf2*)] expression levels ($p < 0.05$). Additionally, the activities of total antioxidant capacity (T-AOC), SOD, peroxidase (POD), and antioxidant genes (*CAT*, *GSH-Px*, *Trx*, and *Hippo*) in the S3B9 treatment of hepatopancreas showed significant improvement ($p < 0.05$). The inclusion of dietary *G. lichenoides* and *B. amyloliquefaciens* resulted in enhanced relative expression of intestinal lipid metabolism genes (fatty acid synthetase (*FAS*), lipophorin receptor (*LR*), fatty acid transport protein 1 (*FATP1*)) and suppressed the expression of the long-chain fatty acid-CoA ligase 4 (*LCL4*) gene. Analysis of microbiota sequencing indicated improvements in composition and structure, with notable increases in Firmicutes at the phylum level and *Vibrio* at the genus level in the S3 group, as well as an increase in *Tenericutes* at the genus level in the S8B11 group. Overall, the inclusion of dietary *G. lichenoides* and *B. amyloliquefaciens* positively impacted the growth, antioxidant capacity, and microbial composition of shrimp, with particular enhancement observed in shrimp fed a supplementary 3% *G. lichenoides* diet.

Keywords: *Penaeus monodon; Gracilaria lichenoides; Bacillus amyloliquefaciens;* growth; antioxidant capacity; intestine microbiota; lipid metabolism

1. Introduction

Penaeus mondon, also known as black tiger shrimp, is the second largest shrimp species in China, with an annual production of over 100,000 tons [1]. *P. monodon*, as a traditional cultured shrimp, has the characteristics of strong adaptability, fast growth, strong disease resistance, delicious flavor, high nutritional value, and resistance to transportation [2]. However, diseases are still the most important constraints affecting the development of shrimp culture. Also, water quality deterioration and serious pollution leads to an increase of the pathogenic microorganisms. Antibiotic use in aquaculture is often applied to shrimp farms to overcome disease and stress problems. In response to these problems, antibiotics are often used in shrimp production and medicine. To combat the adverse effects of antibiotics, researchers have been looking for alternatives. Additionally, consumers demand growing levels of quality and safety for shrimp farming, as well as an absence of antibiotics and pollutants. As a result, shrimp farming strategies must accelerate the search for safe dietary supplements and additives that enhance their health, vital activities, and environmental resilience. Antioxidants and microbes added to feed have been considered a non-antibiotic approach that can improve health and performance [3].

There are several efforts underway to develop products or methods for treating or preventing shrimp farm stress [4]. Macroalgae (seaweeds) are rich in bioactive compounds that can potentially be exploited as functional ingredients in feed additives or supplements for animal health applications [5]. *Gracilaria lichenoides*, belonging to the division Rhodophyta, is an important aquaculture species with good bioremediation potential due to their fast growth rate, high nutrient uptake capacity, and high temperature tolerance [6,7]. A significant portion of the lipid content of seaweeds comes from polyunsaturated fatty acids (PUFAs), especially those of tropical origin [8]. Aside from its polysaccharides and dietary fibers, the seaweeds is known to contain incomparable amounts of minerals, vitamins, and polyphenols [9]. The utilization of seaweed as a dietary supplement for fish and shrimp at low concentrations (2.5–10%) has demonstrated beneficial outcomes, such as enhanced growth performance, increased feed utilization [10–13], and improved immune and stress responses [14,15]. Supplementation of *Sciaenops ocellatus* feed with *Isochrysis galbana* extract, resulted in significantly higher survival rates [16]. Feeding *Oncorhynchus mykiss Gracilaria pygmaea* polysaccharide led to significantly higher final average weight compared to the control group, with feed intake showing a tendency to increase with higher levels of supplementation [17]. Incorporating *Gracilaria gracilis* into the diet of *Sparus aurata* enhanced weight gain rate and feed conversion ratio [18]. The incorporation of *Eucheuma denticulatum* into the diet of *Paralichthys olivaceus* resulted in notable enhancement in the growth rate and feed efficiency of juvenile fish, with a concurrent increase in the accumulation of dorsal muscle omega-3 polyunsaturated fatty acid in experimental groups [19]. Additionally, feeding *Gracilaria persica* to *Acipenser persicus* led to elevated activities of superoxide dismutase

(SOD) and catalase (CAT) [20]. Furthermore, supplementation of *Sargassum aquifolium* in the feed resulted in heightened activities of CAT, SOD, and glutathione (GSH), along with a reduction in malondialdehyde (MDA) content. Notable, *SOD* and *CAT* were significantly upregulated in *Oreochromis niloticus* [21]. *Sargassum horneri* was found to enhance hepatic total antioxidant capacity (T-AOC), SOD, and CAT activities in *Acanthopagrus schlegelii*, with a notable upregulation of *CAT* expression [22]. Similarly, the dietary supplementation of sulphated polysaccharide resulted in a significant increase in hepatic T-AOC, CAT, and SOD activities, as well as marked upregulated $Cu^{2+}/ZnSOD$, *CAT*, and glutathione peroxidase (*GSH-Px*) antioxidant gene expression in *Siganus canaliculatus* [23]. In contrast, some researchers found that seaweeds and their extracts induced several morphological changes in fish guts, inhibiting digestion and nutrient absorption [24–26]. *Gracilaria lemaneiformis* increased the length and evenness of intestinal microvilli in *L. vannamei* [27]. Niu et al. [14] found that the addition of *Undaria pinnatifida* to the feed significantly increased the height and width of folds in shrimp, as well as the lipase activity. Extracts of *Amphiroa Fragilissima* promoted digestive enzyme (protease and amylase) activities in *P. monodon* [28].

Probiotics have been defined as "live microorganisms" which work by rebuilding the gastrointestinal trace (GIT)'s native microflora in response to ingestion of food or water [29]. *Bacillus amyloliquefaciens*, known as probiotic, which is closely related to *Bacillus subtilits*, can improve the intestinal microbial environment, facilitate aquatic animal growth, improve enzyme activity, and boost nonspecific immune responses [30,31]. In our earlier research, a specific probiotic *B. amyloliquefaciens* broke away the healthy *L. vannamei* intestine, which is an endogenous strain that has strong antagonistic effects against seven species of *Vibrio spp.* pathogens in aquatic animals. Besides, it can also improve the non-specific immune function, relieve oxidative stress, and strengthen the ability of shrimp to resist pathogenic bacterial infections. Similar results were also found when a strain of *B. amyloliquefaciens* was isolated from the intestine of healthy *P. clarkia*, exhibiting strong antimicrobial activity against different pathogens. *B. amyloliquefaciens* as a feed additive could improve the intestinal digestive enzyme activity, the innate immune enzyme activities, and the antiviral resistance of *P. clarkia* [32]. *Bacillus* also has beneficial effects on the microbial community and the development of *L. vannamei*. Wang et al. explained the effects of *Bacillus* (10^8~10^9 CFU/g) on the survival rate and molting of *L. vannamei* larvae [33]. *B. amyloliquefaciens* significantly improved the WG, SGR, and FCR of *O. niloticus* [34]. *B. amyloliquefaciens* LSG2-8 added to *Rhynchocypris lagowskii* feed significantly improved FBM, WGR, SGR, and FCR growth indices [35]. *L. vannamei* phenol oxidase activity, phagocytic activity, and scavenging efficiency were significantly increased by feeding higher concentrations of *Bacillus subtilis* E20 [36]. *B. amyloliquefaciens* R8 caused a significant increase in decreased mRNA expression of oxidative stress-related genes (*SOD*, *Gpx*, and *Hsp70*) in zebrafish liver [37].

The dietary role of *G. lichenoides* and *B. amyloliquefaciens*, either alone or in combination, on digestive and metabolic capacity regulatory in *P. monodon* remains uncertain. The potential benefits of incorporating *G. lichenoides* and *B. amyloliquefaciens* into shrimp diets were previously unidentified. Therefore, this study examined the impact of *G. lichenoides* and *B. amyloliquefaciens*, either individually or in combination, on the growth, antioxidant capacity, and intestinal health of *P. mondon*. The aim was to identify the most effective dosage and application method of *G. lichenoides* and *B. amyloliquefaciens* in the diet of *P. mondon*, optimize feed formulation, and enhance feed utilization efficiency and antioxidant stress response in *P. mondon*. Further, the effects of *G. lichenoides* and *B. amyloliquefaciens* on the intestinal microbiota and overall intestinal health of *P. mondon* was explored, with the goal of establishing a theoretical framework for research in black tiger shrimp aquaculture and elucidating their metabolic mechanisms.

2. Materials and Methods

2.1. Ethical Statement

This animal study was conducted under animal use protocol number NHDF2023-12 and was approved by the Laboratory Animal Welfare and Ethics Committee of the South China Sea Fisheries Research Institute, Chinese Academy of Fishery Sciences.

2.2. Experimental Diets Preparation and Nutritional Composition

Basic feed ingredients were purchased from Shandong Wuzhoufeng Agricultural Science and Technology Co., Ltd (Heze, China). *G. lichenoide* was purchased from the breeding farm of Xigao Village, Nanri Town, Xiuyu District, Putian City, Fujian Province. *B. amyloliquefacients* was supplied by our laboratory, isolated from the intestinal tract of healthy *L. vannamei* (Patent No.: ZL201410319443.7). Seven experimental diets were prepared based on the basal diet of black tiger shrimp. The basal diet served as the control, while the other diets included the addition of 3% *G. lichenoide* as S3, the addition of 3% *G. lichenoide* and 10^9 CFU/g *B. amyloliquefacients* as S3B9, the addition of 10^9 CFU/g *B.amyloliquefacients* as S9, the addition of 8% *G. lichenoide* as S8, the addition of 8% *G. lichenoide* and 10^{11} CFU/g *B. amyloliquefacients* as S8B11, and the addition of 10^{11} CFU/g B. amyloliquefacients as B11. The feed formulation and composition are analyzed in Table 1. Feed preparation was done according to the method of feed preparation in our laboratory [38]. The feed was made into pellets with diameters of 1.0 mm and 1.5 mm and were heated in an oven at 90 °C for 90 min, and then air-dried to about 10% moisture and stored in a −20 °C freezer. Moisture, crude protein, crude lipid, and ash of the experimental diets were determined using standard methods [39].

Table 1. Composition and nutrient contents of experimental diets (%).

Items (%)	Control	S3	S3B9	B9	S8	S8B11	B11
Fish meal	31.00	31.00	31.00	31.00	31.00	31.00	31.00
Soybean meal	17.00	17.00	17.00	17.00	17.00	17.00	17.00
Peanut meal	15.00	15.00	15.00	15.00	15.00	15.00	15.00
Wheat flour	20.30	20.00	19.90	20.29	19.50	19.49	20.29
Gracilaria lichenoides	0.00	0.30	0.30	0.00	0.80	0.80	0.00
Bacillus amyloliquefaciens	0.00	0.00	0.01	0.01	0.00	0.01	0.01
Beer yeast	5.00	5.00	5.00	5.00	5.00	5.00	5.00
Krill meal	5.00	5.00	5.00	5.00	5.00	5.00	5.00
Soy lecithin	1.00	1.00	1.00	1.00	1.00	1.00	1.00
Fish oil	1.00	1.00	1.00	1.00	1.00	1.00	1.00
Soybean oil	1.00	1.00	1.00	1.00	1.00	1.00	1.00
Choline chloride (50%)	0.60	0.60	0.60	0.60	0.60	0.60	0.60
Monocalcium phosphate	1.00	1.00	1.00	1.00	1.00	1.00	1.00
Vitamin premix [a]	1.00	1.00	1.00	1.00	1.00	1.00	1.00
Mineral premix [b]	1.00	1.00	1.00	1.00	1.00	1.00	1.00
Ascorbic Phosphate ester	0.10	0.10	0.10	0.10	0.10	0.10	0.10
Sum	100.00	100.00	100.00	100.00	100.00	100.00	100.00
Nutrient levels [c]							
Moisture	9.66	9.97	10.17	9.46	9.86	10.32	9.40
Ash	10.35	11.29	11.53	11.28	13.18	13.64	11.27
Crude protein	40.62	40.73	43.87	41.78	40.15	40.44	39.71
Crude lipid	9.12	8.95	8.73	9.09	9.32	8.99	8.84
Crude fiber	3.94	3.29	4.30	4.52	3.21	3.64	3.73

Note: [a]. Purchased from Guangzhou Feixite Aquatic Technology Co., Ltd (Guangzhou, China).Vitamin premix (g kg^{-1}): VE, 75; VK, 2.5; VB_1, 0.25; VB_2, 1.0; VB_3, 5.0; VB6, 0.75; VB_{12}, 2.5; VA, 2.5; VD, 6.25; folic acid, 0.25; cellulose, 500; inositol, 379; biotin, 2.5. [b]. Purchased from Guangzhou Xinhai Lisheng Biotechnology Co., Ltd (Guangzhou, China). Mineral premix (g kg^{-1}): KCl, 90; NaCl, 40; KI, 0.04; $ZnSO_4{\cdot}7H_2O$, 4; $CuSO_4{\cdot}5H_2O$, 3; $CoSO_4{\cdot}7H_2O$, 0.02; $MnSO_4{\cdot}H_2O$, 3; $FeSO_4{\cdot}7H_2O$, 20; $MgSO_4{\cdot}7H_2O$, 124; $CaCO_3$, 215; $Ca(H_2PO_4)_2{\cdot}2H_2O$, 500. [c]. Nutrient levels in dry-weight feed.

2.3. Shrimp and Culture System

Juvenile black tiger shrimp were obtained from the Shenzhen Experimental Base of the South China Sea Fisheries Research Institute of the Chinese Academy of Fishery Sciences. Shrimp were domesticated with commercial feed for one week before the beginning of the experiment. Shrimp with an initial average weight of (1.04 ± 0.03) g were randomly assigned to 28 fiberglass tanks (500 L, 0.5 m^2 round bottom area, 30 shrimp per tank) in 7 groups of 4 tanks each. During the feeding period, shrimp were fed at 8:00, 17:00, and 22:00 every day, and the daily feeding regimen consisted of providing shrimp with an amount equivalent to 5% of their body weight, closing the screen after each feeding to prevent escape, and adjusting the feeding quantity promptly based on shrimp behavior and prevailing weather conditions. After 1 h of feeding, the remaining shrimp feed in the bait dish was collected, rinsed 2 to 3 times with water, dried at 60 °C, and weighed. The numbers of shrimp deaths and body weight of each tank were recorded. Oxygen was connected to the oxygen supply system by a hose air stone, which was filled day and night. The breeding water was filtered seawater. The rearing water was siphoned off the bottom every day, and 1/3 of the rearing water was changed on a weekly basis. During the experiment, water parameters were maintained as follows: temperature 29.7 ± 0.9 °C, pH 8.0 ± 0.1, salinity 30, dissolved oxygen more than 6.0 mg/L. The feeding trial lasted for 56 days.

2.4. Sampling and Preservation

At the end of the feeding trial, shrimp were fasted for 24 h, and then weighed to determine growth performance. Five shrimp from each tank were randomly selected for whole-body composition analysis. Additionally, five shrimp from each tank were sampled for analysis of enzyme activity and gene expression in the hepatopancreas and intestines. The intestines from two shrimp per tank were extracted at 56 d and stored in Davidson's solution for 24 h. The intestines from five shrimp per tank were extracted and quickly frozen in liquid nitrogen, followed by storage at −80 °C. To reduce the impact of interindividual differences, intestines from each treatment were mixed for the analysis of the intestinal microbiota.

2.5. Measurement and Analysis

2.5.1. Growth Performance

The growth parameters were calculated according to the following formulas:

$$\text{Weight gain rate (WGR, \%)} = 100 \times [\text{FBW (g)} - \text{IBW (g)}]/\text{IBW (g)}$$

$$\text{Specific growth rate (SGR, \%)} = 100 \times [\ln(\text{FBW}) - \ln(\text{IBW})]/56 \text{ days}$$

$$\text{Feed conversion ratio (FCR)} = 100 \times \text{dry feed intake (g)}/\text{body weight gain (g)}$$

$$\text{Survival rate (SR, \%)} = 100 \times \text{final living numbers (shrimp)}/\text{initial numbers (shrimp)}$$

Note: initial weight (IBW), final body weight (FBW)

2.5.2. Activity of Physiological and Biochemical Indexes

The hepatopancreas samples were homogenized in sterilized PBS buffer (0.86%, pH = 7.4; 1: 9, *w*/*v*) by a handheld homogenizer; then, the samples were centrifuged for 10 min (3000 r/min, 4 °C) and the supernatant was fractionated [40]. Hepatopancreas T-AOC, SOD, CAT, GSH-Px, peroxidase (POD) activities, and MDA and total protein contents were determined according to the kit (Nanjing Jiancheng Bioengineering Institute, Nanjing, China). The lipase activity in the stomach and intestine were also analyzed with commercial test kits (Nanjing Jiancheng Bioengineering Institute, Nanjing, China).

2.5.3. RNA Extraction and qPCR Measurement

RNA from hepatopancreas and intestine was extracted using TRIzol reagent (QIAGEN Cat: no. 74104, Guangzhou, China) according to the manufacturer's instructions, and DNase I (QIAGEN Cat: no. 79254) was used to remove contaminated DNA. The primer sequences for the target gene and the internal reference gene *EF-1α* for real-time PCR are shown in Table 2 [41]. All primer pairs for real-time PCR were synthesized by Sangon Biotech (Shanghai) Co., Ltd. (Shanghai, China). The mRNA expression levels of the target genes were quantified using the fluorescence quantitative PCR method described in a previous report. Each sample was tested three times using real-time PCR ($n = 4$). The gene expressions were quantified using the $2^{-\Delta\Delta ct}$ calculation method with an internal reference gene [42].

Table 2. The primers for real-time fluorescence quantification PCR.

cDNA	Forward Primer (5′–3′)	Reverse Primer (5′–3′)	Size of Production	GenBank Nos.
mtMnSOD	TCGCCGCCAGGAGACTCTTC	GGCACAGATGACAGGTTCCAAGG	122	KC461130.1
CAT	GTCCTTCTTCAGCAGCCTCAGTTG	CTTGGCTCGTGGTCAGGTTATCG	155	KR908786.1
GSH-Px	CGTCCGTCCTGGCAATAACTTCG	CTGGCAGCGGCAGTCGTTC	116	JX912159.1
Trx	CCTGAAGGTGGATGTGGACGAATG	AGTGGAATGGATTACTTGTGCTTCTCG	168	JF828310.1
Hippo	TGAGCACAACCAAACCCACCATC	CATCGTCCGACTGTCCACTTCATC	88	Not acquired
Nrf2	CCAACCTCCAGTAACAAGCCAAGAG	TCAACAATTCTGATGAGCACAGCAATG	237	MW390830.1
FAS	GCGTGATAACTGGGTGTCCT	CTTTCAGGCCCTGGATGATA	227	XM030250746.2
LR	GTGGCAGGAATGAGTGTGAGG	GCATACCCTCCAACGCACTC	215	NM009309.2
FATP1	GATGACACGGACATGACTACTG	CTCCTGGCTTCAGCACATTCC	195	NM164934.2
LCL4	CTCCAATCTGTAGTTTAACCAAGTCC	GATGACACGGACATGACTACTG	304	XM006583136.4
EF-1α	AGTATGCTCCTTTTGGACGTTTTGC	CCTTTTCTGCGGCCTTGGTAGTC	120	GU136229.1

Note: *Nrf2*, NF-E2-related factor 2; *Trx*, thioredoxin; *FAS*, fatty acid synthetase; *LR*, lipophorin receptor; *FATP1*, fatty acid transport protein 1; *LCL4*, long-chain fatty acid-CoA ligase 4; *EF-1α*, elongation factor 1α.

2.5.4. HE Stain of the Intestinal Tissue

Two shrimp from each tank had their intestines sampled and preserved in Davidson's solution and ethanol. The tissues were then rinsed, dehydrated, and prepared for staining and photography. Eight samples were randomly selected for measuring the height of the intestinal epithelial cell and intestinal folds height of shrimp.

2.5.5. Intestinal Microbiome Analysis

Genomic DNA was extracted from intestinal microbial samples using the OMEGA Soil DNA Kit (M5635-02) (Omega Bio-Tek, Norcross, GA, USA) and stored at −20 °C. The bacterial 16S rRNA genes V3-V4 region was amplified using primers 338F (5′-ACTCCTACGGGA GGCAGCA-3′) and 806R (5′-GGACTACHVGGGTWTCTAAT-3′), and the resulting PCR products were purified and combined in equal amounts. After quantifying the individuals, the amplicons were pooled and sequenced using the Illumina NovaSeq platform with the NovaSeq 6000 SP Reagent Kit (500 cycles) at Suzhou PANOMIX Biomedical Tech Co., Ltd (Suzhou, China). Microbiota sequencing and analysis were conducted at the same facility. Microbiota sequencing and subsequent analyses were performed at Suzhou Panomix Biomedical Tech Co., Ltd (Suzhou, China).

2.6. Statistical Analysis

The mean ± SD of the data was presented. Statistical analysis was conducted using SPSS 26.0 software (IBM Corporation, Somers, NY, USA) for Windows. Prior to performing ANOVA, the normality and homogeneity of the data were tested using the Kolmogorov-Smirnov test and Levene's test, respectively. Once the data passed the tests, a one-way ANOVA was performed. If there were notable differences, the group means were compared using Duncan's multiple-range test, with a significance level of $p < 0.05$.

3. Results

3.1. Growth Performance

Growth performance of *P. monodon* is presented in Table 3. There were no significant differences in IBW, FCR, and SR among the groups ($p > 0.05$). However, shrimp fed diets with 3% *G. lichenoide* had significantly higher FBW, WGR, and SGR compared to the control group ($p < 0.05$).

Table 3. Effects of different diet treatments on growth performance of shrimp after feeding for 56 days ($n = 4$).

Items	IBW (g)	FBW (g)	WGR (%)	SGR (%)	FCR	SR (%)
Control	1.06 ± 0.02	9.18 ± 0.48 ^{b}	765.56 ± 56.72 ^{b}	3.92 ± 0.12 ^{b}	1.79 ± 0.17	90.00 ± 6.67
S3	1.04 ± 0.06	10.49 ± 0.90 ^{a}	908.94 ± 33.58 ^{a}	4.20 ± 0.06 ^{a}	1.50 ± 0.24	90.00 ± 0.00
S3B9	1.05 ± 0.05	9.47 ± 0.62 ^{ab}	804.95 ± 56.43 ^{b}	4.00 ± 0.12 ^{ab}	1.64 ± 0.14	94.44 ± 1.93
B9	1.04 ± 0.02	8.95 ± 0.56 ^{b}	759.42 ± 41.23 ^{b}	3.91 ± 0.09 ^{b}	1.53 ± 0.30	88.89 ± 3.85
S8	1.02 ± 0.03	8.51 ± 0.45 ^{b}	731.19 ± 49.05 ^{b}	3.85 ± 0.11 ^{b}	1.80 ± 0.10	93.33 ± 0.00
S8B11	1.01 ± 0.04	9.08 ± 1.05 ^{b}	794.15 ± 80.75 ^{b}	3.98 ± 0.17 ^{b}	1.46 ± 0.12	93.33 ± 0.00
B11	1.04 ± 0.04	8.33 ± 0.29 ^{b}	703.55 ± 17.80 ^{b}	3.79 ± 0.04 ^{b}	1.63 ± 0.21	93.33 ± 8.82

Note: Values with different letters in the same column indicate significant differences ($p < 0.05$).

3.2. Whole-Body Composition

Table 4 shows the composition of shrimps' whole-body composition on different diets. There were no significant differences in moisture and protein content among the diet treatments ($p > 0.05$). The lipid content of shrimp in the B9 group was higher than that of shrimp in the B11 group ($p < 0.05$), but not significantly different from shrimp fed other diets ($p > 0.05$). The ash content of shrimp in the B9 group was higher than that of shrimp in the S3B9 group ($p < 0.05$).

Table 4. Body composition of shrimp.

Items	Control	S3	S3B9	B9	S8	S8B11	B11
Moisture (%)	76.51 ± 1.20	77.02 ± 0.45	76.15 ± 0.86	76.76 ± 0.17	76.54 ± 0.47	76.09 ± 0.81	76.00 ± 0.74
Protein (% dry matter)	70.01 ± 1.30	70.37 ± 0.68	70.45 ± 0.35	70.56 ± 1.27	71.09 ± 0.42	71.39 ± 0.87	70.01 ± 0.94
Lipid (% dry matter)	10.40 ± 0.70 ^{ab}	10.49 ± 1.11 ^{ab}	10.73 ± 1.86 ^{ab}	11.08 ± 0.75 ^{a}	9.80 ± 1.06 ^{ab}	9.35 ± 0.92 ^{ab}	9.27 ± 0.76 ^{b}
Ash (% dry matter)	16.61 ± 0.69 ^{ab}	16.11 ± 1.13 ^{ab}	15.67 ± 0.48 ^{b}	16.97 ± 0.56 ^{a}	16.66 ± 0.72 ^{ab}	16.11 ± 0.79 ^{ab}	16.46 ± 0.55 ^{ab}

Note: Values in the same row with different letters are significantly difference ($p < 0.05$).

3.3. Change of Antioxidant Enzyme Activities in the Hepatopancreas

Activities of T-AOC, SOD, GSH-Px, CAT, POD, and MDA content in the hepatopancreas of shrimp for all diet treatments are shown in Figure 1. T-AOC and SOD activities were significantly higher in the S3 and S3B9 groups compared to the other groups ($p < 0.05$). CAT activity was significantly higher in the S3B9 group compared to the B9, S8B11, and B11 groups ($p < 0.05$), but not significantly different from the control, S3, and S8 groups ($p > 0.05$). The GSH-Px activity in shrimp from the S3 group was significantly higher than in the control and B9 groups ($p < 0.05$), but there was no significant difference with the other groups ($p > 0.05$). The POD activity in the S3B9 group was significantly higher than in the control, B9, S8B11, and B11 groups ($p < 0.05$). There was no significant difference in GSH-Px activity among the S3, S3B9, and S8 groups ($p > 0.05$). The MDA content in shrimp from S3 group was significantly higher than in the control, B9, S8B11, and B11 groups ($p < 0.05$). The antioxidant capacity in hepatopancreatic tissues of *P. monodon* was significantly increased in the 3% *G. lichenoide* group (S3) and the 3% *G. lichenoides* and 10^9 CFU/g *B. amyloliquefaciens* (S3B9).

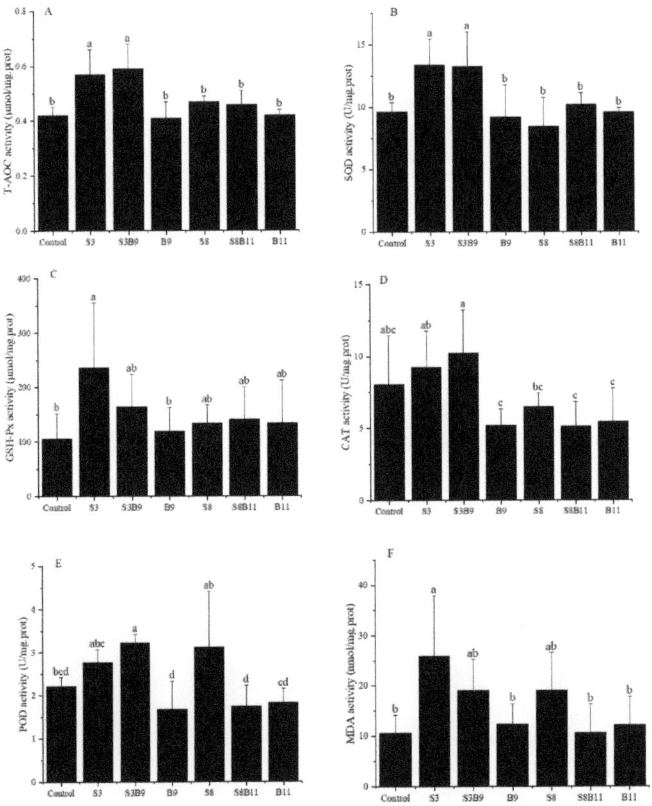

Figure 1. (**A**) T-AOC, (**B**) SOD, (**C**) CAT, (**D**) GSH-Px, (**E**) POD, and (**F**) MDA content in the hepatopancreas of shrimp fed different diets for 56 days. Vertical bars represented the mean ± SD (n = 4) and data indicated with different letters were significantly different ($p < 0.05$) among treatments, the same as below.

3.4. The Relative Expression of Antioxidant Genes in the Hepatopancreas of Shrimp

The relative expression of antioxidant genes in the hepatopancreas are showed in Figure 2. The relative expression level of *mtMnSOD* in the S3 group was significantly higher than that of shrimp in other groups ($p < 0.05$). However, the relative expression of SOD in the S8, S8B11, and B11 groups was significantly lower than that of shrimp in the control group ($p < 0.05$). Regardless of the supplementation of *G. lichenoides* and *B. amyloliquefaciens*, either individually or in combination in the diet, the relative expressions of *CAT* and *GSH-Px* in the hepatopancreas were significantly increased compared to the control group ($p < 0.05$). Furthermore, the *CAT* and *GSH-Px* genes had the highest expression in the S3B9 and S3 groups, respectively. The expression of *Trx* in shrimp increased significantly in the S3, S3B9, and S8B11 groups compared to the control ($p < 0.05$), with the highest levels observed in the S3 group. The expression of *Hippo* in shrimp was significantly higher in the S3 and S3B9 groups compared to the other groups ($p < 0.05$). The *Nrf2* gene relative expression in shrimp was significantly higher in the S3, S8, and S8B11 groups compared to other groups ($p < 0.05$), with the highest levels observed in the S8B11 group. The results indicate a significant increase in antioxidant gene expression in the hepatopancreas of shrimp between the 3% *G. lichenoide* group and the 3% *G. lichenoide* + 10^9 CFU/g *B. amyloliquefaciens* group.

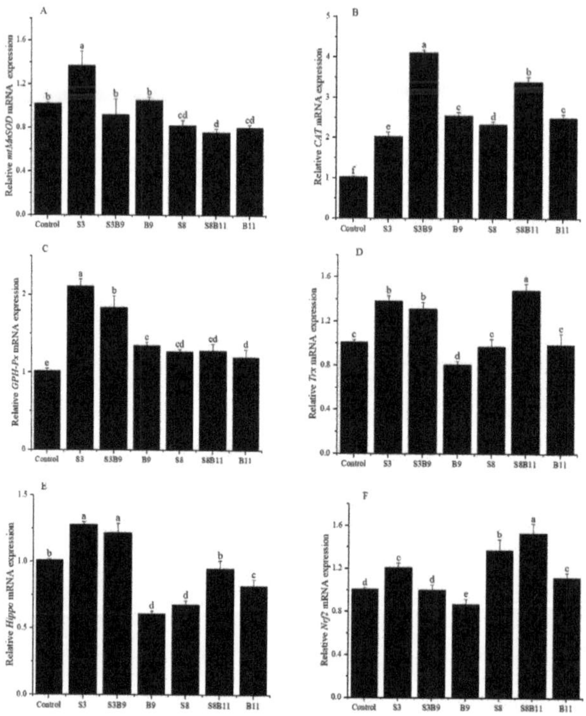

Figure 2. Expression levels of antioxidant genes in the hepatopancreas of shrimp fed different diets for 56 days. (**A**) *mtMnSOD*, (**B**) *CAT*, (**C**) *GSH-Px*, (**D**) *Trx*, (**E**) *Hippo*, (**F**) *Nrf2*. The reference gene is *EF-1α*. Vertical bars represented the mean ± SD ($n = 4$) and data indicated with different letters were sig-nificantly different ($p < 0.05$) among treatments.

3.5. Lipase Activity

The results of stomach and intestinal lipase activity are shown in Figure 3. The lipase activity of intestine and stomach in the experimental groups were significantly higher than those of the control group ($p < 0.05$). Results showed that supplementing the diet with the *G. lichenoide* and *B. amyloliquefacients* improved the stomach and intestinal lipase activity of *P. monodon*.

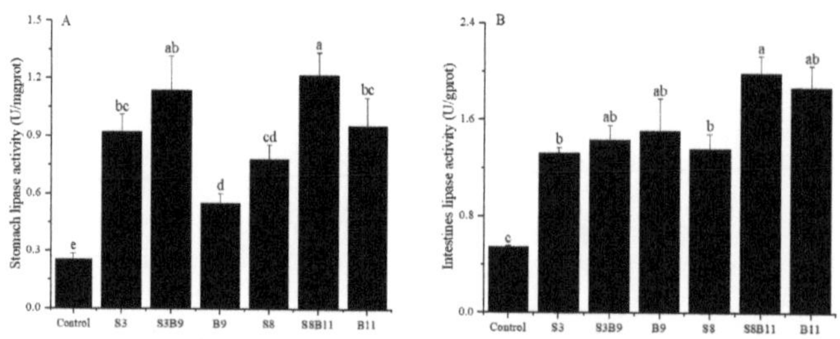

Figure 3. Lipase activity in the stomach and intestine of shrimp fed different diets for 56 days. (**A**) Lipase activity of *P. monodon* in the stomach. (**B**) Lipase activity of *P. monodon* in the intestine. Vertical bars represented the mean ± SD ($n = 4$) and data indicated with different letters were significantly different ($p < 0.05$) among treatments.

3.6. Intestinal Tissue Structure

After staining with HE dye, the intestinal tissue of *P. monodon* showed a higher number of epithelium and neatly arranged brush borders in the experimental group compared to the control group (Figure 4). Vacuoles were almost absent and nuclei in S3 and S8 groups were closely arranged. The height of intestine epithelium in the S3 group was significantly higher than in the S8B11 group ($p < 0.05$). The folds depth in the S3B9, B9, and B11 groups was also significantly higher than in other groups ($p < 0.05$). Overall, supplementing the diet with *B. amyloliquefaciens* (10^9 CFU/g) improved the intestinal tissue structure of *P. monodon*.

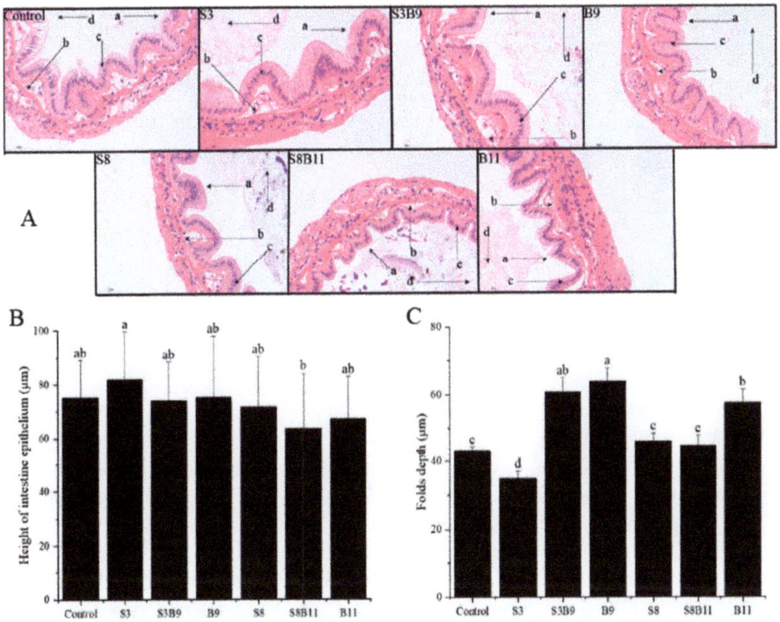

Figure 4. Intestinal structure of *P. monodon* fed the different experimental diets for 56 days. (**A**) Intestine sections of shrimp fed different diets, ×400. (**B**) Height of intestine epithelium. (**C**) Folds depth of intestine in shrimp fed different diets. The letters in figure (**A**) indicate: (a) brush border, (b) epithelium, (c) nuclei, (d) lumen. Vertical bars represented the mean ± SD (n = 4) and data indicated with different letters were significantly different ($p < 0.05$) among treatments.

3.7. Intestinal Lipid Metabolism Genes Expression

The expression of lipid metabolism genes in the intestine is illustrated in Figure 5. The relative expression levels of fatty acid *FAS* and *LR* genes in the shrimp intestine was significantly higher in the groups supplemented with *G. lichenoides* and *B. amyloliquefaciens*, either individually or in combination, compared to the control group ($p < 0.05$). The B9 and S8B11 groups had the highest expression levels of *FAS* and *LR* genes. The relative expression levels of intestinal fatty acid transport protein 1 (*FATP1*) were significantly higher in the B9 and S8B11 groups compared to other groups ($p < 0.05$), while the expression of *FATP1* was lower in the S3, S3B9, S8, and B11 groups compared to the control group ($p < 0.05$). The expression of long-chain fatty acid-CoA ligase 4 (*LCL4*) was lower in all experimental groups with *G. lichenoides* and *B. amyloliquefaciens* supplementation compared to control ($p < 0.05$). The results reveal that adding *B.amyloliquefaciens* (10^9 CFU/g) and 8% *G. lichenoide* + *B. amyloliquefaciens* (10^{11} CFU/g) had a significant impact on the expression of genes involved in shrimp intestinal lipid metabolism.

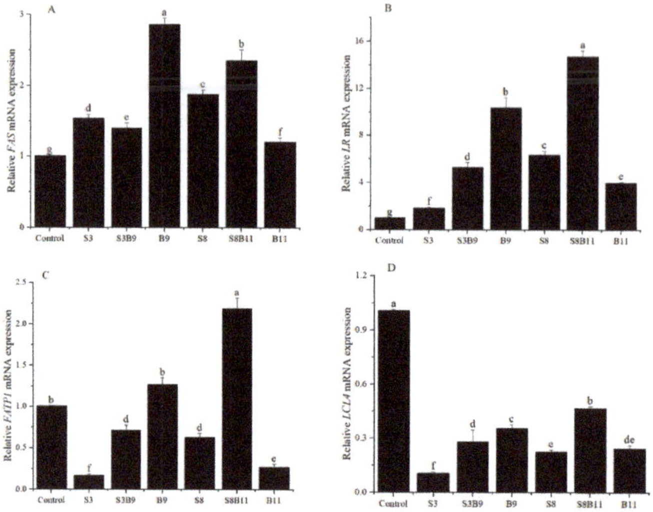

Figure 5. *FAS* (**A**), *LR* (**B**), *FATP1* (**C**), and *LCL4* (**D**) gene expression level in the intestine of shrimp fed different diets for 56 days. The reference gene is *EF-1α*. Vertical bars represented the mean ± SD ($n = 4$) and data indicated with different letters were significantly different ($p < 0.05$) among treatments.

3.8. Intestinal Microbiota Changes

3.8.1. Richness and Diversity

A total of 2,535,118 high-quality sequences were obtained from the intestine microbiota of *P. monodon*. The sequence length was from 159 to 437 bp. The core OUT (operational taxonomic unit) accounted for 394 (Figure 6). The microbiota coverage of each group was all close to 99%, which suggests that the sequencing depth of this intestinal microbiota analysis was sufficient. Thus, the experimental results were representative. The bacterial richness analysis showed that Chao1 index was increased in the S8, B9 and B11 groups compared with the control group, while there was no significant difference among the groups ($p > 0.05$) (Table 5).

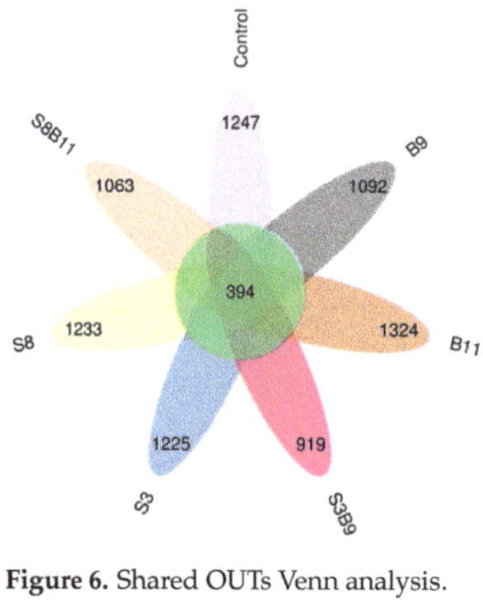

Figure 6. Shared OUTs Venn analysis.

Table 5. Intestinal microbial diversity of *P. monodon* fed the different experimental diets for 56 days.

Alpha Name	Control	S3	S3B9	B9	S8	S8B11	B11
Chao1	1066.90 ± 445.03	1054.39 ± 354.66	876.95 ± 243.56	1139.16 ± 399.85	1229.12 ± 352.32	1052.77 ± 469.62	1298.55 ± 404.43
Simpson	0.94 ± 0.03	0.91 ± 0.05	0.91 ± 0.04	0.94 ± 0.03	0.95 ± 0.02	0.92 ± 0.04	0.93 ± 0.01
Shannon	5.74 ± 0.97	5.39 ± 0.41	5.14 ± 0.94	5.78 ± 0.83	5.94 ± 0.46	5.20 ± 0.97	5.54 ± 0.57
Coverage (%)	99.67 ± 0.17	99.65 ± 0.14	99.73 ± 0.05	99.59 ± 0.15	99.61 ± 0.15	99.61 ± 0.19	99.54 ± 0.13

3.8.2. Analysis of Microbial Community Composition

A total of 43 bacterial phyla were identified in the seven groups. Of the dominant phylum, the relative abundances of Tenericutes in the S8B11 groups were increased compared to the control, S3, S3B9, and S8 groups ($p < 0.05$), yet the relative abundance of Firmicutes in the S3 group was significantly higher than that in the S3B9, B9, S8, and B11 groups ($p < 0.05$) (Figure 7A). As shown in Figure 7B, at the genus level, the relative abundance of *Vibrio* in the S3 group were significantly higher than that in the S3B9 ($p < 0.05$) and had no significant difference with other groups ($p > 0.05$). Moreover, the relative abundance of *Shimia* in the B9 group was significantly higher than that of the control, S3, and S8 groups ($p < 0.05$). However, the relative abundances of *Lactobacillus* in the S3B9, B9, and S8 groups were significantly lower than that of the control group ($p < 0.05$).

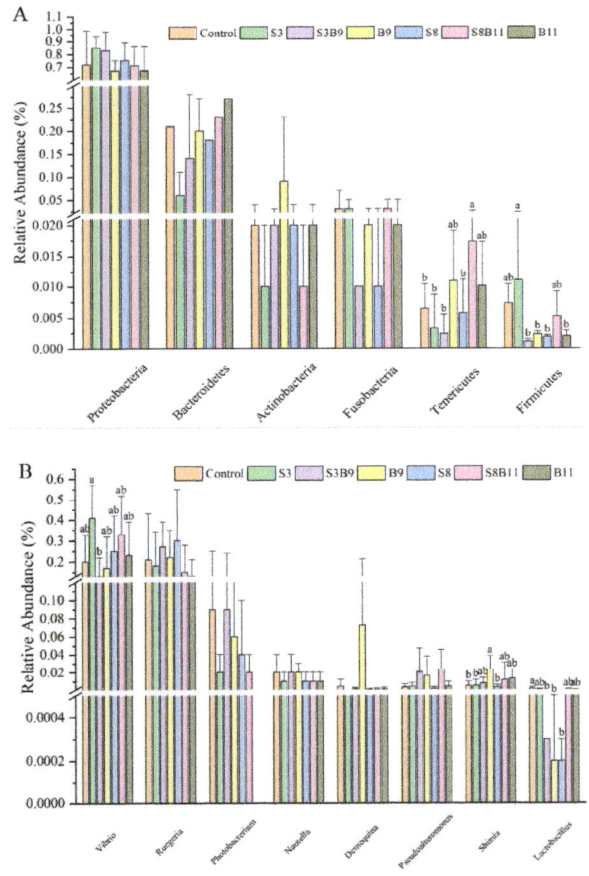

Figure 7. *Cont.*

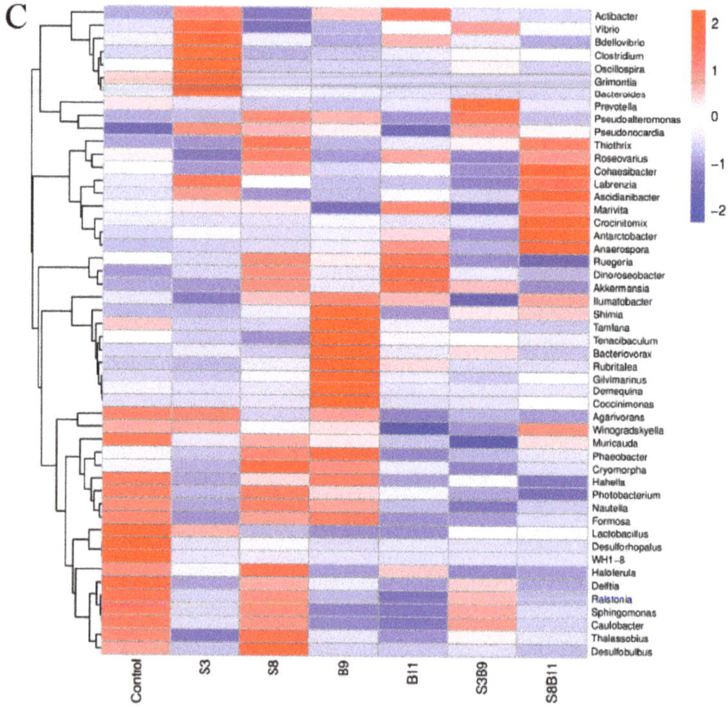

Figure 7. Intestine microbe composition of *P. monodon* that were fed the different experimental diets for 56 days. (**A**) Relative abundance of major bacterial phyla in *P. monodon* fed with the different experimental diets for 56 days. (**B**) Relative abundance of major bacterial genera in *P. monodon* fed with the different experimental diets for 56 days. (**C**) Heat map analysis of intestine microbial on the top 50 genera. Vertical bars represented the mean ± SD (n = 4) and data indicated with different letters were significantly different ($p < 0.05$) among treatments.

The linear discriminant analysis (LDA) effect size (Lefse) package was used to determine the differential abundances of microbiota taxa among the groups. There are 23 bacteria taxa distinguishing among the experimental groups, except the S8 and B9 groups. The relative abundance of 5, 8, 1, 5, and 4 biomarkers (LDA score > 2.0) was higher in the control, S3, B11, S3B9, and S8B11 groups, respectively. LDA scores of Lefse showed that 8 taxa were increased in the S3 group and there were two groups only enriched at genus level, including *Altomonas* and *Clostridium* (Figure 7A). From the results of Lefse cladogram (Figure 8B), 1 phylum, 1 class, 1 order, 4 families, and 2 genera were enriched in the S3, including Firmicutes (from phylum to genus), while 1 phylum, 2 class, and 1 order were enriched in the S8B11 group, including Flavobacteriia (from class to order). At the order level, Acidobacteriales is the possible biomarker for the S3B9 group, while *Lactrobacillales* was found at the control group. However, no suitable biomarker was found in the S8 and B9 groups.

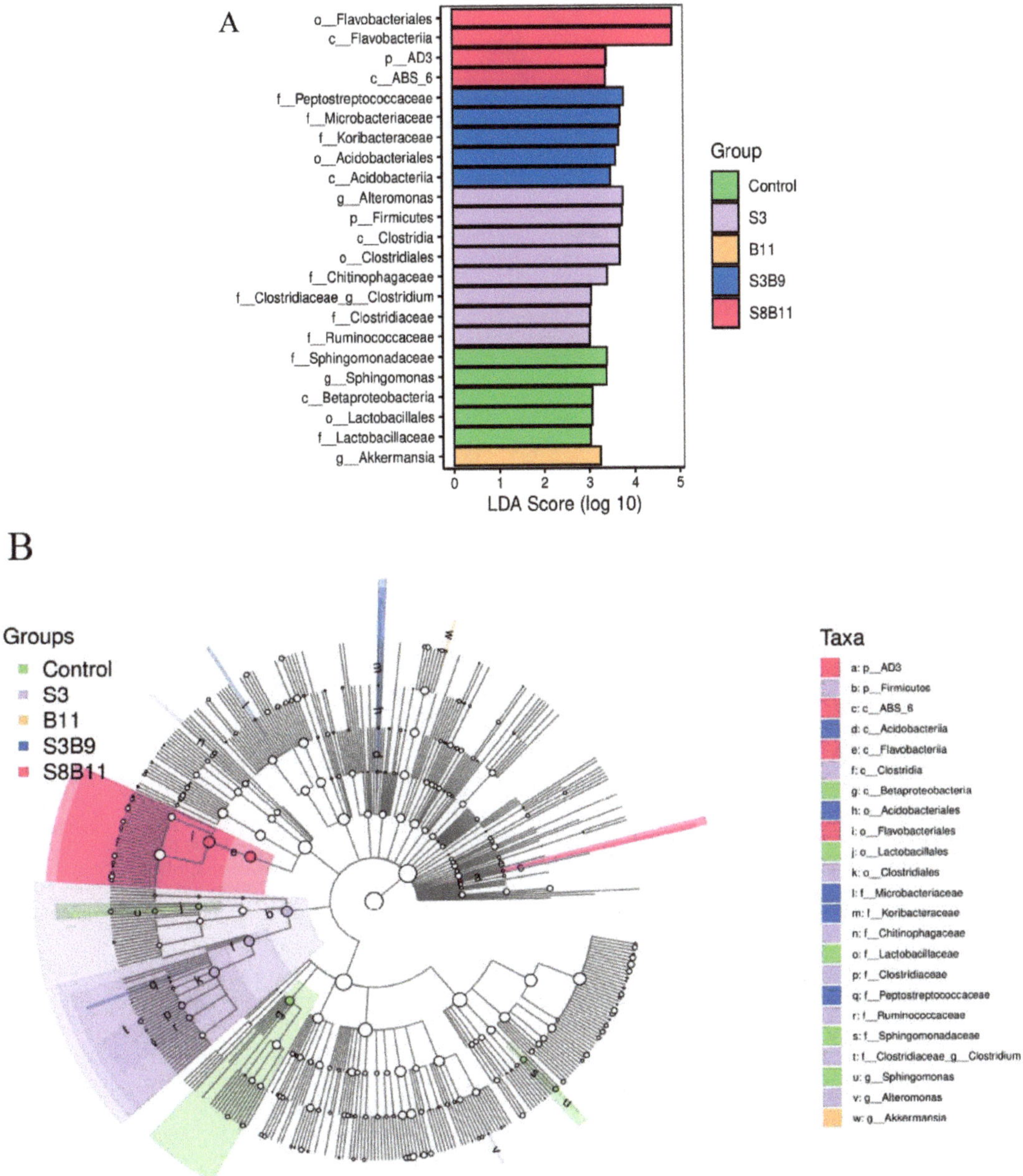

Figure 8. Indicator species analysis of intestinal microbiota of *P. monodon* fed the different experimental diets for 56 days. (**A**) LDA score of Lefse-PICRUSt. The length of the column represents the effect size of bacterial lineages. (**B**) Lefse cladogram. The cladogram was obtained by mapping differences onto a known hierarchical classification tree. Green: bacterial taxa enriched in the control; Purple: bacterial taxa enriched in the S3 group; Orange: bacterial taxa enriched in the B11 group; Blue: bacterial taxa enriched in the S3B9 group; Pink: bacterial taxa enriched in the S8B11; White: no significant differences. Only the taxa that had a linear discriminate analysis (LDA) value above 2.0 are shown.

4. Discussion

The findings of this study demonstrate that the inclusion of a small quantity of seaweed power in the diet resulted in significantly higher FBW, WGR, and SGR in the S3 group compared to the other groups, suggesting a positive impact on the growth of prawns. Niu et al. [14] conducted an experiment in which various concentrations of wakame were administered to spotted prawns, resulting in a significant increase in WG and SGR at the 2% concentration level. Similarly, Jumah et al. [43] supplemented prawn feeds with *k-carrageenan*, resulting in a significant increase in the FBW, WG, and SGR in the experimental group. Anil et al. [44] demonstrated that the inclusion of *Solieriaceae* in the diet of shrimp resulted in increased average daily gain and total weight, suggesting that seaweed have the potential to enhance the growth performance of shrimp. The beneficial effects of seaweeds on shrimp growth can be attributed to their rich content of vitamins, mineral contents, polyunsaturated fatty acids, and amino acids [45]. The active compounds have the ability to modulate the composition of bacterial flora, leading to enhanced intestinal function that facilitates the absorption of carbohydrates and fats in the diet [45]. The growth-promoting impact of seaweeds in the diet can be linked to the digestion and absorption of key nutrients in shrimp feed, which is facilitated by the release of hydrolytic enzymes from shrimps' digestive glands [46]. The results indicate that the S3 group exhibited significantly higher lipase activity compared to the control group, implying that the inclusion of a small quantity of *G. lichenoides* may enhance feed digestion and absorption in shrimp, leading to improved growth performance.

The process of lipid metabolism plays a crucial role in the growth and development of shrimp. Lipid transport, lipolysis, and lipid synthesis collectively govern lipid metabolism [47]. The synthesis of lipids, specifically triglycerides, necessitates the catalytic activity of FAS within the fatty acid synthesis pathway, thereby exerting a significant influence on lipid synthesis [47,48]. Furthermore, a decrease in adipogenesis is indicated by the downregulation of *FAS* [49]. *FATPs* are essential for lipid transport and synthesis, as they exhibit lipo acyl-coenzyme-A synthetase activity and facilitate the transmembrane transport of long-chain fatty acid, ultimately contributing to fat deposition [50]. *LR*s are integral membrane proteins that facilitate the binding of lipoproteins and their corresponding ligands, thereby facilitating the cellular uptake and metabolism of lipoproteins [51]. *LCL*, as a superfamily of membrane proteins, catalyze the activation of fatty acids and the formation of acyl-cofactor thioesters, serving as crucial components in both anabolic and catabolic pathways involving fatty acids [52]. In conclusion, upregulation of the genes FAS, *FATP*, *LR*, and *LCL* has been found to be advantageous for lipid synthesis. The findings of this research indicate that the expression of genes related to fat synthesis, specifically *FAS*, *FATP1*, and *LR*, was increased in the intestines of shrimp in the B9 group compared to the B11 group. This difference in gene expression may account for the significantly higher lipid content observed in the whole body of shrimp of the B9 group compared to the B11 group, suggesting a potential dose-response relationship between *B. amyloliquefaciens* and lipid metabolism. Similar results indicated a rise in muscle and hepatopancreatic fat content in *L. vannamei* following the addition of *B. subtilis* in diet [53], as well as a notable increase in the expression levels of genes associated with lipid synthesis in grass carp after the supplementation *B. subtilis* H2 in diets [54].

The hepatopancreas plays a crucial role as an antioxidant tissue organ in shrimp, as animals generate significant quantities of reactive oxygen species (ROS) under stress conditions, leading to oxidative damage to proteins, lipids, and DNA [49,55]. The measurement of MDA content in tissues and cells serves as an indicator of the level of oxidative damage [56,57]. Following oxidative damage, the antioxidant enzymes within the biological antioxidant enzyme system (SOD, GSH-PX, CAT, POD) work to eliminate excess ROS and safeguard the organism from harm [58]. The magnitude of T-AOC is indicative of the organism's ability to compensate for external stimuli through its antioxidant enzyme system and non-enzymatic system, as well as its metabolic state of free radicals [59]. This measurement can serve as a reflection of the organism's overall antioxidant capacity. Therefore, it is imperative to reduce oxidative damage and enhance the antioxidant capacity of the organism. In the current study, there was a significant increase in T-AOC and SOD enzyme activities in the S3 and S3B9 groups, an increase in MDA and GSH-Px enzyme activities in the S3 group, and a significant increase in POD enzyme activity in the S3B9 group. These results align with the findings of Chen et al. [60], who observed a significant increase in GSH-Px and SOD activities in the hepatopancreas of shrimp fed with *Rhodotorula mucilaginosa* and *B. licheniformis*. Ghasem et al. [61] demonstrated that the activities of GSH-Px and SOD in Nile tilapia were significantly enhanced following the addition of PHDP and *Pediococcus acidilactici* to their diet. Similarly, feeding *Gracilaria tenuistipitata* extract to *L. vannamei* resulted in an increase in SOD activity [62]. Furthermore, the supplementation of dietary with 1 g/kg of seaweed polysaccharides led to elevated levels of T-AOC, SOD, and GSH-Px in the hemolymph of *Fenneropenaeus merguiensis* [63]. The observed phenomenon can be ascribed to the abundance of active biological compounds such as polyphenols, glycosides, anthocyanins, tannins, and thiocarbamates in seaweed. These compounds contribute to the high levels of natural antioxidants that effectively scavenge free radicals, stimulate the production of active antioxidant enzymes, and inhibit oxidative enzymes, thereby mitigating the detrimental effects of free radicals [64]. Additionally, the synthesis of extracellular polysaccharides by prebiotic bacteria plays a role in inhibiting ROS and modulating antioxidant-related genes, ultimately enhancing the antioxidant capacity of the system [65]. The antioxidant capabilities of algae have been evidenced in prior research through a variety of mechanisms, including the reduction of hydrogen peroxide radicals and their conversion into oxygen and water [66]. Furthermore, the observed trends in SOD and GSH-Px activities in the hepatopancreas of shrimp were found to align with their respective mRNA levels.

The Inclusion of 3% *G. lichenoides* in the feed resulted in a significant upregulation of *mtMnSOD, GSH-Px, Trx, Hippo*, and *Nrf2* genes. Furthermore, the addition of 3% *G. lichenoides* in combination with *B. amyloliquefaciens* (10^9 CFU/g) led to an increase in the relative expression levels of *CAT, GSH-Px, Trx*, and *Hippo* genes. This study further examined antioxidant-related genes' expression levels of hepatopancreas in shrimp supplemented with additives, based on the analysis of physiological and biochemical indices. Under conditions of oxidative stress, *Nrf2* levels are maintained through a reduction in ubiquitination and degradation, as well as activation of downstream antioxidant factors [67]. *Nrf2* is regulated by multiple pathways at various stages, forming a dimer with sMAF proteins and controlling the expression of genes containing antioxidant response elements (ARE), such as *GPx, Trx, Prx, TrxR, CAT*, and *SOD* [68]. This process mediates the activation of a diverse array of antioxidant genes through the Keap1-Nrf2- ARE signaling pathway [67,69]. The *SOD* and *CAT* genes in zebrafish was found to be significantly increased by addition of *G. gracilis* [70]. Furthermore, Chen et al. [59] conducted a study which demonstrated that the inclusion of mannan oligosaccharide (MOS) and *B. lincheniformis* in the diet resulted in a significant upregulation of the expression levels of *CAT, GPx*, and *SOD* in *L. vannamei*. The inclusion of 3% *G. lichenoides* in the diet led to an upregulation of *Nrf2, SOD*, and *GSH-Px* gene expression, aligning with the observed antioxidant enzyme activity levels. These findings indicate that even a modest quantity of algal meal can enhance antioxidant enzymes through the stimulation of gene expression. Curcumin was observed to enhance the activities of antioxidant enzyme SOD, CAT, and GPx by upregulating the expression of *Cu/Zn SOD, MnSOD, CAT*, and *GPx* mRNA, which was associated with the *Keap1/Nrf2* signaling pathway in grass carp following infection with *Aeromonas hydrophila* [59]. *Keap1* is capable of regulating the expression of antioxidant genes, such as *CAT* and *GSH-Px*, through modulation of the *Nrf2* signaling pathway. Moreove, a fraction of *Keap1* may potentially increase the efficacy of hepatopancreatic antioxidant enzymes by activating genes through the *Nrf2* signaling pathway [69]. The findings indicated elevated levels of *Trx* expression in the S3, S3B9, and S8B11 groups, aligning with previous research demonstrating the ability of aspalathin to mitigate lead-induced renal oxidative stress in mice through modulation of the *Trx* and *Nrf2* signaling pathways [71]. This mechanism involves the binding of Nrf2 to the antioxidant ARE and subsequent upregulation of *Trx* and *TrxR* genes via the *Trx* signaling pathway, ultimately reducing oxidative stress-induced damage [71]. The inclusion of 8% *G. lichenoides* in combination with *B. amyloliquefaciens* (10^{11} CFU/g) in the diet resulted in an upregulation of *Nrf2*, *CAT*, and *Trx* gene levels. However, there was no notable impact on antioxidant enzyme activities, indicating the necessity for further exploration into the specific mechanisms underlying this effect. The study revealed that the supplementation of *B. amyloliquefaciens* (10^9 CFU/g) in the diet did not have significant alterations in the antioxidant enzyme activities and expression levels of relevant antioxidant genes in *P. monodon*. Conversely, the concurrent administration of modest quantities of algal meal and *B. amyloliquefaciens* induced the activation of the antioxidant response factor *Nrf2*, thereby facilitating the expression of diverse antioxidant enzymes and enhancing the transcription of downstream genes including *Trx*, *SOD*, *CAT*, and *GPx*. It is hypothesized that the coexistence of low levels of *G. lichenoides* and *B. amyloliquefaciens* may result in a heightened presence of natural antioxidants.

In the process of nutrient assimilation, the intestines are essential due to the increased surface area for absorption provided by the folds within them [14]. In the present study, the experimental group demonstrated a greater abundance of intestinal tissue epithelia with well-defined brush borders and densely packed nuclei compared to the control group. Additionally, the study found that the height of intestinal epithelial cells in *P. monodon* fed a diet containing *G. lichenoides* was significantly greater than those in the S8B11 group, indicating that the inclusion of *G. lichenoides* in the diet may enhance the nutrient absorption capacity of the shrimp. Previous research has shown that seaweed (*Enteromorpha*) can increase villus width and lead to a larger villus surface area [63]. Paul et al. [72] illustrated that the inclusion of dietary *K. alvarezii* resulted in a notable enhancement of villus width and crypt depth within the intestine of broiler chicken. The supplementation of seaweeds in cultured animals has been shown to augment intestinal microvilli and increase absorptive surface area, thereby potentially enhancing nutrient utilization and promoting growth. In the current study, the highest values of intestine epithelium in *P. monodon* were observed in the S3 group. Additionally, elevated values of fold depth were noted in the S3B9, B9, and B11 groups, which were significantly greater than those in the control group. Previous research has shown that *B. subtilis* can enhance broiler villus height, villus surface area, and absorbing epithelial cell area [53], while the supplementation of *B. licheniformis* has been found to promote the development of intestinal villi in common carp, with well-developed and well-arranged villi [73]. These findings suggest that *Bacillus* have the potential to improve the structure integrity of intestinal tissue.

The intestinal microbiota serves as the primary locus for organ digestion and absorption, supplying essential nutrients and energy to the host organism while also exerting a significant influence on the growth, health, and developmental processes of shrimp [38]. In this study, there were no significant differences in the diversity and richness of microbial communities. Diversity and richness are comprehensive indicators composed of a large number of bacteria, indicating that the type of feed added does not change the total type and number of intestinal florae [38]. But there are differences in microbial community composition: at the phylum level, Proteobacteria, Bacteroidetes, Actinobacteria, and Fusobacteria were identified as the predominant bacterial taxa in the intestinal microbiota of *P. monodon*. Previous studies have shown that Proteobacteria, whose microbiota is dominated by Gram-negative bacteria, have a high risk of disease and reduce the health of the host [74]. In the study by Wang et al. [75], the microbial community of Proteobacteria in the intestine of *L. vannabaeus* was increased after white spot syndrome virus (WSSV) infection. Actinobacteria have been shown to degrade organic matter, synthesize antibacterial compounds [76], and modulate inflammatory and autoimmune responses through the induction of regulatory T cells, potentially enhancing disease resistance and immunity in shrimp [77,78]. Firmicutes and Bacteroidetes are generally the major microbiota in the gut of humans and other mammals, participating in the digestion and metabolism of food by the host [79], and are generally thought to contribute to improved digestion in aquatic animals [80]. The relative abundance of Firmicutes in the S3 group was significantly higher than that in the S3B9, S8, and B11 groups. Firmicutes is a commonly predominant microbial taxa in the gastrointestinal tract of humans and other mammals, playing a significant role in food digestion and metabolism within the host organism [73,81]. In aquatic animals, these microbial groups are believed to enhance digestive processes. Firmicutes have been shown to facilitate the absorption of fatty acids in host organisms and promote the synthesis of short-chain fatty acids [79], a process closely associated with the development of fatty liver [38]. In light of the alterations observed in the levels of FBW, WGR, SGR, and intestinal lipase activity within the S3 group, the findings indicate that the supplementation of *G. lichenoides* at a low concentration may enhance digestive enzyme activities and growth performance through modulation of the Firmicutes composition. At the taxonomic level of genus, we identified several dominant genera were identified that displayed significant differences among the groups, including *Vibrio*, *Shimia*, and *Lactobacillus*. *Vibrio*, a facultative pathogen, exhibited high enrichment in the intestinal microbiota of aquatic animals [82]. Our findings indicated a significantly relative abundance of *Vibrio* in the S3 group compared to the S3B9. Previous research by Duan et al. [83] supported the enrichment of *Vibrio* in the intestinal microbiota of the fast-growing shrimp, aligning with the results of our study. The growth performance of shrimp in the S3 group was significantly highest.

5. Conclusions

The inclusion of dietary *G. lichenoides* and *B. amyloliquefaciens* positively impacted the growth, antioxidant capacity, and microbial composition of shrimp. Adding 3% *G. lichenoide* had significantly higher FBW, WGR, and SGR; the antioxidant capacity in hepatopancreatic tissues of *P. monodon* was significantly increased in the 3% *G. lichenoide* group and the 3% *G. lichenoide* + 10^9 CFU/g *B. amyloliquefaciens* group; antioxidant gene expression was significantly increased in the hepatopancreas of shrimp between the 3% *G. lichenoide* group and the 3% *G. lichenoide* + 10^9 CFU/g *B. amyloliquefaciens* group; supplementing the diet with the *G. lichenoide* and *B. amyloliquefacients* improved the stomach and intestinal lipase activity of *P. monodon*; supplementing the diet with *B. amyloliquefaciens* (10^9 CFU/g) improved the intestinal tissue structure of *P. monodon*; adding *B. amyloliquefaciens* (10^9 CFU/g) and 8% *G. lichenoide* + *B. amyloliquefaciens* (10^{11} CFU/g) had a significant impact on the expression of genes involved in shrimp intestinal lipid metabolism; and the relative abundance of intestinal microbiome in the control, S3, B11, S3B9, and S8B11 groups had a significant impact. The results indicate that the inclusion of 3% *G. lichenoides* led to specific improvements in shrimp. This research investigated the impacts of *G. lichenoide* and *B. amyloliquefaciens* on growth, antioxidant capacity, and intestinal microbiome, offering a theoretical foundation for further study in this area.

Author Contributions: J.T.: designing and performing experiments and writing the manuscript; Y.W.: designing experiments and revising the manuscript; J.H.: review and revising the manuscript. H.Y.: collecting samples and statistical analyses; Y.D.: review and revising the manuscript; J.W.: review and revising the manuscript; C.Z.: review and revising the manuscript. Z.H.: review and revising the manuscript. All authors have read and agreed to the published version of the manuscript.

Funding: This research was funded the Finance science and technology project of Hainan province (323MS126), Guangdong province Rural Science and Technology Commissioners Project, Guangdong Natural Science Foundation (2019A1515012159), Financial Special Project of the Ministry of Agriculture and Rural Areas (NHYYSWZZZYKZX2020), Central Public-interest Scientific Institution Basal Research Fund, CAFS (2020KX03, 2023TD97), Rural Revitalization Strategy Special Fund Seed Industry Revitalization Project of Guangdong Province (2022SPY02001).

Institutional Review Board Statement: Not applicable.

Informed Consent Statement: Informed consent was obtained from all subjects involved in the study.

Data Availability Statement: Data are contained within the article.

Acknowledgments: We thank the editor and anonymous reviewers for their comments that helped improve this work.

Conflicts of Interest: The authors declare no conflicts of interest.

References

1. Yang, W.Y.; Jiang, S.; Yang, Q.B.; Jiang, S.G.; Huang, J.H.; Yang, L.S.; Chen, X.; Li, Y.D.; Zhou, F.L. Effects of vitamin C on transport of *Penaeus monodon*. *Isr. J. Aquac.-Bamidgeh* **2023**, *75*, 1–10. [CrossRef]
2. Tu, H.T.; Silvestre, F.; Wang, N.; Thome, J.P.; Phuong, N.T.; Kestemont, P. A multi-biomarker approach to assess the impact of farming systems on black tiger shrimp (*Penaeus monodon*). *Chemosphere* **2010**, *81*, 1204–1211. [CrossRef]
3. Xie, J.J.; Chen, X.; Guo, T.Y.; Xie, S.W.; Fang, H.H.; Liu, Z.L.; Zhang, Y.M.; Tian, L.X.; Liu, Y.J.; Niu, J. Dietary values of *Forsythia suspensa* extract in *Penaeus monodon* under normal rearing and *Vibrio parahaemolyticus* 3HP (VP3HP) challenge conditions: Effect on growth, intestinal barrier function, immune response and immune related gene expression. *Fish Shellfish Immunol.* **2018**, *75*, 316–326. [CrossRef]
4. Schleder, D.D.; Peruch, L.G.B.; Poli, M.A.; Ferreira, T.H.; Silva, C.P.; Andreatta, E.R.; Hayashi, L.; do Nascimento Vieira, F. Effect of brown seaweeds on pacific white shrimp growth performance, gut morphology, digestive enzymes activity and resistance to white spot virus. *Aquaculture* **2018**, *495*, 359–365. [CrossRef]
5. Holdt, S.L.; Kraan, S. Bioactive compounds in seaweed: Functional food applications and legislation. *J. Appl. Phycol.* **2011**, *23*, 543–597. [CrossRef]
6. Yang, Y.F.; Fei, X.G.; Song, J.M.; Hu, H.Y.; Wang, G.C.; Chung, I.K. Growth of *Gracilaria lemaneiformis* under different cultivation conditions and its effects on nutrient removal in Chinese coastal waters. *Aquaculture* **2006**, *254*, 248–255. [CrossRef]
7. Xu, Y.; Fang, J.; Wei, W. Application of *Gracilaria lichenoides* (Rhodophyta) for alleviating excess nutrients in aquaculture. *J. Appl. Phycol.* **2008**, *20*, 199–203. [CrossRef]
8. MacArtain, P.; Gill, C.I.R.; Brooks, M.; Campbell, R.; Rowland, I.R. Nutritional value of edible seaweed. *Nutr. Rev.* **2007**, *65*, 535–543. [CrossRef]
9. Burtin, P. Nutritional value of seaweeds. *J. Agric. Food Chem.* **2003**, *2*, 498–503.
10. Moutinho, S.; Linares, F.; Rodríguez, J.L.; Sousa, V.; Valente, L.M.P. Inclusion of 10% seaweed meal in diets for juvenile and on-growing lifestages of senegalese sole (*Solea senegalensis*). *J. Appl. Phycol.* **2018**, *30*, 3589–3601. [CrossRef]
11. Marinho, G.; Nunes, C.; Sousa-Pinto, I.; Pereira, R.; Rema, P.; Valente, L.M.P. The IMTA-cultivated Chlorophyta *Ulva* spp. as a sustainable ingredient in nile tilapia (*Oreochromis niloticus*) diets. *J. Appl. Phycol.* **2013**, *25*, 1359–1367. [CrossRef]
12. Wassef, E.A.; El-Sayed, A.F.M.; Kandeel, K.M.; Sakr, E.M. Evaluation of *Pterocladia* (Rhodophyta) and *Ulva* (Chlorophyta) meals as additives to gilthead seabream *Sparus aurata* diets. *Egypt. J. Aquat. Res.* **2005**, *31*, 321–332.
13. Mustafa, G.; Wakamatsu, S.; Takeda, T.; Umino, T.; Nakagawa, H. Effects of algae meal as feed additive on growth, feed efficiency, andbody composition in red sea bream. *Fish. Sci.* **1995**, *61*, 25–28. [CrossRef]
14. Niu, J.; Chen, X.; Lu, X.; Jiang, S.-G.; Lin, H.-Z.; Liu, Y.-J.; Huang, Z.; Wang, J.; Wang, Y.; Tian, L.-X. Effects of different levels of dietary wakame (*Undaria pinnatifida*) on growth, immunity and intestinal structure of juvenile *Penaeus monodon*. *Aquaculture* **2015**, *435*, 78–85. [CrossRef]
15. Valente, L.M.P.; Araújo, M.; Batista, S.; Peixoto, M.J.; Sousa-Pinto, I.; Brotas, V.; Cunha, L.M.; Rema, P. Carotenoid deposition, flesh quality and immunological response of nile tilapia fed increasing levels of IMTA-cultivated *Ulva* spp. *J. Appl. Phycol.* **2016**, *28*, 691–701. [CrossRef]
16. Lazo, J.P.; Dinis, M.T.; Holt, G.J.; Faulk, C.; Arnold, C.R. Co-Feeding microparticulate diets with algae: Toward eliminating the need of zooplankton at first feeding in larval red drum (*Sciaenops ocellatus*). *Aquaculture* **2000**, *188*, 339–351. [CrossRef]
17. Sotoudeh, E.; Jafari, M. Effects of dietary supplementation with Red Seaweed, *Gracilaria pygmaea*, on growth, carcass composition and hematology of juvenile rainbow trout, *Oncorhynchus Mykiss*. *Aquacult. Int.* **2017**, *25*, 1857–1867. [CrossRef]
18. Passos, R.; Correia, A.P.; Ferreira, I.; Pires, P.; Pires, D.; Gomes, E.; Do Carmo, B.; Santos, P.; Simões, M.; Afonso, C.; et al. Effect on health status and pathogen resistance of gilthead seabream (*Sparus aurata*) fed with diets supplemented with *Gracilaria gracilis*. *Aquaculture* **2021**, *531*, 735888. [CrossRef]
19. Ragaza, J.A.; Koshio, S.; Mamauag, R.E.; Ishikawa, M.; Yokoyama, S.; Villamor, S.S. Dietary supplemental effects of red seaweed *Eucheuma denticulatum* on growth performance, carcass composition and blood chemistry of juvenile flounder, *Paralichthys olivaceus*. *Aquac. Res.* **2015**, *46*, 647–657. [CrossRef]
20. Adel, M.; Omidi, A.H.; Dawood, M.A.O.; Karimi, B.; Shekarabi, S.P.H. Dietary *Gracilaria persica* mediated the growth performance, fillet colouration, and immune response of persian sturgeon (*Acipenser persicus*). *Aquaculture* **2021**, *530*, 735950. [CrossRef]
21. Negm, S.S.; Ismael, N.E.M.; Ahmed, A.I.; Asely, A.M.E.; Naiel, M.A.E. The efficiency of dietary *Sargassum aquifolium* on the performance, innate immune responses, antioxidant activity, and intestinal microbiota of nile tilapia (*Oreochromis niloticus*) raised at high stocking density. *J. Appl. Phycol.* **2021**, *33*, 4067–4082. [CrossRef]
22. Shi, Q.; Yu, C.; Zhu, D.; Li, S.; Wen, X. Effects of dietary *Sargassum horneri* on resisting hypoxia stress, which changes blood biochemistry, antioxidant status, and *Hepatic* HSP mRNA expressions of juvenile Black sea bream *Acanthopagrus Schlegelii*. *J. Appl. Phycol.* **2020**, *32*, 3457–3466. [CrossRef]

23. Bakky, M.A.H.; Tran, N.T.; Zhang, Y.; Hu, H.; Lin, H.; Zhang, M.; Liang, H.; Zhang, Y.; Li, S. Effects of dietary supplementation of Gracilaria lemaneiformis-derived sulfated polysaccharides on the growth, antioxidant capacity, and innate immunity of rabbitfish (*Siganus canaliculatus*). *Fish Shellfish Immunol.* **2023**, *139*, 108933. [CrossRef]
24. Heidarieh, M.; Mirvaghefi, A.R.; Akbari, M.; Farahmand, H.; Sheikhzadeh, N.; Shahbazfar, A.A.; Behgar, M. Effect of dietary ergosan on growth performance, digestive enzymes, intestinal histology, hematological parameters and body composition of rainbow trout (*Oncorhynchus mykiss*). *Fish Physiol. Biochem.* **2012**, *38*, 1169–1174. [CrossRef]
25. Silva, D.M.; Valente, L.M.P.; Sousa-Pinto, I.; Pereira, R.; Pires, M.A.; Seixas, F.; Rema, P. Evaluation of IMTA-produced seaweeds (*Gracilaria*, *Porphyra*, and *Ulva*) as dietary ingredients in Nile tilapia, *Oreochromis niloticus* L., juveniles. Effects on growth performance and gut histology. *J. Appl. Phycol.* **2015**, *27*, 1671–1680. [CrossRef]
26. Araújo, M.; Rema, P.; Sousa-Pinto, I.; Cunha, L.M.; Peixoto, M.J.; Pires, M.A.; Seixas, F.; Brotas, V.; Beltrán, C.; Valente, L.M.P. Dietary inclusion of IMTA-cultivated *Gracilaria vermiculophylla* in rainbow trout (*Oncorhynchus mykiss*) diets: Effects on growth, intestinal morphology, tissue pigmentation, and immunological response. *J. Appl. Phycol.* **2016**, *28*, 679–689. [CrossRef]
27. Yu, Y.-Y.; Chen, W.-D.; Liu, Y.-J.; Niu, J.; Chen, M.; Tian, L.-X. Effect of different dietary levels of *Gracilaria lemaneiformis* dry power on growth performance, hematological parameters and intestinal structure of juvenile pacific White shrimp (*Litopenaeus vannamei*). *Aquaculture* **2016**, *450*, 356–362. [CrossRef]
28. Muttharasi, C.; Gayathri, V.; Muralisankar, T.; Mohan, K.; Uthayakumar, V.; Radhakrishnan, S.; Kumar, P.; Palanisamy, M. Growth Performance, Digestive enzymes and antioxidants activities in the shrimp *Litopenaeus vannamei* fed with *Amphiroa fragilissima* crude polysaccharides encapsulated *Artemia nauplii*. *Aquaculture* **2021**, *545*, 737263. [CrossRef]
29. Fuller, R. Probiotics in man and animals. *J. Appl. Bacteriol.* **1989**, *66*, 365–378.
30. Hong, H.A.; Duc, L.H.; Cutting, S.M. The use of bacterial spore formers as probiotics: Table 1. *FEMS Microbiol. Rev.* **2005**, *29*, 813–835. [CrossRef]
31. Kumar, V.; Roy, S.; Meena, D.K.; Sarkar, U.K. Application of probiotics in shrimp aquaculture: Importance, mechanisms of action, and methods of administration. *Rev. Fish Sci. Aquac.* **2016**, *24*, 342–368. [CrossRef]
32. Xu, L.; Yuan, J.; Chen, X.; Zhang, S.; Xie, M.; Chen, C.; Wu, Z. Screening of intestinal probiotics and the effects of feeding probiotics on the digestive enzyme activity, immune, intestinal flora and WSSV resistance of *Procambarus clarkii*. *Aquaculture* **2021**, *540*, 736748. [CrossRef]
33. Wang, R.; Guo, Z.; Tang, Y.; Kuang, J.; Duan, Y.; Lin, H.; Jiang, S.; Shu, H.; Huang, J. Effects on development and microbial community of shrimp *Litopenaeus vannamei* larvae with probiotics treatment. *AMB Expr.* **2020**, *10*, 109. [CrossRef]
34. Yu, M.; Zhang, Y.; Zhang, D.; Wang, Q.; Wang, G.; Elsadek, M.; Yao, Q.; Chen, Y.; Guo, Z. The effect of adding *Bacillus amyloliquefaciens* LSG2-8 in diets on the growth, immune function, antioxidant capacity, and disease resistance of *Rhynchocypris lagowskii*. *Fish Shellfish Immunol.* **2022**, *125*, 258–265. [CrossRef]
35. Truong Thy, H.T.; Tri, N.N.; Quy, O.M.; Fotedar, R.; Kannika, K.; Unajak, S.; Areechon, N. Effects of the dietary supplementation of mixed probiotic spores of *Bacillus amyloliquefaciens* 54A, and *Bacillus pumilus* 47B on growth, innate immunity and stress responses of striped catfish (*Pangasianodon hypophthalmus*). *Fish Shellfish Immunol.* **2017**, *60*, 391–399. [CrossRef]
36. Tseng, D.-Y.; Ho, P.-L.; Huang, S.-Y.; Cheng, S.-C.; Shiu, Y.-L.; Chiu, C.-S.; Liu, C.-H. Enhancement of immunity and disease resistance in the white shrimp, *Litopenaeus vannamei*, by the probiotic, *Bacillus subtilis* E20. *Fish Shellfish Immunol.* **2009**, *26*, 339–344. [CrossRef]
37. Lin, Y.-S.; Saputra, F.; Chen, Y.-C.; Hu, S.-Y. Dietary administration of *Bacillus amyloliquefaciens* R8 reduces hepatic oxidative stress and enhances nutrient metabolism and immunity against Aeromonas hydrophila and Streptococcus agalactiae in zebrafish (*Danio rerio*). *Fish Shellfish Immunol.* **2019**, *86*, 410–419. [CrossRef]
38. Xu, P.W. Optimum Dietary Lipid Requirement and Short-Chain Fatty Acids Nutritional Function for Juvenile Golden Pompano (*Trachinotus ovatus*). Ph.D. Thesis, Shanghai Ocean University, Shanghai, China, 2022.
39. Baur, F.J.; Ensminger, L.G. The Association of Official Analytical Chemists (AOAC). *J. Am. Oil Chem. Soc.* **1977**, *54*, 171–172. [CrossRef]
40. Ming, J.; Ye, J.; Zhang, Y.; Yang, X.; Shao, X.; Qiang, J.; Xu, P. Dietary optimal reduced glutathione improves innate immunity, oxidative stress resistance and detoxification function of grass carp (*Ctenopharyngodon idella*) against Microcystin-LR. *Aquaculture* **2019**, *498*, 594–605. [CrossRef]
41. Xie, S.; Wei, D.; Liu, Y.; Tian, L.; Niu, J. Dietary fish oil levels modulated lipid metabolism, immune response, intestinal health and salinity stress resistance of juvenile *Penaeus monodon* fed a low fish-meal diet. *Anim. Feed Sci. Technol.* **2022**, *289*, 115321. [CrossRef]
42. Livak, K.J.; Schmittgen, T.D. Analysis of relative gene expression data using real-time quantitative PCR and the $2^{-\Delta\Delta CT}$ method. *Methods* **2001**, *25*, 402–408. [CrossRef]
43. Sun, M.M.; Huang, J.H.; Yang, Q.B.; Zhou, F.L.; Wen, W.G.; Chen, X.; Jiang, S.G. Comparison on characteristics of growth and resistance to ammonia among 13 families of *Penaeus monodon*. *J. Shanghai Ocean. Univ.* **2011**, *20*, 510–516.
44. Anil, K.S.; Balakrishanan, G.; Kanji, J.L.; Jitesh, S.B.; Kumaran, R. Comparison of *Penaeus monodon* (Crustacea, Penaeidae) growth between commercial feed vs. commercial shrimp feed supplemented with *Kappaphycus alvarezii* (Rhodophyta, Solieriaceae) seaweed sap. *AACL Bioflux.* **2011**, *4*, 292–300.
45. Omont, A.; Quiroz-Guzman, E.; Tovar-Ramirez, D.; Peña-Rodríguez, A. Effect of diets supplemented with different seaweed extracts on growth performance and digestive enzyme activities of juvenile white shrimp *Litopenaeus vannamei*. *J. Appl. Phycol.* **2019**, *31*, 1433–1442. [CrossRef]

46. Traifalgar, R.F.; Serrano, A.E.; CoRRE, V.; Kira, H.; TuNG, H.T.; Michael, F.R.; Koshio, S. Evaluation of dietary fucoidan supplementation effects on growth performance and vibriosis resistance of *Penaeus monodon* postlarvae. *Aquacult. Sci.* **2009**, *57*, 167–174.
47. Guo, J.; Fu, Y.; Wu, Z.; Yu, X.; Guo, Y.; Liu, J.; Zhang, W.; Mai, K. Effects of dietary carbohydrate levels on growth performance, body composition, glucose/lipid metabolism and insulin signaling pathway in abalone *Haliotis discus hannai*. *Aquaculture* **2022**, *557*, 738284. [CrossRef]
48. Bai, T. Study on Early Lipid Characteristics and Key Lipid Metabolism Enzymes of *Schizothorax grahami*. Master's Thesis, Guizhou University, Guiyang, China, 2021.
49. Lu, L.; Luo, X.G.; Ji, C.; Liu, B.; Yu, S.X. Effect of manganese supplementation and source on carcass traits, meat quality, and lipid oxidation in broilers. *J. Anim. Sci.* **2007**, *85*, 812–822. [CrossRef]
50. Zhao, L.L.; Wang, D.X.; Liao, Z.B.; Bi, Q.Z.; Ma, Q.; Wei, Y.L.; Liang, M.Q.; Qiao, X.T.; Cheng, Z.Y.; Xu, H.G. Tissue distribution and nutritional regulation of fatty acid transport protein *Scopthalmus maximus* and *Takifugu rubripes*. *Chin. J. Anim. Nutr.* **2022**, *34*, 6620–6633.
51. Li, H.B.; Liu, F.; Fu, J.H.; Feng, L.; Dai, C.G.; Hu, Y. Cloning and spatio-temporal expression of lipophorin receptors genes related to reproduction in the oriental armyworm *Mythimna separata*. *J. Plant Prot.* **2022**, *49*, 741–748.
52. Knights, K.M. Long-chain-fatty-acid coA ligases: The key to fatty acid activation, formation of xenobiotic acyl-coA thioesters and lipophilic xenobiotic conjugates. *Curr. Med. Chem.* **2003**, *3*, 235–244. [CrossRef]
53. Al-Fataftah, A.-R.; Abdelqader, A. Effects of dietary *Bacillus subtilis* on heat-stressed broilers performance, intestinal morphology and microflora composition. *Anim. Feed Sci. Technol.* **2014**, *198*, 279–285. [CrossRef]
54. Shi, F.; Zi, Y.; Lu, Z.; Li, F.; Yang, M.; Zhan, F.; Li, Y.; Li, J.; Zhao, L.; Lin, L.; et al. *Bacillus subtilis* H2 modulates immune response, fat metabolism and bacterial flora in the gut of grass carp (*Ctenopharyngodon idellus*). *Fish Shellfish Immunol.* **2020**, *106*, 8–20. [CrossRef]
55. Zheng, J.; Duan, Y.; Yu, J.; Li, F.; Guo, Q.; Li, T.; Yin, Y. Effects of Long-term protein restriction on meat quality and muscle metabolites of shaziling pigs. *Animals* **2022**, *12*, 2007. [CrossRef]
56. Nordberg, J.; Arnér, E.S. Reactive oxygen species, antioxidants, and the mammalian thioredoxin system. *Free. Radic. Biol. Med.* **2001**, *31*, 1287–1312. [CrossRef]
57. Finkel, T.; Holbrook, N.J. Oxidants, oxidative stress and the biology of ageing. *Nature* **2000**, *408*, 239–247. [CrossRef]
58. Wang, P.; Geng, J.; Gao, J.; Zhao, H.; Li, J.; Shi, Y.; Yang, B.; Xiao, C.; Linghu, Y.; Sun, X.; et al. Macrophage achieves self-protection against oxidative stress-induced ageing through the Mst-Nrf2 axis. *Nat. Commun.* **2019**, *10*, 755. [CrossRef]
59. Abbas, H.H.H.; Authman, M.M.N. Effects of accumulated selenium on some physiological parameters and oxidative stress indicators in tilapia fish (*Oreochromis* spp.). *Am. -Eurasian J. Agric. Environ. Sci.* **2009**, *5*, 219–222.
60. Chen, M.; Chen, X.-Q.; Tian, L.-X.; Liu, Y.-J.; Niu, J. Enhanced intestinal health, immune responses and ammonia resistance in Pacific white shrimp (*Litopenaeus vannamei*) fed dietary hydrolyzed yeast (*Rhodotorula mucilaginosa*) and *Bacillus licheniformis*. *Aquac. Rep.* **2020**, *17*, 100385. [CrossRef]
61. Mohammadi, G.; Hafezieh, M.; Karimi, A.A.; Azra, M.N.; Van Doan, H.; Tapingkae, W.; Abdelrahman, H.A.; Dawood, M.A.O. The synergistic effects of plant polysaccharide and *Pediococcus acidilactici* as a synbiotic additive on growth, antioxidant status, immune response, and resistance of nile tilapia (*Oreochromis niloticus*) against *Aeromonas hydrophila*. *Fish Shellfish Immunol.* **2022**, *120*, 304–313. [CrossRef]
62. Liu, P.-C.; Lin, P.-W.; Huang, C.-L.; Hsu, C.-H.; Chen, J.-C. Long-term administration of diets containing *Gracilaria tenuistipitata* extract induce the expression of immune-related genes and increase the immune response and resistance against *Vibrio harveyi* in white shrimp *Litopenaeus vannamei*. *Gene Rep.* **2019**, *15*, 100378. [CrossRef]
63. Liu, W.-C.; Zhou, S.-H.; Balasubramanian, B.; Zeng, F.-Y.; Sun, C.-B.; Pang, H.-Y. Dietary seaweed (*Enteromorpha*) polysaccharides improves growth performance involved in regulation of immune responses, intestinal morphology and microbial community in banana shrimp *Fenneropenaeus merguiensis*. *Fish Shellfish Immunol.* **2020**, *104*, 202–212. [CrossRef]
64. Arjeh, E.; Akhavan, H.-R.; Barzegar, M.; Carbonell-Barrachina, Á.A. Bio-active compounds and functional properties of pistachio hull: A review. *Trends Food Sci.* **2020**, *97*, 55–64. [CrossRef]
65. Akbary, P.; Ajdari, A.; Ajang, B. Growth, survival, nutritional value and phytochemical, and antioxidant state of *Litopenaeus vannamei* shrimp fed with premix extract of brown *Sargassum ilicifolium, Nizimuddinia zanardini, Cystoseira indica,* and *Padina australis macroalgae*. *Aquac. Int.* **2023**, *31*, 681–701. [CrossRef]
66. Mohammadi, G.; Adorian, T.J.; Rafiee, G. Beneficial effects of *Bacillus subtilis* on water quality, growth, immune responses, endotoxemia and protection against lipopolysaccharide-induced damages in oreochromis niloticus under biofloc technology system. *Aquac. Nutr.* **2020**, *26*, 1476–1492. [CrossRef]
67. Tong, H.D.; Wang, J.M.; Song, Y. The role of Keap1-Nrf2-ARE in organismal defence against oxidative stress injury. *Carcinog.,Teratog. Mutagen.* **2013**, *25*, 71–75.
68. Morilla, M.J.; Ghosal, K.; Romero, E.L. More than pigments: The potential of astaxanthin and bacterioruberin-based nanomedicines. *Pharmaceutics* **2023**, *15*, 1828. [CrossRef]
69. Niemcharoen, S. Effects of Microplastics on Gene Expression to Nonspecific Immune System in Pacific White Shrimp (*Litopenaeus vannamei*). Master's Thesis, Chulalongkorn University, Bangkok, Thailand, 2022.

70. Hoseinifar, S.H.; Yousefi, S.; Capillo, G.; Paknejad, H.; Khalili, M.; Tabarraei, A.; Van Doan, H.; Spanò, N.; Faggio, C. Mucosal immune parameters, immune and antioxidant defence related genes expression and growth performance of zebrafish (*Danio rerio*) fed on *Gracilaria gracilis* powder. *Fish Shellfish Immunol.* **2018**, *83*, 232–237. [CrossRef]
71. Ji, K.; Liang, H.; Ren, M.; Ge, X.; Mi, H.; Pan, L.; Yu, H. The immunoreaction and antioxidant capacity of juvenile blunt snout bream (*Megalobrama amblycephala*) involves the PI3K/Akt/Nrf2 and NF-κB signal pathways in response to dietary methionine levels. *Fish Shellfish Immunol.* **2020**, *105*, 126–134. [CrossRef]
72. Paul, S.S.; Vantharam Venkata, H.G.R.; Raju, M.V.; Rama Rao, S.V.; Nori, S.S.; Suryanarayan, S.; Kumar, V.; Perveen, Z.; Prasad, C.S. Dietary supplementation of extracts of red sea weed (*Kappaphycus alvarezii*) improves growth, intestinal morphology, expression of intestinal genes and immune responses in broiler chickens. *J. Sci. Food Agric.* **2021**, *101*, 997–1008. [CrossRef]
73. Zhang, J.; Huang, M.; Feng, J.; Chen, Y.; Li, M.; Chang, X. Effects of dietary *Bacillus licheniformis* on growth performance, intestinal morphology, intestinal microbiome, and disease resistance in common carp (*Cyprinus carpio* L.). *Aquac. Int.* **2021**, *29*, 1343–1358. [CrossRef]
74. Hasyimi, W.; Widanarni, W.; Yuhana, M. Growth performance and intestinal microbiota diversity in pacific white shrimp *Litopenaeus vannamei* fed with a probiotic bacterium, honey prebiotic, and synbiotic. *Curr. Microbiol.* **2020**, *77*, 2982–2990. [CrossRef]
75. Wang, J.; Huang, Y.; Xu, K.; Zhang, X.; Sun, H.; Fan, L.; Yan, M. White spot syndrome virus (WSSV) infection impacts intestinal microbiota composition and function in *Litopenaeus vannamei*. *Fish Shellfish Immunol.* **2019**, *84*, 130–137. [CrossRef]
76. Liu, L. Effects of Lactobacillus Pentosus Combined with *Arthrospira platensis* on the Growth Performance, Digestive Ability, Immunity, Intestinal Microbiota and Disease Resistance of *Litopenaeus vannamei*. Master's Thesis, Hainan University, Haikou, China, 2022.
77. Binda, C.; Lopetuso, L.R.; Rizzatti, G.; Gibiino, G.; Cennamo, V.; Gasbarrini, A. Actinobacteria: A relevant minority for the maintenance of gut homeostasis. *Dig. Liver Dis.* **2018**, *50*, 421–428. [CrossRef]
78. Sha, Y.; Liu, M.; Wang, B.; Jiang, K.; Qi, C.; Wang, L. Bacterial population in intestines of *Litopenaeus vannamei* fed different probiotics or probiotic supernatant. *J. Microbiol. Biotechnol.* **2016**, *26*, 1736–1745. [CrossRef]
79. Costello, E.K.; Gordon, J.I.; Secor, S.M.; Knight, R. Postprandial remodeling of the gut microbiota in burmese pythons. *ISME J.* **2010**, *4*, 1375–1385. [CrossRef]
80. Yuan, H.; Song, W.; Tan, J.; Zheng, Y.; Wang, H.; Shi, L.; Zhang, S. The effects of dietary protein level on the growth performance, body composition, intestinal digestion and microbiota of *Litopenaeus vannamei* fed *Chlorella Sorokiniana* as the main protein source. *Animals* **2023**, *13*, 2881. [CrossRef]
81. Tian, Z.-K.; Zhang, Y.-J.; Feng, Z.-J.; Jiang, H.; Cheng, C.; Sun, J.-M.; Liu, C.-M. Nephroprotective effect of gastrodin against lead-induced oxidative stress and inflammation in mice by the *GSH*, *Trx*, *Nrf2* antioxidant system, and the *HMGB1* pathway. *Toxicol. Res.* **2021**, *10*, 249–263. [CrossRef]
82. Xiong, J.B.; Wang, K.; Wu, J.F.; Qiuqian, L.L.; Yang, K.L.; Qian, Y.X.; Zhang, D. Changs in intestinal bacterial communities are closely associated with shrimp disease severity. *Appl. Microbiol. Biotechnol.* **2015**, *99*, 6911–6919. [CrossRef]
83. Duan, Y.; Huang, J.; Wang, Y.; Zhang, J. Characterization of bacterial community in intestinal and rearing water of Penaeus monodon differing growth performances in outdoor and indoor ponds. *Aquac. Res.* **2020**, *51*, 4279–4289. [CrossRef]

biology

MDPI

Article

Analysis of Elimination Effects of Inbreeding on Genotype Frequency in Larval Stages of Chinese Shrimp

Qiang Fu [1,2], Jingxin Zhou [1,2], Sheng Luan [1,2], Ping Dai [1,2], Ding Lyu [1,2], Baolong Chen [1,2], Kun Luo [1,2], Jie Kong [1,2] and Xianhong Meng [1,2,*]

1 State Key Laboratory of Mariculture Biobreeding and Sustainable Goods, Yellow Sea Fisheries Research Institute, Chinese Academy of Fishery Sciences, Qingdao 266071, China; oucfuq@163.com (Q.F.); meditationzhou@163.com (J.Z.); luansheng@ysfri.ac.cn (S.L.); daiping@ysfri.ac.cn (P.D.); lvding@ysfri.ac.cn (D.L.); chenbl@ysfri.ac.cn (B.C.); luokun@ysfri.ac.cn (K.L.); kongjie@ysfri.ac.cn (J.K.)

2 Laboratory for Marine Fisheries Science and Food Production Processes, Qingdao Marine Science and Technology Center, Qingdao 266237, China

* Correspondence: mengxianhong@ysfri.ac.cn

Simple Summary: In this research, we established inbred families of *Fenneropenaeus chinensis* with an inbreeding coefficient of 0.25, and studied how inbreeding depression affects the larval development. We examined gene frequency changes in two developmental stages through sequencing. The results showed significant inbreeding depression in larval survival rates during hatching and two major developmental stages, ranging from 24.36% to 45.28%. Sequencing at these stages revealed deviations from Mendelian segregation, with homozygote deficiency and heterozygote excess being the primary selection types. The elimination during the larval stage resulted in more heterozygotes being retained and increased the population's heterozygosity. These findings indicate the presence of large amount of harmful recessive alleles in the Chinese shrimp genome and their potential criteria for elimination. The study provide valuable insights into the mechanisms of inbreeding depression in marine animals and provides guidance for shrimp population breeding strategies.

Abstract: Marine animals possess genomes of considerable complexity and heterozygosity. Their unique reproductive system, characterized by high fecundity and substantial early mortality rates, increases the risk of inbreeding, potentially leading to severe inbreeding depression during various larval developmental stages. In this study, we established a set of inbred families of *Fenneropenaeus chinensis*, with an inbreeding coefficient of 0.25, and investigated elimination patterns and the manifestations of inbreeding depression during major larval developmental stages. Reduced-representation genome sequencing was utilized to explore the genotype frequency characteristics across two typical elimination stages. The results revealed notable mortality in hatching and metamorphosis into mysis and post-larvae stages. Inbreeding depression was also evident during these developmental stages, with depression rates of 24.36%, 29.23%, and 45.28%. Segregation analysis of SNPs indicated an important role of gametic selection before hatching, accounting for 45.95% of deviation in the zoea stage. During the zygotic selection phase of larval development, homozygote deficiency and heterozygote excess were the main selection types. Summation of the two types explained 82.31% and 89.91% of zygotic selection in the mysis and post-larvae stage, respectively. The overall distortion ratio decreased from 22.37% to 12.86% in the late developmental stage. A total of 783 loci were identified through selective sweep analysis. We also found the types of distortion at the same locus could change after the post-larvae stage. The predominant shifts included a transition of gametic selection toward normal segregation and other forms of distortion to heterozygous excess. This may be attributed to high-intensity selection on deleterious alleles and genetic hitchhiking effects. Following larval elimination, a greater proportion of heterozygous individuals were preserved. We detected an increase in genetic diversity parameters such as expected heterozygosity, observed heterozygosity, and polymorphic information content in the post-larvae stage. These findings suggest the presence of numerous recessive deleterious alleles and their linkage and suggest a major role of the partial dominance hypothesis. The results provide valuable insights into the mechanisms of inbreeding depression in marine animals and offer guidance for formulating breeding strategies in shrimp populations.

Keywords: *Fenneropenaeus chinensis*; inbreeding; larval development; segregation distortion; SNPs

1. Introduction

Inbreeding is the mating of individuals with shared ancestry and can lead to a decrease in genetic diversity and the subsequent depression of certain phenotypic traits within a population [1,2]. Inbreeding depression is a matter of great concern across various species and holds significant implications for evolutionary processes, species conservation, and genetic breeding programs [3–5]. The genetic mechanisms of inbreeding depression have long been a subject of debate, with two major hypotheses proposed. The partial dominance hypothesis suggests that detrimental effects arise from the expression of recessive mutant alleles, whereas the overdominance hypothesis posits that heterozygotes possess a selective advantage [6]. Although the importance of these hypotheses remains a topic of discussion, more evidence supports the view that inbreeding depression is predominantly caused by the cumulative effects of deleterious mutations at multiple genetic loci [7].

Empirical evidence consistently indicates that inbreeding primarily impacts fitness-related traits rather than morphological traits [2]. However, the identification of inbreeding depression may fluctuate among studies due to various factors, like varying degrees of inbreeding, genetic diversity, trait types, developmental phases, and environmental factors [8]. The decline in fitness components caused by inbreeding can manifest at various stages of an organism's life history, and lethal and sub-lethal mutations are more evident during early developmental stages, when severe depression is more readily detectable. Consequently, the study of inbreeding during the developmental stage yields more precise and meaningful results, since the early elimination of individuals allows for sampling a more heterogeneous pool of survivors, leading to lower estimates of inbreeding depression [8,9]. Marine animals possess genomes of significant complexity and heterozygosity. Moreover, their great fecundity combined with high larval mortality rates [10] increases the likelihood of inbreeding, potentially resulting in severe inbreeding depression during various larval developmental stages. These characteristics make marine animals ideal models for studying the effects of inbreeding.

Inbreeding depression has been widely reported in aquaculture species [9,11–14]. In shrimp, the impact of inbreeding on typical economic traits and the effect of varying levels of inbreeding have been extensively studied [9,15–17]. However, few studies have been devoted to the specific genetic loci involved and the underlying mechanisms responsible for the elimination effect, particularly during the larval developmental stages. The fundamental principle of inbreeding cannot be completely determined by analysis only at the phenotypic level. The molecular basis of inbreeding can be better understood by studies at the molecular level, which benefit from high-throughput sequencing. In our previous studies, we developed a novel classification method that can discern the selection of different genotypes in segregation distortion [18]. This method can also be applied to investigate the selection effects on genotype frequency during larval stages under inbreeding conditions. By employing this approach, we can gain valuable insights into the criteria for elimination and the genetic mechanisms involved at the molecular level.

Fenneropenaeus chinensis is one of the most valuable mariculture species in China. In 2005, a breeding program was initiated with the aim of improving its growth rate and disease resistance. The breeding strategy was based on large-scale family selection. Despite careful control of inbreeding (<1% per generation), the closed breeding population inevitably experienced growth in inbreeding. Performance traits exhibited inbreeding depression ranging from 4.16% to 4.74% per 10% increase in inbreeding within this population. However, traits related to overall fitness did not show significant decline during the juvenile and adult stages [15]. This inconsistency may be attributed to the early elimination of lethal and sub-lethal mutations during the larval stage, which could mask subsequent declines [8]. Nevertheless, further confirmation is still required.

The objective of this study was to investigate the selection characteristics of inbred families during the larval stages from the perspective of genotype frequency. The results offer insights into the genetic foundation of significant elimination during the larval stage of marine penaeid shrimp. Additionally, they can provide valuable guidance for the control of inbreeding within breeding programs for *F. chinensis*.

2. Materials and Methods

2.1. Research Population and Family Construction

The parental population for this study was derived from G14 of the core breeding population of *F. chinensis*, known as "Huanghai No. 5". Its base population was constructed by crossing six different geographic populations in 2005 and has since been continuously selected using pedigree methods and large-scale family selection. The family construction and tests were conducted at the Marine Genetic Breeding Center of the Chinese Academy of Fishery Sciences in Qingdao, China. After promoting sexual maturity for 60 days, the prawns were mated through artificial insemination following a specific mating design. The control families were established by carefully mating known pedigree males and females based on their estimated breeding values, following optimal contribution selection criteria and with the inbreeding coefficient controlled within 1%. The experimental inbred lines were established by mating full-sibling individuals within the same family, with an inbreeding coefficient of 0.25. The experiment group included three sets of inbreeding families, and each of them comprised five full-sibling F_2 families. For each set, five control families were also established, which were non-inbreeding mated while being half-sibling to the inbreeding families.

After artificial insemination, the parent prawns were transferred to separate 200 L buckets for spawning. Seventy-two hours after fertilization, three 1 mL samples were taken from each family and observed under a microscope to determine the hatching rates. Subsequently, 10,000 nauplii for each family were randomly selected, transferred to larvae-culture buckets, and cultured according to a standardized procedure [19]. The larvae underwent Nauplius stage (N), Zoea stage (Z), Mysis stage (M) and Post-larvae stage (P) over approximately 21 days (Figure 1). At each developmental stage, three 100 mL samples were taken from each family bucket to calculate larval density and the number of survivors. Simultaneously, 100 larvae were randomly sampled from each developmental stage, fixed with 95% alcohol, and stored at −80 °C. At the P5 stage, all post-larvae were collected, and the overall survival rate during the entire larval rearing period was calculated. A total of 1500 larvae were retained and transported to new tanks for further rearing.

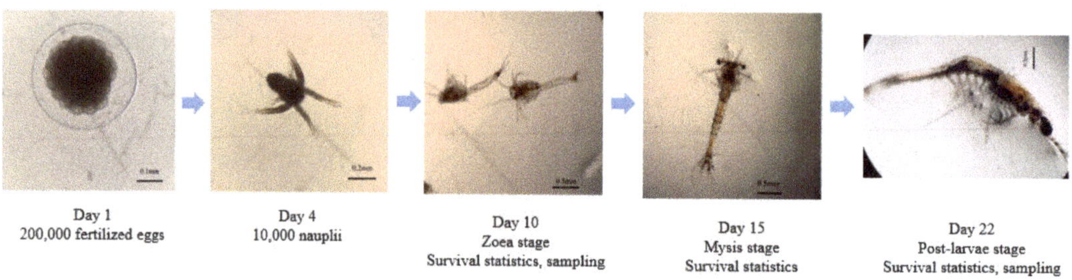

Figure 1. The infographic for larvae development stages.

2.2. Sequencing Library Construction

Out of the 15 F_2 families, the inbred family named 714Y, which exhibited a high mortality rate during larval development, was selected for further genetic analysis. We employed a reduced-representation sequencing strategy (RRS) to analyze the genotype frequency at two different developmental stages. For this purpose, 2b-RAD libraries were prepared for two parents and 36 larvae in zoea and post-larval stages (P3), respectively, following the protocol of Wang et al. [20]. Genomic DNA was extracted using the standard phenol/chloroform method. We constructed standard BsaXI libraries with original adaptors that covered nearly all recognition sites of this Type IIB restriction enzyme in the *F. chinensis* genome. Each library was ligated from multiple restriction enzyme-digested fragments and tagged with a unique barcode during preparation to simplify the pooling process [21]. After quality check, all libraries were subjected to an Illumina Hiseq Xten sequencer (Illumina, San Diego, CA, USA) for 2 × 150 bp paired-end sequencing.

2.3. Data Processing and Genotyping

The raw data were filtered by removing adaptors and reads of poor quality (those with over 15% of bases having a quality value below Q30 or an N base proportion exceeding 8%). The remaining high-quality reads were then applied to de novo 2b-RAD genotyping using the RAD-typing program v1.0 under optimized default parameters for marine animals [22]. A reference of parent-shared representative sites was established based on the male and female parent database and ustacks v2.1.3 software [23], which was utilized for subsequent locus genotyping. High-quality reads from all larvae were genotyped by aligning them to these reference sites and assessed using a likelihood ratio test (−r 0, −M 4, −v 2) [24]. Markers that were polymorphic and heterozygous in at least one parent and could be genotyped in at least 80% of the larvae were considered as qualified segregating markers. To ensure the accuracy of the analyses, genotyping results were further refined by excluding sites with a minor allele frequency (MAF) lower than 0.01, non-biallelic SNP markers, and sites with multiple SNPs within a tag. The resulting markers were retained for subsequent genotype frequency analysis.

2.4. Genotype Frequency Analysis

In each developmental stage, three types of polymorphic markers were observed. *hk* × *hk* represented markers where both parents were heterozygous, and *lm* × *ll* indicated markers where only the male parent was heterozygous. *nn* × *np* denoted markers where only the female parent was heterozygous. These markers were evaluated using the chi-square test to determine if they conformed to the Hardy–Weinberg equilibrium, which was expected to be 1:2:1, 1:1, and 1:1, respectively. Markers that deviated from the expected Mendelian ratio in the *hk* × *hk* type were subjected to further analysis. The gametic or zygotic selection types were determined by two successive chi-square tests (Table 1) following the separation detection model of F_2 generation markers based on the maximum likelihood method [25].

Table 1. Selection type test of deviated *hk* × *hk* markers.

Test Formulae	Test of Significance		
$x_1^2 = \frac{(2np-n)^2+(2nq-n)^2}{n}$	×	√	√
$x_2^2 = \frac{(n_{AA}-np^2)^2}{np^2} + \frac{(n_{Aa}-2npq)^2}{2npq} + \frac{(n_{aa}-nq^2)^2}{nq^2}$	√	√	×
Selection type	Zygotic	Zygotic	Gametic

p = frequency of allele *A*; q = frequency of allele *a*; n = total larvae number that genotyped in each stage; n_{AA}, n_{Aa}, n_{aa} represented the number of the three genotypes; √ means $p < 0.05$ and × means $p > 0.05$.

Regarding the zygotic selection markers, the eliminated genotypes were screened out by the ratio of three genotypes following the classification method we developed [18]. This involved evaluating the ratio between the two homozygotes in the segregating markers (*AA*:*aa*) and the ratio between homozygotes and heterozygotes ((*AA* + *aa*):*Aa*) using chi-square tests in succession (Figure 2). To better understand the severity of marker deviation, we transformed the *p*-value of each distorted marker into its natural logarithm (−Log10 *p*) to conform to a normal distribution. We refer to this normalized *p*-value as the Segregation

Distortion Value (SDV). The higher the SDV, the lower the p-value became, and this indicated more severe segregation distortion at that locus.

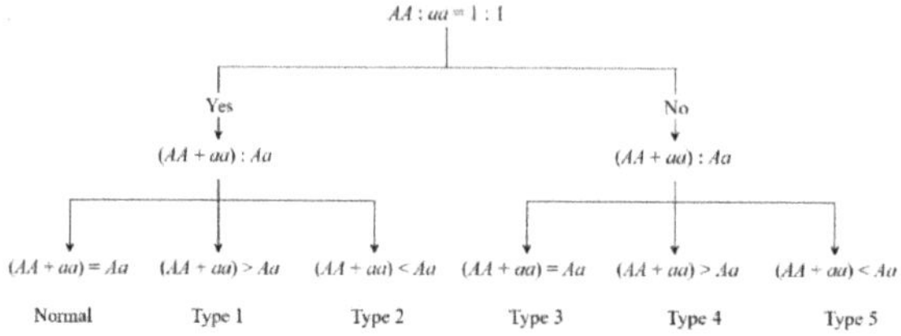

Figure 2. Five classifications of zygotic selection SNPs.

2.5. Comparison of Genetic Parameters and Locus Frequencies

Genetic diversity parameters in the two developmental stages, including the effective number of alleles (*Ne*), expected heterozygosity (*He*), observed heterozygosity (*Ho*), polymorphic information content (*PIC*), nucleotide diversity (P_i), and allele frequencies, were estimated using genepop (Version 1.0.5) [26]. Selective sweep analysis was performed to identify loci under selection from the zoea to the post-larval stage. This method combined the population differentiation coefficient (F_{ST}) and the nucleotide diversity (P_i) to identify strong signals of selection. Loci with a minimum allele frequency (MAF) below 0.05 were further filtered out using vcftools software v0.1.13 in both groups. For the remaining loci, the F_{ST} and P_i values were calculated. The selective sweep region was determined by selecting loci that fell within the top 5% highest F_{ST} region, the bottom 5% P_i region, and the top 5% P_i region. Loci within this intersection region were considered as candidate loci.

2.6. Statistical Analysis

Analysis of variance (ANOVA) was performed to assess the survival rate at different developmental stages, and multiple comparisons were conducted using SPSS 27.0. The significance threshold was set at $p < 0.05$.

3. Results

3.1. Descriptive Statistics of Larval Survival

The hatching rates and survival rates at different larval developmental stages are presented in Table 2. The average hatching rate of the inbred families was significantly lower (24.36% lower) than that of the control families. After normalizing the density of hatched nauplii, the mortality of larvae from the nauplius to the zoea stage and after post-larval metamorphosis was relatively low in both control and inbred groups. However, substantial mortality occurred during the zoea to mysis stage and the mysis to post-larvae stage, with larval survival rates decreasing to 65% and 53% in the control families, respectively. Moreover, the survival performance of the inbred families during these two stages was worse. The inbreeding depression rate of larval survival was 29.23% and 45.28% in the mysis and post-larvae stages, respectively. The results indicated that the predominant larval elimination, as well as the effects of inbreeding depression, occurred during the hatching stages and metamorphosis stages of mysis and post-larvae.

Table 2. Hatching rates and larval survival at different developmental stages.

	Hatching Rates	Larval Survival in Z	Larval Survival in M	Larval Survival in P3	Larval Survival in P10
Control families	0.78 ± 0.07^{a}	0.91 ± 0.04^{a}	0.65 ± 0.08^{a}	0.53 ± 0.09^{a}	0.48 ± 0.08^{a}
Inbred families	0.59 ± 0.14^{b}	0.89 ± 0.06^{a}	0.46 ± 0.12^{b}	0.29 ± 0.14^{b}	0.27 ± 0.13^{b}
Inbreeding depression	24.36%	2.20%	29.23%	45.28%	43.75%

Different superscript letters within columns indicate significant difference ($p < 0.05$).

3.2. Sequencing Data Processing and Marker Development

In total, we obtained 63.11 Gb of clean sequencing data from 74 samples. The average number of reads for the two parents was 67.71 million, and the high-quality reads accounted for 65.08%. As for the 72 larval samples, the average number of sequencing reads was 30.58 million, and the percentage of high-quality reads was 77.05%. The reference tag database, clustering reads from the two parents, included a total of 522,402 tags, which served as a reliable reference for the subsequent genotyping. After aligning the high-quality tags of each larval sample to this reference, we obtained an average of 418,145 unique tags per sample, with an average mapping rate of 63.41%. The sequencing depth for the two parents was 55.56× and 67.15×, respectively, while for the larvae, it ranged from 31.36× to 41.22×, with an average depth of 35.68×. The number of unique tags from each library and their sequencing depth are shown in Figure 3. A total of 31,767 polymorphic SNPs were identified through the RAD-typing program. After filtering, there were 29,435 biallelic SNPs shared by the two developmental stages. These selected segregating markers were then subjected to further analysis.

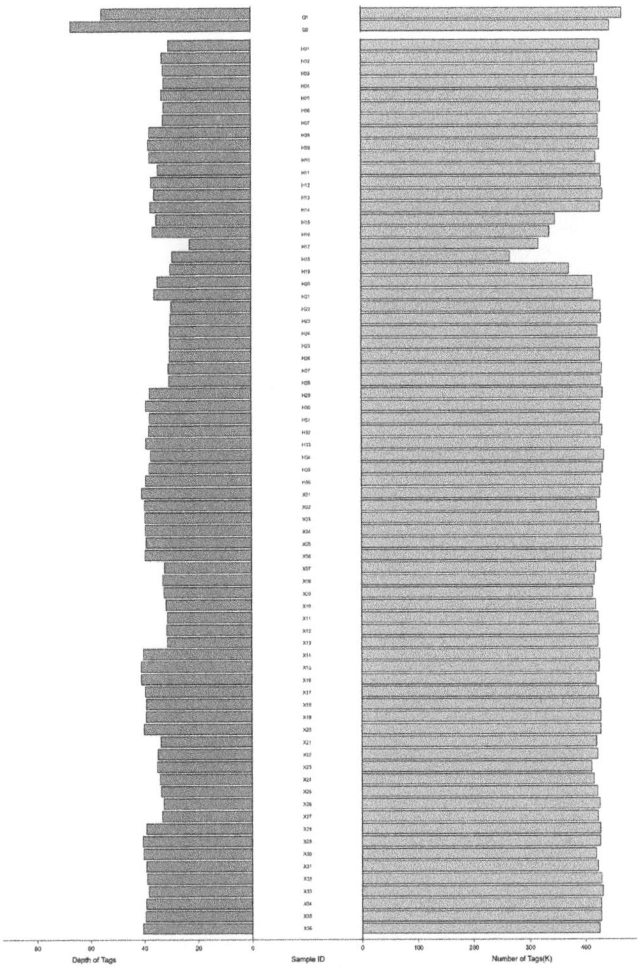

Figure 3. The number of unique tags from each library and their sequencing depth. The vertical axis represents the identification numbers of 74 samples, Q represents the parent sample, H represents the zoea stage, and X represents the post-larvae stage.

3.3. Changes in Genetic Diversity between the Two Stages

Out of all 29,435 segregating markers, the Ts/Tv ratio was 1.38 in all larval samples. Using the major allele at each locus during the zoea stage as a wild type, the average frequency of major alleles at all loci was 75.33% in the zoea stage but decreased to 70.34% in the post-larvae stage. In comparison to the zoea stage, there were 15,306 loci in the post-larvae stage, where the frequency of the wild-type allele changed by more than 5%. Among these, 11,094 loci had a decrease in wild-type allele frequency, accounting for 37.69% of all loci; while 4212 loci had an increase in wild-type allele frequency, accounting for 14.31% of all loci. This seems to be a type of balancing selection, where heterozygotes are preferred during larval selection. From the perspective of genetic diversity parameters, observed heterozygosity was higher than expected heterozygosity in both stages, and both exhibited an increasing trend (Table 3). From the zoea to post-larvae stage, the expected heterozygosity increased from 0.36 to 0.38, and the observed heterozygosity increased from 0.47 to 0.50. Other genetic diversity parameters, such as the effective number of alleles, polymorphic information content, and nucleotide diversity, also showed increases.

Table 3. Genetic diversity parameters in the two developmental stages.

Developmental Stage	*He*	*Ho*	*PIC*	*Ne*	P_i
Zoea	0.3641	0.4735	0.2935	1.6075	0.3703
Post-larvae	0.3826	0.5087	0.3060	1.6489	0.3886

He: expected heterozygosity; *Ho*: observed heterozygosity; *PIC*: polymorphic information content; *Ne*: effective number of alleles; P_i: nucleotide diversity.

The average F_{ST} value estimated from the shared SNPs between the two stages was 0.0043, indicating a fundamental consistency in their genetic structure. A total of 783 loci were identified based on the F_{ST} and P_i values (Figure 4). These loci served as important candidates under selection from the zoea to post-larvae stages.

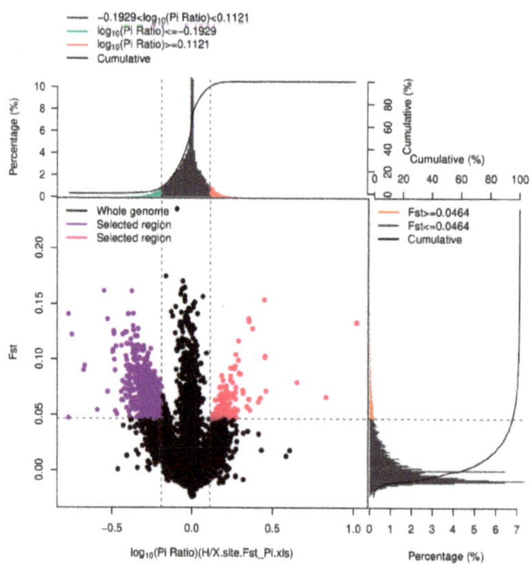

Figure 4. Selective sweep analysis of candidate loci under selection from zoea to post-larvae. The *x*-axis represents the Log10 (P_i ratio) values, while the *y*-axis represents the F_{ST} values, corresponding to the frequency distribution plot located above and to the right, respectively. The topmost green and red regions represent the bottom and top 5% regions selected based on P_i values, while the orange region on the right indicates the top 5% regions selected based on F_{ST} values. The overlapping red and purple region of the scatter plot in the middle represents the intersection of F_{ST} and P_i and is considered to represent the candidate loci.

3.4. Overall Manifestation of Segregation Distortion in Larvae

Among the 29,435 common SNPs shared between the two stages, 22.37% exhibited significant deviation from the expected Mendelian segregation in the zoea stage, while the proportion of distorted markers decreased to 12.86% in the post-larvae stage (Table 4). Three types of segregating markers showed distinct proportions and degrees of distortion. The proportion of distorted markers was higher in the *hk* × *hk* type compared to markers with only one heterozygous parent. No significant difference was observed between the *ll* × *lm* and *nn* × *np* types. This trend was particularly evident in the post-larvae stage.

Table 4. Ratios of distorted markers in different types of segregating markers between the two developmental stages.

Marker	Type Specification	Number	Ratios of Distortion in Zoea	Ratios of Distortion in Post-Larvae
hk × *hk*	Biparental heterozygous	5582	27.73%	20.71%
lm × *ll*	Male heterozygous	11,793	21.06%	11.38%
nn × *np*	Female heterozygous	12,060	26.84%	10.67%
Total		29,435	22.37%	12.86%

The analysis of markers from the *hk* × *hk* type revealed that the proportion of gametic selection was 45.95% in the zoea stage and decreased to 29.28% in the post-larvae stage. Type 2 loci with heterozygote excess and type 5 loci with homozygote deficiency were the predominant types of zygotic selection in both stages. The sum of the two types accounted for 82.13% of zygotic selection in the zoea stage and 89.91% in the post-larvae stage (Table 5). In the zoea stage (Figure 5), the average SDR of type 2 markers was 6.07, significantly higher than the average SDR of 3.42 for type 5 markers ($p < 0.001$). The characteristic of SDR in the post-larvae stage was the same. These results indicated a greater selection pressure on loci with heterozygote advantage compared to those exhibiting homozygous deletion.

Table 5. Classifications of zygotic selection SNPs in the two developmental stages.

Classification	Normal	Type 1	Type 2	Type 3	Type 4	Type 5
Number in zoea	3915	96	494	57	8	246
Ratio in zoea	-	10.65%	54.83%	6.33%	0.89%	27.30%
Number in post-larvae	4349	38	554	49	1	230
Ratio in post-larvae	-	4.36%	63.53%	5.62%	0.11%	26.38%
Possible explanation	Mendelian segregation	Heterozygote deficiency	Heterozygote excess	Impact of partial recessive deleterious		Homozygous deficiency

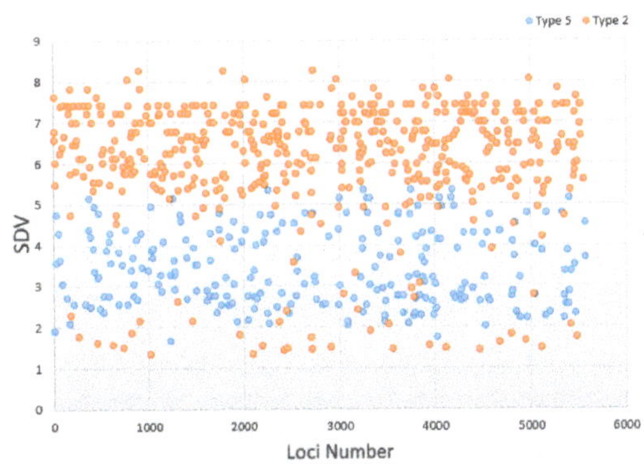

Figure 5. The SDR distribution of the two primary distorted markers in the zoea stage. The ordinate value refers to segregation distortion values (SDVs) of these markers. Type 2 and type 5 markers refer to those classifications of zygotic selection in Table 5.

For single-parent heterozygous loci, the results from Table 4 indicate a noticeable decrease in their deviation ratios from the zoea stage to the post-larvae stage. The specific elimination types are presented in Table 6. In both stages, the heterozygotes rather than the homozygotic types were predominant in suffering from selection. Among all deviated markers, the proportions of the *lm* and *np* types under elimination were 94.16% and 92.99% in the zoea stage, respectively; in the post-larvae stage, these proportions decreased to 85.02% and 84.69% for the *lm* and *np* types, respectively. Mutant alleles of *m*/*p* possibly showed a disadvantage in gametes or zygotes and were more likely to be eliminated, and thus these loci were considered as dominant. This result further supported the partial dominance hypothesis.

Table 6. Elimination type of single-parent heterozygous loci in the two developmental stages.

Type	Genotype under Elimination	Number/Ratio in Zoea	Number/Ratio in Post-Larvae
lm × *ll*	*lm*	2339/94.16%	1141/85.02%
	ll	145/5.84%	201/14.98%
nn × *np*	*np*	2373/92.99%	1090/84.69%
	nn	179/7.01%	197/15.31%

3.5. Change in Distorted Types

We further compared the types of distorted loci between the two developmental stages and observed potential changes in the types of distortion at the same locus as the larvae were eliminated. Out of 5582 *hk* × *hk* parental heterozygous loci, 3589 loci showed normal segregation in both stages. The five main types of changes in distorted loci are shown in Figure 6. Among these, 326 loci shifted from normal segregation to distorted segregation, and 652 loci shifted from gametic selection to normal segregation. A total of 83 loci shifted from homozygous deficiency to heterozygous excess, and 554 loci shifted from other types of distortion to heterozygous excess. In addition, 230 loci shifted from other types of distortion to homozygous deficiency.

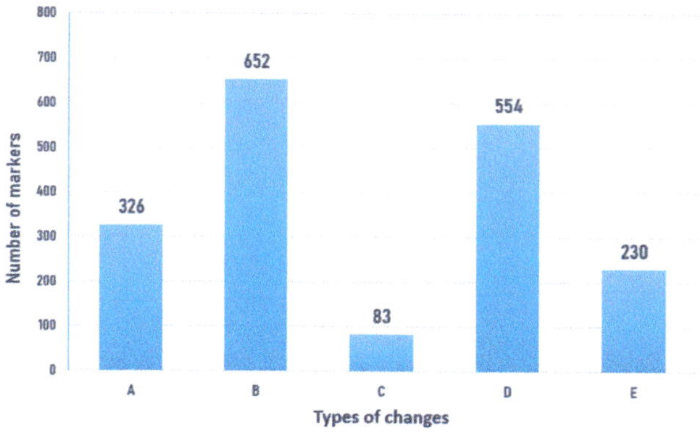

Figure 6. The main five types of changes in segregation distortion. A: normal segregation to distorted segregation; B: gametic selection to normal segregation; C: homozygous deficiency to heterozygous excess; D: other types of distortion to heterozygous excess; E: other types of distortion to homozygous deficiency.

4. Discussion

Compared to terrestrial animals, marine animals have a distinct reproductive system, characterized by high fecundity, rapid development through a planktonic larval stage, and significant early mortality rates [27]. This adaptation enables them to survive in fluctuating ocean environments. Hedgecock [28] introduced the concept of Sweepstakes

Reproductive Success (SRS) to describe this unique reproductive strategy, where only a few individuals succeed while many others do not. SRS is based on the idea of a wide variation in individual reproductive success during competitive life cycles [10], providing an opportunity for inbreeding effects to influence selection outcomes and potentially serve as a significant driving force for natural selection. In this study, we investigated the eliminated characteristics of the main larval developmental stage of Chinese shrimp and the effects of inbreeding. The results of genotype segregation at two substantial elimination stages indicated that the selection after inbreeding resulted in significant heterozygote excess and homozygous deficiency. More heterozygous individuals were retained due to the selection process.

It has been reported that natural elimination mainly occurred in the larval developmental stages in the life history of *F. chinensis*, and the inbreeding depression in mortality rates was not evident after the post-larvae stage [15]. Our study further revealed that elimination mainly occurred in hatching (from fertilized eggs to nauplius), the zoea to mysis period, and the mysis to post-larvae period (Table 2). The results of depression rates in the inbred group indicated that a decrease of larval mortality rates due to inbreeding was also primarily observed during these three stages. The depression rate (F = 0.25) of hatching as well as survival in the mysis and post-larvae stages was 24.36%, 29.23% and 45.28%, respectively. These levels were much higher than the depression of body weight observed in P140 (4.16% per 10% increase in F) [15] as well as in adult shrimp (2.19% per 10% increase in F) of *L. vannamei* [16]. In contrast, the decline in mortality rates for juvenile and adult shrimp was significantly lower than that during the larval stage, which was 1.77% and 0.009% per 10% increase in F in the above two studies. Similar results were also found in inbreeding studies of *E. carinicauda* [17]. This aligned with the empirical evidence that inbreeding depression is severe in traits that are closely related to fitness, such as infant survival [29], and was consistent with studies in other marine animals [30]. Based on the above, in the subsequent genotype frequency study, we focused on the zoea and post-larvae stages. These two stages can help us understand how the genotype frequency of each locus changes after the first major elimination and the three main elimination events.

Until the cost of resequencing becomes affordable for non-model organisms, the RRS technique remains an economical and efficient method for obtaining genotypes at specific loci. In this research, 2b-RAD was applied, since this technology featured even and tunable genome coverage in order to provide reliable and flexible large-scale SNPs [20]; it has successfully been applied to the construction of high-resolution linkage maps of *C. farreri* [22] and *F. chinensis* [31]. Sequencing produced an average of 67.71 and 35.68 million reads for parental and progeny libraries, respectively. After quality filtering, more than 65.08% of parental reads and 77.05% of progeny reads remained. Compared to our initial application of this technique in constructing the first high-density genetic map of *F. chinensis*, there was a decrease in the proportion of high-quality reads (85% in that research), possibly due to the use of isolength restriction site–associated DNA (isoRAD) [21] in this sequencing. However, due to a significant increase in sequencing output, the sequencing depth of the parents and progenies (61× and 36×) also significantly increased. This depth far exceeded the suggested depth (20× and 15× for parents and progenies through simulation analysis, with a genotyping accuracy > 96%) [24]. There was no difference in the high-quality reference assembled from the two parents. Therefore, we believe that the markers we developed were reliable.

Segregation distortion is a common occurrence in genetic studies. Deviations from Mendelian segregation ratios have been reported in various marine animals, including prawn [32–34]. Typically, only a few markers show deviation, and they are excluded from the genetic analysis. However, segregation distortion was an indication of the linkage between molecular markers and distorting factors such as recessive detrimental genes [35]. It can therefore be utilized to study the quantity of inbreeding loads as well as the selected alleles and genotypes. A few days after hatching, in the zoea stage, 54.05% of the parental heterozygous loci were of the zygotic selection type, while 45.95% resulted from gamete abortion. This suggested that both zygotic and gametic selection were present and had a significant impact on segregation distortion. Further analysis revealed that within the zygotic selection type SNPs, 27.30% exhibited homozygous deficiency and 54.83% showed heterozygous excess. This indicated that overdominance and partial dominance hypotheses both played a crucial role in explaining inbreeding depression. The above evidence points to the fact that deleterious recessive mutations are abundant in shrimps and they possibly play a major role in inbreeding depression.

In the two larval developmental stages, the overall segregation distortion ratios were 22.37% and 12.86%, respectively. These figures were similar to those reported in mapping populations of marine animals [33,34,36,37] but were lower than the 50.77% observed in another study on a Chinese shrimp family [18]. In that study, adult shrimps from a full-sibling family underwent artificial selection, with an intensity exceeding 3%. The high selection pressure could potentially explain the substantial severe deviation. Our findings also revealed a decrease in segregation distortion ratios from the initial discernible elimination to the third main elimination event. In addition, we observed a reduction in gametic segregation distortion ratios, decreasing from 45.95% in the zoea stage to 29.28% in the post-larval stage. Given that gametic selection was essentially completed before hatching, this decline suggested the potential selective elimination of another wild-type allele at the zygotic stage, if not due to sampling error. Due to constraints in experimental design and sequencing costs, we did not establish replicate groups in the sequencing of samples from the two developmental stages. This could potentially introduce some sampling errors into the statistical analysis of genotype frequencies. To minimize the risk of such errors, we did not select larvae samples based on size, with the aim of preserving their size distribution within a normal range.

These seemingly perplexing results can be accounted for by considering the pseudo-overdominance effect. The pseudo-overdominance effect, reviewed by Charlesworth and Willis [7], is primarily caused by selection on neutral loci due to the genetic hitchhiking effect [38,39]. We also found and elaborated on evidence of this effect in a separate study within the mapping population of Chinese shrimp [18]. The shrimp genome exhibits high heterozygosity and leads to the presence of numerous deleterious recessive alleles. Inbreeding depression thus manifests particularly severely in shrimp and the majority of marine animals [27]. Due to the presence of numerous deleterious loci and their linkage relationships, neutral loci might also undergo significant selection pressure because of their linkage with detrimental loci during ongoing elimination process in the larval stage. This could explain the reduced gametic and overall segregation distortion ratios. We speculated that, following the pseudo-overdominance effect, subsequent selection might target neutral homozygotes and lead to a decrease in the proportion of normal alleles, which was a form of balancing selection. In the comparison between two stages, we identified 783 loci through selective sweep analysis. These loci might potentially be neutral loci caused by genetic hitchhiking effects, so we did not conduct further gene functional and enrichment analysis. However, they at least verify that different loci may be under selection during different developmental stages. Under this selection pattern, the elimination effect of detrimental alleles at the gametic stage was weakened. The gametic selection loci could possibly become normally segregated. Supporting evidence could be seen in the change of distortion types between the two stages (Figure 6). There were 683 gametic selection loci whose distortion types changed in the post-larvae stage. The vast majority (95.46%) of them shifted to normal segregation loci. Additionally, 554 loci shifted from other types to heterozygote excess, indicating a reduction in homozygotes for both alleles due to continuous selection pressure. Notably, as observed in the results of single parental heterozygous selection loci (Table 6), the significant increase in the elimination rate of wild-type homozygotes *ll* and *nn* in the post-larval stage provided further supporting evidence to some extent.

The Ts/Tv ratio observed in our study was 1.38 in all larvae and was higher than that reported in other species. For instance, scallops [40] exhibited ratios ranging from 1.07 to 1.13, while fish [41] showed ratios ranging from 1.22 to 1.24. This may be attributed to the conversion deviation strategy proposed by Wakeley [42], which is a potential mechanism to counteract negative effects during long-term evolutionary processes. Our findings might be consistent with this concept, particularly considering the high reproductive capacity and elimination rate observed in crustaceans.

The impact of inbreeding on populations over the long term leads to an increase in the frequency of homozygous genotypes and a decrease in heterozygous frequencies. This is a phenomenon well-documented in most model organisms [7] and shrimps [17]. Our study focuses on the changes in genotype frequency influenced by inbreeding during the larval development stages in one generation. Through the analysis of genetic diversity parameters in two stages, it was observed that all five parameters calculated based on different genotype frequencies exhibited varying degrees of increase in the post-larval stage (Table 3), aligning with the decrease in segregation distortion ratios. This illustrated, from another perspective, that the selection process resulted in a higher proportion of heterozygotes, which supported a significant role of overdominance but could be readily explicable by deleterious mutations [7]. The above results also gave a possible interpretation of why it was difficult to explain the large effect of inbreeding on fitness in model species without a significant contribution from variability as maintained by selection [43], as well as why mutation due to deleterious alleles was not the sole source of the variation, which can only explain only about 60% of the observed inbred load for fitness components [44].

5. Conclusions

Noticeable purging selection and inbreeding depression occur in certain larval developmental stages of Chinese shrimp. Homozygote deficiency and heterozygote excess are the two major types of zygotic selection. Possibly owing to the genetic hitchhiking effect of deleterious loci, the distortion types of the same locus can change over different developmental stages. Nature selection tends to retain more heterozygotes after larval developmental stages. The results provide valuable insights into the mechanisms of inbreeding depression in marine animals, which possess genomes of considerable complexity and heterozygosity and reproductive characteristics with high elimination rates. It also offers guidance for formulating breeding strategies in marine economic animal and marine biodiversity conservation.

Author Contributions: Conceptualization, Q.F.; methodology, Q.F. and S.L.; formal analysis, P.D. and D.L.; investigation, J.Z.; resources, B.C.; data curation, K.L.; writing—original draft preparation, Q.F.; writing—review and editing, Q.F. and X.M.; supervision, J.K. and X.M.; funding acquisition, Q.F. and J.K. All authors have read and agreed to the published version of the manuscript.

Funding: This work was funded by the National Key R&D Program of China (No. 2018YFD0901302), the National Natural Science Foundation of China (No. 31602151), the China Agriculture Research System of MOF and MARA (CARS-48), the Science and Technology Innovation Team project of the Chinese Academy of Fishery Sciences (No. 2020TD26), and the open competition program of the top ten critical priorities of Agricultural Science and Technology Innovation for the 14th Five-Year Plan of Guangdong Province (No. 2022SDZG01).

Data Availability Statement: The data presented in this study are available on request from the corresponding author due to the protection of breeding population information.

Conflicts of Interest: The authors declared no conflicts of interest.

References

1. Bierne, N.; Beuzart, I.; Vonau, V.; Bonhomme, F.; Bédier, E. Microsatellite-associated heterosis in hatchery-propagated stocks of the shrimp *Penaeus stylirostris*. *Aquaculture* **2000**, *184*, 203–219. [CrossRef]
2. Falconer, D.S. Introduction to quantitative genetics. *Population* **1962**, *17*, 1529. [CrossRef]
3. Keller, L.F. Inbreeding and its fitness effects in an insular population of song sparrows (*Melospiza melodia*). *Evolution* **1998**, *52*, 240–250. [CrossRef] [PubMed]
4. Henter, H.J. Inbreeding depression and haplodiploidy: Experimental measures in a parasitoid and comparisons across diploid and haplodiploid insect taxa. *Evolution* **2003**, *57*, 1793–1803. [CrossRef] [PubMed]

5. Haag, C.R.; Ebert, D. Genotypic selection in *Daphnia* populations consisting of inbred sibships. *J. Evol. Biol.* **2010**, *20*, 881–891. [CrossRef] [PubMed]
6. Wright, S. *Evolution and the Genetics of Populations*; University of Chicago Press: Chicago, IL, USA, 1978; Volume 1. [CrossRef]
7. Charlesworth, D.; Willis, J.H. The genetics of inbreeding depression. *Nat. Rev. Genet.* **2009**, *10*, 783–796. [CrossRef] [PubMed]
8. Keller, L. Inbreeding effects in wild populations. *Trends Ecol. Evol.* **2002**, *17*, 230–241. [CrossRef]
9. Moss, D.R.; Arce, S.M.; Otoshi, C.A.; Doyle, R.W.; Moss, S.M. Effects of inbreeding on survival and growth of Pacific white shrimp *Penaeus (Litopenaeus) vannamei*. *Aquaculture* **2007**, *272* (Suppl. S36), S30–S37. [CrossRef]
10. Hedgecock, D.; I Pudovkin, A. Sweepstakes Reproductive Success in Highly Fecund Marine Fish and Shellfish: A Review and Commentary. *Bull. Mar. Sci.* **2011**, *87*, 971–1002. [CrossRef]
11. Wang, S.; Hard, J.J.; Utter, F. Salmonid inbreeding: A review. *Rev. Fish Biol. Fish.* **2002**, *11*, 301–319. [CrossRef]
12. Bucklin, K.A. Analysis of the Genetic Basis of Inbreeding Depression in the Pacific Oyster Crassostrea Gigas. Ph.D. Thesis, University of California Davis, Davis, CA, USA, 2002; 140p.
13. Evans, F.; Matson, S.; Brake, J.; Langdon, C. The effects of inbreeding on performance traits of adult Pacific oysters (*Crassostrea gigas*). *Aquaculture* **2004**, *230*, 89–98. [CrossRef]
14. Keys, S.J.; Crocos, P.J.; Burridge, C.Y.; Coman, G.J.; Davis, G.P.; Preston, N.P. Comparative growth and survival of inbred and outbred *Penaeus* (*Marsupenaeus*) japonicus, reared under controlled environment conditions: Indications of inbreeding depression. *Aquaculture* **2004**, *241*, 151–168. [CrossRef]
15. Luo, K.; Kong, J.; Luan, S.; Meng, X.-H.; Zhang, T.-S.; Wang, Q.-Y. Effect of inbreeding on survival, WSSV tolerance and growth at the postlarval stage of experimental full-sibling inbred populations of the Chinese shrimp *Fenneropenaeus chinensis*. *Aquaculture* **2014**, *420–421*, 32–37. [CrossRef]
16. Ríos-Pérez, L.d.L.; Campos-Montes, G.R.; Martínez-Ortega, A.; Castillo-Juárez, H.; Montaldo, H.H. Inbreeding Effects on Body Weight at Harvest Size and Grow-out Survival Rate in a Genetic Selected Population of Pacific White Shrimp *Penaeus (Litopenaeus) vannamei*. *J. World Aquac. Soc.* **2015**, *46*, 53–60. [CrossRef]
17. Wang, J.; Li, J.; Ge, Q.; Chen, Z.; Li, J. Effects of Inbreeding on Genetic Characteristic, Growth, Survival Rates, and Immune Responses of a New Inbred Line of *Exopalaemon carinicauda*. *Int. J. Genom.* **2020**, *2020*, 5735968. [CrossRef] [PubMed]
18. Fu, Q.; Meng, X.; Luan, S.; Chen, B.; Cao, J.; Li, X.; Kong, J. Segregation distortion: High genetic load suggested by a Chinese shrimp family under high-intensity selection. *Sci. Rep.* **2020**, *10*, 21820. [CrossRef] [PubMed]
19. Sui, J.; Luan, S.; Luo, K.; Meng, X.; Cao, B.; Liu, N.; Li, W.; Lu, X.; Chai, Z.; Kong, J. Genetic parameters and response to selection of harvest body weight of the Chinese shrimp *Fenneropenaeus chinensis* after five generations of multi-trait selection. *Aquaculture* **2015**, *452*, 134–141. [CrossRef]
20. Wang, S.; Meyer, E.; McKay, J.K.; Matz, M.V. 2b-RAD: A simple and flexible method for genome-wide genotyping. *Nat. Methods* **2012**, *9*, 808 810. [CrossRef] [PubMed]
21. Wang, S.; Liu, P.; Lv, J.; Li, Y.; Cheng, T.; Zhang, L.; Xia, Y.; Sun, H.; Hu, X.; Bao, Z. Serial sequencing of isolength RAD tags for cost-efficient genome-wide profiling of genetic and epigenetic variations. *Nat. Protoc.* **2016**, *11*, 2189–2200. [CrossRef]
22. Jiao, W.; Fu, X.; Dou, J.; Li, H.; Su, H.; Mao, J.; Yu, Q.; Zhang, L.; Hu, X.; Huang, X.; et al. High-Resolution Linkage and Quantitative Trait Locus Mapping Aided by Genome Survey Sequencing: Building Up an Integrative Genomic Framework for a Bivalve Mollusc. *DNA Res.* **2014**, *21*, 85–101. [CrossRef]
23. Catchen, J.; Hohenlohe, P.A.; Bassham, S.; Amores, A.; Cresko, W.A. Stacks: An analysis tool set for population genomics. *Mol. Ecol.* **2013**, *22*, 3124–3140. [CrossRef] [PubMed]
24. Fu, X.; Dou, J.; Mao, J.; Su, H.; Jiao, W.; Zhang, L.; Hu, X.; Huang, X.; Wang, S.; Bao, Z. RADtyping: An Integrated Package for Accurate De Novo Codominant and Dominant RAD Genotyping in Mapping Populations. *PLoS ONE* **2013**, *8*, e79960. [CrossRef] [PubMed]
25. Lorieux, M.; Perrier, X.; Goffinet, B.; Lanaud, C.; de León, D.G. Maximum-likelihood models for mapping genetic markers showing segregation distortion. 2. F2 populations. *Theor. Appl. Genet.* **1995**, *90*, 81–89. [CrossRef] [PubMed]
26. Rousset, F. genepop'007: A complete re-implementation of the genepop software for Windows and Linux. *Mol. Ecol. Resour.* **2008**, *8*, 103–106. [CrossRef] [PubMed]
27. Winemiller, K.O.; Rose, K.A.; Pedersen, E.J.; Goto, D.; Gaeta, J.W.; Hansen, G.J.; Sass, G.G.; Zanden, M.J.V.; Cichosz, T.A.; Rypel, A.L.; et al. Patterns of Life-History Diversification in North American Fishes: Implications for Population Regulation. *Can. J. Fish. Aquat. Sci.* **1992**, *49*, 2196–2218. [CrossRef]
28. Hedgecock, D. Does variance in reproductive success limit effective population sizes of marine organisms. *Genet. Evol. Aquat. Org.* **1994**, *122*, 122–134.
29. DeRose, M.A.; Roff, D.A. A comparison of inbreeding depression in life-history and morphological traits in animals. *Evolution* **1999**, *53*, 1288–1292. [CrossRef]
30. Zheng, H.; Li, L.; Zhang, G. Inbreeding depression for fitness-related traits and purging the genetic load in the hermaphroditic bay scallop *Argopecten irradians irradians* (Mollusca: Bivalvia). *Aquaculture* **2012**, *366–367*, 27–33. [CrossRef]
31. Meng, X.; Fu, Q.; Luan, S.; Luo, K.; Sui, J.; Kong, J. Genome survey and high-resolution genetic map provide valuable genetic resources for *Fenneropenaeus chinensis*. *Sci. Rep.* **2021**, *11*, 7533. [CrossRef]
32. Plough, L.V.; Hedgecock, D. Quantitative Trait Locus Analysis of Stage-Specific Inbreeding Depression in the Pacific Oyster *Crassostrea gigas*. *Genetics* **2011**, *189*, 1473–1486. [CrossRef]

33. Tian, M.; Li, Y.; Jing, J.; Mu, C.; Du, H.; Dou, J.; Mao, J.; Li, X.; Jiao, W.; Wang, Y.; et al. Construction of a High-Density Genetic Map and Quantitative Trait Locus Mapping in the Sea Cucumber *Apostichopus japonicus*. *Sci. Rep.* **2015**, *5*, 14852. [CrossRef]
34. Yu, Y.; Zhang, X.; Yuan, J.; Li, F.; Chen, X.; Zhao, Y.; Huang, L.; Zheng, H.; Xiang, J. Genome survey and high-density genetic map construction provide genomic and genetic resources for the Pacific White Shrimp *Litopenaeus vannamei*. *Sci. Rep.* **2015**, *5*, 15612. [CrossRef] [PubMed]
35. Lyttle, T.W. Segregation distorters. *Annu. Rev. Genet.* **1991**, *25*, 511–581. [CrossRef] [PubMed]
36. Wilson, K.; Li, Y.; Whan, V.; Lehnert, S.; Byrne, K.; Moore, S.; Pongsomboon, S.; Tassanakajon, A.; Rosenberg, G.; Ballment, E.; et al. Genetic mapping of the black tiger shrimp *Penaeus monodon* with amplified fragment length polymorphism. *Aquaculture* **2002**, *204*, 297–309. [CrossRef]
37. Li, Y.; Byrne, K.; Miggiano, E.; Whan, V.; Moore, S.; Keys, S.; Crocos, P.; Preston, N.; Lehnert, S. Genetic mapping of the kuruma prawn *Penaeus japonicus* using AFLP markers. *Aquaculture* **2003**, *219*, 143–156. [CrossRef]
38. Smith, J.M.; Haigh, J. The hitch-hiking effect of a favourable gene. *Genet. Res.* **1974**, *23*, 23–35. [CrossRef] [PubMed]
39. Harr, B.; Kauer, M.; Schlötterer, C. Hitchhiking mapping: A population-based fine-mapping strategy for adaptive mutations in *Drosophilamelanogaster*. *Proc. Natl. Acad. Sci. USA* **2002**, *99*, 12949–12954. [CrossRef] [PubMed]
40. Wang, C.; Liu, B.; Chen, M.; Ning, J.; Lu, X.; Wang, C. Mutations in Growth-Related Genes Induced by EMS Treatment in Scallops. *Front. Genet.* **2022**, *13*, 879844. [CrossRef]
41. Su, X.-L.; Zhao, S.-S.; Xu, W.-J.; Shuang, L.; Zheng, G.-D.; Zou, S.-M. Efficiently whole-genomic mutagenesis approach by ARTP in blunt snout bream (*Megalobrama amblycephala*). *Aquaculture* **2022**, *555*, 738241. [CrossRef]
42. Wakeley, J. The excess of transitions among nucleotide substitutions: New methods of estimating transition bias underscore its significance. *Trends Ecol. Evol.* **1996**, *11*, 158–162. [CrossRef]
43. Charlesworth, B.; Charlesworth, D. The genetic basis of inbreeding depression. *Genet. Res.* **2000**, *74*, 329–340. [CrossRef] [PubMed]
44. Charlesworth, B.; Hughes, K.A. The maintenance of genetic variation in life-history traits. In *Evolutionary Genetics: From Molecules to Morphology*; Cambridge University Press: Cambridge, UK, 2000; pp. 369–392.

biology

MDPI

Article

Statocyst Ultrastructure in the Norwegian Lobster (*Nephrops norvegicus*)

Marta Solé

Laboratory of Applied Bioacoustics, Universitat Politècnica de Catalunya, BarcelonaTech. Rambla Exposició s/n, 08800 Vilanova i la Geltrú, Barcelona, Spain; marta.sole@upc.edu

Simple Summary: In this study, we describe the structure of the Norwegian lobster (*Nephops norvegicus*) statocyst using scanning and transmission electron microscopy. These results contribute to our understanding of sound perception systems in an additional crustacean species in order to favor actions for marine noise pollution mitigation to protect the marine fauna.

Abstract: Statocyst anatomy and fine morphology in Norwegian lobster (*Nephrops norvegicus*) are studied for the first time using scanning and transmission electron microscopy. *N. norvegicus* exhibits sensory setae projecting from the statocyst inner cavity floor into a mass of sand granules (statoconia) embedded in a gelatinous substance. The setae are distributed in four areas: a curved field made up of an inner single row and an outer double row that run on a circle around the medial and lateral rim of the central depression, a small setal field in the posterior part, a large setal field, opposite to the small field, and a short row, running internally and lying parallel to the inner single row, next to the small setal field. A study of the fine morphology of the statocyst sensory setae shows that the structure of the setae in the different areas is similar, with a bulb (the proximal portion of the sensillum), a setal shaft, a tooth (the smooth portion of the bulb), a fulcrum (a transverse fold), and filamentous hairs. The hair cells are firmly implanted within the cuticular layer. Although the type of innervation of the statocyst was not determined in the present study, the close taxonomic position of the lobster to that of the crayfish and crab would suggest that the setae in *N. norvegicus* are pure mechanoreceptors rather than sensory cells.

Keywords: statocyst; hair cell; crustaceans; *Nephrops norvegicus*; scanning electron microscopy; transmission electron microscopy

1. Introduction

Marine invertebrates, in general, and crustaceans, in particular, are known to use acoustic signals for intra- and interspecific communication. Although sound production was evidenced in two crustacean groups, barnacles (*Cirripeda*) and decapods (*Eumalacostraca*), sound detection is widespread among crustaceans [1–3].

Three systems make crustaceans able to detect mechanical disturbance in water/sediment associated with sound waves: a pair of statocysts, chordotonal organs linked to the joints of antennae or legs, and internal and external sensilla [2,4]. To be able to orientate itself in the 3D marine environment, marine invertebrates need a gravity receptor system. These receptors, the statocysts, are common in different groups of marine invertebrates (cephalopods [5], crustaceans [6–8], cnidarians [9], and gastropods [10]). Statocysts are fundamental for the regulation of vital invertebrate behavior, including locomotion, posture, balance, and movement in the water column [2,11,12]. In addition, invertebrate statocysts detect sound particle motion rather than the sound pressure and are involved in underwater hearing [12,13]. Lovell et al. found that a sound-evoked response was generated in the statocyst after both ablation of the organ and section of the innervating nerve [7].

In crustaceans, the statocyst consists of a sac-like epidermal invagination of the cuticle located in the basal segment of each antennule in decapods and in the uropod or telson of the tail in mysids and isopods. The basic statocyst structure has similar features in all crustacean species [8]. The crustacean statocyst shows an inner sensory epithelium arranged in two to four rows of hair cells (setae) polarized towards the center of the organ and an overlying statolith, made of agglomerated sand granules [13], which stimulates the sensory setae [13,14]. When the animal changes its position, the statolith deflects the setae, and as a consequence, the afferent neurons innervating the statocyst are stimulated, promoting different responses [11,15] through compensatory movements of the appendages and the body [16].

The structure and morphology of the statocyst sensory epithelium vary depending on the crustacean taxonomical group [11,17]. The statocyst of the crayfish *Orconectes limosus* shows four separate fields of setae: a lateral group in two semicircles, a fusiform medial group, and a single row of proximal setae [18]. The statocyst of the Australian crayfish (*Cherax destructor*) is formed by three fields of setae: a curved field of two setal rows forming a semicircle around the medial and the posterior rim of the central depression, a large triangular lateral field of setae, and a smaller triangular setal field on the medial side of the depression [8]. The blue crab (*Callinectes sapidus*) statocyst exhibits three groups of mechanosensory setae (thread, free hook, and statolith) located in the statocyst canals [19]. Smaller species show smaller statocysts with different setal distribution. Lovell et al. described the fine structure of the prawn *Palaemon serratus*, which exhibits a regular row of setae dividing further into two irregular rows of setae, slightly curved and oriented towards the center of the crescent [7]. The hermit crab *Dardanus calidus* statocyst exhibits four groups of setae distributed along the statocyst inner walls (a curved field made up of two setal rows forming a semicircle around the central depression, a lateral semicircle of setae, a smaller medial semi-circular field, and a single row of setae in the center of the cavity) [20]. The setae in the telson statocyst were studied in some crustacean species [21,22].

Providing precise information on auditory systems in diverse marine species is of importance because of the rising concerns regarding the acoustic impact of anthropogenic noise on marine animals [2]. This concerns have promoted research regarding (i) crustaceans' capacity to perceive strong vibrations transmitted through a solid [23], including loud anthropogenic noise generated in the marine environment [2,6]; (ii) the structure and physiology of crustacean sensory sound perception systems.

Although both structure and function of some crustacean species' statocysts are well understood, in the Norwegian lobster (*N. norvegicus*), a species of important commercial interest, they have not been previously described. *N. norvegicus* is widely distributed on soft sediment, commonly at depths of 200–800 m, throughout the North-East Atlantic, from Iceland in the north to Morocco in the south, including the Mediterranean and Adriatic waters. Here, we examine and describe for the first time the micromorphology and ultrastructure of *N. norvegicus* statocyst sensory epithelium using scanning (SEM) and transmission (TEM) electron microscopy. This study aims to better understand the sound perception system in an additional crustacean species in order to favor actions for marine noise pollution mitigation to protect the marine fauna.

2. Materials and Methods

2.1. Animals

Thirty adult Norwegian lobsters (*N. norvegicus*) of mixed sex, ranging in total length from 13 cm (77 g) to 18 cm (120 g), from South West Scotland, were used for this study (January–March 2023). The animals were kept for five days, prior to the start of the analysis process, in the LAB's (Laboratory of Applied Bioacoustics, 41°12′57.1″N 1°43′59.0″E) maintenance system, a closed circuit of recirculating water (at 15 °C, salinity 35, and natural oxygen pressure) consisting of 2 mechanically filtered fiberglass-reinforced plastic tanks of 2000 L connected to each other. These maintenance facilities included a physicochemical self-filtration system with activated carbon and sand, driven by a circulation pump.

2.2. Scanning Electron Microscopy (SEM) Ultrastructural Analysis

For the SEM analysis, we used the 40 statocysts from 20 animals of both sexes. The lobsters were anaesthetized and sacrificed with an overdose of 2-phenoxyethanol. The statocysts were dissected and placed in a solution of 2.5% S-carboxymethyl–l-cysteine in sodium chloride to hydrolyze the mucus surrounding the statocyst hair sensilla in order to eliminate the statoconia (sand grains). After 2 min, the solution was removed, and the samples were fixed and processed by routine procedures for analysis by SEM. Fixation was performed in glutaraldehyde 2.5% for 24–48 h at 4 °C. The samples were dehydrated in graded alcohol solutions and critical-point dried with liquid carbon dioxide in a Leica Em CPD300 unit (Leica Mycrosystems, Vienna, Austria). The dried samples were mounted on specimen stubs with double-sided tape. The mounted tissues were gold-coated with a Quorum Q150R S sputter coated unit (Quorum Technologies, Ltd., Lewes, UK) and viewed with a variable-pressure microscope (Hitachi High Technologies Co., Ltd., Tokyo, Japan) at an accelerating voltage of 5 kV at the Institute of Marine Sciences of the Spanish Research Council (CSIC).

2.3. Setae Counting and Measurements

The number of setae was counted in each zone of the statocyst, and the results were averaged to obtain a mean number of setae for each of the five zones described.

To describe the diameter and length of the inner (ir) and outer (or) rows, these lengths were measured in 10 setae from the ir and or rows of the 40 statocysts (20 animals), and the results were averaged to obtain the mean size of the setae.

2.4. Transmission Electron Microscopy (TEM) Ultrastructural Analysis

For the TEM analysis, we used 5 animals (10 statocysts) of both sexes. The lobsters were anaesthetized and sacrificed with an overdose of 2-phenoxyethanol. The statocysts were dissected, fixed, and processed by routine procedures for analysis by TEM. Fixation was performed in 2.5% glutaraldehyde–2% paraformaldehyde for 24 h at 4 °C. Subsequently, the samples were osmicated in 1% osmium tetroxide, dehydrated in acetone, and embedded in Spurr. To orient the specimens properly, semithin sections (1 mm) were cut transversally or tangentially with a glass knife, stained with methylene blue, covered with Durcupan, and observed on an Olympus CX41 microscope. Ultrathin (around 100 nm) sections of the samples were then obtained by using a diamond knife (Diatome) with an Ultracut Ultramicrotome from Reichert-Jung. The sections were double-stained with uranyl acetate and lead citrate and viewed with a Jeol JEM 1010 microscope at 80 kV. Images were obtained with a Bioscan camera model 792 (Gatan, Pleasanton, CA, USA) at the University of Barcelona technical services.

3. Results

3.1. Scanning Electron Microscopy

The statocyst, located on the basal segment of the antennule, appeared as a cup-like invagination of the cuticle forming a closed oval cavity (Figure 1). The ventral floor of the cavity had a depression that exhibited an inner cuticular sensory epithelium formed of hairs (mechanosensory setae) polarized towards the center and an overlying statoconia (consisting of sand granules cemented together by tegmental gland secretions (Figures 1D and 2A), which stimulates the sensory hair cells (Figures 2 and 3). After removing the sand granules (Figure 2B), the setae distribution within the statocyst was visible. A curved field consisting of an inner single row ($\overline{X}$ = 145, Figure 2B–D and Figure 3A–C) and an outer double row ($\overline{X}$ = 95, Figure 2A,D and Figure 3A) run on a circle around the medial and lateral rim of the central depression. On the posterior side, it merged into a small triangle of setae ($\overline{X}$ = 45, Figure 2C). A short row ($\overline{X}$ = 10, Figures 2D and 3B) was shown to run internally in parallel to the inner single row only next to the small setal field. Opposite this small field, on the anterior side of the depression, there was a large setal triangular field ($\overline{X}$ = 125, Figure 2) ($\overline{X}$:

setae mean number for N = 20). Table 1 reports the setae mean number for each statocyst region.

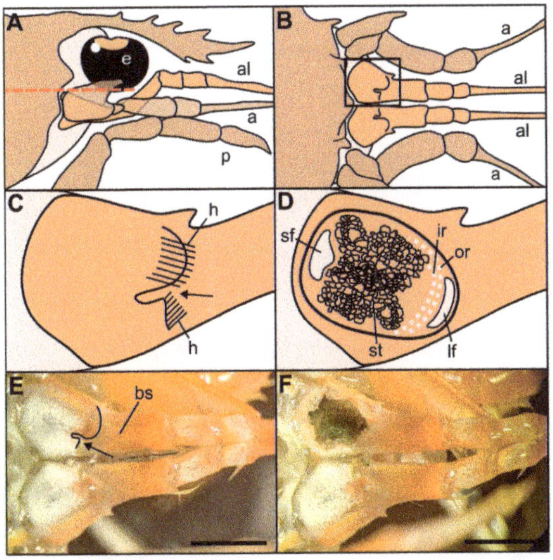

Figure 1. Drawings of the location and morphology of the statocyst organs in Norwegian lobster *Nephrops norvegicus*. (**A**) Right lateral view of the region of the statocysts, with the eye (e), antennule (al), antenna (a), pereiopod (p). (**B**) Dorsal view of the statocyst region after removal of the tissues that cover it dorsally, above the red dashed line in (**A**). (**C**) Detail from the inset in (**B**). Dorsal view of the closed statocyst in the basal segment of the antennule. Arrow: opening of the statocyst organ covered by hairs (h) on two sides. (**D**) Dorsal view of the opened statocyst in the basal segment of the antennule, showing the statolith (st) and the setae distribution: large setal field (lf), small setal field (sf), outer row setae (or), inner row setae (ir). (**E**) Dorsal view of the left and right antennules with the covered statocyst. Basal segment (bs). Arrow: statocyst's opening. (**F**) Dorsal view of the left and right antennules with the left statocyst exposed. Note the statolith made of sand particles inside the statocyst. Scale bar: (**E**,**F**) = 5 mm.

Table 1. Summary of setae mean number in each statocyst region.

$\overline{X}$ (Setae Mean Number for N = 20)		Figure
Inner single row (ir)	145	Figures 2B–D and 3A,C
Outer double row (or)	95	Figures 2A–D and 3A
Large setal field (lf)	125	Figure 2B
Small setal field (sf)	45	Figure 2A–D
Short internal row	10	Figures 2D and 3B

The inner row setae appeared overlaid by the statolith and hook-shaped, curving to the center (Figures 2 and 3). The outer double row showed upright setae free of sand grains (Figures 2 and 3). The structure of the setae in the two rows was similar, showing a bulb (the proximal portion of the sensillum) with a diameter of 30 μm at its widest point, a setal shaft extending for 370 μm (inner row) and 170 μm (outer row) into the statocyst lumen (setae mean measures for N = 20), a tooth (the smooth portion of the bulb), a fulcrum (a transverse fold), and filamentous hairs, which, in the inner row, showed attached statoconia (grain sands) (Figure 3C,D).

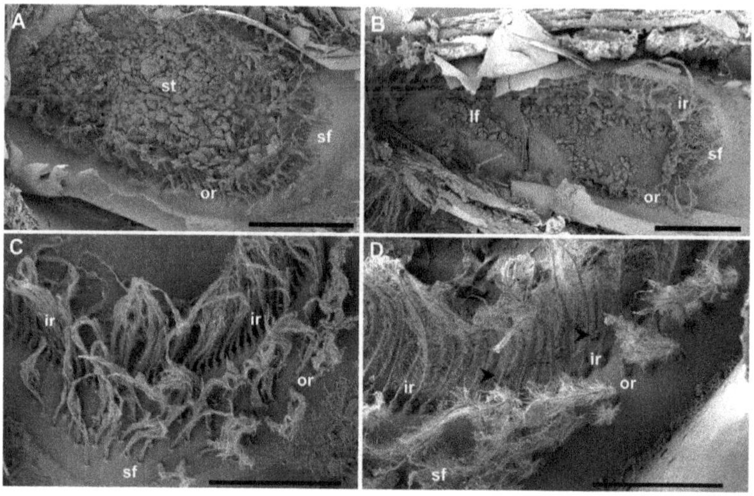

Figure 2. SEM. *N. norvegicus* statocyst structure and setae distribution. (**A**) Dorsal view of the statocyst with the statolith (st) covering all the sensilla. (**B**) Dorsal view of the statocyst after removing the statolith. The different areas are visible: inner single row (ir), outer double row (or), large setal field (lf), small setal field (sf). (**C**) Small setal field (sf) converging with the outer row (or). The inner row (ir) is visible behind it. (**D**) The arrowheads mark the short row that runs internally in parallel to the inner single row (ir) next to the small setal field (sf). Scale bars: (**A**,**B**) = 1 mm. (**C**) = 500 µm. (**D**) = 300 µm.

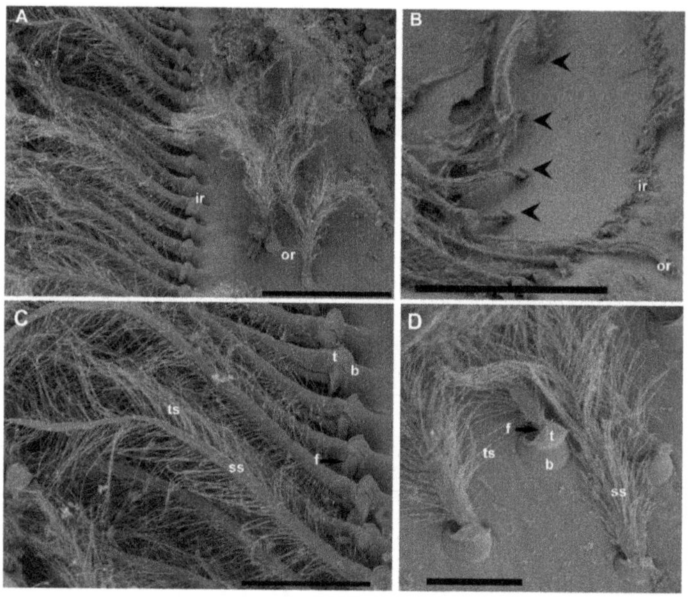

Figure 3. SEM. *N. norvegicus* statocyst setae structure. (**A**) Detail of the setae in the regular single inner row (ir) and the double outer row (or). (**B**) The arrowheads mark the short row visible after the removal of the inner row. (**C**) Inner row setae structure with a bulb (the proximal portion of the sensillum, b), a setal shaft (ss), a tooth (the smooth portion of the bulb, t), a fulcrum (a transverse fold, f), and filamentous thread-like strands (ts). (**D**) Setae structure in the small field. The setae in this area are similar to those described in (**C**) but with a shorter setal shaft (ss). Scale bars: (**A**) = 200 µm. (**B**) = 300 µm. (**C**) = 100 µm. (**D**) = 50 µm.

3.2. *Transmission Electron Microscopy*

In the sensory epithelium (Figure 4), the statocyst hair cells appeared buttressed by dark supporting cells. The medial portion of the hair cells showed a clear cytoplasm containing large nuclei (Figure 4B). Towards the hair cell base, membranous junction complexes were shown to attach neighboring hair cells together. In addition, a root was found to anchor the cells into the cuticular layer.

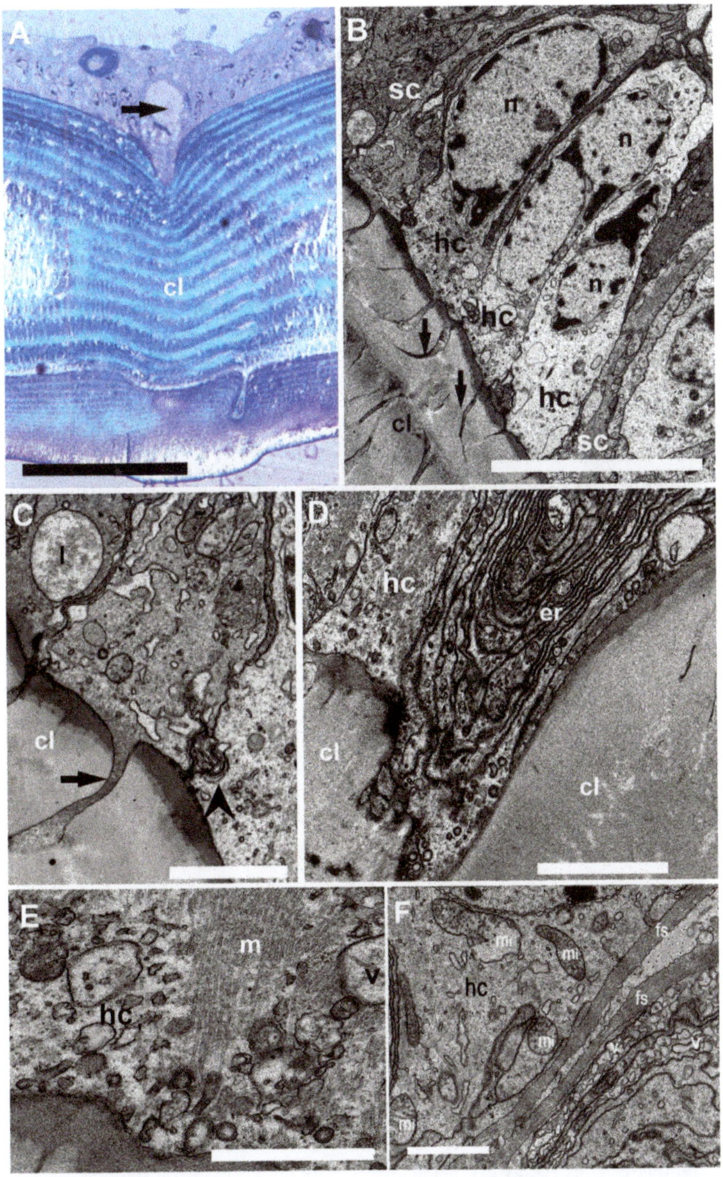

Figure 4. Light microscopy (**A**) and TEM (**B–F**). Inner structure of *N. Norvegicus* statocyst sensory epithelium. (**A**) The arrow marks the hair cell anchored in the cuticular layer (cl). (**B**) View of the sensory epithelium showing the hair cells (hc) buttressed by dark supporting cells (sc). Note the

prominent nuclei (n) and the cytoplasm in the hair cells. The arrow marks the "anchor root" of the hair cell base in the cuticular layer (cl). **(C)** Detail of the anchoring structure of a hair cell (arrow). The arrowhead marks a complex membranous junction between two hair cells (lysosome (l)). **(D)** Abundant endoplasmic reticulum (er) is visible in a hair cell (hc) basal portion. **(E)** A bundle of microtubules (m) in relation with vesicles (v) in the basal portion of a hair cell. **(F)** Basal pole of a hair cell containing abundant mitochondria (mi). Note also fibrous strands (fs) and numerous vesicles (v). Scale bars: **(A)** = 100 µm. **(B)** = 10 µm. **(C,D,F)** = 2 µm. **(E)** = 1000 nm.

The basal portion of the hair cells appeared to enter the cuticular layer and contain abundant endoplasmic reticulum and mitochondria. At this level, fibrous strands and numerous vesicles were shown.

4. Discussion

The statocyst is a crucial sensory system, the starting point of the sound perception process in marine invertebrates [12,13]. The outcome of this work is a description of the statocyst morphology in a previously undescribed Astacidea member. Astacidea is an infraorder of decapod crustaceans, including lobster and crayfish species. Although some previous works described the statocyst structure of some crayfish species, this is the first description of *N. norvegicus* statocyst, which allows a comparison to be made between some species.

The general morphology of the statocyst of *N. norvegicus* appeared to be similar to that of crayfish species previously described (*Procambarus clarkii* [24], *Faxonius limosus* [18], *Cherax destructor* [8]). Hertwig et al. [18] showed that all the setae of the statocyst capsule in *O. limosus* were morphologically identical. In *C. destructor* [8], the statocyst setae were of a single type too, and their morphology was closely similar to that of *O. limosus* setae [8]. However, despite these morphological similarities, the physiological responses of statocyst setae can differ [25]. Our results showed similarities in morphology between *N. norvegicus* setae and the previously described setae of the crayfish [8,18], even though these animals inhabit different ecosystems. In the next sections, we will attempt a further comparison on the arrangement of these elements in a variety of species with differing lifestyles, which has the potential to reveal principles of statocyst structure and function.

Hertwig et al. [18] found four distinct distribution areas of setae in *O. limosus* statocyst. Findley et al. [8] were able only to clearly detect three areas in *C. destructor* statocyst. Our results showed the presence of four areas, resembling the description of *O. limosus*. The inability to identify a fourth area in *C. destructor* might be due to the fact that the posterior line of eight setae (described in *O. limosus*) was not evident in *C. destructor*, but it is possible that these setae are included in the outer curved group. *N. norvegicus* exhibited similar numbers of setae as those in *C. destructor* and *O. limosus* in the different fields, confirming the similarity of the statocyst structure among *Astacidea* members, independently of their habitat.

In *N. norvegicus*, the curved and small fields appeared to occupy a similar position relative to the statolith, the same as that observed in *O. limosus* and *C. destructor*, and had approximately the same number of setae. As a consequence, it is probable that they serve similar functions about position information. Similarly, as in *C. destructor* and *O. limosus*, in lobsters (including our studied species, *N. norvegicus*) the lateral and posterior field respond to body roll, whereas the anterior field of setae responds mostly to acceleration [11].

It was hypothesized in different species that statolith-free setae are most suited for detecting angular accelerations [11,18,25]. In *N. norvegicus*, the double external row of setae is free from the statocyst; therefore, these setae could accomplish this function. Further research on behavioral aspects like movement in the water column or escape is necessary to determine the relation between these setae and the organization of the statocyst.

A description of the ultrastructure of the statocyst of the shrimp *Palaemon serratus* [6] confirms the specificity of the statocyst depending on its taxonomic group. *Palaemon* is a small species and, in addition to the obvious difference in the size of the setae with respect to *N. norvegicus*, its statocyst consists of vertical cellular projections arranged in a single row of hair cells oriented towards a common crescent-shaped central region and covered by a statolith (sand granules).

A comparison with taxonomically distant crustacean groups would lead us to see even greater differences, such as the simple crescent-shaped distribution of the few setae on the statocysts of the decapod Portunidae crabs (e.g., blue crab *Callinectes sapidus*) [14,19]. Interestingly, a recent description of the statocyst in the hermit crab *Dardanus calidus* (decapod, Anomura), taxonomically distant from *N. norvegicus*, showed four setal zones (medial, lateral, ventral, and caudal) as described in the present study, but compared to *N. norvegicus*, their distribution is different, and only the medial group of setae is in contact with the statolith [20].

The present TEM investigations essentially revealed that the statocyst hair cells have a robust microtubular cytoskeleton, are firmly anchored into the cuticular layer, and are strongly attached to one another. These morphological characteristics typically correspond to the organization of receptors devoted to the reception of mechanical vibrations. They actually attest to the ability of the organ to withstand water movements to some extent.

In decapod crustaceans, the setae have different functions depending on the species. In the crayfish *O. limosus* [18] and in the blue crab *C. sapidus* [14], the setae have a purely mechanical function. Shearing of the setae in one direction causes a mechanical constraint on a lever spring attached to the base of the setae, which in turn stimulates a sensory afferent neuron [12,26,27]. Similarly, in the crayfish *Astacus fluviatis*, setae stimulation causes mechanical stress in a chorda thread connected to a few bipolar afferent neurons [12,26,27]. In contrast, in the statocyst of the shrimp *P. serratus*, the setae have a mechanosensory function. The setae have a root buried in the cuticular layer that establishes synaptic contacts with infracuticular bipolar afferent neurons [6]. In the present study, a connection of the setae with neurons could not be observed. Based on the taxonomic proximity of the lobster to the crayfish and crab, the setae in *N. norvegicus* could have a mechanical function rather than a sensory function. Further ultrastructural research using TEM is needed to clarify the mode of connection between setae and afferent neurons in the statocyst of *N. norvegicus*.

5. Conclusions

N. norvegicus statocyst was described for first time. The statocyst setae appeared distributed in four areas (a curved field consisting of an inner single row and an outer double row that run on a circle around the medial and lateral rim of the central depression, a small setal field in the posterior part, a large setal field, opposite to the small field, and a short row running internally and lying parallel to the inner single row). The structure of the seta consists of a bulb (the proximal portion of the sensillum), a setal shaft, a tooth (the smooth portion of the bulb), and a fulcrum (a transverse fold), and filamentous hairs. The hair cells are firmly implanted within the cuticular layer. The data presented here contribute to the necessary knowledge of the sensory systems of crustaceans, as an initial step for the evaluation of the long-term effects of intense low-frequency sounds on their hearing ability. Although the type of innervation of the statocyst was not considered in the present study, the close taxonomic position of the lobster to that of the crayfish and crab would suggest that the setae in *N. norvegicus* are pure mechanoreceptors rather than sensory cells.

Funding: This work was funded as part of the SATURN project, which received funding from the European Union's Horizon2020 research and innovation program under grant agreement no. 101006443.

Institutional Review Board Statement: Although there are no legal requirements for studies involving crustaceans in Spain, the experimental protocol strictly followed the necessary precautions to comply with the current ethical and welfare considerations when dealing with vertebrates and cephalopods in scientific experimentation (Royal Decree 1386/2018 of November 19; Directive 2010/63/EU). This study was also carefully analyzed and approved by the Ethical Committee for

Scientific Research of the Universitat Politècnica de Catalunya, BarcelonaTech (UPC) (approval code B9900085). In this study, we limited the number of animals used in the experiments.

Informed Consent Statement: Not applicable.

Data Availability Statement: Data are contained within the article.

Acknowledgments: I would like to thank Marc Lenoir from INSERM (Montpellier, France) for his generosity and valuable comments on the manuscript draft, which greatly improved it, and Steffen De Vreese for the diagram of the statocyst location and morphology (Figure 1). I would like to thank José Manuel Fortuño from the Institut de Ciències del Mar (CSIC) for assistance in obtaining the SEM images and the staff of Unitat de microscòpia electronica from the Hospital Clínic (Universitat de Barcelona) for assistance and advice in obtaining the TEM images.

Conflicts of Interest: The author declares no conflicts of interest.

References

1. Edmonds, N.J.; Firmin, C.J.; Goldsmith, D.; Faulkner, R.C.; Wood, D.T. A review of crustacean sensitivity to high amplitude underwater noise: Data needs for effective risk assessment in relation to UK commercial species. *Mar. Pollut. Bull.* **2016**, *108*, 5–11. [CrossRef] [PubMed]
2. Sole, M.; Kaifu, K.; Mooney, T.A.; Nedelec, S.L.; Olivier, F.; Radford, A.N.; Vazzana, M.; Wale, M.A.; Semmens, J.M.; Simpson, S.D.; et al. Marine invertebrates and noise. *Front. Mar. Sci.* **2023**, *10*, 1129057. [CrossRef]
3. Radford, C.A.; Stanley, J.A. Sound detection and production mechanisms in aquatic decapod and stomatopod crustaceans. *J. Exp. Biol.* **2023**, *226*, jeb243537. [CrossRef] [PubMed]
4. Breithaupt, T. Sound Perception in Aquatic Crustaceans. In *The Crustacean Nervous System*; Springer: Berlin/Heidelberg, Germany, 2002; pp. 548–558. [CrossRef]
5. Solé, M.; Lenoir, M.; Durfort, M.; López-Bejar, M.; Lombarte, A.; van der Schaar, M.; André, M. Does exposure to noise from human activities compromise sensory information from cephalopod statocysts? *Deep. Sea Res. Part II Top. Stud. Oceanogr.* **2013**, *95*, 160–181. [CrossRef]
6. Lovell, J.M.; Findlay, M.M.; Moate, R.M.; Yan, H.Y. The hearing abilities of the prawn *Palaemon serratus*. *Comp. Biochem. Physiol.—A Mol. Integr. Physiol.* **2005**, *140*, 89–100. [CrossRef] [PubMed]
7. Lovell, J.M.; Moate, R.M.; Christiansen, L.; Findlay, M.M. The relationship between body size and evoked potentials from the statocysts of the prawn *Palaemon serratus*. *J. Exp. Biol.* **2006**, *209*, 2480–2485. [CrossRef] [PubMed]
8. Finley, L.; Macmillan, D. The Structure and Growth of the Statocyst in the Australian Crayfish *Cherax destructor*. *Biol. Bull.* **2000**, *199*, 251–256. [CrossRef] [PubMed]
9. Solé, M.; Lenoir, M.; Fontuño, J.M.; Durfort, M.; van der Schaar, M.; André, M. Evidence of Cnidarians sensitivity to sound after exposure to low frequency noise underwater sources. *Sci. Rep.* **2016**, *6*, 37979. [CrossRef] [PubMed]
10. Solé, M.; Fortuño, J.-M.; van der Schaar, M.; André, M. An acoustic treatment to mitigate the effects of the apple snail on agriculture and natural ecosystems. *J. Mar. Sci. Eng.* **2021**, *9*, 969. [CrossRef]
11. Cohen, M.J. The function of receptors in the statocyst of the lobster *Homarus americanus*. *J. Physiol.* **1955**, *130*, 9–34. [CrossRef] [PubMed]
12. Budelmann, B.U. Hearing in Non-arthropod Invertebrates. In *The Evolutionary Biology of Hearing*; Webster, D.B., Fay, R.R., Eds.; Springer: Berlin/Heidelberg, Germany, 1992; pp. 141–155.
13. Budelmann, B.U. Hearing in Crustacea. In *The Evolutionary Biology of Hearing*; Webster, D.B., Fay, R.R., Eds.; Springer: Berlin/Heidelberg, Germany, 1992; pp. 131–139. [CrossRef]
14. Cate, H.S.; Roye, D.B. Ultrastructure and physiology of the outer row statolith sensilla of the blue crab *Callinectes sapidus*. *J. Crustac. Biol.* **1997**, *17*, 398–411. [CrossRef]
15. Breithaupt, T.; Tautz, J. Vibration sensitivity of the crayfish statocyst. *Naturwissenschaften* **1988**, *75*, 310–312. [CrossRef]
16. Patton, M.L.; Grove, R.F. Statolith hair movements and the regulation of tonic gravity reflexes in the lobster *Homarus americanus*. *Comp. Biochem. Phys.* **1992**, *101A*, 259–268. [CrossRef] [PubMed]
17. Sekiguchi, H.; Terazawa, T. Statocyst of *Jasus edwardsii pueruli* (*Crustacea, Palinuridae*), with a review of crustacean statocysts. *Mar. Freshw. Res.* **1997**, *48*, 715–719. [CrossRef]
18. Hertwig, I.; Schneider, H.; Hentschel, J. Light and electron microscopic analysis of the statocyst of the American crayfish *Orconectes limosus* (Crustacea, Decapoda). *Zoomorphology* **1991**, *110*, 189–202. [CrossRef]
19. Solé, M.; De Vreese, S.; Sánchez, A.M.; Fortuño, J.-M.; van der Schaar, M.; Sancho, N.; André, M. Cross-sensory interference assessment after exposure to noise shows different effects in the blue crab olfactory and sound sensing capabilities. *Sci. Total Environ.* **2023**, *873*, 162260. [CrossRef] [PubMed]
20. Solé, M.; De Vreese, S.; Fortuño, J.-M.; van der Schaar, M. Artificial sound impact could put at risk hermit crabs and their symbiont anemones. *Sci. Total Environ.* **2023**, *900*, 165756. [CrossRef] [PubMed]
21. Letourneau, J.G. Addition of sensory structures and associated neurons to the crayfish telson during development. *J. Comp. Physiol.* **1976**, *110*, 13–23. [CrossRef]

22. Schmitz, B. Post embryonic development of the crayfish *Procambarus clarkii* and its tailfan mechanosensory system. In *Nervous Systems: Principles of Design and Function*; Singh, P.N., Ed.; Wiley Eastern: Hoboken, NJ, USA, 1992; pp. 69–90.
23. Roberts, L.; Laidre, M.E. Finding a home in the noise: Cross-modal impact of anthropogenic vibration on animal search behaviour. *Biol. Open* **2019**, *8*, bio041988. [CrossRef]
24. Takahata, M.; Hisada, M. Functional polarization of statocyst receptors in the crayfish *Procambarus clarkii Girard*. *J. Comp. Physiol.* **1979**, *130*, 201–207. [CrossRef]
25. Patton, M.L.; Grove, R.F. The response of statocyst receptors of the lobster *Homarus americanus* to movements of statolith hairs. *Comp. Biochem. Physiol.* **1992**, *101*, 249–257. [CrossRef] [PubMed]
26. Janse, C. The function of statolith-hair and free-hook-hair receptors in statocyst of the crab *Scylla serrata*. *J. Comp. Physiol.* **1980**, *137*, 51–62. [CrossRef]
27. Schöne, H.; Steinbrecht, R.A. Fine structure of statocyst receptor of *Astacus fluviatis*. *Nature* **1968**, *220*, 184–186. [CrossRef] [PubMed]

biology

MDPI

Article

Effects of Different Salinity Stress on the Transcriptomic Responses of Freshwater Crayfish (*Procambarus clarkii*, Girard, 1852)

Lei Luo [1,2,3], Li-Shi Yang [1,2,4,*], Jian-Hua Huang [1,2], Shi-Gui Jiang [1,2], Fa-Lin Zhou [1,2], Yun-Dong Li [1,2], Song Jiang [1,2] and Qi-Bin Yang [4]

1. Key Laboratory of South China Sea Fishery Resources Exploitation and Utilization, Ministry of Agriculture and Rural Affairs, South China Sea Fisheries Research Institute, Chinese Academy of Fishery Sciences, Guangzhou 510300, China; leiluo2019@163.com (L.L.); huangjianhua@scsfri.ac.cn (J.-H.H.); jiangsg@21cn.com (S.-G.J.); zhoufalin@aliyun.com (F.-L.Z.); liyd2019@163.com (Y.-D.L.); tojiangsong@163.com (S.J.)
2. Shenzhen Base of South China Sea Fisheries Research Institute, Chinese Academy of Fishery Sciences, Shenzhen 518108, China
3. School of Life Sciences, Sun Yat-Sen University, Guangzhou 510275, China
4. Key Laboratory of Efficient Utilization and Processing of Marine Fishery Resources of Hainan Province, Sanya Tropical Fisheries Research Institute, Sanya 572018, China; yangqibin1208@163.com

* Correspondence: yangls2016@163.com; Tel.: +86-2084451448

Simple Summary: *Procambarus clarkii* is an economic freshwater aquaculture species which is popular with consumers for its delicious flavor and high protein content. Salinization of freshwater ecosystems is an increasingly pressing global issue that poses a significant threat to aquaculture. Salinity is an important environmental factor directly affecting the metabolism, growth, reproduction, and physiological processes of aquatic animals. In this study, crayfish were subjected to acute low salt (6 ppt) and high salt (18 ppt) stress and investigated using transcriptome sequencing technology. The response of the crayfish to different salinity stresses, especially immunity, metabolism, ion transport, and osmoregulation, was analyzed to illustrate the resistance mechanism of crayfish facing salt stress. The results of this study are intended to deepen our understanding of the mechanisms by which freshwater organisms respond to salinity stress and provide useful references for the healthy culture of crayfish and the utilization of saline soils.

Abstract: Salinization of freshwater ecosystems is a pressing global issue. Changes in salinity can exert severe pressure on aquatic animals and jeopardize their survival. *Procambarus clarkii* is a valuable freshwater aquaculture species that exhibits some degree of salinity tolerance, making it an excellent research model for freshwater aquaculture species facing salinity stress. In the present study, crayfish were exposed to acute low salt (6 ppt) and high salt (18 ppt) conditions. The organisms were continuously monitored at 6, 24, and 72 h using RNA-Seq to investigate the mechanisms of salt stress resistance. Transcriptome analysis revealed that the crayfish responded to salinity stress with numerous differentially expressed genes, and most of different expression genes was observed in high salinity group for 24h. GO and KEGG enrichment analyses indicated that metabolic pathways were the primary response pathways in crayfish under salinity stress. This suggests that crayfish may use metabolic pathways to compensate for energy loss caused by osmotic stress. Furthermore, gene expression analysis revealed the differential expression of immune and antioxidant-related pathway genes under salinity stress, implying that salinity stress induces immune disorders in crayfish. More genes related to cell proliferation, differentiation, and apoptosis, such as the Foxo, Wnt, Hippo, and Notch signaling pathways, responded to high-salinity stress. This suggests that regulating the cellular replication cycle and accelerating apoptosis may be necessary for crayfish to cope with high-salinity stress. Additionally, we identified 36 solute carrier family (SLC) genes related to ion transport, depicting possible ion exchange mechanisms in crayfish under salinity stress. These findings aimed to establish a foundation for understanding crustacean responses to salinity stress and their osmoregulatory mechanisms.

Keywords: salinity; transcriptome; mechanism; osmotic mediation; *Procambarus clarkii*

1. Introduction

Freshwater ecosystems provide a wide range of natural resources for fisheries, aquaculture, production, and recreational activities, which are essential for human survival [1]. However, human activities such as urbanization, commercial pollution, and hydrological disturbances severely affect freshwater ecosystems such as rivers, lakes, and wetlands [2,3]. Salinization also exerts pressure on freshwater ecological resources. Global warming has led to rising sea and estuarine levels, and the continued intrusion of saline water into the upper reaches of rivers is one of the main causes of increased salinity in freshwater ecosystems [4–6]. Furthermore, human industrial activities, road de-icing salts, and agricultural irrigation have accelerated freshwater resource salinization [7,8]. Groundwater salt rises to the surface via capillary action in arid and semiarid areas, and salts deposited on the soil surface after water evaporation also cause soil salinization [9]. Freshwater ecosystem salinization is a global problem affecting over 100 countries and poses a significant threat to agriculture, negatively impacting biodiversity, species ecosystems, food security, and human economic welfare [10]. Therefore, studying the response of freshwater species to saline stress and the effective exploitation of saline soils are urgent and essential tasks for scientists.

Salinity is an important environmental factor affecting metabolism, growth, reproduction, and physiological processes [11–13]. Studies have shown that salinity in aquaculture waters above the species acclimation threshold will lead to histopathological changes in organisms and can also cause immune disorders, osmoregulatory imbalances, disturbances in the gut microbiota, and increased mortality, among other harmful effects [14–16]. Aquatic animals respond to the osmotic stress caused by increased salinity through heightened energy metabolism [17]. However, excessive energy metabolism alters oxidative levels, producing reactive oxygen species (ROS) accumulation, possibly triggering oxidative stress injury [18]. Furthermore, salinity stress affects the ion transport processes and transport-related enzyme activities, leading to disturbances in the osmotic regulation of salt ions in organisms [19]. Among all aquatic animals, crustaceans are the most sensitive to salinity fluctuations due to their osmoregulatory system's inefficiency, especially in narrowly saline species [20]. Although the effects of salinity fluctuations on crustaceans have become a popular topic, most of them are only detected using macroscopic indicators (e.g., enzyme activities and changes in gene expression) [21–23], and the microscopic internal regulatory mechanisms are still unclear. Therefore, elucidating these regulatory mechanisms will help us understand whether crustaceans can cope with further salinity fluctuations, is important for screening salt-tolerant species of aquatic animals, and provides a scientific basis for exploring the aquaculture of aquatic species in saline soils.

High-throughput RNA sequencing technology has been widely used in various organisms and has been successfully used to analyze many aquatic species in response to salinity stress. Jiang et al. established a transcriptome database of pufferfish under long-term low-salinity stress and screened three solute carrier family (SLC) genes related to ion transport [24]. Liu et al. analyzed the liver transcript levels of halibut under low-salinity stress and showed that salinity stress affects the liver lipid metabolism of halibut [25]. Through transcriptome sequencing, Chen et al. identified various genes involved in osmoregulation under different salinity stresses in *Oratosquilla oratoria* [26]. Although considerable progress has been made in the studying of transcript levels in salinity-stressed aquatic animals, relatively few studies have been conducted on freshwater, narrowly saline crustaceans, especially at the transcript level.

Procambarus clarkii is a freshwater crayfish native to the United States and Mexico and was introduced to China from Japan in the 1920s [27]. Its delicious meat and high protein content has made it widely welcomed by consumers. Crayfish production was ~2.89 million tons in 2022 in China, making it the largest crustacean aquaculture species in China [28]. Therefore, in the present study, crayfish were subjected to salinity stress and investigated using transcriptome sequencing technology. The response of the crayfish to different salinity stresses, especially immunity, metabolism, ion transport, and osmoregulation, was analyzed to illustrate the mechanism of crayfish resistance to salinity. The results of this study are intended to deepen our understanding of the mechanisms by which freshwater organisms respond to salinity stress and provide useful references for the healthy culture of crayfish and the utilization of saline soils.

2. Materials and Methods

2.1. Experimental Animal

Healthy *P. clarkii*, with a body length of 6.0 ± 0.5 cm and a weight of 15.0 ± 2.0 g, were purchased from Jian Li, China. They were temporarily reared in the recirculating water system of the Shenzhen Test Base of the South China Sea Fisheries Research Institute, Chinese Academy of Fishery Sciences (Shenzhen, China), and fed 3% of their body weight with commercial feed (China Hubei Tianbang Feed, Tianmen, China). The water temperature was maintained at 25 ± 1 °C, with a pH of 7.5 and dissolved oxygen levels of 6.5 ± 0.5 mg/L.

2.2. Exposure and Sample Collection

According to the safe salinity (6 ppt) of *P. clarkii* (body weight about 15 g) and our previous results of the 96 h semi-lethal salinity experiment (21 ppt) [29,30], the crayfish were marked with different labels as follows: control group (NC, 0 ppt), low-salt group (LS, 6 ppt), and high-salt group (HS, 18 ppt). A total of 270 healthy crayfish were selected and divided into three groups of 90 crayfish each. The experimental salt water was prepared in proportions of seawater and fresh water, and the water was not replaced during the test. The hepatopancreas tissues of three crayfish of each group were dissociated from the carapace at 6 h, 24 h, 72 h, respectively, and frozen in liquid nitrogen for transcriptome sequencing.

2.3. RNA Extraction, Library Preparation, and Sequencing

The Trizol reagent kit (Invitrogen, Carlsbad, CA, USA) was used to extract total RNA following the manufacturer's protocol. RNA quality was assessed using an Agilent 2100 Bioanalyzer (Agilent Technologies, Palo Alto, CA, USA) and RNase-free agarose gel electrophoresis. The NEBNext® Ultra™ RNA Library Prep Kit (Ipswich, MA, USA) was used to construct mRNA-seq libraries for all the samples, which were then sequenced on an Illumina HiSeqTM 4000 by Gene Denovo Biotechnology Co. (Guangzhou, China). High-quality reads were produced by fastp software (version 0.18.0) to remove reads that contained adapters [31], poly nucleotides (N), and low-quality bases. Thus, contigs and unigenes were obtained by a short reads assembling program—Trinity software (version: r20140413p1) [32].

2.4. Data Analysis

2.4.1. Gene Differential Expression and Functional Annotation

The unigene expression was calculated and normalized using RPKM (reads per kb per million reads) [33]. Differential expression analysis of RNAs was performed between the two groups using the DESeq2_1.20.0 software (and between the two samples using edgeR) [34,35]. Genes were considered differentially expressed if they had a false discovery rate (FDR) parameter of less than 0.05 and an absolute fold change of ≥2. All unigenes were used to analyze gene function with Blastx alignment based on the following databases: NCBI non-redundant protein (Nr) database, Swiss-Prot protein database, Kyoto Encyclopedia of Genes, Genomes (KEGG) database, COG/KOG database, and Gene Ontology (GO) database.

2.4.2. GO, KEGG Enrichment, and Gene Expression Pattern Analysis

To explore how *P. clarkii* responds to salinity fluctuations, we performed GO classification and KEGG pathway annotation of all the differentially expressed genes obtained. Firstly, we used the Blast2GO program (v2.3.5) to identify the characteristics and biochemical metabolic pathways of the DEGs and their products, and the threshold parameter for significant differences was $Q \leq 0.05$. Subsequently, we mapped the unigenes to the KEGG database to find the pathways to which they belonged and further determined the unigenes of the ion transport-related pathway.

The expression levels were then normalized to different values to illustrate the expression pattern of the DEGs. In brief, the expression level of each sample was normalized to 0, log2 (v1/v0), log2 (v2/v0), and the values were clustered by Short Time-series Expression Miner software 1.3.8 (STEM 1.3.8). The cluster profile was constructed by three rules as follows: At first, the maximum unit change between the different time points would be 1. Then, the maximum output profiles should be set to 20 (similar expression pattern profiles should be merged). At last, the minimum ratio of the fold change in the DEGs should be more than 2.0. The profiles with p-value ≤ 0.05 were considered as significant differences.

2.4.3. Ion Transport-Related Gene Selection and Function Prediction

To understand the differential expression of ion transport-related genes at seven salinity time points, based on the detailed information of the Nr-annotated DEGs, we selected unigenes whose annotation information was functionally annotated as "ion channel" and "transporter". Next, the functional characteristics of all ion transport genes were described using the GeneCard database (http://www.genecards.org/, accessed on 22 August 2023).

2.5. Verification by Quantitative Real-Time PCR (qRT-PCR)

To ensure the reliability of the RNA sequencing results, we randomly selected 9 DEGs for verification by quantitative RT-PCR at two different salinity time points. Using 18S rRNA as a reference gene, Primer Premier 6.0 software was used to design the gene-specific primer pairs (Supplementary Materials Table S1). The experiments were performed on a Roche Light Cycler 480 thermal cycler (Roche Applied Science, Penzberg, Germany). The total volume of the reaction solution was 12.5 μL containing 1 μL cDNA, 0.5 μL upstream and downstream primers, 6.25 μL 2×SYBR Premix ExTaq, and 4.25 μL sterile water. qRT-PCR was performed as follows: Five qPCR replicates for each gene, with each replicate using independently prepared RNA samples, and each sample from three technical replicates. The experimental procedure was 95 °C for 2 min. Running conditions were 95 °C for 2 min, 1 cycle, then 95 °C for 15 s, and 60 °C for 1 min, 40 cycles. Amplicons were verified by melting curve analysis and relative expression was determined using the comparative CT method ($2^{-\Delta\Delta CT}$) [36].

2.6. Statistical Analysis

All biological experiments were repeated three times independently, and the data analyzed were expressed as mean ± standard deviation (SD). Significant differences were analyzed using Duncan's multiple comparison test followed by GraphPad Prism software 9. $p < 0.05$ was defined as statistically significant.

3. Results

3.1. RNA-Seq Data

A total of 884,590,210 raw reads were obtained from seven groups of crayfish hepatopancreas treated at different salinities and time points (Table 1). After data filtering, we obtained 882,552,854 clean reads, accounting for an average of 99.76% of the raw reads. The average Q20% and Q30% were 96.62% and 91.18%, respectively, and the average GC content was 51.84% (Table 1). Clean reads were assembled using Trinity software, and 52,533 unigenes were generated with an average length of 1000 bp, the maximum length of 32,384 bp, and a minimum length of 201 bp. The number and length of N50 were 8596 and 1688 bp, respectively, with an average GC content of 43.18% (Table 2). Based on the protein databases, the gene annotation results showed that 16,957 unigenes were annotated. Of these, 16,893, 8908, 12,210, 8234, and 4816 unigenes showed significant matches with sequences in the Nr, Swiss-Prot, KEGG, COG/KOG, and GO databases, respectively (Supplementary Materials Table S2).

Table 1. Sequencing sample data and quality control.

Sample	Raw Data	Clean Data (%)	Q20 (%)	Q30 (%)	GC (%)
NC-1	43,079,546	42,982,162 (99.77%)	96.51%	90.75%	51.68%
NC-2	38,593,886	38,496,476 (99.75%)	96.17%	90.06%	51.31%
NC-3	43,843,540	43,725,484 (99.73%)	96.80%	91.58%	51.91%
LS-6h-1	44,961,252	44,810,234 (99.66%)	96.41%	90.82%	51.87%
LS-6h-2	40,666,280	40,580,188 (99.79%)	96.61%	91.15%	52.15%
LS-6h-3	45,471,462	45,335,054 (99.70%)	96.56%	91.11%	50.69%
LS-24h-1	36,343,018	36,273,260 (99.81%)	96.75%	91.39%	52.11%
LS-24h-2	37,333,288	37,262,908 (99.81%)	96.61%	91.05%	51.67%
LS-24h-3	43,514,756	43,385,520 (99.70%)	96.48%	91.04%	50.90%
LS-72h-1	41,937,250	41,860,300 (99.82%)	96.72%	91.26%	50.97%
LS-72h-2	39,298,830	39,218,856 (99.80%)	96.96%	91.88%	52.03%
LS-72h-3	39,176,456	39,097,798 (99.80%)	96.30%	90.59%	52.02%
HS-6h-1	45,496,514	45,380,462 (99.74%)	96.26%	90.52%	50.76%
HS-6h-2	37,971,054	37,894,032 (99.80%)	96.69%	91.36%	53.46%
HS-6h-3	44,680,448	44,583,988 (99.78%)	96.63%	90.88%	51.51%
HS-24h-1	40,967,824	40,872,802 (99.77%)	96.24%	90.51%	52.03%
HS-24h-2	41,390,576	41,287,814 (99.75%)	96.22%	90.49%	52.74%
HS-24h-3	46,509,192	46,401,214 (99.77%)	98.00%	94.48%	52.18%
HS-72h-1	41,216,070	41,140,894 (99.82%)	96.42%	90.79%	52.28%
HS-72h-2	48,341,970	48,254,082 (99.82%)	96.89%	91.75%	52.41%
HS-72h-3	43,796,998	43,709,326 (99.80%)	96.70%	91.35%	52.01%

Table 2. Quality statistics of unigenes assembly.

Genes Number	GC (%)	N50 Number	N50 Length	Max Length	Min Length	Average Length	Total Bases
52,533	43.18	8596	1688	32,394	201	1000	52,553,090

3.2. *Differential Expression Responses of P. clarkii Exposed to Different Salinity Gradients*

To investigate the genes that respond to changes in salinity, we compared DEG counts between the six salinity treatment groups and the control group. Compared with the control group, 292 (184 up and 108 down), 622 (280 up and 342 down), 614 (425 up and 189 down), 324 (214 up and 110 down), 2545 (1368 up and 1177 down), and 613 (299 up and 314 down) unigenes were obtained at LS-6h, LS-24, LS-72, HS-6h, HS-24h, and HS-72h, respectively (Figure 1). We then performed GO term enrichment analysis on the DEGs from the six treatment groups to explore how salinity fluctuations affected physiological processes in crayfish (Figure 2). The results indicated that differentially expressed genes were mainly enriched in biological processes (mainly single organisms and metabolic processes), cellular components (mainly cells and cell parts), and molecular functions (mainly catalytic activity and binding).

KEGG enrichment analysis revealed the top 20 pathways in which the DEGs were enriched (Figure 3). Of the three low-salt treatment groups, DEGs were mainly enriched in pyruvate metabolism (Q = 0.010) and glycolysis/gluconeogenesis (Q = 0.035) pathways within the LS-6h group. The DEGs in the LS-24h group were primarily enriched in the pentose and glucuronate interconversion (Q = 0.000), ascorbate and aldarate metabolism (Q = 0.000), drug metabolism—cytochrome P450 (Q = 0.000), metabolism of xenobiotics by cytochrome P450 (Q = 0.000), and retinol metabolism (Q = 0.000) pathways. However, fewer DEGs were significantly enriched in the LS-72h treatment group, mainly in drug metabolism—other enzymes (Q = 0.033). In the high-salinity treatment groups, the DEGs were significantly enriched in 19 pathways at HS-24h, including the pentose and glucuronate interconversions (Q = 0.000) and metabolic pathways (Q = 0.000). Furthermore, DEGs in the ECM–receptor interaction (Q = 0.002) and ascorbate and aldarate metabolism (Q = 0.028) pathways were significantly enriched in the HS-6h and HS-72h treatment groups.

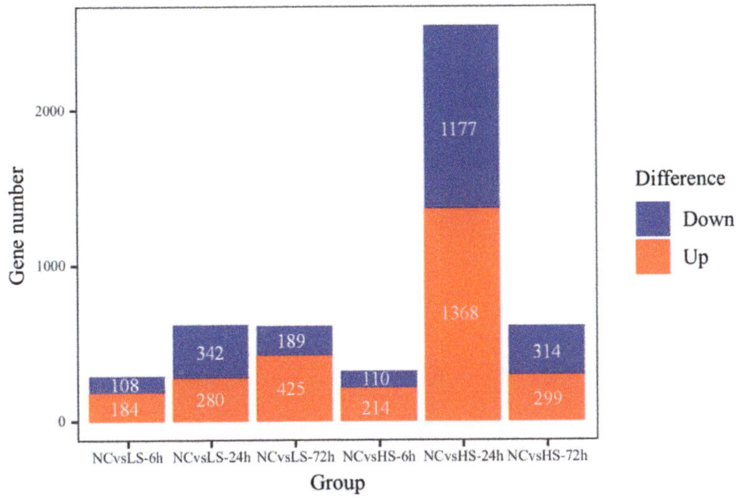

Figure 1. Statistics of DEGs in different salinity treatment groups.

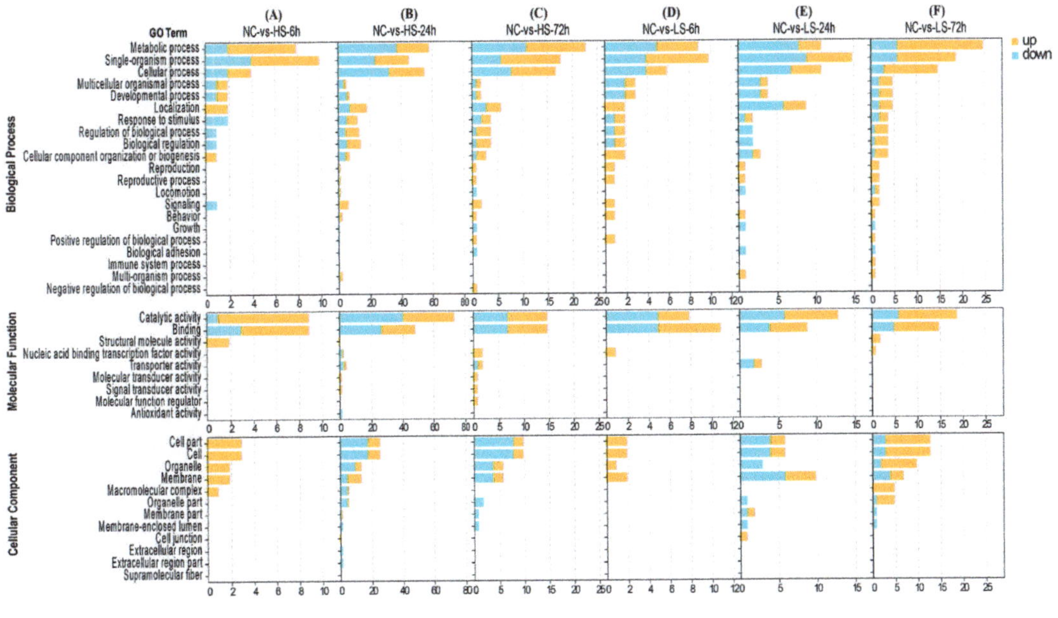

Figure 2. GO functional annotations of DEGs for *P. clarkii* exposed to salinity fluctuation; (**A**): NCvsHS-6h, (**B**): NCvsHS-24h, (**C**): NCvsHS-72h; (**D**): NCvsLS-6h, (**E**): NCvsLS-24h, (**F**): NCvsLS-72h.

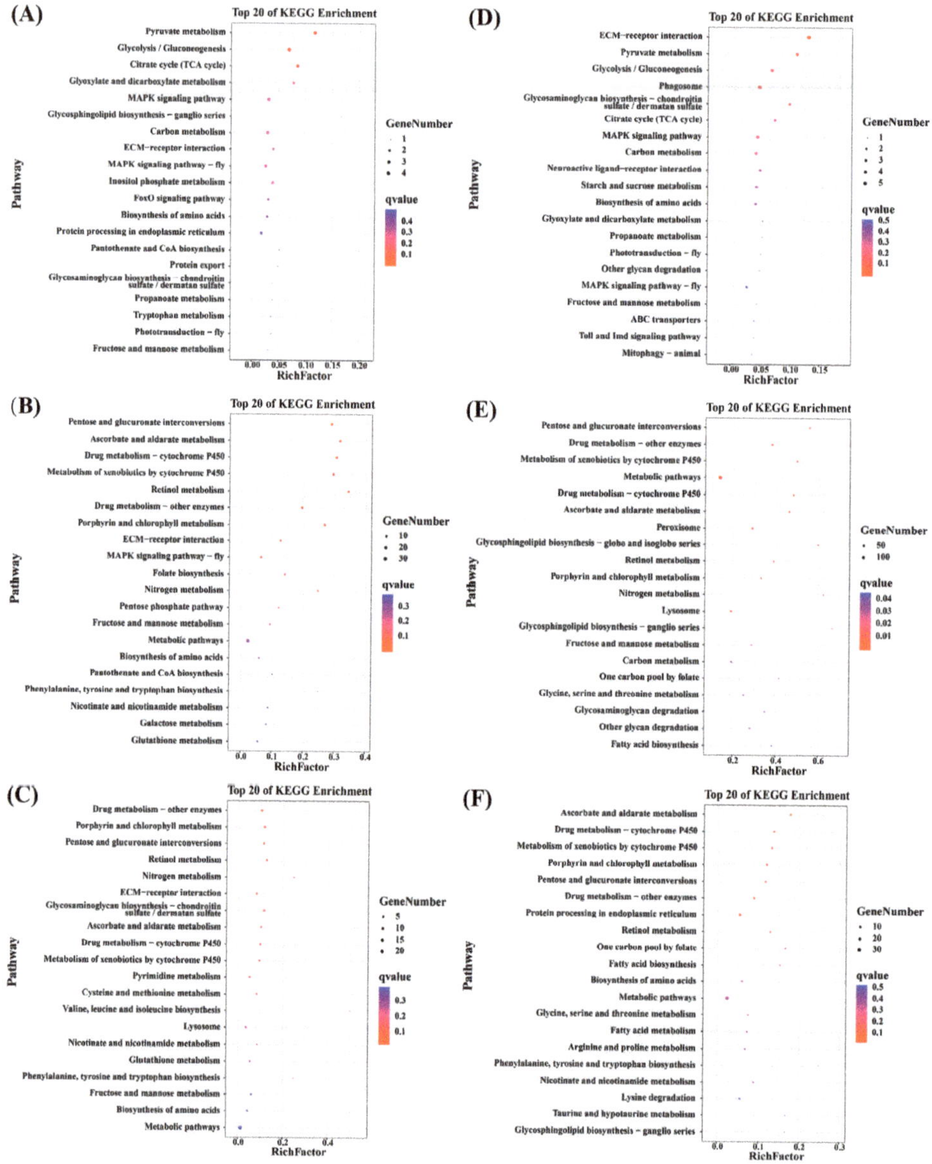

Figure 3. KEGG enrichment analysis results of DEGs of *P. clarkii* under salt stress; (**A**): NCvsLS-6h, (**B**): NCvsLS-24h, (**C**): NCvsLS-72h; (**D**): NCvsHS-6h, (**E**): NCvsHS-24h, (**F**): NCvsHS-72h.

3.3. Trend Analysis of DEGs under Salinity Treatment

To visually observe the expression patterns of transcripts in relation to salinity, we analyzed the expression trends of genes in the different salinity treatment groups and performed hierarchical clustering of genes enriched in different pathways (Figures 4 and 5).

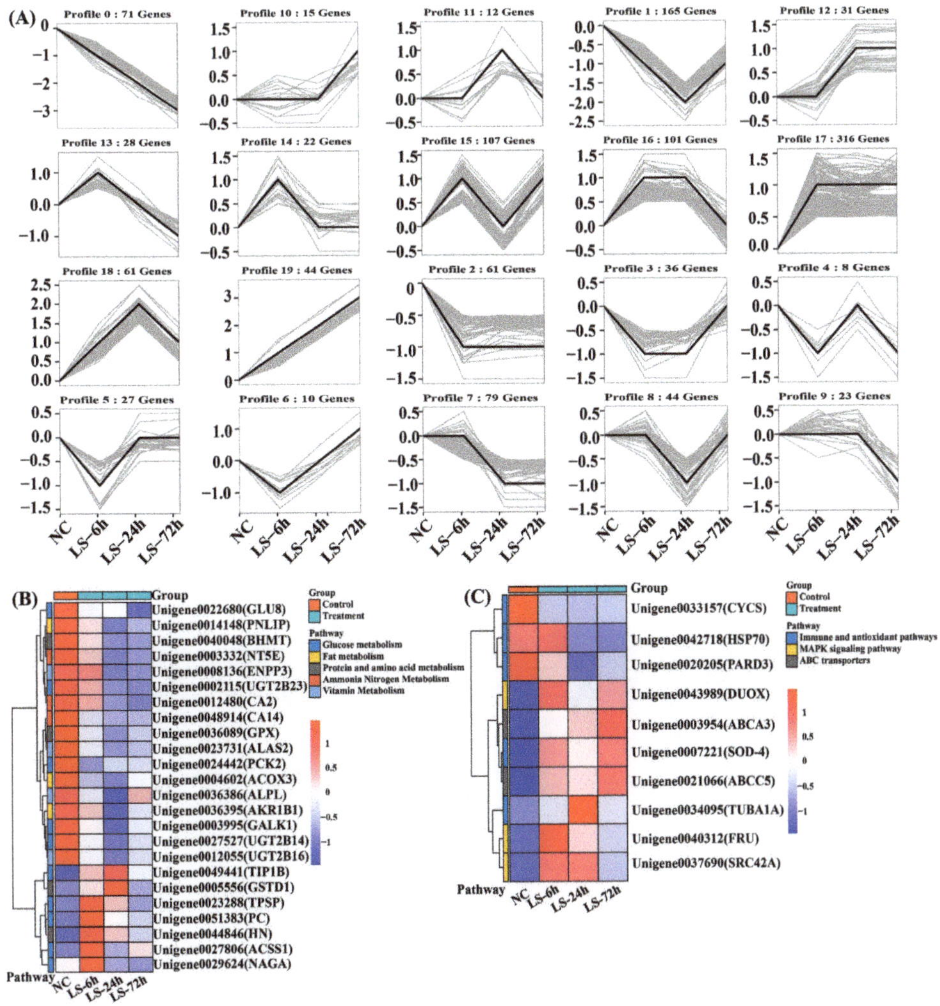

Figure 4. Analysis of gene expression patterns under low salt stress based on normalization of gene expression trends. (**A**): gene expression trend analysis; (**B**): gene expression clustering pattern analysis of metabolic pathways; (**C**): gene expression clustering pattern analysis of other signaling pathways. (The color scale on the right ranges from the lowest (blue) to the highest (red) expression level). The detailed gene information in the right range is shown in Supplementary Materials Table S3.

Based on the normalization of gene expression trends, 20 expression patterns of 1231 genes were identified under low salt stress (Figure 4A). Among them, seven categories of gene expression patterns showed an overall post-change upregulation trend with increasing salinity stress duration, including profiles of 17 (316 genes), 15 (107 genes), 18 (61 genes), 19 (44 genes), 12 (31 genes), 10 (15 genes), and 6 (10 genes). Similarly, the following seven categories of gene expression patterns showed an overall post-change downregulation trend: profile 1 (165 genes), profile 7 (79 genes), profile 2 (61 genes), profile 13 (28 genes), profile 9 (23 genes), profile 4 (8 genes), and profile 0 (7 genes). In addition, six categories of genes, profile 16 (101 genes), profile 8 (44 genes), profile 3 (36 genes), profile 5 (27 genes), profile 14 (22 genes), and profile 11 (12 genes), showed an overall unchanged expression trend. Furthermore, 36 genes that were significantly differentially expressed in different pathways and signaling pathways were identified using cluster analysis. These included metabolic pathways (glucose, lipid, protein, vitamin, ammonia, and nitrogen metabolism) (Figure 4B), immune and antioxidant pathways, and ABC transporter signaling pathways (Figure 4C).

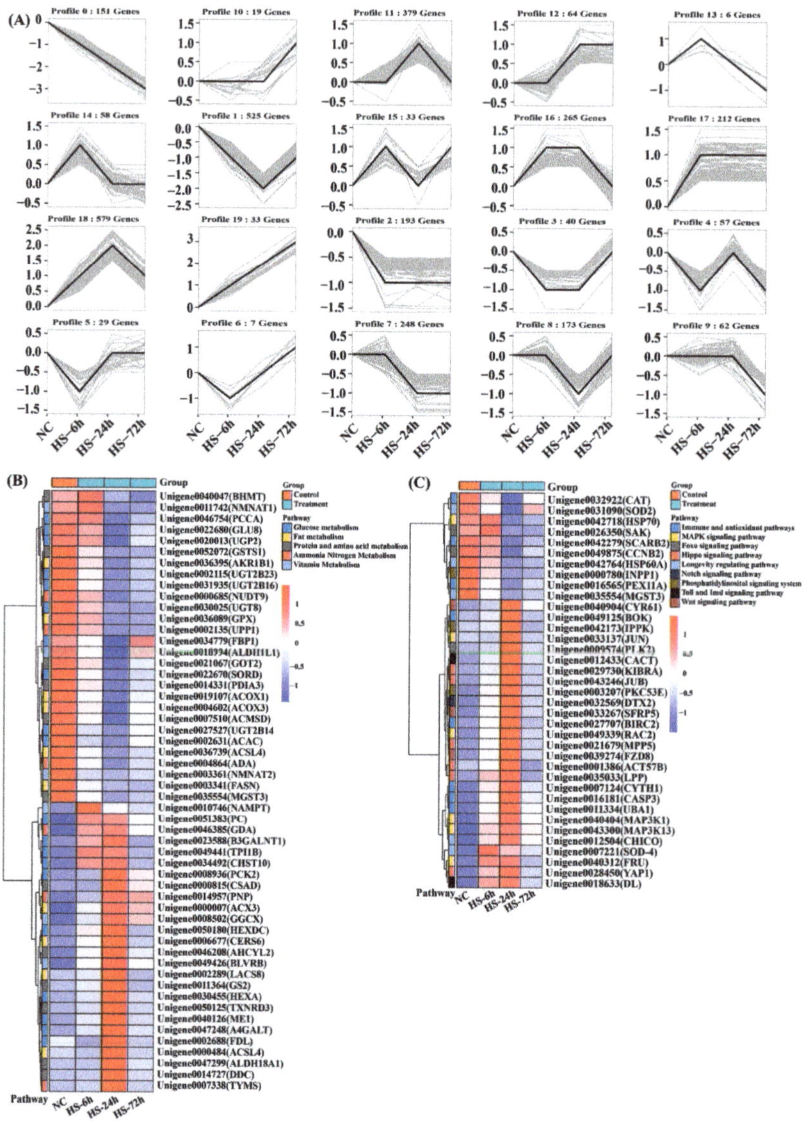

Figure 5. Analysis of gene expression patterns under high salt stress based on normalization of gene expression trends. (**A**): gene expression trend analysis; (**B**): gene expression clustering pattern analysis of metabolic pathways; (**C**): gene expression clustering pattern analysis of other signaling pathways. (The color scale on the right ranges from the lowest (blue) to the highest (red) expression level). The detailed gene information in the right range is shown in Supplementary Materials Table S4.

Similarly, 3133 genes with 20 different expression trends were characterized under high salt stress (Figure 5A). Of these, seven categories of gene expression patterns showed a general up-regulation trend, including profiles 18 (579 genes), 17 (212 genes), 12 (64 genes), 19 (33 genes), 15 (33 genes), 10 (19 genes), and 6 (7 genes). Profiles 1 (525 genes), 7 (248 genes), 2 (193 genes), 0 (151 genes), 9 (62 genes), 4 (57 genes), and 13 (6 genes) showed an overall post-change downregulation trend. Moreover, six gene categories, Profile 11 (379 genes), Profile 16 (265 genes), Profile 8 (173 genes), Profile 14 (58 genes), Profile 3 (40 genes), and Profile 5 (29 genes) showed overall unchanged expression trends. A total of 96 significant differentially expressed genes were clustered into 14 pathway classes. In addition to the metabolic, immune, and antioxidant pathways (Figure 5B), several important signaling pathways were included (Foxo, Hippo, MAPK, Notch, Wnt Toll, and Imd signaling pathways, phosphatidylinositol signaling system, and longevity regulating pathway) (Figure 5C).

3.4. Analysis of Genes Related to Ion Exchange in *P. clarkii*

Based on the screening of transcript data and gene annotation, we identified 36 solute carrier family (SLC) genes related to ion transport that were differentially expressed in the six salinity treatment groups (Table 3 and Figure 6). These genes include SLC41A1 (Magnesium ion transporter), SLC40A1 (Iron ion transporter), SLC39A11/9/2 (Zinc transporter), SLC30A9/7/6/5/2/1 (Zinc transporter), SLC26A11/6 (chloride–bicarbonate exchanger), SLC22A8/6/5/3/15B (Organic cation transporter), SLC13A5 (Na^+-dependent carboxylate and sulfate transporter), SLC12A9/6/2 (Cation-coupled Cl-Cotransporter), SLC10A/6/3/2 (Ileal sodium/bile acid cotransporter), SLC9A9/8/3R1/2 (Sodium/hydrogen exchanger), SLC8A2/1 (Sodium–calcium exchanger), SLC5A3/1 (Sodium-dependent vitamin transporter), and SLC4A10/1 (Sodium bicarbonate cotransporter).

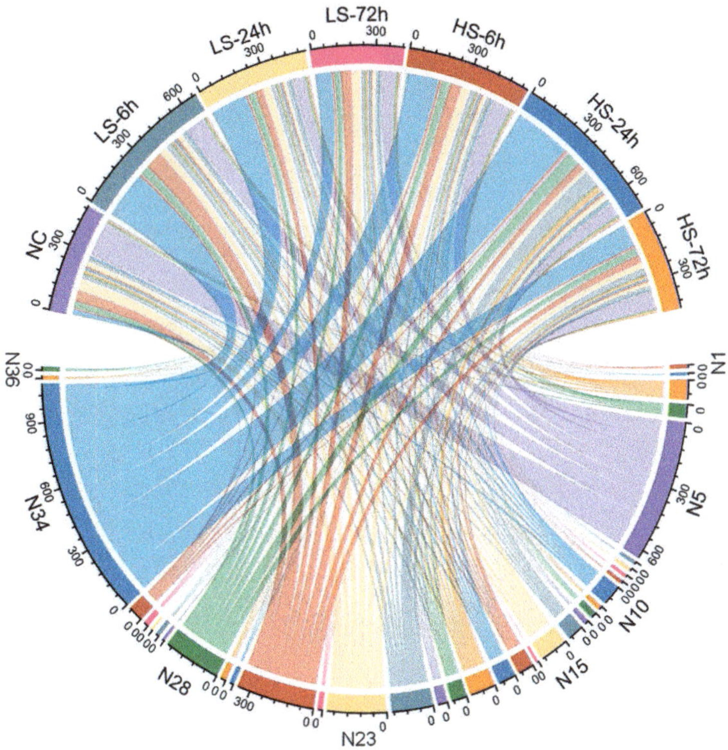

Figure 6. Chordal graph of 36 solute carrier family genes under different salinity treatments. N1–N36 represent the genes of Number 1–36 in Table 3, the genes correspond to the treatment groups one by one, the wider the chord width, the higher the gene expression.

Table 3. The 36 SLC family (solute carrier family) genes related to ion transport were differentially expressed in six salinity groups.

Number	Symbol	Description	Rpkm (Mean)						
			NC	LS-6h	LS-24h	LS-72h	HS-6h	HS-24h	HS-72h
1	SLC41A1	Solute carrier family 41 member 1 (Magnesium ion transporter)	1.21	1.11	1.96	0.96	1.12	3.69	0.83
2	SLC40A1	Solute carrier family 40 member 1 (Iron ion transporter)	1.50	0.76	0.39	0.86	0.89	0.61	0.68
3	SLC39A9	Solute carrier family 39 member 9 (Zinc transporter ZIP9)	9.99	11.04	11.65	11.11	12.65	12.88	13.12
4	SLC39A11	Solute carrier family 39 member 11 (Zinc transporter ZIP11)	6.87	7.49	8.40	8.30	8.70	8.62	6.25
5	SLC39A1	Solute carrier family 39 member 1 (Zinc transporter ZIP1)	169.84	96.51	63.39	68.11	90.07	62.21	83.81
6	SLC30A9	Solute carrier family 30 member 9 (Zinc transporter 9)	1.37	1.12	1.24	1.58	1.24	1.10	1.10
7	SLC30A7	Solute carrier family 30 member 7 (Zinc transporter 7)	1.96	1.11	1.14	1.67	1.46	2.03	1.44
8	SLC30A6	Solute carrier family 30 member 6 (Zinc transporter 6)	0.89	1.05	1.06	0.94	1.11	1.38	0.93
9	SLC30A5	Solute carrier family 30 member 5 (Zinc transporter 5)	0.85	1.20	1.20	0.94	1.31	1.30	1.36
10	SLC30A2	Solute carrier family 30 member 2 (Zinc transporter 2)	14.44	12.71	10.37	12.75	14.66	13.42	9.99
11	SLC30A1	Solute carrier family 30 member 1 (Zinc transporter 1)	2.88	2.39	1.52	1.82	2.47	1.63	1.70
12	SLC26A6	Solute carrier family 26 member 6 (chloride-bicarbonate exchanger)	2.50	2.82	2.77	2.97	4.08	4.83	2.83
13	SLC26A11	Solute carrier family 26 member 11 (chloride-bicarbonate exchanger)	1.89	2.31	2.34	2.35	3.34	2.39	1.55
14	SLC22A8	Solute carrier family 22 member 8 (Organic cation transporter)	9.04	10.59	12.87	8.80	11.79	4.89	11.21
15	SLC22A6	Solute carrier family 22 member 6 (Organic cation transporter)	22.11	24.18	14.19	15.98	17.75	19.86	14.23
16	SLC22A5	Solute carrier family 22 member 5 (Organic cation transporter)	0.79	0.86	0.76	0.63	0.68	0.54	0.47
17	SLC22A3	Solute carrier family 22 member 3 (Organic cation transporter)	8.86	11.13	8.70	6.63	9.70	12.58	10.05
18	SLC22A15B	Solute carrier family 22 member 15B (Organic ion transporter)	11.15	11.08	9.40	11.31	11.34	10.87	10.04
19	SLC13A5	Solute carrier family 13 member 5 (Na^+-dependent carboxylate and sulfate transporter)	15.67	12.67	9.84	13.95	11.28	34.18	15.09
20	SLC12A9	Solute carrier family 12 member 9 (Cation-coupled Cl^- cotransporter)	10.11	9.59	10.29	9.13	11.05	8.53	6.02
21	SLC12A6	Solute carrier family 12 member 6 (Cation-coupled Cl^- cotransporter)	5.00	4.57	8.13	5.15	7.42	5.91	4.65
22	SLC12A2	Solute carrier family 12 member 2 ($Na^+/K^+/2Cl^-$ cotransporter)	13.86	25.66	17.36	11.79	17.32	74.29	26.06
23	SLC10A6	Solute carrier family 10 member 6 (Ileal sodium/bile acid cotransporter)	56.69	37.14	40.56	46.31	43.40	16.17	27.24
24	SLC10A3	Solute carrier family 10 member 3 (Ileal sodium/bile acid cotransporter)	3.08	2.52	3.42	1.81	2.65	2.11	1.50
25	SLC10A2	Solute carrier family 10 member 2 (Sodium/bile acid cotransporter)	51.12	89.37	36.31	50.32	54.28	34.86	36.86
26	SLC9A8	Solute carrier family 9 member 8 (Sodium/hydrogen exchanger 8)	1.12	1.71	1.74	1.75	1.84	3.16	1.95
27	SLC9A9	Solute carrier family 9 member 9 (Sodium/hydrogen exchanger 9)	2.64	2.88	3.02	2.51	3.28	4.20	2.28
28	SLC9A3R1	Solute carrier family 9 member 3 regulator 1 (Na^+/H^+exchange regulatory cofactor NHE-RF1)	29.13	40.50	32.40	28.00	37.29	58.53	32.90
29	SLC9A2	Solute carrier family 9 member 2 (Na^+/H^+ exchanger)	0.13	0.34	0.27	0.16	0.54	5.04	0.56
30	SLC8A2	Solute carrier family 8 member 2 (Sodium-calcium exchanger)	0.68	1.60	1.72	1.22	2.14	4.33	1.48
31	SLC8A1	Solute carrier family 8 member 1 (Sodium-calcium exchanger)	0.31	0.63	0.38	0.62	0.46	0.50	0.46
32	SLC5A8	Solute carrier family 5 member 8 (Sodium-dependent vitamin transporter)	2.32	2.59	2.65	2.35	2.78	3.90	1.77
33	SLC5A3	Solute carrier family 5 member 3 (Sodium/myo-inositol cotransporter)	5.42	10.56	8.26	7.93	11.63	12.35	8.77
34	SLC5A1	Solute carrier family 5 member 1 (Sodium/glucose cotransporter)	23.56	227.19	181.73	82.43	160.33	283.00	110.62
35	SLC4A1	Solute carrier family 4 member 1 (Sodium bicarbonate cotransporter)	1.68	1.79	1.72	1.50	1.84	1.56	1.00
36	SLC4A10	Solute carrier family 4 member 10 (Sodium bicarbonate cotransporter)	2.04	2.40	2.23	1.81	2.60	1.29	1.37

3.5. Quantitative Real-Time PCR

To confirm the RNA-Seq results, nine DEGs were randomly selected for RT-qPCR validation in two salinity groups (Figure 7). Among these, five genes were upregulated (*SLC5A1*, *SOD-4*, *DUOX*, *CHER*, and *TPI1B*) (Figure 7A,C,E,F,H), and four genes were downregulated (*AO2*, *HSP70*, *RGN*, and *SORD*) (Figure 7B,D,G,I). The gene expression trend was similar to that of the transcriptome sequencing results, indicating that the sequencing results were reliable.

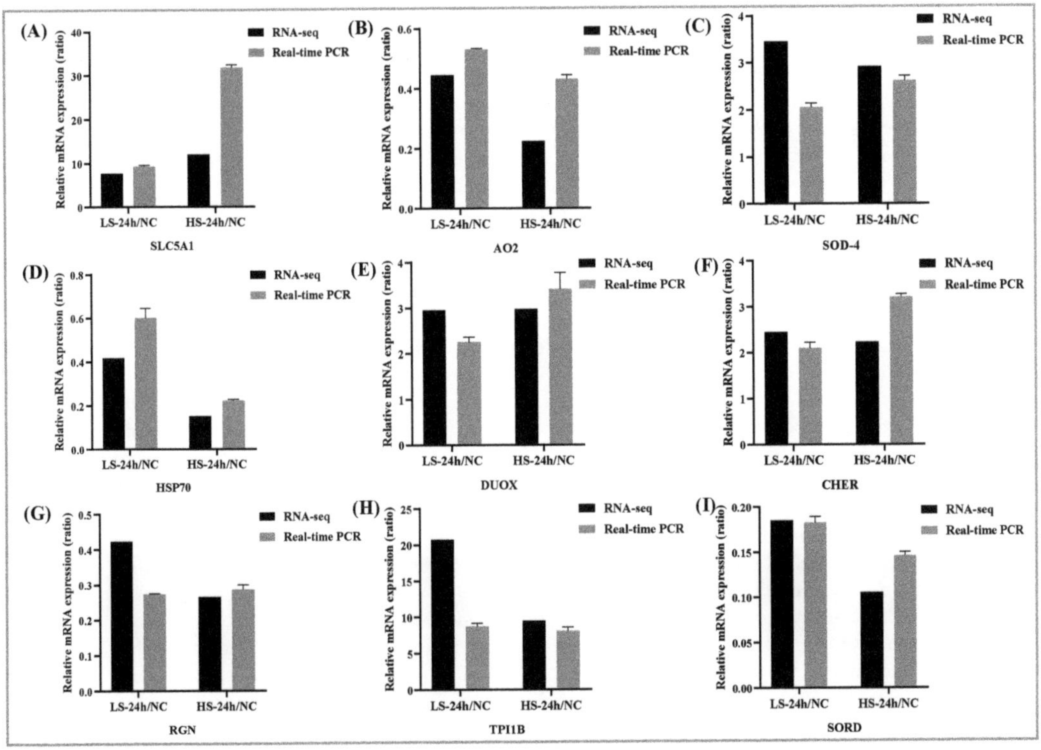

Figure 7. Validation of DEGs with qRT-PCR. Relative mRNA expression level < 1 indicates downregulation, and relative mRNA expression level > 1 indicates upregulation. (**A**): *SLC5A1*, (**B**): *AO2*, (**C**): *SOD-4*, (**D**): *HSP70*, (**E**): *DUOX*, (**F**): *CHER*, (**G**): *RGN*, (**H**): *TPI1B*, (**I**): *SORD*. The data of this study are presented as mean ± SD of three parallel measurements.

4. Discussion

Salinity is an important environmental factor, and changes in salinity affect aquatic animal growth and development, metabolism, immunity, and numerous physiological processes [11,12]. In their natural habitats, crustaceans are constantly exposed to a wide range of salinities, making their survival extremely difficult because they lack effective osmoregulators [37]. Understanding the mechanisms by which crustaceans respond to environmental salinity fluctuations is an increasingly important scientific issue. Therefore, in the present study, a freshwater economic crayfish species was used as a research model, and its response pathways to different salinities and ion osmoregulation mechanisms were investigated by transcriptome sequencing.

In this study, we set a low salt concentration of 6 ppt and a high salt concentration of 18 ppt stress on crayfish. The results showed that the differentially expressed genes showed a tendency to increase and then decrease with the prolongation of time under both high and low salt stress, with the most remarkable response to the low salt stress of 24 h. Although crayfish temporarily adapted to these stresses, they still posed significant stress challenges, which explains why crayfish only survive for a short period in many lake inlets [38]. To further determine which pathways were enriched for functional genes under salinity stress, we performed GO functional enrichment analyses on six comparator groups. These showed that the upregulated and downregulated genes were significantly enriched in metabolic processes, cellular components, catalytic activity, and binding processes. Crustaceans respond to environmental stress by compensating for energy loss through metabolic pathways [39–41]. A study of the response of mud crabs to salinity fluctuations also showed that enzyme catalysis regulates ionic changes and osmotic pressure [42]. Numerous ion transporters bind and transport salt ions to alleviate the difference in salt ion concentration inside and outside the cell membrane [19]. Similarly, KEGG enrichment analyses showed that metabolic pathways responded most strongly to salinity stress, which was consistent with the results of the GO analyses, revealing that metabolic processes are important in the response to salinity fluctuations in crayfish. Notably, differentially expressed genes were also significantly enriched in ECM–receptor interactions under high salinity stress (Q = 0.002). Therefore, we hypothesized that high salt levels induce a cellular stress response that activates functional genes to regulate cellular responses and ultimately prevent cell death [43,44].

Under varying salinity pressures, crayfish require the synergistic expression of numerous genes to respond to the hypertonic external environment. In our study, 1132 genes were categorized into 20 different expression patterns under low salt (6 ppt) stress, while there were 3133 genes under high salt (18 ppt) stress. This demonstrates that high salt levels induce a more dramatic response and that crayfish require more genes to be differentially expressed to cope with osmotic stress [15]. Gene clustering analyses indicated that metabolic pathways were important in the response to salinity stress in crayfish. This was consistent with our GO and KEGG enrichment analyses, suggesting that metabolic pathways may be necessary for crayfish to compensate for energy loss due to hypertonic environments. Furthermore, changes in immune- and antioxidant-related genes were also found in the salinity results, possibly suggesting that salinity stress leads to immune disorders in crayfish; similar findings have been reported in various aquatic animals [15,45,46]. MAPKs are key signal transducers from the cell's surface to the inside of the cell and are activated to control many key processes in cell physiology, including cell growth, differentiation, and adaptation to environmental pressures [47]. Our results showed that the differential expression of genes related to the MAPK signaling pathway suggests that the MAPK signaling pathway may be essential for organisms to cope with environmental stress. Two ATP-binding protein genes were differentially expressed under low-salinity stress, implying that crayfish must rapidly translocate ATP to counteract hyperosmotic stress [48]. Notably, crayfish require additional signaling pathways to respond to high-salinity stress. The FoxO signaling pathway is an important biological regulatory network, and the activation of the FoxO signaling pathway can have important roles in metabolic regulation, disease genesis, and cell proliferation [49]. We discovered that FoxO signaling pathway genes were differentially expressed at different time points during high salinity stress, suggesting that high salinity stress may lead to reduced metabolism and increased fatty acid oxidation and glycolysis in crayfish to counteract metabolic diseases; a similar report has been published on *Oreochromis mossambicus* [50]. The Hippo, Notch, and phosphatidylinositol signaling pathways play important roles in the cell differentiation, proliferation, and apoptosis pathways [51–53]. The related pathway genes in our results were also differentially expressed at different time points under high salinity stress, suggesting that high salinity stress influences the survival of cells, and that the organism adjusts the cell cycle and promotes cell apoptosis to adapt to the hyperosmotic environment outside the cell membrane. Additionally, the Wnt signaling pathway and Toll and IMD signaling pathway genes were also differentially expressed, which are related to intercellular communication and pathogen recognition, and are similarly expressed in many aquatic animals during increased environmental stress [54–56]. Notably, longevity regulatory pathway genes were differentially expressed under high salt stress, which is rare in crustaceans. The longevity regulatory pathway is a complex regulatory network involving numerous signaling pathways, such as the mitochondrial, insulin, and reproductive pathways [57]. It has been reported in higher animals [58] but is less frequently reported in aquatic animals in response to environmental stress. Further, our results revealed that most genes under high salt stress had a significant degree of differential change (sharply upregulated or downregulated) at HS-24h and a tendency to regress at HS-72h. This suggests that high salinity treatment of the crayfish for a period of time (at 24 h) resulted in peak stress, and that the crayfish required more rapid changes in gene expression to regulate body functions in response to the hypertonic environment and produced transient acclimatization and survival, followed by gene expression callbacks. However, failure to acclimatize to high salinity may have led to mortality in the crayfish [59]. Similar findings were reported for the transcript of *Oreochromis niloticus* in response to salinity stress [60].

In aquatic animals, osmotic homeostasis under salinity stress is regulated by multiple inorganic ion exchanges both inside and outside the cell membrane [61,62]. This study revealed the differential expression of genes related to ion exchange in crayfish under low and high salinity gradients in seven groups. To visualize the ion regulatory roles of these genes, we simplified all the genes related to ion exchange into a single diagram (Figure 8). Sodium (Na^+) and chloride ions (Cl^-) account for more constituents of the hemolymph of crustaceans [11,42,62]. Thus, when crayfish are exposed to salinity stress, osmotic pressure is balanced by the preferential uptake or secretion of Na^+ and Cl^-. Our results revealed that SLC12A2/6/9, which are mainly responsible for the transport of sodium chloride ions [63], were differentially expressed under salinity stress; in particular, SLC12A2 was significantly upregulated at HS-24h, suggesting that crayfish under acute salinity stress may first reduce salt loss through the active uptake of extramembrane sodium chloride and potassium ions. Notably, the upregulated expression of SLC9A9/8/2/3R1 may generate a steep electric potential difference by enhancing H^+ release (hydration), which, in turn, promotes Na^+ uptake [64]. Under low-salinity conditions, we found that the expression of SLC8A2/1 increased. New evidence suggests that SLC8A2/1 is involved in the removal (Ca^{2+}) from cells, followed by the release of only $2Na^+$ [65]. We also found that SLC4A10/1 and SLC26A11/6 were initially upregulated and subsequently downregulated under salinity stress. Earlier studies have suggested that SLC4A10/1 may increase the co-transport of Na^+ and bicarbonate (HCO_3^-), allowing more Na^+ to be transported into the plasma [66]. Simultaneously, SLC26A11/6 can promote Cl^- uptake and HCO_3^- secretion [67]. Specifically, a high positive transepithelial potential would be generated in the plasma owing to the rapid influx of large amounts of Na^+ [68]. To drive Na^+ out of the plasma, SLC4A10/1 was downregulated at HS-72h to slow this trend [62,66]. Additionally, seven sodium-dependent biomolecule co-transporter genes (SLC13/10/5 family) were also identified, among which SLC5A1 (sodium-dependent glucose transporter) expression was significantly elevated under salinity stress, which once again corroborated that rapid transmembrane transport and constant catabolism of glucose to provide ATP quickly are essential for crayfish to cope with osmotic stress [68]. Finally, several genes involved in the transport of Fe^{2+}, Zn^{2+}, Mg^{2+}, and other cations were identified. Among them, Mg^{2+}, Fe^{2+}, and Zn^{2+} may activate other ion channels by altering catalytic processes or cellular structures [69,70]. Previous studies have suggested that some inorganic cations may play a role in the regulation of cell morphology during osmotic control, indirectly influencing ion secretion and uptake [71]. Overall, we hypothesized that changes in the salt stress levels dominate the ion transport patterns of *P. clarkii*. The synergistic action of various ion transport channels triggers a complex regulatory strategy that ultimately improves tolerance to salt stress.

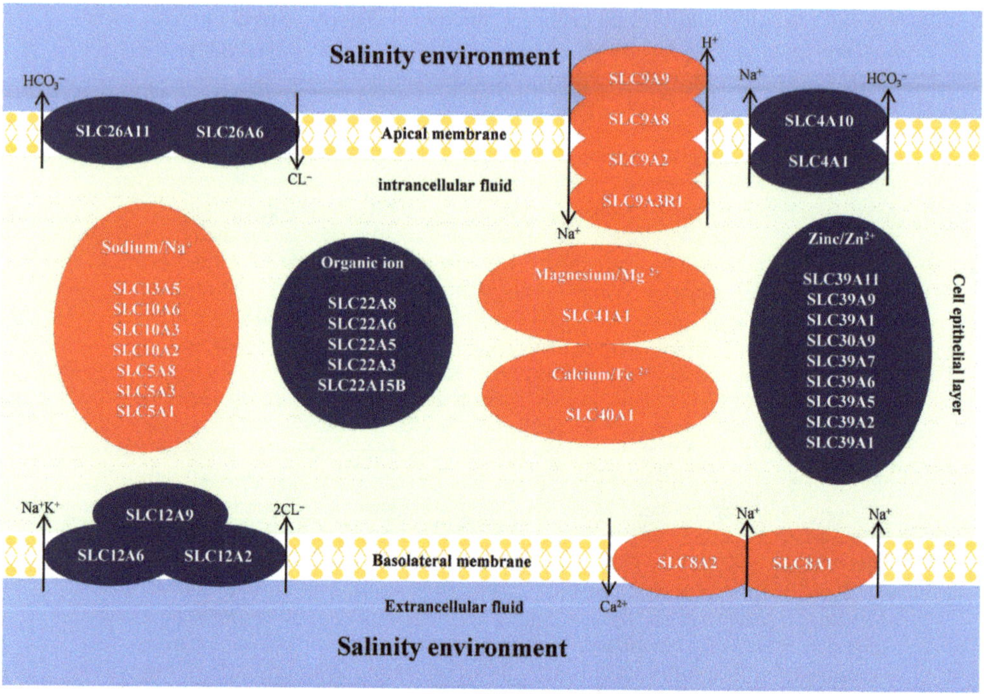

Figure 8. Schematic of ion exchange-related genes in *P. clarkii* exposed to salinity environment.

5. Conclusions

Freshwater salinization is a pressing global environmental problem. Excessive salinity severely affects the metabolism, osmoregulation, behavior, and reproduction of organisms, ultimately leading to a serious loss of biodiversity in freshwater ecosystems. In this study, RNA-Seq analyses were performed on *P. clarkii* under acute high and low salt stress, and more differentially expressed genes were identified under high-salt stress than under low salt stress. This evidence suggests that high salt concentrations may induce more intense osmotic stress than low salt concentrations. Our results also suggest that salinity stress alters the rates of glucose, lipid, and amino acid metabolism and may ultimately enhance osmoregulation in crayfish by regulating the rates of energy metabolism and ion transmembrane transport. Moreover, immune- and antioxidant-related pathway genes were differentially expressed under salinity stress, suggesting that salinity stress induces immune disorders in crayfish. Importantly, cell proliferation, differentiation, apoptosis, and related signaling pathways (Foxo, Wnt, Hippo, Notch signaling pathway, etc.) responded to high salinity stress in crayfish compared to low-salinity stress, revealing that regulating the cell replication cycle and accelerating apoptosis may be necessary for crayfish to respond to high salinity stress. We identified 36 solute carrier family (SLC) genes related to ion transport, indicating possible ion exchange mechanisms in crayfish exposed to salinity stress. The results of the study elaborate on the response mechanism of crayfish to salinity, provide a molecular theoretical basis for the domestication of crayfish for low-salt adaptation, and provide data support for the development and utilization of saline soils.

Supplementary Materials: The following supporting information can be downloaded at: https://www.mdpi.com/article/10.3390/biology13070530/s1, Table S1: Gene Primers used for quantity PCR; Table S2: Statistics of unigenes annotation; Table S3: Detailed DEGs information under low salt stress in Figure 4; Table S4: Detailed DEGs information under high salt stress in Figure 5.

Author Contributions: L.L., J.-H.H., S.-G.J. and L.-S.Y. conceived and designed the experiments; L.L., L.-S.Y., Q.-B.Y., F.-L.Z. and J.-H.H. performed the experiments; S.J. and Y.-D.L. contributed to sample collection and data analysis; L.L. analyzed the data and wrote the paper; L.-S.Y. assisted with writing and proofreading. The entire study has not been published, accepted for publication, or under consideration for publication elsewhere. All authors have read and agreed to the published version of the manuscript.

Funding: This study was funded by the Shenzhen Science and Technology Plan of China (JCYJ20180504170233070), Natural Science Foundation of Hainan Province of China (324MS130), China Agriculture Research System-48 (CARS-48-3), Rural Revitalization Program of Guangdong Province (2022-SPY-02-001), Central Public-interest Scientific Institution Basal Research Fund, CAFS (no. 2023TD34) and National Key Research and Development Program of China (2022YFD2400104).

Institutional Review Board Statement: All experiments in this study were approved by the Animal Care and Use Committee of the South China Sea Fisheries Research Institute, Chinese Academy of Fishery Sciences (no. SCSFRI96-253) and performed according to the regulations and guidelines established by this committee.

Informed Consent Statement: Not applicable.

Data Availability Statement: Data are contained within the article or Supplementary Materials.

Acknowledgments: We thank all the staff in Laboratory of Aquaculture and Genetic breeding, South China Sea Fisheries Research Institute, for their help during the experiments. We greatly appreciate the technical assistance provided by Wen-Ya Wei for his help with qRT-PCR assays. Special thanks to Dong-Liang Liu for his help in preparing the materials for this experiment.

Conflicts of Interest: The authors declare no conflicts of interest.

References

1. Dudgeon, D.; Arthington, A.H.; Gessner, M.O.; Kawabata, Z.I.; Knowler, D.J.; Lévêque, C.; Naiman, R.J.; Prieur-Richard, A.H.; Soto, D.; Stiassny, M.L.J.; et al. Freshwater Biodiversity: Importance, Threats, Status and Conservation Challenges. *Biol. Rev. Camb. Philos. Soc.* **2006**, *81*, 163–182. [CrossRef] [PubMed]
2. Dudgeon, D. Multiple Threats Imperil Freshwater Biodiversity in the Anthropocene. *Curr. Biol.* **2019**, *29*, R960–R967. [CrossRef] [PubMed]
3. Reid, A.J.; Carlson, A.K.; Creed, I.F.; Eliason, E.J.; Gell, P.A.; Johnson, P.T.J.; Kidd, K.A.; MacCormack, T.J.; Olden, J.D.; Ormerod, S.J.; et al. Emerging Threats and Persistent Conservation Challenges for Freshwater Biodiversity. *Biol. Rev.* **2019**, *94*, 849–873. [CrossRef] [PubMed]
4. Jeppesen, E.; Søndergaard, M.; Pedersen, A.R.; Jürgens, K.; Strzelczak, A.; Lauridsen, T.L.; Johansson, L.S. Salinity Induced Regime Shift in Shallow Brackish Lagoons. *Ecosystems* **2007**, *10*, 47–57. [CrossRef]
5. Gardner, A.S.; Moholdt, G.; Cogley, J.G.; Wouters, B.; Arendt, A.A.; Wahr, J.; Berthier, E.; Hock, R.; Pfeffer, W.T.; Kaser, G.; et al. A Reconciled Estimate of Glacier Contributions to Sea Level Rise: 2003 to 2009. *Science* **2013**, *340*, 852–857. [CrossRef] [PubMed]
6. Müller, B.; Gächter, R. Increasing Chloride Concentrations in Lake Constance: Characterization of Sources and Estimation of Loads. *Aquat. Sci.* **2012**, *74*, 101–112. [CrossRef]
7. Cañedo-Argüelles, M.; Kefford, B.J.; Piscart, C.; Prat, N.; Schäfer, R.B.; Schulz, C.J. Salinisation of Rivers: An Urgent Ecological Issue. *Environ. Pollut.* **2013**, *173*, 157–167. [CrossRef]
8. Kaushal, S.S.; Likens, G.E.; Mayer, P.M.; Shatkay, R.R.; Shelton, S.A.; Grant, S.B.; Utz, R.M.; Yaculak, A.M.; Maas, C.M.; Reimer, J.E.; et al. The Anthropogenic Salt Cycle. *Nat. Rev. Earth Environ.* **2023**, *4*, 770–784. [CrossRef] [PubMed]
9. Cañedo-Argüelles, M.; Kefford, B.; Schäfer, R. Salt in Freshwaters: Causes, Effects and Prospects—Introduction to the Theme Issue. *Philos. Trans. R. Soc. B Biol. Sci.* **2019**, *374*, 20180002. [CrossRef]
10. Arora, S.; Singh, A.K.; Singh, Y.P. *Bioremediation of Salt Affected Soils: An Indian Perspective*; Springer: Berlin/Heidelberg, Germany, 2017; ISBN 9783319482576.
11. Rind, K.; Beyrend, D.; Blondeau-Bidet, E.; Charmantier, G.; Cucchi, P.; Lignot, J.H. Effects of Different Salinities on the Osmoregulatory Capacity of Mediterranean Sticklebacks Living in Freshwater. *J. Zool.* **2017**, *303*, 270–280. [CrossRef]
12. Thomas, D.; Rekha, M.U.; Jani, J.R.; Sreekanth, G.B.; Thiagarajan, G.; Subburaj, R.; Kailasam, M.; Vijayan, K.K. Effects of Salinity Amendments on the Embryonic and Larval Development of a Tropical Brackishwater Ornamental Silver Moony Fish, *Monodactylus argenteus* (Linnaeus, 1758). *Aquaculture* **2021**, *544*, 737073. [CrossRef]
13. Qin, Y.; Jiang, S.; Huang, J.; Zhou, F.; Yang, Q.; Jiang, S.; Yang, L. C-Type Lectin Response to Bacterial Infection and Ammonia Nitrogen Stress in Tiger Shrimp (*Penaeus Monodon*). *Fish Shellfish Immunol.* **2019**, *90*, 188–198. [CrossRef] [PubMed]
14. Chen, J.; Zhou, F.; Huang, J.; Jiang, S. Ammonia and salinity tolerance of *Penaeus monodon* across eight breeding families. *SpringerPlus* **2016**, *5*, 171. [CrossRef] [PubMed]

15. Xiao, Y.; Zhang, Y.; Xu, W.; Chen, D.; Li, B.; Cheng, Y.; Guo, X. The Effects of Salinities Stress on Histopathological Changes, Serum Biochemical Index, Non-Specific Immune and Transcriptome Analysis in Red Swamp Crayfish *Procambarus clarkii*. *Sci. Total Environ.* **2022**, *840*, 156502. [CrossRef] [PubMed]
16. Zhao, Y.; Luo, L.; Liu, D.L.; Huang, J.H.; Jiang, S.G.; Zhou, F.L.; Yang, Q.B.; Jiang, S.; Li, Y.D.; Tan, L.Q.; et al. Effect of Acute Salinity Stress on Metabolism, Antioxidant Status, and Histological Structure of *Procambarus clarkii*. *Aquac. Res.* **2023**, *2023*, 2748257. [CrossRef]
17. Jacquin, L.; Petitjean, Q.; Côte, J.; Laffaille, P.; Jean, S. Effects of Pollution on Fish Behavior, Personality, and Cognition: Some Research Perspectives. *Front. Ecol. Evol.* **2020**, *8*, 86. [CrossRef]
18. Bal, A.; Panda, F.; Pati, S.G.; Das, K.; Agrawal, P.K.; Paital, B. Modulation of Physiological Oxidative Stress and Antioxidant Status by Abiotic Factors Especially Salinity in Aquatic Organisms: Redox Regulation under Salinity Stress. *Comp. Biochem. Physiol. Part C Toxicol. Pharmacol.* **2021**, *241*, 108971. [CrossRef] [PubMed]
19. Lou, F.; Gao, T.; Han, Z. International Journal of Biological Macromolecules Effect of Salinity Fl Uctuation on the Transcriptome of the Japanese Mantis Shrimp *Oratosquilla oratoria*. *Int. J. Biol. Macromol.* **2019**, *140*, 1202–1213. [CrossRef] [PubMed]
20. Castaño-Sánchez, A.; Hose, G.C.; Reboleira, A.S.P.S. Salinity and Temperature Increase Impact Groundwater Crustaceans. *Sci. Rep.* **2020**, *10*, 12328. [CrossRef]
21. Pan, L.Q.; Zhang, L.J.; Liu, H.Y. Effects of Salinity and PH on Ion-Transport Enzyme Activities, Survival and Growth of *Litopenaeus vannamei* Postlarvae. *Aquaculture* **2007**, *273*, 711–720. [CrossRef]
22. Fabri, L.M.; Moraes, C.M.; Costa, M.I.C.; Garçon, D.P.; Fontes, C.F.L.; Pinto, M.R.; McNamara, J.C.; Leone, F.A. Salinity-Dependent Modulation by Protein Kinases and the FXYD2 Peptide of Gill (Na^+, K^+)-ATPase Activity in the Freshwater Shrimp *Macrobrachium amazonicum* (Decapoda, Palaemonidae). *Biochim. Biophys. Acta-Biomembr.* **2022**, *1864*, 183982. [CrossRef] [PubMed]
23. Koyama, H.; Kamiya, K.; Sasaki, Y.; Yamakawa, R.; Kuniyoshi, H.; Piyapattanakorn, S.; Watabe, S. Cloning of Glutamine Synthetase Gene from Abdominal Muscle of Kuruma Shrimp *Marsupenaeus japonicus* and Its Expression Profile. *Fish. Sci.* **2023**, *89*, 215–222. [CrossRef]
24. Jiang, J.L.; Xu, J.; Ye, L.; Sun, M.L.; Jiang, Z.Q.; Mao, M.G. Identification of Differentially Expressed Genes in Gills of Tiger Puffer (*Takifugu rubripes*) in Response to Low-Salinity Stress. *Comp. Biochem. Physiol. Part B Biochem. Mol. Biol.* **2020**, *243*, 110437. [CrossRef] [PubMed]
25. Liu, Z.; Ma, A.; Yuan, C.; Zhao, T.; Chang, H.; Zhang, J. Transcriptome Analysis of Liver Lipid Metabolism Disorders of the Turbot *Scophthalmus maximus* in Response to Low Salinity Stress. *Aquaculture* **2021**, *534*, 736273. [CrossRef]
26. Lou, F.; Gao, T.; Han, Z. Transcriptome Analyses Reveal Alterations in Muscle Metabolism, Immune Responses and Reproductive Behavior of Japanese Mantis Shrimp (*Oratosquilla oratoria*) at Different Cold Temperature. *Comp. Biochem. Physiol.-Part D Genom. Proteom.* **2019**, *32*, 100615. [CrossRef] [PubMed]
27. Huang, J.; Tang, S.; Cai, F.; Lin, Y.; Wu, Z. Microsatellite Evidence of Dispersal Mechanism of Red Swamp Crayfish (*Procambarus clarkii*) in the Pearl River Basin and Implications for Its Management. *Sci. Rep.* **2017**, *7*, 8272. [CrossRef] [PubMed]
28. Ministry of Agriculture and Rural Affairs of China. *China Fishery Statistical Yearbook*; Ministry of Agriculture and Rural Affairs of China: Beijing, China, 2023; p. 34. (In Chinese)
29. Li, H.T.; Zhou, W.Z.; Gao, H.L.; Zhang, G. Combined toxicity test of salinity and alkalinity on *Procambarus clarkii*. *Aquaculture* **2006**, *27*, 1. (In Chinese)
30. Liu, D.L.; Huang, J.H.; Yang, L.S.; Tan, L.Q. Effects of multiple environmental factors on molting death of *Procambarus clarkii* and countermeasures. *S. China Fish. Sci.* **2020**, *16*, 29–35. (In Chinese) [CrossRef]
31. Chen, S.; Zhou, Y.; Chen, Y.; Gu, J. Fastp: An Ultra-Fast All-in-One FASTQ Preprocessor. *Bioinformatics* **2018**, *34*, i884–i890. [CrossRef]
32. Grabherr, M.G.; Haas, B.J.; Yassour, M.; Levin, J.Z.; Thompson, D.A.; Amit, I.; Adiconis, X.; Fan, L.; Raychowdhury, R.; Zeng, Q.; et al. Full-Length Transcriptome Assembly from RNA-Seq Data without a Reference Genome. *Nat. Biotechnol.* **2011**, *29*, 644–652. [CrossRef]
33. Mortazavi, A.; Williams, B.A.; McCue, K.; Schaeffer, L.; Wold, B. Mapping and Quantifying Mammalian Transcriptomes by RNA-Seq. *Nat. Methods* **2008**, *5*, 621–628. [CrossRef] [PubMed]
34. Love, M.I.; Huber, W.; Anders, S. Moderated Estimation of Fold Change and Dispersion for RNA-Seq Data with DESeq2. *Genome Biol.* **2014**, *15*, 550. [CrossRef]
35. Robinson, M.D.; McCarthy, D.J.; Smyth, G.K. EdgeR: A Bioconductor Package for Differential Expression Analysis of Digital Gene Expression Data. *Bioinformatics* **2009**, *26*, 139–140. [CrossRef]
36. Livak, K.J.; Schmittgen, T.D. Analysis of Relative Gene Expression Data Using Real-Time Quantitative PCR and the $2^{-\Delta\Delta CT}$ Method. *Methods* **2001**, *25*, 402–408. [CrossRef]
37. Chen, K.; Li, E.; Li, T.; Xu, C.; Wang, X.; Lin, H.; Qin, J.G.; Chen, L. Transcriptome and Molecular Pathway Analysis of the Hepatopancreas in the Pacific White Shrimp *Litopenaeus vannamei* under Chronic Low-Salinity Stress. *PLoS ONE* **2015**, *10*, 0131503. [CrossRef] [PubMed]
38. Bissattini, A.M.; Traversetti, L.; Bellavia, G.; Scalici, M. Tolerance of Increasing Water Salinity in the Red Swamp Crayfish *Procambarus clarkii* (Girard, 1852). *J. Crustac. Biol.* **2015**, *35*, 682–685. [CrossRef]

39. Luo, L.; Huang, J.H.; Liu, D.L.; Jiang, S.G.; Zhou, F.L.; Jiang, S.; Yang, Q.B.; Li, Y.D.; Li, T.; Tan, L.Q.; et al. Transcriptome Reveals the Important Role of Metabolic Imbalances, Immune Disorders and Apoptosis in the Treatment of *Procambarus clarkii* at Super High Temperature. *Comp. Biochem. Physiol.-Part D Genom. Proteom.* **2021**, *37*, 100781. [CrossRef]
40. Ye, L.; Jiang, S.; Zhu, X.; Yang, Q.; Wen, W.; Wu, K. Effects of salinity on growth and energy budget of juvenile *Penaeus monodon*. *Aquaculture* **2009**, *290*, 140–144. [CrossRef]
41. Luo, L.; Huang, J.H.; Liu, D.L.; Jiang, S.G.; Zhou, F.L.; Jiang, S.; Yang, Q.B.; Li, Y.D.; Li, T.; Tan, L.Q.; et al. Comparative Transcriptome Analysis of Differentially Expressed Genes and Pathways in *Procambarus clarkii* (Louisiana Crawfish) at Different Acute Temperature Stress. *Genomics* **2022**, *114*, 110415. [CrossRef]
42. Wang, H.; Tang, L.; Wei, H.; Lu, J.; Mu, C.; Wang, C. Transcriptomic Analysis of Adaptive Mechanisms in Response to Sudden Salinity Drop in the Mud Crab, *Scylla paramamosain*. *BMC Genom.* **2018**, *19*, 421. [CrossRef]
43. Lee, M.H.; Atkinson, S.; Murphy, G. Identification of the Extracellular Matrix (ECM) Binding Motifs of Tissue Inhibitor of Metalloproteinases (TIMP)-3 and Effective Transfer to TIMP-1. *J. Biol. Chem.* **2007**, *282*, 6887–6898. [CrossRef] [PubMed]
44. Freedman, B.R.; Bade, N.D.; Riggin, C.N.; Zhang, S.; Haines, P.G.; Ong, K.L.; Janmey, P.A. The (Dys)Functional Extracellular Matrix. *Biochim. Biophys. Acta-Mol. Cell Res.* **2015**, *1853*, 3153–3164. [CrossRef] [PubMed]
45. Cao, Q.; Wang, H.; Fan, C.; Sun, Y.; Li, J.; Cheng, J.; Chu, P.; Yin, S. Environmental Salinity Influences the Branchial Expression of TCR Pathway Related Genes Based on Transcriptome of a Catadromous Fish. *Comp. Biochem. Physiol.-Part D Genom. Proteom.* **2021**, *38*, 100815. [CrossRef] [PubMed]
46. He, Z.; Shou, C.; Han, Z. Transcriptome Analysis of Marbled Rockfish *Sebastiscus marmoratus* under Salinity Stress. *Animals* **2023**, *13*, 400. [CrossRef] [PubMed]
47. Fang, J.Y.; Richardson, B.C. Biological Effects of the MAPK Pathways. *Lancet Oncol.* **2005**, *6*, 322–327. [CrossRef] [PubMed]
48. Wang, L.; Wang, W.N.; Liu, Y.; Cai, D.X.; Li, J.Z.; Wang, A.L. Two Types of ATPases from the Pacific White Shrimp, *Litopenaeus vannamei* in Response to Environmental Stress. *Mol. Biol. Rep.* **2012**, *39*, 6427–6438. [CrossRef] [PubMed]
49. Gutekunst, J.; Andriantsoa, R.; Falckenhayn, C.; Hanna, K.; Stein, W.; Rasamy, J.; Lyko, F. Clonal Genome Evolution and Rapid Invasive Spread of the Marbled Crayfish. *Nat. Ecol. Evol.* **2018**, *2*, 567–573. [CrossRef] [PubMed]
50. Su, H.; Ma, D.; Fan, J.; Zhong, Z.; Li, Y.; Zhu, H. Metabolism Response Mechanism in the Gill of *Oreochromis Mossambicus* under Salinity, Alkalinity and Saline-Alkalinity Stresses. *Ecotoxicol. Environ. Saf.* **2023**, *251*, 114523. [CrossRef] [PubMed]
51. Fu, M.; Hu, Y.; Lan, T.; Guan, K.L.; Luo, T.; Luo, M. The Hippo Signalling Pathway and Its Implications in Human Health and Diseases. *Signal Transduct. Target. Ther.* **2022**, *7*, 376. [CrossRef]
52. Bray, S.J. Notch Signalling: A Simple Pathway Becomes Complex. *Nat. Rev. Mol. Cell Biol.* **2006**, *7*, 678–689. [CrossRef]
53. Powis, G.; Phil, D. Inhibitors of Phosphatidylinositol Signalling as Antiproliferative Agents. *Cancer Metastasis Rev.* **1994**, *13*, 91–103. [CrossRef] [PubMed]
54. Jeffries, K.M.; Connon, R.E.; Verhille, C.E.; Dabruzzi, T.F.; Britton, M.T.; Durbin-Johnson, B.P.; Fangue, N.A. Divergent Transcriptomic Signatures in Response to Salinity Exposure in Two Populations of an Estuarine Fish. *Evol. Appl.* **2019**, *12*, 1212–1226. [CrossRef] [PubMed]
55. Zhou, Q.L.; Wang, L.Y.; Zhao, X.L.; Yang, Y.S.; Ma, Q.; Chen, G. Effects of Salinity Acclimation on Histological Characteristics and MiRNA Expression Profiles of Scales in Juvenile Rainbow Trout (*Oncorhynchus mykiss*). *BMC Genom.* **2022**, *23*, 300. [CrossRef] [PubMed]
56. Tang, D.; Shi, X.; Guo, H.; Bai, Y.; Shen, C.; Zhang, Y.; Wang, Z. Comparative Transcriptome Analysis of the Gills of *Procambarus clarkii* Provides Novel Insights into the Immune-Related Mechanism of Copper Stress Tolerance. *Fish Shellfish Immunol.* **2020**, *96*, 32–40. [CrossRef] [PubMed]
57. Lu, J.Y.; Simon, M.; Zhao, Y.; Ablaeva, J.; Corson, N.; Choi, Y.; Yamada, K.L.Y.H.; Schork, N.J.; Hood, W.R.; Hill, G.E.; et al. Comparative Transcriptomics Reveals Circadian and Pluripotency Networks as Two Pillars of Longevity Regulation. *Cell Metab.* **2022**, *34*, 836–856.e5. [CrossRef] [PubMed]
58. Liu, L.l.; Cao, Z.h.; He, C.l.; Zhong, Y.c.; Liu, W.y.; Zhang, P.; Yang, F.; Xu, Y.j. Ferric Ion Induction of Triggering Receptor Expressed in Myeloid Cells-2 Expression and PI3K/Akt Signaling Pathway in Preosteoclast Cells to Promote Osteoclast Differentiation. *Orthop. Surg.* **2020**, *12*, 1304–1312. [CrossRef] [PubMed]
59. Prymaczok, N.C.; Pasqualino, V.M.; Viau, V.E.; Rodríguez, E.M.; Medesani, D.A. Involvement of the Crustacean Hyperglycemic Hormone (CHH) in the Physiological Compensation of the Freshwater Crayfish *Cherax quadricarinatus* to Low Temperature and High Salinity Stress. *J. Comp. Physiol. B Biochem. Syst. Environ. Physiol.* **2016**, *186*, 181–191. [CrossRef]
60. Xu, Z.; Gan, L.; Li, T.; Xu, C.; Chen, K.; Wang, X.; Qin, J.G.; Chen, L.; Li, E. Transcriptome Profiling and Molecular Pathway Analysis of Genes in Association with Salinity Adaptation in Nile Tilapia *Oreochromis niloticus*. *PLoS ONE* **2015**, *10*, e0136506. [CrossRef] [PubMed]
61. Evans, D.H.; Piermarini, P.M.; Choe, K.P. The Multifunctional Fish Gill: Dominant Site of Gas Exchange, Osmoregulation, Acid-Base Regulation, and Excretion of Nitrogenous Waste. *Physiol. Rev.* **2005**, *85*, 97–177. [CrossRef]
62. Velotta, J.P.; McCormick, S.D.; O'Neill, R.J.; Schultz, E.T. Relaxed Selection Causes Microevolution of Seawater Osmoregulation and Gene Expression in Landlocked Alewives. *Oecologia* **2014**, *175*, 1081–1092. [CrossRef]
63. Arroyo, J.P.; Kahle, K.T.; Gamba, G. The SLC12 Family of Electroneutral Cation-Coupled Chloride Cotransporters. *Mol. Asp. Med.* **2013**, *34*, 288–298. [CrossRef] [PubMed]

64. Xu, H.; Ghishan, F.K.; Kiela, P.R. SLC9 Gene Family: Function, Expression, and Regulation. *Compr. Physiol.* **2018**, *8*, 555–583. [CrossRef] [PubMed]
65. Khananshvili, D. The SLC8 Gene Family of Sodium-Calcium Exchangers (NCX)-Structure, Function, and Regulation in Health and Disease. *Mol. Asp. Med.* **2013**, *34*, 220–235. [CrossRef] [PubMed]
66. Remigante, A.; Spinelli, S.; Pusch, M.; Sarikas, A.; Morabito, R.; Marino, A.; Dossena, S. Role of SLC4 and SLC26 Solute Carriers during Oxidative Stress. *Acta Physiol.* **2022**, *235*, e13796. [CrossRef] [PubMed]
67. Lam, S.H.; Lui, E.Y.; Li, Z.; Cai, S.; Sung, W.K.; Mathavan, S.; Lam, T.J.; Ip, Y.K. Differential Transcriptomic Analyses Revealed Genes and Signaling Pathways Involved in Iono-Osmoregulation and Cellular Remodeling in the Gills of Euryhaline Mozambique Tilapia, *Oreochromis mossambicus*. *BMC Genom.* **2014**, *15*, 921. [CrossRef] [PubMed]
68. Vallaeys, L.; Van Biervliet, S.; De Bruyn, G.; Loeys, B.; Moring, A.S.; Van Deynse, E.; Cornette, L. Congenital Glucose-Galactose Malabsorption: A Novel Deletion within the SLC5A1 Gene. *Eur. J. Pediatr.* **2013**, *172*, 409–411. [CrossRef] [PubMed]
69. Marshall, W.S. Mechanosensitive Signalling in Fish Gill and Other Ion Transporting Epithelia. *Acta Physiol.* **2011**, *202*, 487–499. [CrossRef] [PubMed]
70. Feeney, G.P.; Zheng, D.; Kille, P.; Hogstrand, C. The Phylogeny of Teleost ZIP and ZnT Zinc Transporters and Their Tissue Specific Expression and Response to Zinc in Zebrafish. *Biochim. Biophys. Acta-Gene Struct. Expr.* **2005**, *1732*, 88–95. [CrossRef]
71. Voss, J.G. Effect of Inorganic Cations on Bactericidal Activity of Anionic. *J. Bacteriol.* **1963**, *86*, 207–211. [CrossRef]

biology

MDPI

Article

Corazonin Stimulates Ecdysteroid Synthesis during the Molting Process of the Swimming Crab, *Portunus trituberculatus*

Xi Xie, Jun Zhang, Shisheng Tu, Qi Zhou and Dongfa Zhu *

Key Laboratory of Aquacultural Biotechnology Ministry of Education, School of Marine Sciences, Ningbo University, Ningbo 315832, China; xiexi@nbu.edu.cn (X.X.)
* Correspondence: zhudongfa@nbu.edu.cn

Simple Summary: Functional studies of neuropeptides are of great significance in revealing the neuroendocrine regulation mechanism in crustaceans, and in guiding and developing artificial regulation techniques in economic species. In this paper, we investigated the effect of a crustacean neuropeptide corazonin (Crz) on ecdysteroid synthesis based on the high expression of Crz receptor in the Y-organ, the main site of ecdysteroid synthesis. By using Crz treatment and CrzR silencing, we found that Crz/CrzR signaling could cause up-regulation of ecdysteroid synthesis genes and an increase in ecdysteroid levels both in vitro and in vivo. In addition, it was also observed that Crz/CrzR signaling affects the expression level of ecdysis-triggering hormones. Our results provide new insights into the understanding of Crz function in crustaceans.

Abstract: The neuropeptide corazonin (Crz) exerts diverse physiological effects in insects, yet its role in crustaceans remains elusive. The abundant expression of Crz receptor (CrzR) in the Y-organs of several crustaceans suggests a potential involvement of Crz in regulating ecdysteroid synthesis. In this study, we examined the effects of PtCrz on ecdysteroid synthesis during the molting period of *Portunus trituberculatus* through PtCrz treatments and PtCrzR silencing. Our results showed that PtCrz peptide stimulates ecdysteroid levels and the gene expression involved in ecdysteroidogenesis both in vitro and in vivo, whereas dsPtCrzR treatments had opposite effects on ecdysteroid levels and associated gene expression. Thus, our study suggests that PtCrz may modulate ecdysteroid synthesis via Y-organ-expressed PtCrzR. Furthermore, we also discovered the involvement of PtCrz/PtCrzR signaling in regulating *PtETH* expression. Notably, the inhibitory effect of dsPtCrzR on ecdysteroid synthesis or *PtETH* expression can be reversed by PtCrz treatment, suggesting the potential existence of multiple receptors for PtCrz. This study provides new insights into the function of crustacean Crz and, for the first time, elucidates the presence of a neuropeptide that can stimulate ecdysteroid synthesis in crustaceans.

Keywords: corazonin; corazonin receptor; *Portunus trituberculatus*; ecdysteroid synthesis; molting

1. Introduction

Corazonin (Crz) is a peptidergic neurohormone widely distributed among arthropods. Initially discovered as a cardioacceleratory peptide in the cockroach *Periplaneta americana* [1], its function as such is not universally observed across all species [2,3]. Crz may also play a role in nutritional stress signaling [4], behavior modulation [5–7], and reproductive physiology [6,8,9] among other functions, which may vary among species. All Crz prepropeptides contain a predicted signal peptide, a conserved mature peptide of 11 amino acids, and an additional sequence known as Crz-precursor-related peptide (CrzRP) [10,11]. Among the six isoforms of mature Crzs, [Arg7]-corazonin is the most common form found in insects [12] and the sole form present in crustaceans [13].

The diverse physiological effects of Crz are primarily mediated by its membrane receptor (CrzR), a member of the G-protein coupled receptor (GPCR) family. In *Drosophila melanogaster*, gene silencing of CrzR often leads to altered physiological states, including changes in stress responses, feeding behavior, and energy metabolism; all of these phenotypes are associated with Crz action [14,15]. Structurally homologous to the mammalian gonadotropin-releasing hormone (GnRH) receptor [16], CrzR has been identified and characterized in various insects [5,17–21]. Upon activation, CrzR releases secondary messengers such as cAMP and Ca^{2+}, thereby regulating cellular physiological functions [19]. The tissue distribution of CrzR also varies among different species, which may reflect the variable functions of Crz. Notably, in the tobacco hornworm *Manduca sexta*, the high expression of CrzR in Inka cells is crucial for initiating ecdysis as its activation by Crz leads to the release of ecdysis-triggering hormone (ETH) [5].

In recent years, advances in transcriptomics have facilitated the identification of Crz and CrzR sequences across various crustacean species [3,22–25]. However, a comprehensive understanding of the Crz/CrzR signaling functions within crustaceans remains incomplete. Crz had no effect on heart activity, blood glucose levels, lipid mobilization, or pigment distribution in chromatophores in the green shore crab *Carcinus maenas* [3]. By contrast, it influences pigment migration in the crayfish *Procambarus clarkia* [26], and some molt behavior in the redclaw crayfish *Cherax quadricarinatus* [27]. In a previous investigation, we discovered that the Crz/CrzR signaling system may also be involved in the vitellogenesis process of *Portunus trituberculatus*, thereby uncovering a novel function for this pathway [10]. It is noteworthy that that study not only revealed that *P. trituberculatus* Crz is predominantly derived from neural tissue, but also discovered a high expression of *CrzR* in the Y-organ, a primary site for crustacean ecdysteroid biosynthesis [10]. Extremely high *CrzR* levels were also observed in *C. maenas* Y-organ; however, it is paradoxical that while application of the Crz peptide had no significant impact on ecdysteroid biosynthesis in *C. maena* (with exception of slight stimulation during early postmolt) [3], it did induce an increase in expression of several ecdysteroidogenesis genes in our cultured Y-organ explant [10].

Although it is common for the Crz/CrzR action to vary across species, our aforementioned studies were specifically conducted during the period of ovarian development. In this current study, we utilized both in vitro and in vivo Crz treatments, as well as CrzR RNA interference (RNAi), to investigate whether Crz/CrzR signaling stimulates ecdysteroid synthesis during the molting period. Furthermore, given the findings in *Manduca sexta* and *Cherax quadricarinatus* indicating potential involvement of Crz/CrzR signaling in ETH action [5,27], we further explored their potential regulatory role on *ETH* expression.

2. Materials and Methods

2.1. Experimental Animals

Juvenile male crabs (40–70 g of bodyweight) were purchased from Xianxiang, Ningbo, in August 2021 and reared at the aquaculture base of the Institute of Marine and Fisheries, Ningbo, China. The crabs were kept in large tanks with continuous seawater flow and fed live razor clams once a day. Different stages of the molt cycle, including postmolt (stages A and B), intermolt (stage C), and premolt (stage D_0, D_1, D_2, D_3, and D_4), were determined by observing the hardness of the carapace and the formation of new epidermis as described previously [28]. All crabs were anaesthetized on ice for 10 min prior to sacrifice, and tissues including the Y-organ, eyestalk, brain, and thoracic ganglion were dissected and stored in RNA preservation fluid (Cwbiotech, Taizhou, China) at −80 °C until RNA extraction.

2.2. In Vitro Treatments

The mature PtCrz peptide (pQTFQYSRGWTN-NH2) was synthesized by Sangon Biotech (Shanghai, China), with a purity of 98%. The cDNA fragments of the PtCrzR transmembrane region (565 bp) and the green fluorescent protein (GFP, 568 bp) were cloned into the pMD19-T vector (Takara, Kyoto, Japan), respectively, and the corresponding dsRNAs were prepared as previously described [10]. Since the hemolymph ecdysteroid titer generally begins to rise during the premolt stages during a natural molt, Y-organs from premolt D_0 crabs were utilized for in vitro experiments. For PtCrz treatments, Y-organ explants were exposed to synthetic PtCrz peptide at final concentrations of 10^{-5} M, 10^{-6} M, and 10^{-7} M after pre-incubation for 1 h. The control group received an equivalent volume of crab saline. In the case of dsPtCrzR treatments, Y-organ explants from the same stage were treated with dsPtCrzR, dsGFP, and dsCrzR supplemented with PtCrz at 10^{-6} M (the dosage was chosen according to the results of the above PtCrz treatment experiments). The dosage of dsRNA was set at 5 μg per plate. Following an additional incubation period of 8 h, all tissue explants and corresponding culture mediums were collected for further analysis of gene expression and ecdysteroid levels.

2.3. In Vivo Treatments

Crabs in the premolt stage at D_0 and D_2 were selected for PtCrz in vivo injection. Considering that hemolymph volumes are estimated to be approximately 27% of body mass [29], the dosage of PtCrz injection (3–4.2 × 10^{-4} M in 50 μL) was calculated to maintain a hemolymph concentration of added Crz at 10^{-6} M. The control group received an equivalent volume of crab saline. Crabs at premolt D_0 were utilized for in vivo RNA interference, with groups corresponding to those used in the in vitro experiments, including dsGFP, dsPtCrzR, and dsPtCrzR + Crz groups. The dosage of dsRNA was standardized at 3 μg/g bodyweight. All injections were administered at the base of the last walking foot using a 1 mL syringe. Hemolymph and tissue samples were collected 24 h post-injection for further analysis of gene expression and ecdysteroid levels.

2.4. Determination of Ecdysteroid Levels

The ecdysteroid levels in culture medium and hemolymph samples were quantified using the 20-Hydroxyecdysone Enzyme Immunoassay Kit (ARBOR ASSAYS, Ann Arbor, MI, USA). Briefly, 0.5 mL of culture medium and hemolymph was extracted, added to 1.5 mL of methanol, vortexed thoroughly, and then centrifuged at 4 °C for 15 min at 12,000× *g*. The resulting supernatant was transferred to a new sterile centrifuge tube, dried using a nitrogen blower, and stored at −80 °C. Prior to use, the dried powder was reconstituted in diluted Assay Buffer (1:5 with sterile water) and analyzed according to the provided kit protocols.

2.5. Gene Expression Analysis

The relative gene expression levels were determined using quantitative reverse-transcription PCR (qRT-PCR) with specific primers (Table 1). The qRT-PCR was performed on the ABI 7500 qPCR instrument (Thermofisher, Waltham, MA, USA) following the manufacturer's instructions of the SYBR® Premix Ex Taq™ II Kit (Takara). PCR conditions included an initial denaturation at 95 °C for 2 min, followed by 40 cycles of amplification at 95 °C for 15 s and extension at 56 °C for 20 s. Additional melting curve analysis was conducted to confirm product specificity, with temperature increasing from 55 to 95 °C at a rate of 0.2 °C/s. Amplification efficiencies were assessed using standard curve analysis based on a five-point, tenfold dilution series of cDNA. Each sample was analyzed in triplicate for technical replicates. Normalization of target gene expression was achieved using β-actin, and relative mRNA expression levels were calculated utilizing the comparative Ct ($2^{-\Delta\Delta Ct}$) method [30].

Table 1. The qRT-PCR primers used in this study.

Gene	Primer (5′-3′)	GenBank Accession No.
PtCrz	Forward: CAGTTGTGGTGCTCGTTGCC Reverse: GCTCAGCGGACCTTTTTCG	OL694705
PtCrzR	Forward: GTCATCTGCTGGACTCCCTACTAC Reverse: CGGGTTGACGAGACTGTTGG	OL694706
PtSpo	Forward: GTTTTGGCTCCCGCAACTA Reverse: TGTCGTCGGTGAGGCTTGT	KM030021

Table 1. *Cont.*

Gene	Primer (5′-3′)	GenBank Accession No.
PtSad	Forward: CAGATATGGGCAGATTCATCG Reverse: AAGGCGTCATCCAGGCAC	KM596851
PtDib	Forward: GGCAAACACTGGTGGGAACT Reverse: ACCCTTCACGCCTCATCTTG	KM880023
PtETH	Forward: ATGCTCTCTGTTCTGGACTCAAG Reverse: TCACTTCTGCAGGTAACGCA	MT890695
β-actin	Forward: CGAAACCTTCAACACTCCCG Reverse: CGGGTTGACGAGACTGTTGG	FI641977

2.6. Statistical Analysis

The data were subjected to normality testing using the Kolmogorov–Smirnov and Cochran tests prior to all statistical analyses. Non-normally distributed data were assessed using the nonparametric Mann–Whitney test, while normally distributed data were analyzed using one-way ANOVA followed by Tukey's test or the student's *t*-test (SPSS 24.0 software, IBM, Armonk, NY, USA). Statistical significance was considered at $p < 0.05$ and denoted with letters or asterisks in all cases.

3. Results

3.1. Expression Profiles of *PtCrz* and *PtCrzR* during the Molt Cycle

Based on previous reports of tissue distribution [10], we conducted an analysis of the variations in *PtCrz* expression levels in the eyestalk, brain, and thoracic ganglion, as well as *PtCrzR* expression in the Y-organ and eyestalk throughout the molt cycle. While the expression patterns of *PtCrz* in these three nerve tissues were not identical, they all exhibited a gradual increase from postmolt (stage A), reaching a peak at premolt D_2 substage, followed by a rapid decline to a minimum at premolt D_4 (Figure 1A). A similar trend was observed for *PtCrzR* expression in the Y-organ; however, the peak of *PtCrzR* expression in the eyestalk occurred at premolt D_1 (Figure 1B).

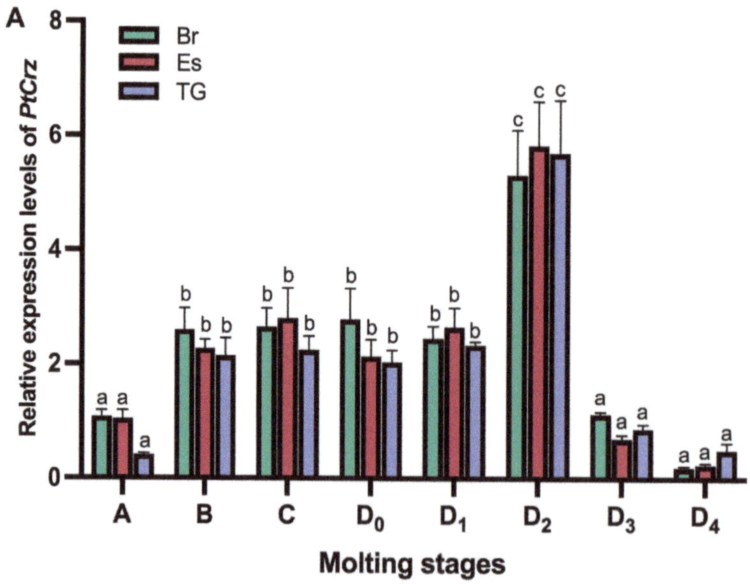

Figure 1. *Cont.*

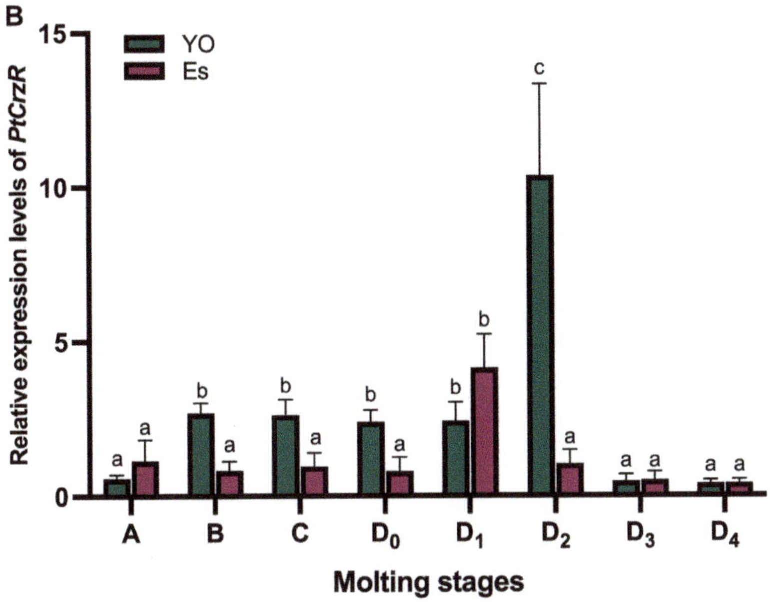

Figure 1. Expression profiles of PtCrz (**A**) and PtCrzR (**B**) during the molt cycle of *P. trituberculatus*. The relative mRNA levels of PtCrz were examined in brain (Br), eyestalk (Es), and thoracic ganglion (TG), while PtCrzR were examined in Y-organ (YO) and eyestalk (Es). The molting stages includes postmolt (stage A and B), intermolt (stage C), and premolt (stage D_0, D_1, D_2, D_3, and D_4). Bars represent mean + SEM (n = 5). Different letters indicate statistically significant differences within each tissue ($p < 0.05$).

3.2. *Effects of PtCrz Peptide and dsPtCrzR on Ecdysteroid Synthesis In Vitro*

We investigated the effects of varying concentrations (10^{-7} M, 10^{-6} M, and 10^{-5} M) of PtCrz peptide on ecdysteroid synthesis in the Y-organ cultured in vitro. Our findings revealed that exposure to 10^{-6} M and 10^{-5} M PtCrz resulted in significantly elevated levels of ecdysteroids (124.2 ± 2.5 pg/mL and 125.2 ± 2.0 pg/mL, respectively) in the culture medium compared to the control group (97.1 + 2.5 pg/mL) (Figure 2A). By contrast, the ecdysteroids levels in 10^{-7} M PtCrz treatment (102.8 + 1.4 pg/mL) showed no significant difference with the control group. Furthermore, these treatments effectively stimulated the expression of *PtCrzR* and induced the expression of two ecdysteroid synthesis pathway genes (*PtSpo* and *PtSad*), while showing no discernible effect on another ecdysteroidogenesis gene, *PtDib* (Figure 2B). On the other hand, treatment with dsPtCrzR led to a marked decrease in *PtCrzR* within the Y-organ, resulting in opposing effects on ecdysteroid levels (98.0 ± 2.3 pg/mL for dsGFP and 67.3 ± 1.2 pg/mL for dsCrzR) and *PtSpo* and *PtSad* expression compared to PtCrz treatments (Figure 2C,D). However, it was observed that such effects could be recovered by administration of PtCrz peptide.

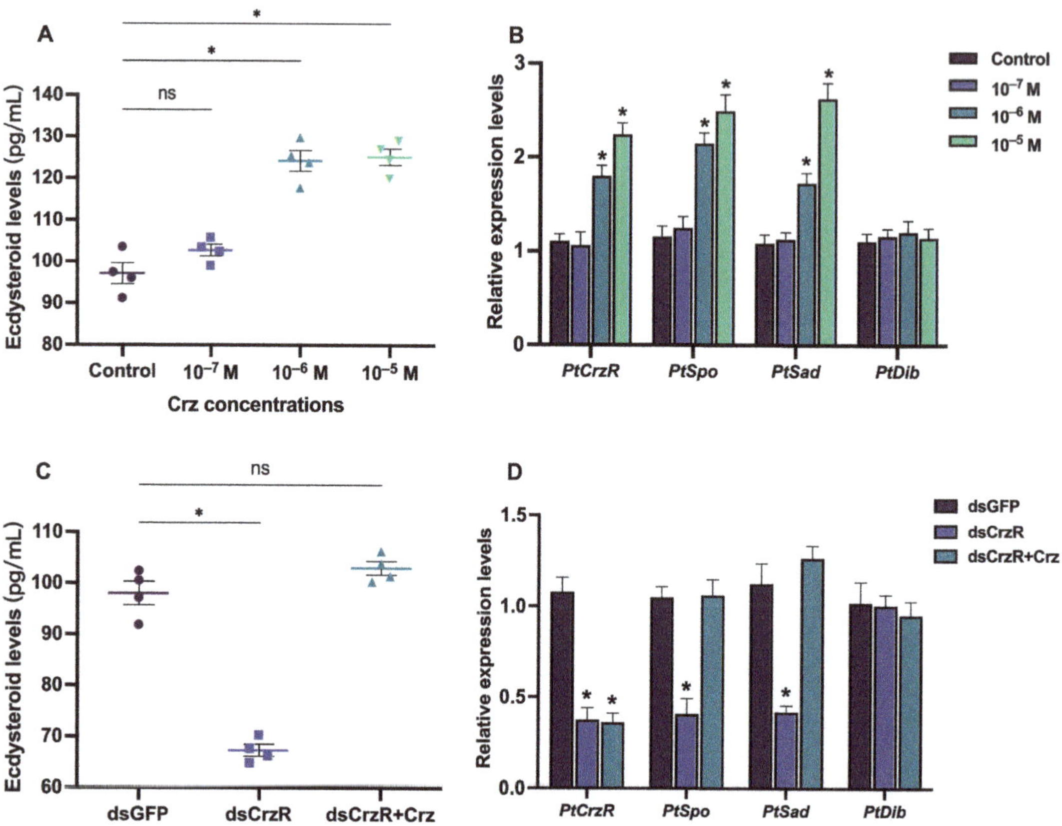

Figure 2. Effects of PtCrz peptide and dsPtCrzR on ecdysteroid synthesis in vitro. The ecdysteroid levels in Y-organ culture medium (**A**) and the *PtCrzR*, *PtSpo*, *PtSad*, and *PtDib* expression levels (**B**) when treated with different concentrations of Crz peptides. The ecdysteroid levels in Y-organ culture medium (**C**) and the *PtCrzR*, *PtSpo*, *PtSad*, and *PtDib* expression levels (**D**) when treated with different dsGFP, dsCrzR, and dsCrzR supplemented with PtCrz at 10^{-6} M. Bars represent mean ± SEM (n = 4). The asterisk indicates a statistically significant difference when compared with control or dsGFP group ($p < 0.05$), while "ns" means no significant difference.

3.3. Effects of PtCrz Peptide and dsPtCrzR on Ecdysteroid Synthesis In Vivo

Similar to the findings observed in in vitro experiments, PtCrz peptide was found to stimulate hemolymph ecdysteroid levels (Figure 3A). The hemolymph ecdysteroid titer increased from 827.5 ± 120.6 pg/mL of the crab saline group to 1179.0 ± 141.5 pg/mL of the Crz injection group in the premolt D_0 stage, and from 1198.8 ± 92.3 pg/mL to 2855.7 ± 409.9 pg/mL in the premolt D_2 stage. In addition, the expression of *PtCrzR*, *PtSpo*, and *PtSad* during both premolt D_0 and premolt D_2 stages were also upregulated after Crz treatment (Figure 3B). Notably, the upregulation of PtCrzR was more pronounced during premolt D_2, exhibiting an approximately 6-fold increase compared to the crab saline group, while only showing an approximately 2-fold increase during premolt D_0. Treatment with dsPtCrzR in premolt D_0 resulted in contrasting effects on ecdysteroid levels (764.4 ± 93.0 pg/mL for dsGFP and 457.9 ± 88.4 pg/mL for dsCrzR) and associated gene expression, which could be rescued by administration of PtCrz (Figure 4).

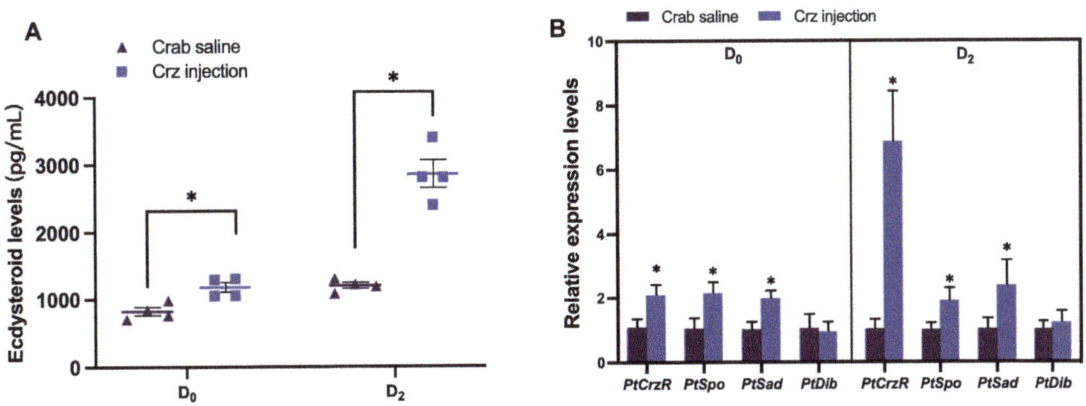

Figure 3. Effects of PtCrz peptide injection on ecdysteroid synthesis in vivo. (**A**) Induced hemolymph ecdysteroid levels by injection with PtCrz peptide in premolt D_0 and D_2, respectively. (**B**) The change in expression levels of *PtCrzR*, *PtSpo*, *PtSad*, and *PtDib* in Y-organ after PtCrz peptide treatments in two stages. Bars represent mean ± SEM (n = 4). The asterisk indicates a statistically significant difference when compared with crab saline group ($p < 0.05$).

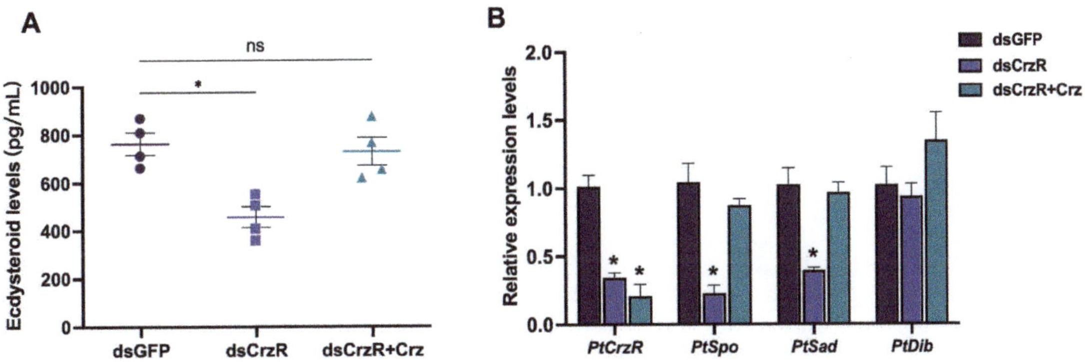

Figure 4. Effects of dsPtCrzR on ecdysteroid synthesis in vivo. The hemolymph ecdysteroid levels (**A**) and the *PtCrzR*, *PtSpo*, *PtSad*, and *PtDib* expression levels (**B**) when treated with different dsGFP, dsCrzR, and dsCrzR supplemented with PtCrz at 10^{-6} M. Bars represent mean ± SEM (n = 4). The asterisk indicates statistically significant different when compared with control group ($p < 0.05$), while "ns" means no significant difference.

3.4. *Effects of PtCrz/PtCrzR Signaling on PtETH Expression*

The expression of *PtETH* in nerve tissues, including the brain, eyestalk, and thoracic ganglion, was detected in the aforementioned in vivo experiments. It was observed that *PtETH* expression in these three tissues was upregulated by PtCrz peptide (Figure 5) and downregulated by dsPtCrzR (Figure 6). Furthermore, the inhibitory effects of dsPtCrzR could be rescued by PtCrz injection.

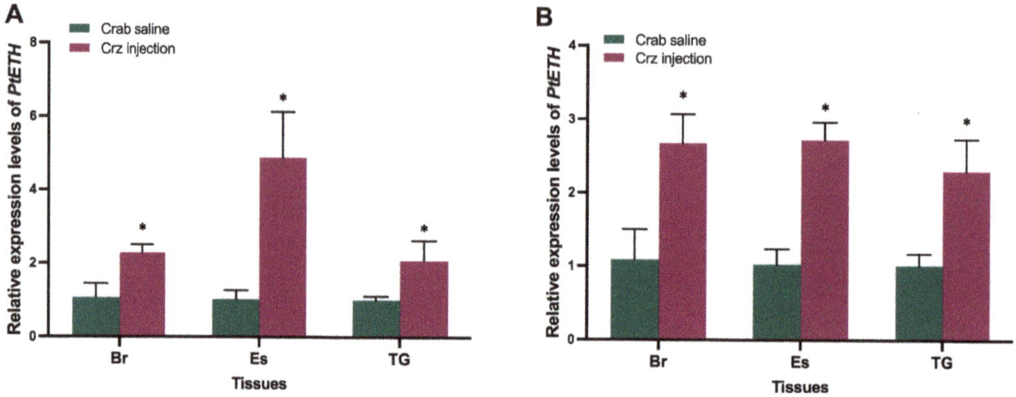

Figure 5. Effects of PtCrz peptide on *PtETH* expression. The *PtETH* expression was examined in brain (Br), eyestalk (Es), and thoracic ganglion (TG) in treatments including PtCrz injection in premolt D_0 (**A**) and premolt D_2 (**B**). Bars represent mean ± SEM (n = 4). The asterisk indicates a statistically significant difference when compared with crab saline group ($p < 0.05$).

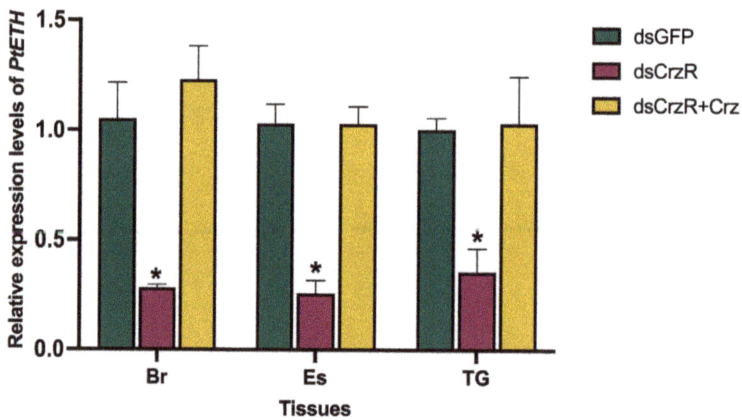

Figure 6. Effects of dsPtCrzR injection on *PtETH* expression. The *PtETH* expression was examined in brain (Br), eyestalk (Es), and thoracic ganglion (TG). Bars represent mean ± SEM (n = 4). The asterisk indicates a statistically significant difference when compared with dsGFP group ($p < 0.05$).

4. Discussion

In this study, we investigated the effect of PtCrz on ecdysteroid synthesis during the molting period of *P. trituberculatus* through neuropeptide treatments and RNA silencing of its receptor. Combined with our previous research [10], we suggest that PtCrz plays a regulatory role in ecdysteroidogenesis in *P. trituberculatus*, both during ovarian development and molting process. While our results indicate the involvement of PtCrzR in mediating the effects of PtCrz, it is plausible to consider the potential involvement of other Crz receptors, as evidenced by the recovery of dsPtCrzR effects following PtCrz treatment. Notably, there are two types of CrzRs expressed in the Y-organ of the blackback crab *Gecarcinus lateralis*, belonging to GPCR A6 and A7, respectively [24]. Further phylogenetic analysis showed that PtCrzR is more closely related to GPCR A6 type (Supplementary Material), while the A7 type of CrzR was not found in our transcriptomic data. Certainly, given that the dsRNA utilized in this investigation is designed for gene knockdown rather than gene knockout, it is also plausible that the impact of Crz may not be fully attenuated. Consequently, further research is warranted to validate the aforementioned hypothesis.

Our findings are in contrast with the observed effects of Crz in the green shore crab, *C. maenas*, where the Crz peptide only weakly stimulated ecdysteroid synthesis during early postmolt [3]. As previously mentioned, the function of the Crz peptide may vary between species. However, this inconsistency is intriguing considering that CrzR was highly expressed in the Y-organ of both species and that both PtCrzR and CmCrzR clustered together in phylogenetic analysis (Supplementary Material). It should be noted that the concentration of Crz administered in our study was substantially higher (10^{-6} M in this study vs. 5×10^{-8} M in *C. maenas*), but both levels exceeded their respective EC_{50} values for CrzR activation in each species. Thus, the discrepancy cannot be readily attributed to the amount of Crz applied.

Instead, the variation in responses may be attributed to intrinsic functional disparities within the receptors themselves. For instance, PtCrzR is clustered together with GlCrzR-A6, but their expression patterns differ throughout the molt cycle: *PtCrzR* is significantly upregulated in premolt D_2, while *GlCrzRA6* shows high expression only in postmolt stages [24], indicating potential divergent functional roles for the same type of CrzR across species. In fact, although not examined, the expression of *CmCrzR* is high, at least in the intermolt, the period from which its tissue distribution was analyzed [3]. This is very different from the expression pattern of *GlCrzRA6* and possibly also *PtCrzR*. On the other hand, the expressions of *PtCrz* and *PtCrzR* during the molt cycle exhibit a strong correlation with ecdysteroidogenesis, as hormone levels peak in premolt D_3 [31] following the peak expressions of *PtCrz* and *PtCrzR*.

To the best of our knowledge, this study represents the first documentation of a neuropeptide's involvement in promoting ecdysteroid synthesis in crustaceans. PtCrz may work in conjunction with classical molt-inhibiting hormone (MIH), which belongs to the crustacean hyperglycemic hormone (CHH) family of neuropeptides [32]. It is generally accepted that MIH signaling suppresses ecdysteroidogenesis through the cyclic nucleotide second messengers cAMP and cGMP [33]. Therefore, PtCrz/PtCrzR may potentially utilize alternative GPCR mediated pathways, such as the phospholipase C (PLC)/PKC pathway, which has been suggested to promote ecdysteroid synthesis and secretion by the Y-organ [11]. The decrease in hemolymph levels of MIH during late intermolt triggers ecdysteroid synthesis in the Y-organ, while high levels of MIH reappear during mid-premolt, where it is no longer sensitive to the Y-organ [33]. In contrast, the Y-organ appears to be more sensitive to PtCrz during premolt D_2, as evidenced by PtCrz inducing greater fold changes in PtCrzR expression and ecdysteroid levels compared to premolt D_0. However, given the expression of multiple neuropeptide receptors in the crustacean Y-organ [24,34], it is still open for discussion whether Crz serves as the primary regulator in stimulating ecdysteroidogenesis. It is worth noting that ecdysteroid synthesis in Drosophila is governed by a complex network of multiple neuropeptides [35], and a similar scenario can be anticipated in crustaceans.

In order to stimulate ecdysteroid levels, the Crz signaling system may target the genes that encode enzymes involved in the ecdysteroid biosynthetic pathway. This is supported by the altered expression of two such genes, *PtSpo* and *PtSad*, in both Crz treatment and CrzR RNAi. The classical ecdysteroid biosynthesis pathway involves seven enzymes, five of which are cytochrome P450 monooxygenases (CYPs). Their coding genes are collectively known as the Halloween genes, including neverland (nvd), non-molting glossy/shroud (sro), spook (spo, CYP307A1), phantom (phm, CYP306A1), disembodied (dib, CYP302A1), shadow (sad: CYP315A1), and shade (shd: CYP314A1) [36]. *PtSpo*, *PtDib*, and *PtSad* were selected for expression analysis in this study because their dsRNA can downregulate hemolymph ecdysteroid levels [37]. In contrast to *PtSpo* and *PtSad*, *PtDib* expression was not sensitive to Crz treatment or CrzR RNAi, suggesting that Crz/CrzR signaling only targets certain Halloween genes. Interestingly, eyestalk ablation, a common method to induce hemolymph ecdysteroid levels, induces the expression of *PtDib* and *PtSad* but not

PtSpo [37]. This again suggests a complex neuropeptide system in the control of crustacean ecdysteroid synthesis.

In the tobacco hornworm *Manduca sexta*, Crz was able to stimulate early secretion of the ecdysis-triggering hormone (ETH), which in turn initiated the onset of the ecdysis behavioral sequence [5]. Crz may play a similar role in crustaceans, as demonstrated in the redclaw crayfish *Cherax quadricarinatus*, where Crz injection induced typical molting responses, albeit in a slower and more asynchronous manner than that observed with ETH injection. Our results showed that Crz treatments at two premolt stages significantly promoted *PtETH* expression in three neural tissues, suggesting a potential function in controlling ETH action. However, neither PtCrz nor PtETH induced any putative molting responses in vivo (data is being prepared and scheduled for future published). Therefore, the physiological significance of this result remains to be investigated. It has recently been proposed that ETH may possess functions beyond behavioral regulation, as evidenced by its RNA silencing in the mud crab *Scylla paramamosain* leading to widespread mortality and perturbations in numerous metabolic pathways-related genes [38,39].

5. Conclusions

In conclusion, the current study demonstrated that PtCrz plays a stimulatory role in ecdysteroid synthesis and may be involved in the regulation of PtETH expression. CrzR is involved in both actions of Crz, but may not be the only Crz receptor. Our results provide new insights into the understanding of Crz function in crustaceans.

Supplementary Materials: The following supporting information can be downloaded at: https://www.mdpi.com/article/10.3390/biology13080630/s1, Figure S1: Phylogenetic analysis of CrzRs with RPCHR and ACPR. The following sequences were used: *Portunus trituberculatus* CrzR (PtCrzR, OL694706), *Rhodnius prolixus* CrzR1 and CrzR2 (RpCrzR1 and RpCrzR2, AND99324 and AND99325), *Carcinus maenas* CrzR and RPCHR (CmCrzR and CmRPCHR, AVA26879 and MF974387), *Daphnia pulex* RPCHR (DpRPCHR, ACD75498), *Ixodes scapularis* APCR1-4 (IsACPR1-4, AHB51763, AHB51764, AHB51765, AHB51766). *Tribolium castaneum* ACPR (TcACPR, NP001280549). *Nasonia vitripennis* ACPR and ACPRX1-X4 (NvACPR and NvACPRX1-4, NP001164571, XP031779949, XP031779950, XP031779951, XP031779952).

Author Contributions: Conceptualization, X.X.; Data curation, S.T.; Formal analysis, J.Z.; Funding acquisition, D.Z.; Investigation, J.Z.; Methodology, S.T.; Software, Q.Z.; Supervision, D.Z.; Validation, J.Z. and Q.Z.; Writing—original draft, X.X.; Writing—review & editing, X.X. All authors have read and agreed to the published version of the manuscript.

Funding: This research was supported by the National natural Science Foundation of China (Nos. 32373124 and U23A20248), and the K. C. Wong Magna Fund in Ningbo University.

Institutional Review Board Statement: In China, ethical approval is not required for experiments on crabs. All the experiments comply with the requirements of the governing regulation for the use of experimental animals in Zhejiang Province (Zhejiang provincial government order No. 263, released on 17 August 2009, effective from 1 October 2010) and the Animal Care and Use Committee of Ningbo University.

Informed Consent Statement: Not applicable.

Data Availability Statement: The data that support the findings of this study are available from the corresponding author upon reasonable request.

Conflicts of Interest: The authors declare no conflicts of interest.

References

1. Veenstra, J.A. Isolation and structure of Corazonin, a cardioactive peptide from the American cockroach. *FEBS Lett.* **1989**, *250*, 231–234. [CrossRef]
2. Hillyer, J.F.; Estévez-Lao, T.Y.; Funkhouser, L.J.; Aluoch, V.A. *Anopheles gambiae* Corazonin: Gene structure, expression and effect on mosquito heart physiology. *Insect. Mol. Biol.* **2012**, *21*, 343–355. [CrossRef]

3. Alexander, J.L.; Oliphant, A.; Wilcockson, D.C.; Audsley, N.; Down, R.E.; Lafont, R.; Webster, S.G. Functional characterization and signaling systems of Corazonin and red pigment concentrating hormone in the green shore crab, *Carcinus maenas*. *Front. Neurosci.* **2018**, *11*, 752. [CrossRef] [PubMed]
4. Veenstra, J.A. Does Corazonin signal nutritional stress in insects? *Insect. Biochem. Mol. Biol.* **2009**, *39*, 755–762. [CrossRef] [PubMed]
5. Kim, Y.J.; Spalovská-Valachová, I.; Cho, K.H.; Zitnanova, I.; Park, Y.; Adams, M.E.; Žitňan, D. Corazonin receptor signaling in ecdysis initiation. *Proc. Natl. Acad. Sci. USA* **2004**, *101*, 6704–6709. [CrossRef]
6. Gospocic, J.; Shields, E.J.; Glastad, K.M.; Lin, Y.; Penick, C.A.; Yan, H.; Mikheyev, A.S.; Linksvayer, T.A.; Garcia, B.A.; Berger, S.L.; et al. The Neuropeptide Corazonin controls social behavior and caste identity in ants. *Cell* **2017**, *170*, 748–759. [CrossRef]
7. Varga, K.; Nagy, P.; Arsikin Csordás, K.; Kovács, A.L.; Hegedűs, K.; Juhász, G. Loss of Atg16 delays the alcohol-induced sedation response via regulation of Corazonin neuropeptide production in Drosophila. *Sci. Rep.* **2016**, *6*, 34641. [CrossRef] [PubMed]
8. Ben-Menahem, D. GnRH-Related Neurohormones in the fruit fly *Drosophila melanogaster*. *Int. J. Mol. Sci.* **2021**, *22*, 5035. [CrossRef]
9. Cheng, J.; Zhao, P.; Zhu, L.; Zhu, F.; Tian, Z.; Shen, Z.; Liu, X.; Liu, X. Corazonin signaling modulates the synthetic activity of male accessory gland in *Grapholita molesta*. *Int. J. Biol. Macromol.* **2022**, *216*, 446–455. [CrossRef]
10. Tu, S.; Ge, F.; Han, Y.; Wang, M.; Xie, X.; Zhu, D. Putative role of corazonin in the ovarian development of the swimming crab *Portunus trituberculatus*. *Front. Mar. Sci.* **2022**, *9*, 976754. [CrossRef]
11. Mykles, D.L. Signaling pathways that regulate the crustacean molting gland. *Front. Endocrinol.* **2021**, *12*, 674711. [CrossRef] [PubMed]
12. Predel, R.; Neupert, S.; Russell, W.K.; Scheibner, O.; Nachman, R.J. Corazonin in insects. *Peptides* **2007**, *28*, 3–10. [CrossRef] [PubMed]
13. Veenstra, J.A. Similarities between decapod and insect neuropeptidomes. *PeerJ* **2016**, *4*, e2043. [CrossRef]
14. Kubrak, O.I.; Lushchak, O.V.; Zandawala, M.; Nässel, D.R. Systemic Corazonin signalling modulates stress responses and metabolism in *Drosophila*. *Open Biol.* **2016**, *6*, 160152. [CrossRef]
15. Sha, K.; Choi, S.-H.; Im, J.; Lee, G.G.; Loeffler, F.; Park, J.H. Regulation of ethanol-related behavior and ethanol metabolism by the Corazonin neurons and Corazonin receptor in *Drosophila melanogaster*. *PLoS ONE* **2014**, *9*, e87062. [CrossRef]
16. Buckley, S.J.; Nguyen, T.V.; Cummins, S.F.; Elizur, A.; Fitzgibbon, Q.P.; Smith, G.S.; Mykles, D.L.; Ventura, T. Evaluating conserved domains and motifs of decapod gonadotropin-releasing hormone G protein-coupled receptor superfamily. *Front. Endocrinol.* **2024**, *15*, 1348465. [CrossRef] [PubMed]
17. Cazzamali, G.; Saxild, N.P.E.; Grimmelikhuijzen, C.J.P. Molecular cloning and functional expression of a Drosophila Corazonin receptor. *Biochem. Biophys. Res. Commun.* **2002**, *298*, 31–36. [CrossRef]
18. Belmont, M.; Cazzamali, G.; Williamson, M.; Hauser, F.; Grimmelikhuijzen, C.J.P. Identification of four evolutionarily related g protein-coupled receptors from the malaria mosquito *Anopheles gambiae*. *Biochem. Biophys. Res. Commun.* **2006**, *344*, 160–165. [CrossRef]
19. Yang, J.; Huang, H.; Yang, H.; He, X.; Jiang, X.; Shi, Y.; Alatangaole, D.; Shi, L.; Zhou, N. Specific activation of the g protein-coupled receptor bngr-a21 by the neuropeptide Corazonin from the silkworm, *Bombyx mori*, dually couples to the Gq and Gs signaling cascades. *J. Biol. Chem.* **2013**, *288*, 11662–11675. [CrossRef] [PubMed]
20. Oryan, A.; Wahedi, A.; Paluzzi, J.P.V. Functional characterization and quantitative expression analysis of two GnRH-related peptide receptors in the mosquito, *Aedes aegypti*. *Biochem. Biophys. Res. Commun.* **2018**, *497*, 550–557. [CrossRef]
21. Sha, K.; Conner, W.C.; Choi, D.Y.; Park, J.H. Characterization, expression, and evolutionary aspects of Corazonin neuropeptide and its receptor from the house fly, *Musca domestica* (Diptera: Muscidae). *Gene* **2012**, *497*, 191–199. [CrossRef] [PubMed]
22. Buckley, S.J.; Fitzgibbon, Q.P.; Smith, G.G.; Ventura, T. In silico prediction of the G-Protein coupled receptors expressed during the metamorphic molt of *Sagmariasus Verreauxi* (Crustacea: Decapoda) by mining transcriptomic data: RNA-Seq to repertoire. *Gen. Comp. Endocrinol.* **2016**, *228*, 111–127. [CrossRef] [PubMed]
23. Tu, S.; Xu, R.; Wang, M.; Xie, X.; Bao, C.; Zhu, D. Identification and characterization of expression profiles of neuropeptides and their GPCRs in the swimming crab, *Portunus trituberculatus*. *PeerJ* **2021**, *9*, 12179. [CrossRef] [PubMed]
24. Tran, N.M.; Mykles, D.L.; Elizur, A.; Ventura, T. Characterization of G-Protein coupled receptors from the blackback land crab *Gecarcinus lateralis* Y organ transcriptome over the molt cycle. *BMC Genom.* **2019**, *20*, 74. [CrossRef] [PubMed]
25. Nguyen, T.V.; Rotllant, G.E.; Cummins, S.F.; Elizur, A.; Ventura, T. Insights into sexual maturation and reproduction in the norway lobster (*Nephrops norvegicus*) via in silico prediction and characterization of neuropeptides and G protein-coupled receptors. *Front. Endocrinol.* **2018**, *9*, 430. [CrossRef] [PubMed]
26. Porras, M.G.; De Loof, A.; Breuer, M.; Aréchiga, H. Corazonin promotes tegumentary pigment migration in the crayfish *Procambarus clarkii*. *Peptides* **2003**, *24*, 1581–1589. [CrossRef] [PubMed]
27. Minh Nhut, T.; Mykles, D.L.; Elizur, A.; Ventura, T. Ecdysis triggering hormone modulates molt behaviour in the redclaw crayfish *Cherax quadricarinatus*, providing a mechanistic evidence for conserved function in molt regulation across pancrustacea. *Gen. Comp. Endocrinol.* **2020**, *298*, 113556. [CrossRef]
28. Shen, J.; Zhu, D.; Hu, Z.; Qi, Y.; Wang, C. Molt staging in the swimming crab *Portunus trituberculatus*. *J. Fish. China* **2011**, *35*, 1481–1487. [CrossRef]
29. Reddy, P.R.; Nagaraju, G.P.C.; Reddy, P.S. Involvement of methyl farnesoate in the regulation of molting and reproduction in the freshwater crab *Oziotelphusa senex senex*. *J. Crus. Biol.* **2004**, *24*, 511–515. [CrossRef]

30. Livak, K.J.; Schmittgen, T.D. Analysis of relative gene expression data using real-time quantitative PCR and the $2^{-\Delta\Delta CT}$ method. *Methods* **2001**, *25*, 402–408. [CrossRef] [PubMed]
31. Wang, C.J.; Zhu, D.F.; Qi, Y.Z.; Hu, Z.H.; Xie, X.; Shen, J. Molt-Inhibiting hormone levels and ecdysteroid titer during a molt cycle of *Portunus trituberculatus*. *Acta Hydrobio. Sin.* **2013**, *37*, 22–28. [CrossRef]
32. Webster, S.G.; Keller, R.; Dircksen, H. The CHH-Superfamily of multifunctional peptide hormones controlling crustacean metabolism, osmoregulation, moulting, and reproduction. *Gen. Comp. Endocrinol.* **2012**, *175*, 217–233. [CrossRef]
33. Mykles, D.L.; Chang, E.S. Hormonal control of the crustacean molting gland: Insights from transcriptomics and proteomics. *Gen. Comp. Endocrinol.* **2020**, *294*, 113493. [CrossRef] [PubMed]
34. Techa, S.; Thongda, W.; Bunphimpapha, P.; Ittarat, W.; Boonbangyang, M.; Wilantho, A.; Ngamphiw, C.; Pratoomchat, B.; Nounurai, P.; Piyapattanakorn, S. Isolation and functional identification of secretin family G-Protein coupled receptor from Y-organ of the mud crab, *Scylla Olivacea*. *Gene* **2023**, *848*, 146900. [CrossRef] [PubMed]
35. Kannangara, J.R.; Christen, K.M.; Warr, C.G. Regulation of ecdysone production in Drosophila by neuropeptides and peptide hormones. *Open Biol.* **2021**, *11*, 200373. [CrossRef]
36. Mykles, D.L. Ecdysteroid metabolism in crustaceans. *J. Steroid. Biochem. Mol. Biol.* **2011**, *127*, 196–203. [CrossRef] [PubMed]
37. Xie, X.; Liu, Z.; Liu, M.; Tao, T.; Shen, X.; Zhu, D. Role of Halloween genes in ecdysteroids biosynthesis of the swimming crab (*Portunus trituberculatus*): Implications from RNA interference and eyestalk ablation. *Comp. Biochem. Physiol. A Mol. Integr. Physiol.* **2016**, *199*, 105–110. [CrossRef]
38. Zhao, Y.F.; Wen, Q.Q.; Ao, C.M.; Wang, W.; Shi, L.L.; Wang, C.G.; Chan, S.F. Ecdysis triggering hormone, eclosion hormone, and crustacean cardioactive peptide play essential but different roles in the molting process of mud crab, *Scylla paramamosain*. *Front. Mar. Sci.* **2022**, *9*, 855391. [CrossRef]
39. Chan, S.F.; Wen, Q.Q.; Ao, C.M.; Wang, W.; Wang, C.G.; Zhao, Y.F. Transcriptome responses of RNAi-mediated ETH knockdown in *Scylla paramamosain* at different premolt substages. *Front. Endocrinol.* **2022**, *13*, 917088. [CrossRef]

biology

MDPI

Article

Comparative Study of Nutritional Composition, Physiological Indicators, and Genetic Diversity in *Litopenaeus vannamei* from Different Aquaculture Populations

Yundong Li [1,2,3,4,†], Siyao Cao [1,2,†], Shigui Jiang [1], Jianhua Huang [1,4], Qibin Yang [1,4], Song Jiang [1], Lishi Yang [1] and Falin Zhou [1,2,3,4,*]

1 South China Sea Fisheries Research Institute, Chinese Academy of Fishery Sciences, Key Laboratory of South China Sea Fishery Resources Exploitation and Utilization, Ministry of Agriculture and Rural Affairs, Guangzhou 510300, China; liyd2019@163.com (Y.L.); cao13793933909@163.com (S.C.); jiangsg@21cn.com (S.J.); huangjianhua@scsfri.ac.cn (J.H.); yangqibin1208@163.com (Q.Y.); tojiangsong@163.com (S.J.); yangls2016@163.com (L.Y.)

2 College of Fisheries and Life Sciences, Dalian Ocean University, Dalian 116023, China

3 Key Laboratory of Efficient Utilization and Processing of Marine Fishery Resources of Hainan Province, Sanya Tropical Fisheries Research Institute, Sanya 572018, China

4 Shenzhen Base of South China Sea Fisheries Research Institute, Chinese Academy of Fishery Sciences, Shenzhen 518108, China

* Correspondence: zhoufalin@alinyun.com

† These authors contributed equally to this work.

Simple Summary: This study investigates the nutritional quality and genetic diversity of Pacific white shrimp (*Litopenaeus vannamei*) from three different aquaculture farms in Guangdong Province, China. We assessed the shrimp's muscle tissue for key nutrients like protein, fat, and amino acids, alongside a detailed genetic analysis. Our findings reveal that the nutritional content is consistent across all three populations, indicating the reliable protein quality and essential nutrient levels critical for shrimp health and growth. However, significant differences in genetic diversity were observed, with one population exhibiting reduced genetic diversity and signs of inbreeding. These results underscore the necessity of implementing effective breeding strategies to enhance genetic diversity and ensure the sustainable development of *L. vannamei* in aquaculture. This study provides essential insights for optimizing breeding practices to maintain the long-term health and productivity of this vital species in the aquaculture industry.

Abstract: This study aimed to evaluate the quality and genetic diversity of farmed *Litopenaeus vannamei* across three distinct populations from Maoming City (MM), Zhanjiang City (ZJ), and Yangjiang City (YJ) in Guangdong Province. Muscle tissues from *L. vannamei* were analyzed for phenotypic traits, conventional nutrients, amino acids, and fatty acids, while genetic diversity was assessed using whole genome sequencing techniques. The analysis revealed that the crude protein content in shrimp across the three populations ranged from 20.87 to 21.95 g/100 g, crude fat content ranged from 0.90 to 1.50 g/100 g, essential amino acid content ranged from 5.55 to 5.86 g/100 g, total amino acid content ranged from 14.73 to 15.27 g/100 g, total fatty acid content ranged from 682.73 to 793.97 mg/100 g, total antioxidant capacity (T-AOC) ranged from 2.68 to 2.72 μmol/g, superoxide dismutase (SOD) activity ranged from 1021.97 to 1057.21 U/g, and catalase (CAT) activity ranged from 78.65 to 81.33 μmoL/min. No significant differences were observed in ash and crude fat levels among conventional nutrients, nor in the biochemical indexes T-AOC, CAT, and SOD. Genetic analysis showed that the single nucleotide polymorphism density (SNP/Kb) ranged from 15.323 to 17.461, nucleotide diversity (π) ranged from 2.98×10^{-5} to 15.84×10^{-5}, polymorphism information content (PIC) ranged from 0.300 to 0.317, heterozygosity (Ho) ranged from 0.033 to 0.048, and inbreeding coefficients (FIS) ranged from 0.834 to 0.887. The genetic differentiation index (FST) values among the three populations ranged from 0.056 to 0.106. This study provides an evaluation of the germplasm resources and genetic diversity of farmed *L. vannamei*, offering insights for the efficient management and sustainable utilization of this species' germplasm resources.

Keywords: *Litopenaeus vannamei*; germplasm resources; population; genetic diversity

1. Introduction

Litopenaeus vannamei, classified within the phylum Arthropoda, subphylum Crustacea, class Malacostraca, order Decapoda, and family Penaeidae, is a species of considerable global importance in aquaculture [1]. Known as the Pacific white shrimp, it is highly valued for its rapid growth and broad environmental adaptability, making it a cornerstone of the aquaculture industry [2]. Although *L. vannamei* is noted for its growth rate and environmental adaptability, it does not possess uniform resistance to all diseases. In practice, this species remains vulnerable to various diseases, which can significantly impact its growth and production efficiency [3]. However, the domestic aquaculture environment has deteriorated, leading to frequent disease outbreaks [4]. Concurrently, the expansion of aquaculture scale and the indiscriminate selection of shrimp broodstock—often directly from ponds—have resulted in reduced genetic diversity and inbreeding depression [5]. These practices have decreased production efficiency, severely impacting the conservation and utilization of superior genetic resources and overall production development [6]. Therefore, while breeding strategies aim to improve production efficiency, they also raise concerns about diminishing genetic diversity within shrimp populations.

Genetic diversity is fundamental for maintaining species adaptability and long-term survival and is a critical criterion for evaluating natural biological resources. Low genetic diversity can result in inbreeding depression, characterized by slow growth, reduced immunity, and increased susceptibility to diseases [7]. In the context of commercial *L. vannamei* farming, excessive reliance on a limited number of genotypes can reduce the genetic diversity of the entire population, thereby increasing vulnerability to environmental changes and disease outbreaks. Ensuring the sustainable development of *L. vannamei* aquaculture requires balancing breeding practices with the conservation of genetic diversity to maintain robust and adaptable shrimp populations [8].

This study aims to compare the phenotypic differences, nutritional composition, physiological–biochemical indicators, and genetic diversity among three different populations of *L. vannamei* using whole-genome sequencing technology. By examining the muscle nutritional components and physiological–biochemical indexes, we aim to understand how genetic diversity influences their growth and health. This comprehensive analysis seeks to elucidate the genetic, nutritional, and physiological differences among these populations, providing scientific evidence for future breeding programs to ensure the sustainable development of the *L. vannamei* aquaculture industry. Furthermore, the findings offer new perspectives and methods to optimize farming practices, thereby enhancing the efficiency of *L. vannamei* cultivation. This research also serves as a reference for other aquaculture species' genetic diversity conservation and breeding strategies.

2. Materials and Methods

2.1. Test Materials

In September 2022, a total of three geographically distinct populations of cultured *L. vannamei* were sampled from representative aquaculture farms in Southeast China. Specifically, 30 samples were collected from each of the following locations: Maoming City (LvA), Zhanjiang City (LvB), and Yangjiang City (LvC). For simplicity, these populations are identified by letter codes corresponding to their collection sites. The sampling locations and their corresponding codes are illustrated in Figure 1A. Detailed data regarding the cultured *L. vannamei* samples from each site are provided in Table 1.

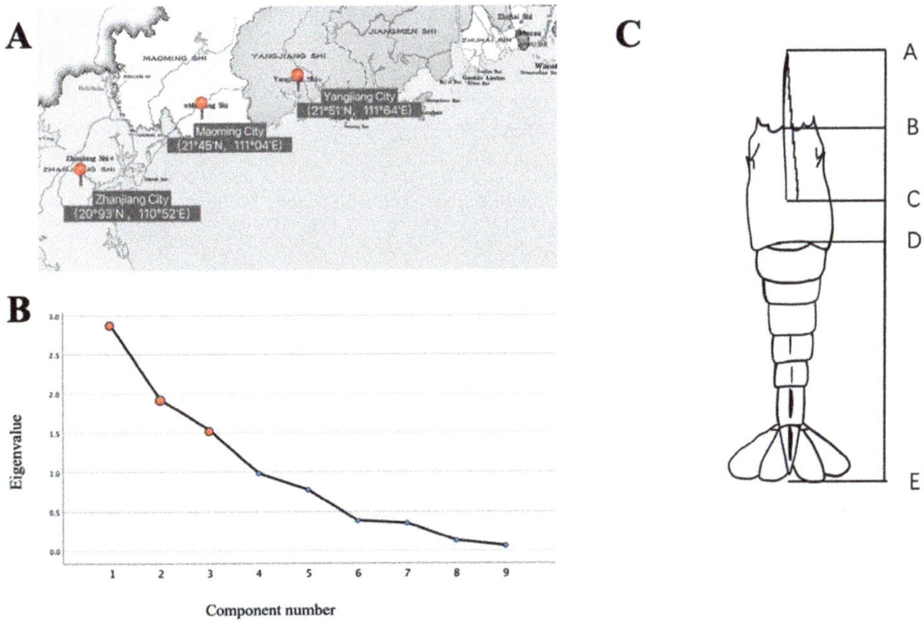

Figure 1. Illustration of the study area and data analysis. (**A**) Geographic location of the three sampling sites for *Litopenaeus vannamei*. (**B**) Principal component analysis of eight characteristics of three populations of *Litopenaeus vannamei* in a gravel map (red dots represent three principal components with characteristic values greater than 1). (**C**) Schematic diagram of the measurement of some morphological characteristics of *Litopenaeus vannamei*. AC: frontal angle length (FAL); BD: cephalothorax length (CL); BE: body length (BL); AE: total length (TL).

Table 1. Information on samples of *Litopenaeus vannamei*.

Survey Point	Sample Name	Quantities	Average Weight (g)	Average Armor Length (mm)	Month	Water Temperature (°C)	Salinity (ppt)	Cultivation Environment
Maoming City	LvA	30	59.28 ± 6.39	143.47 ± 3.12	3 to 4	27.5	30	Natural bottom ponds
Zhanjiang City	LvB	30	60.01 ± 6.63	142.87 ± 2.50	3 to 4	27.4	29	Concrete bottoms
Yangjiang City	LvC	30	60.32 ± 8.94	145.17 ± 4.32	3 to 4	27.4	30	Canvas bottoms

The three farms employ distinct aquaculture practices tailored to their specific environments. The Maoming farm (LvA) utilizes open-air earthen ponds, drawing seawater from the nearby Beibu Gulf. This farm focuses on maintaining moderate stocking densities from 30–35 initially to 10–15 shrimp per square meter during the growth phase, and provides high-protein (35–40%) feed with 5–8% fat and 15–18% energy content, sourced mainly from fish meal, soybean meal, and lecithin to support optimal growth conditions. In contrast, the Zhanjiang farm (LvB) operates a semi-closed recirculating system with concrete ponds, maintaining lower stocking densities ranging from 25–30 initially to 8–12 shrimp per square meter. It uses an automated feeding system with imported feed consisting of 40% protein, 8–10% fat, and 18–20% energy to ensure consistent shrimp quality. The Yangjiang farm (LvC) employs modern plastic-lined ponds, utilizing seawater treated through natural sedimentation. This farm emphasizes controlled stocking densities from 20–25 initially to 5–10 shrimp per square meter and uses nutrient-enriched feed with 40% protein, 10–12% fat, and 20–22% energy, including high proportions of fish meal, soybean meal, lecithin, and probiotics to promote uniform growth. All farms implement stringent parent shrimp management practices, sourcing shrimp from diverse genetic pools to avoid inbreeding, with comprehensive monitoring and documentation to ensure that genetic diversity is maintained.

2.2. Measurement Methods

The shrimp were weighed accurately using an electronic balance (accuracy of 0.01 g). Eighty-two phenotypic traits of shrimp were measured accurately using vernier calipers (accuracy of 0.01 mm) (Figure 1B). These traits included total length (TL), body length (BL), cephalothorax length (CL), frontal angle length (FAL), number of frontal teeth (NFT), length of the second leg (LSL), and the distance between the bases of the fifth step foot (Step 5 Interbasal distance, SID). Measurements such as the first touch (FT) and the second whip (SW) were also recorded. The surface moisture of the samples was absorbed by gauze before weighing [9]. The measurement of phenotypic traits in *L. vannamei* was conducted to assess physical and morphological differences among the three populations. These traits, including total length and body length, serve as indicators of growth performance and health. By analyzing these characteristics, we aim to link phenotypic variation with genetic diversity and environmental factors, providing insights into the factors that influence shrimp quality and development in aquaculture.

2.3. Biochemical Composition Analysis

The abdominal muscles of the shrimp were taken to determine the moisture, crude ash, crude protein, crude fat, and total sugar contents in the muscles. Moisture was determined by the electrothermal constant temperature drying method (GB5009.3-2016) [10]; ash was determined by the Mafu furnace ashing method (GB/T6438-92) [11]; crude protein was determined by the Kjeldahl nitrogen determination method (GB/T6432-94) [12]. Crude fat was determined by the acid hydrolysis test method (GB5009.6-2016) [13]. Total sugar was determined by ultraviolet–visible spectrophotometry (GB/T9695.31-2008) [14]. Apart from the moisture content, the specific determination steps of the remaining indicators followed the AOAC guidelines (1984).

2.4. Nutritional Quality Assessment

Seventeen common amino acids were determined by the hydrochloric acid hydrolysis method (GB/T5009.124-2016) [15]. The Hitachi LA8080 amino acid autoanalyzer(Hitachi High-Tech Corporation, Tokyo, Japan) determined the amino acid content, and tryptophan was not determined because it was des I'm sure to keep ittroyed in the process of hydrolysis. The fatty acid profile was determined by gas chromatography (GB/T5009.168-2016) [16].

2.5. Antioxidant Enzyme Activity Assays

The abdominal muscles of the shrimp were used to determine the activities of superoxide dismutase (SOD), catalase (CAT), and total antioxidant capacity.

Superoxide dismutase (SOD) activity is defined as the amount of enzyme per milliliter of reaction solution corresponding to 50% inhibition of superoxide dismutase as one unit of enzyme activity (U/mg). This indicator was determined by referring to the instruction manual of the WST-8 method activity assay kit(JianCheng, Nanjing, China) [17]. The activity of the WST-8 method was determined according to the instructions for the WST-8 Activity Kit.

Catalase (CAT) activity was defined as one unit of enzyme activity (U/mg) per milliliter of tissue protein breaking down 1 μmol of H_2O_2 per minute.

Total antioxidant capacity (T-AOC) is defined as the total antioxidant level consisting of various antioxidant substances, antioxidant enzymes, etc. This index was determined by referring to the instructions of the total antioxidant capacity (T-AOC) assay kit using the ABTS method.

2.6. Genetic Diversity

Genomic DNA was extracted from the tail muscle of *L. vannamei* using the phenol-chloroform method. DNA integrity was assessed via 1% agarose gel electrophoresis, while concentration and purity were measured using UV spectrophotometry(Shimadzu Corporation, Kyoto, Japan), ensuring that DNA concentration exceeded 12.5 ng/μL. For sequencing library preparation, the DNA was fragmented into smaller pieces using sonication. Libraries were then constructed according to the Illumina TruSeq protocol, which involved end-repair to blunt the DNA ends, A-tailing to add an adenine overhang, adapter ligation to attach sequencing adapters, and PCR amplification to enrich the library. The quality and quantity of the constructed libraries were evaluated using a bioanalyzer and quantitative PCR (qPCR) to ensure suitability for sequencing.

High-throughput sequencing was performed on the Illumina platform(Illumina, Inc., San Diego, CA, USA), generating raw reads. Initial quality control of these reads was conducted using FastQC to identify any sequencing biases or errors. The reads were aligned to the reference genome using the Burrows–Wheeler Aligner (BWA)(Wellcome Trust Sanger Institute, Cambridge, UK), an efficient and accurate alignment tool [18]. Variant calling was performed using the Genome Analysis Toolkit (GATK)(Broad Institute, Cambridge, MA, USA), which identified single nucleotide polymorphisms (SNPs) and insertions/deletions (indels), producing an initial Variant Call Format (VCF) file [19]. To ensure the accuracy of the variant data, the VCF file was filtered using VCF tools, applying stringent criteria for variant quality and coverage depth to exclude low-confidence variants. The remaining high-confidence variants were functionally annotated using ANNOVAR(University of Southern California, Los Angeles, CA, USA), providing insights into the potential effects of these variants on gene function [20]. For population genetic analysis, genotype data underwent further quality control using PLINK(Broad Institute, Cambridge, MA, USA), with checks for missing data, minor allele frequency thresholds, and Hardy–Weinberg equilibrium [21]. Population structure was analyzed using ADMIXTURE(Broad Institute, Cambridge, MA, USA) and Principal Component Analysis (PCA) to identify genetic clusters and assess potential population stratification. Linkage disequilibrium (LD) was calculated using Haploview, offering insights into the genetic linkage between markers [22]. Selection signals were detected using iHS (Integrated Haplotype Score)(Pritchard Lab, University of Chicago, Chicago, IL, USA) and XP-EHH (Cross-population Extended Haplotype Homozygosity) (Sabeti Lab, Harvard University, Cambridge, MA, USA) methods [23].

2.7. Data Analysis

The experimental data were expressed as mean ± standard deviation, and one-way ANOVA (one-way ANOVA) was performed with SPSS 27.0 statistical software, and Duncan's multiple comparisons analysis was performed if there were significant differences ($p < 0.05$) [24]. The proportional parameters were input into SPSS 27.0 statistical software to obtain the contribution rate and cumulative contribution rate of each principal component through the factor analysis of SPSS, and then to obtain which of the nine proportional parameters played a decisive role [25]; using the DNASP5.0 software, the analysis of nucleotide diversity (π) was computed [26]; the PopGene32 software was used to calculate the completion of the observed heterozygosity (Ho) [27]; and the polymorphic information content (PIC) of each locus in each population was calculated using PIC_CALC0.6 software [28]. The inbreeding coefficient (FIS) and genetic differentiation index (FST) were carried out using Arlequin3.5 [29]; gene flow (Nm) was calculated according to the formula: Nm = 0.25 (1 − FST)/FST [30], and the genetic distance between populations (DR) was calculated according to the formula: −In (1 − FST) [31].

3. Results

3.1. Analysis of Phenotypic Traits

A one-way ANOVA was conducted to analyze eight morphometric traits across three groups of *L. vannamei* (Table 2). The results showed variations in these traits among the populations, but none of the differences were statistically significant ($p > 0.05$). Notably, the second whip (SW) length and cephalothorax length (CL) of the LvB population were markedly larger than those of the other populations, with the LvA population exhibiting the smallest values. In the Principal Component Analysis (Table 3), it was noted that the initial principal component accounted for 31.756% of the variance, predominantly representing metrics such as body length (BL) about the length of the second leg (LSL), body length (BL) compared to the distance between the bases of the fifth step foot (SID), body length (BL) juxtaposed with cephalothorax length (CL), and body length (BL) relative to frontal angle length (FAL). This refers to principal component 1 primarily reflecting the morphological aspects associated with body length (BL), suggesting considerable morphological distinctions across the three *L. vannamei* populations. The second principal component contributed 21.383% to the variation, signifying a focus on metrics like the first touch (FT) length per cephalothorax length (CL), the first touch (FT) length in relation to body length (BL), and other relevant parameters. Thus, principal component 2 predominantly highlights the morphological characteristics of the cephalothorax length (CL), suggesting that differences in this area significantly contribute to the overall morphological variations observed among the three populations. The third principal component, responsible for 17.065% of the variation, predominantly emphasizes metrics such as the second whip (SW) length per cephalothorax length (CL) and the second whip (SW) length relative to body length (BL). Principal component 3 primarily mirrors the morphological attributes associated with the second whip (SW) length. The cumulative contribution of the three principal components amounts to 70.204%. Notably, the three principal components with eigenvalues surpassing 1 were derived from the nine indicators in Figure 1C. These findings highlight that body length (BL), carapace length (CL), and second whip (SW) length vary significantly among *L. vannamei* populations. These morphological features are potentially important in distinguishing populations, suggesting that they may be important clues for the further investigation of genetic and ecological differences among populations.

Table 2. Results of analysis of variance (ANOVA) for eight quantifiable traits in three groups of *Litopenaeus vannamei*.

Variant	LvA	LvB	LvC
Body length/second step foot length	6.0533 ± 0.3866^{ab}	6.1767 ± 0.32129^{a}	5.9167 ± 0.35631^{b}
Body length/basal distance of fifth footsteps	11.9667 ± 1.0083^{a}	11.9533 ± 0.6622^{a}	12.0300 ± 0.84431^{a}
Body length/cephalothoracic armor length	3.7500 ± 0.2418^{ab}	3.8867 ± 0.1717^{a}	3.6000 ± 0.2084^{b}
Body length/frontal horn length	8.5767 ± 0.76954^{ab}	8.9167 ± 0.6097^{a}	8.5033 ± 0.7573^{b}
Frontal horn length/cephalothoracic armor length	0.4367 ± 0.4901^{a}	0.4367 ± 0.4901^{a}	0.4267 ± 0.4498^{a}
First whip length/cephalothoracic armor length	1.5733 ± 0.2586^{a}	1.6467 ± 0.2713^{a}	1.5000 ± 0.31948^{a}
Second whip length/cephalothoracic armor length	0.2900 ± 0.10939^{a}	0.4267 ± 0.06400^{b}	0.3967 ± 0.0556^{b}
First whip length/body length	0.4167 ± 0.6989^{a}	0.4200 ± 0.0610^{a}	0.4067 ± 0.0785^{a}
Second whip length/body length	0.0800 ± 0.4068^{a}	0.100 ± 0.0000^{b}	0.1000 ± 0.0000^{b}

Note: Superscript letters (a, b, ab) within the same row denote statistically significant differences between groups. Values sharing the same letter are not significantly different ($p > 0.05$).

3.2. Conventional Nutrients

Fresh muscle samples′ protein and fat contents from *L. vannamei* stocks LvA, LvB, and LvC were analyzed (Figure 2). LvA exhibited a crude protein content of 20.87 ± 0.64 g/100 g and a crude fat content of 1.50 ± 0.10 g/100 g. LvB showed a slightly higher protein content at 21.95 ± 1.65 g/100 g and a lower fat content of 1.05 ± 0.15 g/100 g. LvC had a protein content of 21.75 ± 1.70 g/100 g and a fat content of 0.90 ± 0.10 g/100 g. The conventional nutrient contents, including crude ash and fat, were consistent among the three stocks, with no significant differences observed ($p > 0.05$). This suggests that, despite minor variations in protein and fat content, the overall nutritional composition across the three populations remains comparable, with no statistically significant differences observed.

Table 3. Characteristic vectors of eight traits of three populations of *Litopenaeus vannamei* on three principal components and the contribution of the principal components.

Morphological Proportionality Traits	Principal Component		
	1	2	3
Body length/second step foot length	0.339	0.012	−0.118
Body length/basal distance of fifth footsteps	0.214	−0.049	−0.060
Body length/cephalothoracic armor length	0.286	0.088	−0.053
Body length/frontal horn length	0.301	−0.057	0.118
Frontal horn length/cephalothoracic armor length	−0.113	0.176	−0.248
First whip length/cephalothoracic armor length	0.035	0.485	0.009
Second whip length/cephalothoracic armor length	0.010	0.047	0.463
First whip length/body length	−0.048	0.482	0.003
Second whip length/body length	−0.163	0.010	0.555
Contribution of each principal component (%)	31.756	21.383	17.065
Cumulative contribution (%)	31.756	53.139	70.204

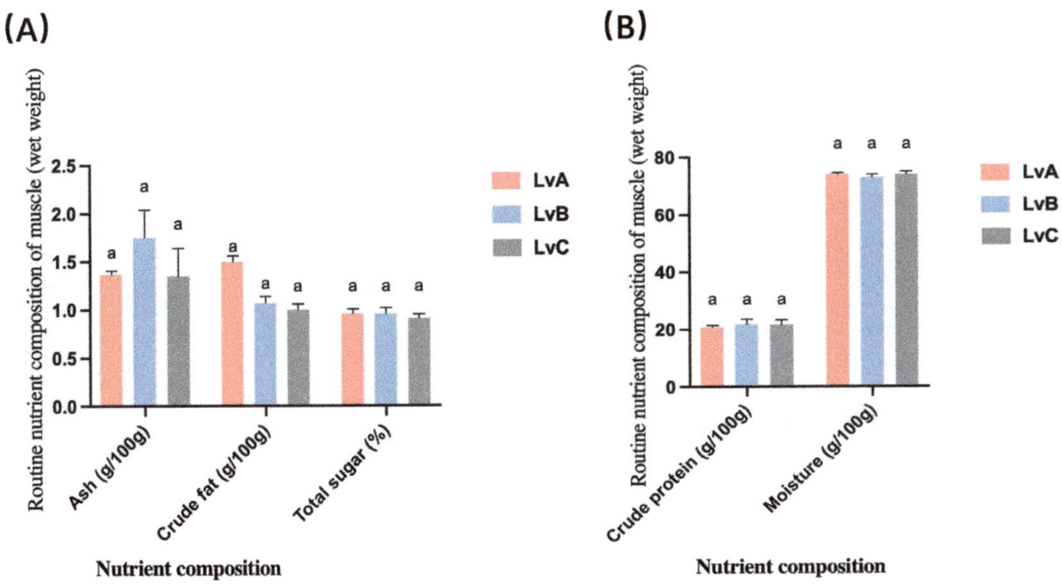

Figure 2. Routine nutrient composition of muscle (wet weight) of three groups of *Litopenaeus vannamei*. (**A**) The contents of ash, crude fat, and total sugar in three populations. (**B**) The water content and crude protein content of the three populations. Superscript "a" indicates no statistically significant difference ($p > 0.05$) between the groups.

3.3. Amino Acid Composition and Content

Our comprehensive analysis detected a total of 17 amino acids in the three populations of *L. vannamei*, reflecting a thorough assessment of the species' amino acid profile. This detailed analysis included seven essential amino acids (EAA: Thr, Val, Met, Phe, Ile, Leu, Lys), although tryptophan was not detected due to disruption during acid hydrolysis. Additionally, two semi-essential amino acids (SEAA: His, Arg) and eight non-essential amino acids (NEAA: Asp, Ser, Glu, Gly, Ala, Cys, Pro, Tyr) were identified (Table 4). This comprehensive dataset provides a wealth of information on the amino acid composition of *L. vannamei*, offering valuable insights for understanding its nutritional quality and supporting further research on its aquaculture management and breeding.

Table 4. Amino acid composition and content (g/100 g, wet weight) of muscle from three *Litopenaeus vannamei* groups.

Amino Acids	LvA	LvB	LvC
Aspartic acid Asp @	1.54 ± 0.09 a	1.53 ± 0.04 a	1.50 ± 0.05 a
Threonine Thr *	0.60 ± 0.03 a	0.59 ± 0.01 a	0.57 ± 0.02 a
Serine Ser	0.53 ± 0.03 a	0.53 ± 0.01 a	0.50 ± 0.02 a
Glutamic acid Glu @	2.43 ± 0.15 a	2.45 ± 0.06 a	2.45 ± 0.06 a
Glycine Gly @	1.04 ± 0.06 a	1.08 ± 0.03 a	1.04 ± 0.03 a
Alanine Ala @	0.96 ± 0.04 a	0.97 ± 0.02 a	0.96 ± 0.03 a
Cystine Cys	0.22 ± 0.02 a	0.17 ± 0.02 b	0.16 ± 0.01 b
Val *	0.73 ± 0.04 a	0.73 ± 0.02 a	0.70 ± 0.02 a
Methionine Met *	0.41 ± 0.02 a	0.38 ± 0.01 b	0.35 ± 0.01 b
Isoleucine Ile *	0.66 ± 0.03 a	0.65 ± 0.01 a	0.63 ± 0.02 a
Leucine Leu *	1.25 ± 0.07 a	1.22 ± 0.03 a	1.18 ± 0.04 a
Tyr tyrosine	0.48 ± 0.02 a	0.45 ± 0.01 a	0.41 ± 0.01 b
Phenylalanine Phe *	0.67 ± 0.03 a	0.60 ± 0.01 b	0.61 ± 0.02 b
Lys *	1.21 ± 0.06 a	1.20 ± 0.03 a	1.20 ± 0.03 a
Histidine His &	0.33 ± 0.02 a	0.32 ± 0.01 a	0.31 ± 0.01 a
Arginine Arg &	0.95 ± 0.05 a	0.95 ± 0.03 a	0.91 ± 0.03 b
Proline Pro	1.27 ± 0.08 a	1.30 ± 0.06 a	1.25 ± 0.06 a
EAA	5.86 ± 0.31 a	5.70 ± 0.13 b	5.55 ± 0.16 c
SEAA	1.28 ± 0.07 a	1.27 ± 0.04 a	1.22 ± 0.03 a
NEAA	8.13 ± 0.47 a	8.17 ± 0.21 a	7.97 ± 0.28 b
DAA	5.97 ± 0.34 a	6.04 ± 0.15 a	5.95 ± 0.17 a
TAA	15.27 ± 0.84 a	15.13 ± 0.38 a	14.73 ± 0.47 a
EAA/TAA	0.38 ± 0.00 a	0.37 ± 0.00 a	0.37 ± 0.00 a
EAA/NEAA	0.72 ± 0.00 a	0.70 ± 0.00 a	0.70 ± 0.00 a

Note: * indicates essential amino acids; @ indicates flavoring amino acids; & indicates semi-essential amino acids; EAA: total essential amino acids; SEAA: total semi-essential amino acids; NEAA: total non-essential amino acids; DAA: total flavoring amino acids. Superscript letters (a, b, c) within the same row indicate statistically significant differences between groups ($p < 0.05$). Values sharing the same letter are not significantly different.

3.4. Fatty Acid Composition and Content

A total of 16 fatty acids were detected in the three populations of *L. vannamei*, including six saturated fatty acids (SFA), two monounsaturated fatty acids (MUFA), and five polyunsaturated fatty acids (PUFA) (Table 5). The total content of saturated fatty acids ranged from 267.17 to 317.83 mg/100 g; the total content of monounsaturated fatty acids ranged from 10.43 to 12.53 mg/100 g; and the total content of polyunsaturated fatty acids ranged from 242.42 to 277.10 mg/100 g.

3.5. Analysis of Physiological Indicators

The total antioxidant capacity (T-AOC) of *L. vannamei* LvA was 2.68 ± 0.19 μMol/g; the activity range of superoxide dismutase (SOD) was 1035.72 ± 132.34 U/g; the activity range of catalase (CAT) was 78.65 ± 17.49 μMol/L/(ming). The total antioxidant capacity (T-AOC) of LvB was 2.68 ± 0.17 μMol/g; the activity range of superoxide dismutase (SOD) was 1021.97 ± 127.25 U/g; the activity range of catalase (CAT) was 81.33 ± 23.52 μMol/L/(ming). The total antioxidant capacity (T-AOC) of LvC was 2.72 ± 0.18 μMol/g; the activity range of superoxide dismutase (SOD) was 1057.21 ± 126.62 U/g; the activity range of catalase (CAT) was 80.14 ± 19.55 μMol/L/(ming). There were no significant differences in physiological and biochemical indicators among the groups (Figure 3).

Table 5. Fatty acid composition and content of muscle fatty acids (mg/100 g, wet weight) of three groups of *Litopenaeus vannamei*.

Fatty Acids	LvA	LvB	LvC
Palmitic acid C16:0	168.30 ± 2.69 [a]	138.53 ± 15.23 [b]	148.87 ± 9.56 [ab]
Palmitoleic acid C16:1	5.77 ± 0.12 [a]	4.10 ± 0.60 [b]	5.13 ± 0.42 [c]
C17:0 heptadecanoic acid	10.67 ± 0.12 [a]	9.03 ± 1.03 [b]	9.90 ± 0.56 [ab]
Stearic acid C18:0	126.43 ± 2.27 [a]	109.67 ± 13.27 [b]	114.70 ± 6.42 [b]
Oleic acid C18:1 n9c	123.07 ± 2.04 [a]	106.27 ± 13.55 [b]	104.53 ± 3.40 [b]
Linoleic acid C18:2n6c	160.57 ± 2.57 [a]	138.77 ± 17.04 [b]	140.23 ± 8.30 [b]
Arachidic acid C20:0	7.80 ± 0.10 [a]	5.13 ± 4.46 [b]	7.40 ± 0.40 [a]
Linolenic acid C18:3n3	9.33 ± 0.15 [a]	7.57 ± 0.97 [b]	7.70 ± 0.60 [b]
Eicosatetraenoic acid C20:1	6.60 ± 0.10 [a]	6.07 ± 0.87 [a]	6.07 ± 0.55 [a]
Eicosadienoic acid C20:2	15.33 ± 0.40 [a]	14.03 ± 1.68 [a]	14.40 ± 0.87 [a]
Docosanoic acid C22:0	9.53 ± 0.15 [a]	8.60 ± 1.61 [a]	8.73 ± 0.49 [a]
Arachidonic acid ARAC20:4n6	31.57 ± 0.45 [a]	27.17 ± 3.63 [b]	28.97 ± 1.46 [b]
Erucic acid C22:1n9	3.90 ± 0.00 [a]	4.37 ± 0.42 [b]	6.47 ± 0.45 [b]
XXIII carbonic acid C23:0	1.12 ± 1.92 [a]	1.13 ± 1.96 [a]	1.13 ± 1.96 [a]
Eicosapentaenoic acid C20:5n3(EPA)	54.80 ± 0.92 [a]	47.37 ± 4.97 [b]	51.10 ± 2.97 [ab]
C22:6n3(DHA)	60.30 ± 0.95 [a]	54.93 ± 5.51 [b]	60.30 ± 3.06 [a]
Total fatty acid content TOTAL	793.97 ± 12.94 [a]	682.73 ± 85.68 [b]	715.63 ± 42.04 [c]
Total saturated fatty acids ΣSFA	317.83 ± 5.32 [a]	267.17 ± 36.44 [b]	285.97 ± 19.03 [b]
Monounsaturated fatty acids ΣMUFA	10.50 + 0.10 [b]	10.43 + 1.25 [b]	12.53 + 0.76 [a]
Polyunsaturated fatty acids ΣPUFA	277.10 ± 4.49 [a]	242.42 ± 28.63 [b]	251.60 ± 13.75 [b]
DHA ± EPA	115.10 ± 1.87 [a]	103.30 ± 10.47 [b]	111.40 ± 5.86 [a]
n-3 series polyunsaturated fatty acidsn-3ΣPUFA	124.43 ± 2.02 [a]	109.87 ± 11.43 [b]	119.10 ± 6.46 [b]
n-6 series polyunsaturated fatty acidsn-6ΣPUFA	192.13 ± 3.01 [a]	165.93 ± 20.67 [b]	169.20 ± 9.42 [b]

Note: Superscript letters (a, b, ab, c) within the same row indicate statistically significant differences between groups ($p < 0.05$). Values sharing the same letter are not significantly different.

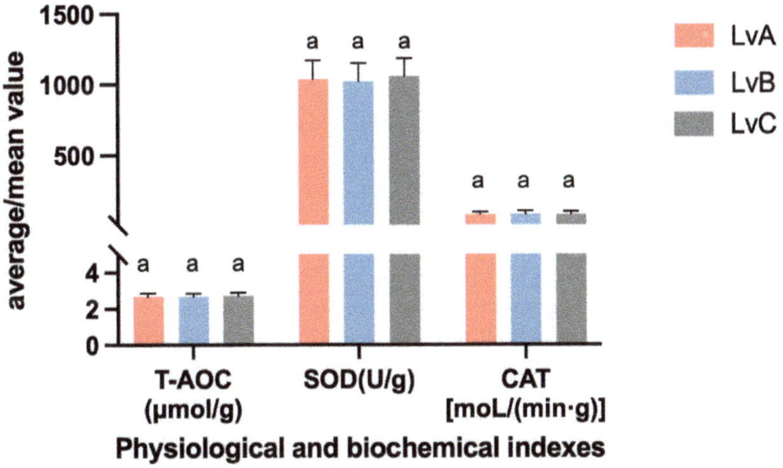

Figure 3. Physiological and biochemical indexes of the three populations of *Litopenaeus vannamei*. Superscript "a" indicates no statistically significant difference ($p > 0.05$) between the groups.

3.6. Genetic Diversity and Population Structure Analysis

Statistical Analysis of Genetic Diversity Parameters

In an in-depth assessment of the genetic diversity of the three populations of *L. vannamei* (LvA, LvB, and LvC) (Table 6), we found significant differences in SNP density and nucleotide diversity among the three populations. Although the observed heterozygosity was relatively consistent among the three, the nucleotide diversity of LvC (15.84×10^{-5})

was significantly higher than that of LvA (3.11×10^{-5}) and LvB (2.98×10^{-5}), suggesting that the LvC population may have richer genetic diversity.

Table 6. Statistics on population genetic diversity parameters of three populations of *Litopenaeus vannamei*.

Population	SNP Number	SNP Density (SNP/Kb)	Nucleotide Diversity (π)	Polymorphism Information Content (PIC)	Observed Heterozygosity (Ho)	Inbreeding Coefficient (FIS)
LvA	29047925	17.461	3.11×10^{-5}	0.303 ± 0.125	0.048 ± 0.001	0.835
LvB	25491738	15.323	2.98×10^{-5}	0.317 ± 0.124	0.048 ± 0.001	0.834
LvC	26677995	16.036	15.84×10^{-5}	0.300 ± 0.130	0.033 ± 0.014	0.887

In the analysis of genetic differentiation among populations (Table 7), the genetic differentiation index between LvB and LvC (FST = 0.106) was the highest among the three, indicating that there was a large genetic isolation between the two populations. In contrast, the genetic connection between LvA and LvB was relatively closer (FST = 0.056). In addition, gene flow analysis further revealed genetic isolation between populations, especially the lowest gene flow observed between LvB and LvC (Nm = 2.108). These findings highlight the need to strengthen genetic monitoring of *L. vannamei* populations to ensure the diversity of genetic resources and the maintenance of population health, especially in the face of dual environmental and economic pressures. Through the effective management and use of these genetic data, we can lay a solid foundation for future breeding activities.

Table 7. Genetic differentiation index (FST), inter-population genetic distance (DR), and gene flow (Nm) statistics of *Litopenaeus vannamei*.

Population 1	Population 2	Population Differentiation Coefficient (FST)	Genetic Distance (DR)	Gene Flow (Nm)
LvA	LvB	0.056	0.057	4.214
LvA	LvC	0.084	0.088	2.726
LvB	LvC	0.106	0.112	2.108

4. Discussion

4.1. Analysis of Differences in Phenotypic Traits of Different Populations of Litopenaeus vannamei

The results of the morphometric analysis of *L. vannamei* suggest a high degree of uniformity in the measured proportions among the different groups. Significant morphological differences may not be present due to the groups being cultivated under similar aquaculture conditions. Additionally, selective breeding strategies might have been consistent, focusing on common important traits such as growth, which could reduce morphological differences among the groups.

Although of morphometric proportions did not reveal significant differences, principal component analysis (PCA) was applied to the three groups of *L. vannamei* and successfully identified three main components. These components predominantly reflect the shrimp's total and second antennal lengths. These findings indicate that these traits play a significant role in distinguishing between shrimp groups and can be considered important discriminators. Total length, as a fundamental indicator of growth and development, directly reflects the health and growth rate of an individual. Carapace length is not only crucial for morphological assessment but also relates to predation behavior and immune function. The length of the second antenna plays a key role in the sensory and locomotive abilities of shrimp, potentially affecting their adaptability and behavioral performance. The cumulative contribution of these three principal components was 70.20%, similar to the results of Li Xiaoshuang et al. [32], who reported a cumulative contribution rate of 72.58% from four principal components in a morphological variation analysis of six groups of *L. vannamei*.

This similarity suggests that these components can explain a significant portion of the morphological variation and are highly representative. These traits can be utilized not only to distinguish between these three groups but also in identifying other populations. Therefore, future breeding and management efforts should integrate genetic diversity with key morphological traits. By ensuring breeding objectives are met while enhancing genetic diversity protection, sustainable development and the health of *L. vannamei* populations can be ensured.

4.2. Analysis of Differences in the Routine Nutrient Composition of the Muscles of L. vannamei from Different Stocks

The evaluation of the nutritional value of aquatic products often relies on assessing the protein and fat content present in the muscle tissue [17]. In this study, the research findings indicate that the average crude protein content in the muscle tissue of *L. vannamei* across three populations exceeds 21.52%, with no significant differences observed between them. Comparatively, this protein content surpasses that of *Mozambican Penaeus Monodon* (18.67%), *Macrobrachium rosenbergii* (18.27%) [33], *Marsupenaeus japonics* (17.74%) [34], *M. nipponenses* (17.71%), and *Exopalaemon modestus* (17.60%) [35]. These results align with the research conducted by Chen Xiaohan et al. [36] in 2001, showcasing *L. vannamei*'s protein content of 21.57%. Furthermore, the fat content in the three populations of *L. vannamei* falls within the range of 0.90% to 1.50%, consistent with *Exopalaemon modestus* (1.43%) [35] and *Macrobrachium nipponense* (1.19%) [34]. However, it is lower than *Marsupenaeus japonics* (2.08%) [34], *Macrobrachium rosenbergii* (1.97%) [33], and *Cambarus clarkii* (1.70%) [37]. The lower fat content of *L. vannamei* positively influences its acceptance among consumers, as low-fat content is generally perceived as healthier. This finding highlights the unique nutritional profile of *L. vannamei*, further emphasizing its value in commercial aquaculture. The study demonstrates that *L. vannamei* has significant advantages in terms of muscle crude protein and fat contents, particularly its superior protein content, enhancing its market competitiveness. These findings not only help elevate the status of *L. vannamei* in the aquaculture industry but also provide essential scientific evidence for future selective breeding and nutritional improvement. This study showed no statistically significant differences in the conventional nutritional components across the three muscle groups ($p > 0.05$). This may be related to the uniformity of feed provided during the cultivation process [38]. Future research should continue to focus on the variations in the nutritional composition of *L. vannamei* under different environmental and feeding conditions to optimize farming strategies and maximize its nutritional and economic value.

4.3. Analysis of Differences in Muscle Amino Acid Composition of Different Stocks of Litopenaeus vannamei

Most aquatic animals contain high-quality animal proteins [39]. Protein quality mainly depends on the protein content, essential amino acid content, and the proportion and digestibility and the type and content of amino acids play an important role in the growth and development of shrimp as well as in the evaluation of their quality, which is the main factor determining the nutritional value [40,41]. This study analyzes the amino acid content of three populations of *L. vannamei*, revealing no significant differences among the populations. This indicates a high degree of consistency in amino acid composition across different populations, despite certain genetic variations and environmental differences. Such consistency may stem from a similar genetic basis among the populations, particularly concerning genes related to amino acid metabolism, and from cultivation under similar conditions and feed formulations. Consequently, the environmental impact on amino acid composition is minimal [42]. This stability is crucial for ensuring that *L. vannamei* consistently provides high-quality protein across different aquaculture environments.

According to the standards for high-quality food protein set by the Food and Agriculture Organization of the United Nations (FAO) and the World Health Organization (WHO) [43], good-quality proteins should not only contain a complete range of essential amino acids but also have the right or correct proportion of essential amino acids to each other [44,45]. The content and proportion of essential amino acids are used as indicators. The nutritional value of shrimp protein was evaluated using the content and ratio of essential amino acids as indicators. According to the essential amino acid model recommended by FAO/WHO, the ratio of EAAs/TAAs should be greater than or equal to 0.40, and the EAAs/NEAAs ratio should be greater than or equal to 0.06 [46]. It can be determined from Table 5 that the ratio of EAAs/TAAs in shrimp from *L. vannamei* was 0.37–0.38, which was very close to 0.40, and the ratio of EAAs/NEAAs in shrimp was 0.70–0.72, which was relatively close to 0.60. These results suggest that the high-quality protein in shrimp meat or by-products is associated with the ratios of essential amino acids to total amino acids (EAAs/TAAs) and essential amino acids to non-essential amino acids (EAAs/NEAAs). *L. vannamei* exhibits the advantage of high-quality protein and a balanced amino acid composition, facilitating efficient protein synthesis and growth. This amino acid balance is crucial for enhancing shrimp growth performance and health, reinforcing its status as a premium aquaculture product [47].

4.4. Analysis of Differences in Muscle Fatty Acid Composition of Different Populations of *Litopenaeus vannamei*

The composition and content of fatty acids in food is an essential indicator for evaluating the nutritional value of lipids [46]. Saturated fatty acids play an important role in the energy metabolism of shrimp. However, humans' excessive intake of saturated fatty acids is detrimental to human health and may increase the risk of cardiovascular disease. Oleic acid in monounsaturated fatty acids is also a major source of energy in the growth and metabolism of shrimp [46]. Polyunsaturated fatty acids (PUFAs) are bioactive substances with unique physiological functions, and their nutritional value is positively correlated with the content of unsaturated fatty acids [46]. The nutritional value of PUFAs is positively correlated with their content [48].

This study conducted a comparative analysis of the fatty acid composition in three populations of *L. vannamei* (LvA, LvB, and LvC). Significant differences were observed among the populations regarding total fatty acid content and specific fatty acid types. The LvA population exhibited the highest total fatty acid content, particularly in polyunsaturated fatty acids (PUFAs) and essential omega-3 fatty acids, such as eicosapentaenoic acid (EPA) and docosahexaenoic acid (DHA). These fatty acids are crucial for the growth and health of *L. vannamei* and significantly enhance their nutritional value. The content of n-3 polyunsaturated fatty acids (PUFA) ranged from 109.87 to 124.43 mg/100 g, and the content of n-6 polyunsaturated fatty acid (PUFA) ranged from 165.93 to 192.13 mg/100 g. Significant differences ($p < 0.05$) were observed in C16:0, C18:1n9c, C22:1n9, and C20:5n3 among the different groups. These findings illustrate the diversity in fatty acid profiles among the studied populations, potentially reflecting variations in their diet, genetic makeup, and environmental conditions. In particular, the LvA population had the highest total content of polyunsaturated fatty acids (PUFAs) and essential omega-3 fatty acids such as EPA and DHA, which are essential for shrimp growth and health, significantly increasing their nutritional value. A possible explanation is that LvA populations live in lagoons with natural substrates and have access to a more diverse range of natural food resources. Combined with the higher proportions of fishmeal, soybean meal, and phospholipids in their feed, these rich nutrient sources promote the synthesis of key healthy fatty acids such as EPA and DHA. This comprehensive feed and natural food supply not only provides balanced nutrition, but also contributes to the diversity and enrichment of fatty acids.

In contrast, the LvB and LvC populations were in an environment with artificial concrete and canvas substrates. Although the crude protein and energy content of the feed was higher, the limitations of natural food may have affected the synthesis of specific fatty acids, especially polyunsaturated fatty acid content. This difference in environment and feed may result in these populations being lower in key fatty acids than LvA. Although high fat and energy in the diet is beneficial for the rapid growth of shrimp, this is not necessarily effective in promoting the production of essential fatty acids, indicating that optimizing the quality and quantity of fatty acids is critical. The contents of EPA and DHA in the muscle of the *Marsupenaeus japonicas* accounted for 1.70% and 10.92% of the total fatty acid content, respectively. In the present study, the muscle fatty acid contents of EPA, DHA, and EPA + DHA were higher in *L. vannamei*, ranging from 6.9 to 7.1%, 7.6 to 8.4%, and 14.49 to 15.56%, respectively, which were significantly higher than those of shrimp species such as *M. nipponenses* (EPA0.05%), *Exopalaemon modestus* (EPA0.04%), and *Monopterus albus* (EPA 0.74%, DHA 1.33%) and *snakehead* (DHA 1.4%) [49]. As a result, *L. vannamei* is a high-quality shrimp with a perfect fatty acid profile. Although the amino acid composition of the three populations of *L. vannamei* is consistent, the difference in fatty acid composition provides a direction for optimizing aquaculture and nutrition management strategies. Breeding environment and feed formula play a key role in optimizing the fatty acid composition of shrimp. The future aquaculture management and nutrition strategy should focus on improving the nutritional value and market competitiveness of shrimp by adjusting the key ingredients in the feed and optimizing environmental conditions. The scientific design of feed nutrition and comprehensive consideration of natural and artificial food sources will help to improve the health level and nutritional quality of shrimp products. In addition, by increasing the contents of polyunsaturated fatty acids (PUFAs) and omega-3 fatty acids, the nutritional status of vulnerable groups can be further improved, and the overall benefits of aquaculture can be enhanced.

4.5. Analysis of Differences in Physiological Indexes of Different Stocks of Litopenaeus vannamei

In this study, the physiological indicators of *L. vannamei*, including total antioxidant capacity (T-AOC), superoxide dismutase (SOD) activity, and catalase (CAT) activity, showed no significant differences between the groups. This indicates that the environmental and physiological conditions of the shrimp in each group were largely consistent throughout the experiment, without significant environmental stress or other external disturbances. This observation suggests that the farming conditions provided during the experiment did not induce oxidative stress responses or significant physiological pressure, thus ensuring the consistency of the experimental conditions. Antioxidant enzymes, such as SOD and CAT, are key indicators used to assess the antioxidant capacity of organisms and are typically employed to evaluate the response to oxidative stress [50]. These enzymes help neutralize reactive oxygen species (ROS) to prevent oxidative damage, thereby maintaining the redox balance within cells [51]. T-AOC, as a comprehensive indicator of antioxidant capacity, reflects the overall antioxidant status of the organism [52]. In this study, the stable activity of these antioxidant enzymes between the groups indicates that the shrimp populations did not experience significant oxidative stress throughout the experiment. This result further confirms the controlled and consistent experimental conditions. Since oxidative stress is often associated with environmental stress, stocking density, or pathological conditions, the stability in antioxidant enzyme activity rules out these factors as potential influences on the experimental results. In particular, in aquaculture, changes in antioxidant indicators are often early signs of stress response. The absence of such changes in this study indicates that the experimental conditions were relatively ideal and stable for the shrimp populations.

Moreover, the findings regarding antioxidant enzyme activity provide important insights into explaining the observed differences in fatty acid composition. Although there were differences in fatty acid composition among the groups, these differences cannot be attributed to oxidative stress or physiological pressure. Lipid peroxidation is a common cause of changes in fatty acid composition. However, given the stability of antioxidant enzyme activity during the experiment, it is reasonable to infer that the differences in fatty acid composition were likely caused by other factors, such as feed composition or slight variations in farming conditions, rather than by oxidative stress. In summary, the stability of the antioxidant physiological indicators demonstrates that the shrimp populations maintained good health throughout the experiment, without significant environmental or physiological stress. This result provides a solid foundation for the analysis of differences in fatty acid composition, ruling out oxidative stress as a potential factor influencing changes in fatty acids.

4.6. Analysis of Differences in Genetic Diversity of Different Stocks of *Litopenaeus vannamei*

Genetic diversity is the cornerstone of a species' ability to adapt to intricate and fluctuating environments, ensuring survival and evolution [53]. The extent of genetic variation significantly impacts a species' evolutionary capacity and its resilience to environmental shifts. Parameters such as Polymorphic Information Content (PIC) are vital in assessing populations' genetic diversity [54]. PIC represents the probability that alleles from offspring and their parents originate from the same source. PIC values exceeding 0.5 are considered highly polymorphic, while values between 0.25 and 0.5 are moderately polymorphic, and values below 0.25 are categorized as low polymorphism. In the study of the *L. vannamei* populations, the PIC values for the LvA, LvB, and LvC populations were 0.303 ± 0.125, 0.317 ± 0.124, and 0.300 ± 0.130, respectively, indicating moderate polymorphism levels across the three populations.

The single nucleotide polymorphism (SNP) density and the nucleotide diversity (π) for all three populations were low, suggesting limited genetic diversity. The observed heterozygosity (Ho) was consistent in the LvA and LvB populations (0.048 ± 0.001), but significantly lower in the LvC population (0.033 ± 0.014), indicating a higher degree of inbreeding in LvC. This was further confirmed by the inbreeding coefficient (FIS), which was highest in the LvC population (0.887), underscoring the limited genetic diversity issue. Analysis of the genetic differentiation index revealed a certain degree of genetic differentiation among the populations. The genetic differentiation coefficient (FST = 0.106) and genetic distance (DR = 0.112) between the LvB and LvC populations were the highest, indicating significant genetic differences between these two populations. This is consistent with the results of Ren et al. (FST = 0.09) [55]. Comparatively, the genetic differentiation between LvA and the other two populations was lower but still notable. Gene flow (Nm) data indicated that gene flow was highest between LvA and LvB (4.214), suggesting frequent genetic exchange between these two populations. Conversely, gene flow between LvC and the other two populations was lower, with the lowest gene flow observed between LvC and LvB (2.108), further supporting the conclusion of limited genetic diversity in the LvC population. These findings highlight the need to strengthen genetic monitoring of *L. vannamei* populations to ensure the diversity of genetic resources and the maintenance of population health, especially in the face of dual environmental and economic pressures. Through the effective management and use of these genetic data, we can lay a solid foundation for future breeding activities. Scholars Crawford and Littlepohn [56] emphasize that differentiation time predominantly influences inter-population variations, with genetic distance serving as an objective metric to gauge differentiation time and genetic diversity within populations. In this study, the populations with the most significant genetic distance were LvB and LvC, with a measured value of 0.112. In contrast, the genetic distance between LvA and LvB was the smallest, at 0.057, indicating a close genetic relationship between LvA and LvB. At the same time, LvB and LvC exhibited a distant genetic relationship. This study revealed significant genetic diversity and structure differences among the three *L. vannamei* populations. The LvC population exhibited lower genetic diversity and a higher inbreeding coefficient, indicating genetic uniformity and a potentially higher risk of inbreeding depression. Lower genetic diversity may reduce the population's ability to adapt to environmental changes, leading to weaker resilience and long-term cultivation performance [57]. To enhance the genetic diversity of these populations, future breeding programs should increase gene flow between populations. Additional genetic resources can increase genetic variation and improve inbreeding conditions, ensuring the health and sustainability of cultured *L. vannamei* populations.

5. Conclusions

This study presents a comprehensive comparative analysis of the phenotypic characteristics, nutritional composition, physiological biochemistry, and genetic diversity among three populations of *L. vannamei*. The results revealed no significant differences in muscle crude protein content, fat content, or amino acid content among the three populations, indicating a high degree of consistency and substantial nutritional value. Principal component analysis identified body length, carapace length, and the length of the second antenna as critical distinguishing features among these populations, with a cumulative contribution rate of 70.20%. Genetic diversity analysis showed that the LvC population exhibited a higher inbreeding coefficient and lower observed heterozygosity, highlighting issues of constrained genetic diversity. Furthermore, the genetic differentiation coefficient and genetic distance indicated the most significant genetic differentiation between LvB and LvC, whereas LvA exhibited higher gene flow with the other two populations, suggesting more frequent genetic exchanges. In summary, this study found no significant differences in nutritional composition or morphological traits among the three populations but did identify significant differences in genetic diversity. These findings underscore the importance of enhancing gene flow between populations and optimizing breeding strategies to improve genetic diversity and adaptability. Such measures are crucial for ensuring the sustainable development of *L. vannamei*.

Author Contributions: F.Z. conceived the project and supervised the work. Y.L. and S.C. performed the bioinformatics analysis and prepared the manuscript and the tables and figures. S.J. (Shigui Jiang), J.H. and S.C. conducted the experiment. L.Y., S.J. (Song Jiang), and Q.Y. collected the samples and performed sequencing. All authors have read and agreed to the published version of the manuscript.

Funding: This research was funded by National Key R & D Program of China (2022YFD2401900, 2022YFD2400104); the Central Public-interest Scientific Institution Basal Research Fund, CAFS (2023TD34, 2021SD13); the China Agriculture Research System (CARS-48); the Guangdong Basic and Applied Basic Research Foundation (2023A1515012410); the Hainan Provincial Natural Science Foundation of China (323MS127, 322RC806); the project funded by the China Postdoctoral Science Foundation (2023T160178); the Hainan Province postdoctoral research funding project (2022-BH-13, 2022-BHMS-03); the project of Sanya Yazhou Bay Science and Technology City (SCKJ-JYRC-2022-01).

Institutional Review Board Statement: The animal study was reviewed and approved by the Animal Care and Use Committee of the South China Sea Fisheries Research Institute.

Informed Consent Statement: Not applicable.

Data Availability Statement: No new data were created or analyzed in this study.

Conflicts of Interest: The authors declare no conflicts of interest.

References

1. Chen, X.M. Analyzation of Single Nucleotide Polymorphisms in CAT Gene and SOD Gene and Associated with Resistant Hypoxia Traits in *Litopenaeus Vannamei*. Master's Thesis, Guangdong Ocean University, Zhanjiang, China, 2017.
2. Li, M.; Wen, H.; Huang, F.; Wu, M.; Yu, L.; Jiang, M.; Lu, X.; Tian, J. Role of arginine supplementation on muscular metabolism and flesh quality of Pacific white shrimp (*Litopenaeus vannamei*) reared in freshwater. *Front. Nutr.* **2022**, *9*, 980188. [CrossRef] [PubMed]
3. Santos, C.A.; Andrade, S.C.S.; Freitas, P.D. Identification of SNPs potentially related to immune responses and growth performance in *Litopenaeus vannamei* by RNA-seq analyses. *PeerJ* **2018**, *6*, e5154. [CrossRef] [PubMed]
4. Luo, Z.; Yu, Y.; Zhang, Q.; Bao, Z.; Xiang, J.; Li, F. Comparative Transcriptome Analysis Reveals the Adaptation Mechanism to High Salinity in *Litopenaeus vannamei*. *Front. Mar. Sci.* **2022**, *9*, 864338. [CrossRef]
5. Lind, C.E. Population Genetics, Phylogeography and the Effects of Aquaculture on Genetic Diversity of the Silver-Lipped Pearl Oyster, *Pinctada Maxima* (Jameson). Ph.D. Thesis, James Cook University, Douglas, Australia, 2009.
6. Zhao, Y.; Zhu, X.; Li, Z.; Xu, W.; Dong, J.; Wei, H.; Li, Y.; Li, X. Genetic diversity and structure of Chinese grass shrimp, *Palaemonetes sinensis*, inferred from transcriptome-derived microsatellite markers. *BMC Genet.* **2019**, *20*, 75. [CrossRef] [PubMed]
7. Teletchea, F. Fish domestication in aquaculture: 10 unanswered questions. *Anim. Front.* **2021**, *11*, 87–91. [CrossRef] [PubMed]
8. Hsu, T.-H.; Gwo, J.-C. Genetic diversity and stock identification of small abalone (*Haliotis diversicolor*) in Taiwan and Japan. *PLoS ONE* **2017**, *12*, e0179818. [CrossRef]

9. Yang, M.; Ding, F.J. Correlation Analysis between Phenotypic Traits and Body Weight of *Macrobrachium rosenbergii* at Different Month Ages. *J. Jiangsu Ocean. Univ. Nat. Sci. Ed.* **2021**, *30*, 1–9.
10. *GB5009.3-2016*; Determination of Moisture in Food. National Standards of the People's Republic of China: Beijing, China, 2016.
11. *GB/T 6438-92*; Determination of Crude Ash in Feedstuffs. National Standards of the People's Republic of China: Beijing, China, 1992.
12. *GB/T 6432-94*; Determination of Crude Protein by Kjeldahl Method. National Standards of the People's Republic of China: Beijing, China, 1994.
13. *GB 5009.6-2016*; National Food Safety Standard-Determination of Crude Fat in Food. National Health and Family Planning Commission: Beijing, China, 2016.
14. *GB/T 9695.31-2008*; Determination of Total Sugar in Food by Ultraviolet-Visible Spectrophotometry. National Standards of the People's Republic of China: Beijing, China, 2008.
15. *GB/T 5009.124-2016*; National Food Safety Standard-Determination of Amino Acids in Foods by Hydrochloric Acid Hydrolysis. National Health and Family Planning Commission: Beijing, China, 2016.
16. *GB/T 5009.168-2016*; National Food Safety Standard-Determination of Fatty Acids in Foods by Gas Chromatography. National Health and Family Planning Commission: Beijing, China, 2016.
17. Zhou, F.; Xiao, J.; He, Z.; Zhang, H.; He, F.; Wu, J. The Muscular Proximate Composition and Activities of Digestive and Immune Enzymes in Different Sizes of Mozambican Tiger Shrimp *Penaeus monodon*. *Fish. Sci.* **2013**, *32*, 653–656.
18. Chen, S.; Huang, T.; Zhou, Y.; Han, Y.; Xu, M.; Gu, J. AfterQC: Automatic filtering, trimming, error removing and quality control for fastq data. *BMC Bioinform.* **2017**, *18*, 80. [CrossRef]
19. Stuart, K.C.; Edwards, R.J.; Sherwin, W.B.; Rollins, L.A. Contrasting Patterns of Single Nucleotide Polymorphisms and Structural Variation Across Multiple Invasions. *Mol. Biol. Evol.* **2023**, *40*, msad046. [CrossRef]
20. Xiong, X.; Samollow, P.B.; Cao, W.; Metz, R.; Zhang, C.; Leandro, A.C.; VandeBerg, J.L.; Wang, X. Genetic and genomic architecture in eight strains of the laboratory opossum *Monodelphis domestica*. *G3 Genes Genomes Genet.* **2022**, *12*, jkab389. [CrossRef] [PubMed]
21. Liu, J.; Xu, L.; Ding, X.; Ma, Y. Genome-Wide Association Analysis of Reproductive Traits in Chinese Holstein Cattle. *Genes* **2023**, *15*, 12. [CrossRef] [PubMed]
22. Park, N.; Kim, H.; Oh, J.; Kim, J.; Heo, C.; Kim, J. PAPipe: A Pipeline for Comprehensive Population Genetic Analysis. *Mol. Biol. Evol.* **2024**, *41*, msae040. [CrossRef] [PubMed]
23. Zhao, X.; Zheng, T.; Gao, T.; Song, N. Whole-genome resequencing reveals genetic diversity and selection signals in warm temperate and subtropical *Sillago sinica* populations. *BMC Genom.* **2023**, *24*, 547. [CrossRef] [PubMed]
24. Ren, S.; Zhao, L.; Wu, X.; Ying, Z.; Ling, Y.; Cai, Z.; Ye, J.; Li, Q.; Tong, W.; Weng, X.; et al. Effects of Different Feeding Strategies on Growth, Digestion and Antioxidant capacity of Chinese Hook Snout Carp *Opsariicjthys bidens*. *J. Fish. Sci.* **2024**, 1–7.
25. Lu, C.; Yang, W.; Feng, Y.; Liu, Y.; Li, T.; Li, L.; Cao, Y.; Xu, C. Effect of Heating Temperature on Volatile Components of Sweet Potato Caramel Flavor. *Flavour Fragr. Cosmet.* **2024**, *2*, 133–141.
26. Yu, H.; Liu, F.; Zhu, G.; Ren, X.; Liu, X.; Ling, W.; Su, S. Analysis of Genetic Diversity of *P. clarkii* Culture Population in Chuzhou Based on Mitochondrial COI Gene. *J. Anhui Agric. Sci.* **2022**, *50*, 111–113+125.
27. Zhou, Z.; Zhang, L.; Huo, Y.; Li, H.; He, X.; Zhang, Y. Morphological differences and genetic diversity of *Paratapes undulatus* from an area of southern Fujian and Beihai sea, Guangxi. *J. Appl. Oceanogr.* **2023**, *42*, 581–590.
28. Wang, J.; Wang, Q.; Qin, Z.; Chen, Y.; Li, J.; Li, J. Development of SSR markers from genomic data for *Litopenaeus vannamei* and analysis of genetic diversity in different cultured populations. *J. Fish.* **2023**, *47*, 64–74.
29. Luo, X.; Qin, W.; Liang, H.; Sha, H.; Zou, G. Genetic monitoring of Changfeng silver carp based on microsatellite. *J. Hydrobiol.* **2022**, *46*, 725–734.
30. Yang, X.; Xia, M.; Jiang, J.; Wang, J.; Chen, N.; Liao, M.; Lian, Y. Genetic diversity of *Pseudorasbora elongata* based on mitochondrial Cyt b gene and D-loop region sequences. *J. Fish. Sci.* **2023**, *30*, 1031–1041.
31. Zhao, M.; Li, Q.; Qin, Q.; Cai, Y.X.; Xu, Z.Q.; Wang, M.H.; Chen, M.H.; Chen, X.H.; Ge, J.C.; Bian, W.J. Genetic diversity based on microsatellite markers in five channel catfish (*Ictalurus punctatus*) populations. *Fish. Sci.* **2011**, *32*, 24–30.
32. Li, X.; Cui, Q.; Yuan, C. Morphological Differences and Discriminant Analysis of Six Populations of *Litopenaeus vannamei*. *J. Hydroecology* **2010**, *31*, 87–91.
33. Yao, G.; Li, X. Determination of Nutritional Components of *Macrobrachium rosenbergii*. *Fish. Sci. Technol. Inf.* **1981**, *4*, 23+32.
34. Xu, X.; Liu, X.; Yan, B.; Xu, J.; Xu, G.; Shao, Y.; Wu, J. Nutritional Component Analysis and Quality Evaluation of *Penaeus japonicus*. *Food Sci.* **2011**, *32*, 297–301.
35. Zhang, T.; Lin, H.; Ge, J.; Lu, Q.; Peng, G.; Li, X. Analysis of Muscle Nutritional Quality of Wild Populations of *Macrobrachium nipponense* and *Exopalaemon modestus*. *Feed. Res.* **2008**, *1*, 59–63.
36. Chen, X.; Chen, Q.; Xie, D. Evaluation of Meat Yield and Muscle Nutritional Value of *Litopenaeus vannamei*. *Fish. Sci. Technol. Inf.* **2001**, *4*, 165–168.
37. Li, L. Analysis of Muscle Nutritional Components of *Macrobrachium nipponense* and *Procambarus clarkii* in Nanwan Reservoir. *Water Conserv. Fish.* **2005**, *3*, 28–29.
38. Huang, Y.; Yang, X.; Wang, Y.; Li, Y.; Zhou, K.; Sun, Z.; Me, Z.; Lu, J.; Lai, Q. Analysis and quality evaluation of nutrient components in muscle of *Litopenaeus vannamei* of two saline—*Alkali* aquaculture and mariculture. *Fish. Inf. Strategy* **2023**, *38*, 60–66.

39. Mai, K.; Wei, Y.; Wang, J.; Zhang, W. Main nutrient compositions of Antarctic krill and its application in aquafeeds. *J. Ocean. Univ. China Nat. Sci. Ed.* **2016**, *46*, 1–15.
40. Anzani, C.; BoukidM, F.; Drummond, L.; Mullen, A.M.; Álvarez, C. Optimising the use of proteins from rich meat co-products and non-meat alternatives: Nutritional, technological and allergenicity challenges. *Food Res. Int.* **2020**, *137*, 109575. [CrossRef] [PubMed]
41. Shan, F.; Liu, B.; Sun, C.; Zhou, Q.; Song, C.; Sun, H.; Yang, Z. Effect of Feed Vitamin E Content on Growth Performance, Serum Biochemical Indicators and Muscle Quality of *Macrobrachium rosenbergii*. *J. Anim. Nutr.* **2019**, *31*, 214–225.
42. Tang, J.; Qin, B.; Ye, J.; Dai, Y. Effects of stocking modes on growth traits and muscle nutritional composition of *Macrobrachium nipponense*. *J. Zhejiang Agric.* **2024**, *36*, 254–263.
43. FAO/WHO. *Protein Quality Evaluation: Report of the Joint FAO/WHO Expert Consultation, Bethesda, MD, USA, 4–8 December 1989*; Food and Agriculture Organization of the United Nations: Rome, Italy, 1991.
44. Fauconneau, B.; Alami-Durante, H.; Laroche, M.; Marcel, J.; Vallot, D. Growth and meat quality relations in carp. *Aquaculture* **1995**, *129*, 265–297. [CrossRef]
45. Wang, L.; Chen, Z.Z.; Leng, X.J.; Gao, J.Z.; Sun, P.Y.; Qu, H.C.; Zhang, X.X.; Song, X.L. Comparison of muscle composition of wild and cultured discus fishes *Symphysodon* spp. *J. Shanghai Ocean. Univ.* **2016**, *25*, 719–725.
46. Liu, Z.; Liu, Q.; Zhang, D.; Wei, S.; Sun, Q.; Xia, Q.; Shi, W.; Ji, H.; Liu, S. Comparison of the Proximate Composition and Nutritional Profile of Byproducts and Edible Parts of Five Species of Shrimp. *Foods* **2021**, *10*, 2603. [CrossRef]
47. Lin, J.W.; Zhang, C.X.; Sun, Y.Z.; Zhai, S.W.; Song, K.; Ye, J. Complete replacement of fish meal by poultry by-product meal in practical diets supplemented with coated amino acids for Pacific white shrimp *Litopenaeus vannamei*. *Chin. J. Anim. Nutr.* **2015**, *27*, 1709–1721.
48. Shi, R.H. Genetic and Molecular Analysis of Fatty Acid Quality Traits of the Pacific Oyster. Ph.D. Thesis, University of Chinese Academy of Sciences (Institute of Oceanology, Chinese Academy of Sciences), Beijing, China, 2022.
49. Huang, B.X.; Wang, D.W.; Wang, J.F. Progress in the research on polyunsaturated fatty acids. *Agric. Eng. Technol. Process. Agric. Prod. Prog. Biotechnol.* **2009**, *8*, 26–30.
50. Aydın Yaz, Y.; Yıldırım, N.; Yaz, Y.; Tekin, N.; İnal, M.; Şahin, F.M. Role of Oxidative Stress in Pseudoexfoliation Syndrome and Pseudoexfoliation Glaucoma. *Turk. J. Ophthalmol.* **2019**, *49*, 61–67. [CrossRef] [PubMed]
51. Attia, H.G.; El-Morshedy, S.M.; Nagy, A.M.; Ibrahim, A.M.; Aleraky, M.; Abdelrahman, S.S.; Osman, S.M.; Alasmari, S.M.; El Raey, M.A.; Abdelhameed, M.F. Citrus clementine Peel Essential Oil Ameliorates Potassium Dichromate-Induced Lung Injury, Insights into the PI3K/AKT Pathway. *Metabolites* **2024**, *14*, 68. [CrossRef] [PubMed]
52. Ren, Z.; Fan, Y.; Zhang, Z.; Chen, C.; Chen, C.; Wang, X.; Deng, J.; Peng, G.; Hu, Y.; Cao, S.; et al. Sodium selenite inhibits deoxynivalenol-induced injury in GPX1-knockdown porcine splenic lymphocytes in culture. *Sci. Rep.* **2018**, *8*, 17676. [CrossRef] [PubMed]
53. O'Connell, M.; Wright, J.M. Microsatellite DNA in Fishes. *Rev. Fish Biol. Fish.* **1997**, *7*, 331–363. [CrossRef]
54. Li, J.L.; Li, H.X.; Tang, Y.K.; Yu, J.H.; Yu, F. Genetic difference analysis of two genetically improved farmed tilapia populations by using microsatellite marker. *J. South. Agric.* **2015**, *46*, 138–143.
55. Ren, S.; Mather, P.B.; Tang, B.; Hurwood, D.A. Levels of genetic diversity and inferred origins of *Penaeus vannamei* culture resources in China: Implications for the production of a broad synthetic base population for genetic improvement. *Aquaculture* **2018**, *491*, 221–231. [CrossRef]
56. Crawford, A.M.; Littlejohn, R.P. The use of DNA markers in deciding conservation priorities in sheep and other livestock. *Anim. Genet. Resour. Inf.* **1998**, *23*, 21–26. [CrossRef]
57. Hu, Y.; Li, H.; Xia, J. Population Structure, Genetic Diversity, and Conservation Strategies of a Commercially Important Sleeper Fish, *Odontobutis potamophilus* (Gobiiformes: Odontobutidae) Based on Gene-Capture Data. *Front. Genet.* **2022**, *13*, 843848. [CrossRef]

biology

MDPI

Article

Development of DNA Markers for Acute Hepatopancreatic Necrosis Disease Tolerance in *Litopenaeus vannamei* through a Genome-Wide Association Study

Sukhuman Whankaew [1], Phassorn Suksri [2], Ammara Sinprasertporn [3], Jumroensri Thawonsuwan [3] and Ponsit Sathapondecha [2,*]

1 Faculty of Technology and Community Development, Thaksin University, Phatthalung Campus, Phatthalung 93210, Thailand; sukhuman.wha@gmail.com
2 Center for Genomics and Bioinformatics Research, Division of Biological Sciences, Faculty of Science, Prince of Songkla University, Hat Yai, Songkhla 90110, Thailand; phassorn.tua5122@gmail.com
3 Songkhla Aquatic Animal Health Research and Development Center, Department of Fisheries, Songkhla 90110, Thailand; naphanang@gmail.com (A.S.); pjumroensri@gmail.com (J.T.)
* Correspondence: ponsit.s@psu.ac.th

Simple Summary: Acute hepatopancreatic necrosis disease (AHPND) is one of the most serious diseases, leading to massive levels of death in marine shrimp. To achieve sustainable prevention, molecular breeding is a promising approach that requires DNA markers for genetic selection. The aim of this study was to identify DNA markers associated with the AHPND-tolerant phenotype in *Litopenaeus vannamei* using DArT sequencing and a genome-wide association study. The post-larval shrimp were examined for infection with *Vibrio parahaemolyticus*, which causes AHPND (Vp_{AHPND}), and examined for AHPND susceptibility and tolerance. DNA markers were identified in 93 individuals, including single-nucleotide polymorphisms (SNPs) and insertions/deletions (InDels). We found a number of filtered SNPs and InDels, while only four SNPs and 17 InDels were significantly associated with the trait of AHPND tolerance. This information could be useful as a genetic tool for the selection of AHPND-tolerant shrimp.

Abstract: Shrimp aquaculture is facing a serious disease, acute hepatopancreatic necrosis disease (AHPND), caused by *Vibrio paraheamolyticus* (Vp_{APHND}). For sustainable shrimp aquaculture, massive losses of shrimp infected with Vp_{APHND} must be prevented. Research and selection of shrimp tolerant to Vp_{APHND} infection is a sustainable approach to reducing the risk of AHPND. This study focused on the identification and development of potential DNA markers associated with AHPND using DArT sequencing (DArTSeq) and a genome-wide association study. Three populations of post-larval *Litopenaeus vannamei* were immersed in Vp_{APHND} to collect susceptible (D) and tolerant (S) samples. The 45 D and 48 S shrimp had their genotypes analyzed using DArTSeq. A total of 108,983 SNPs and 17,212 InDels were obtained from the DArTseq data, while the biallelic 516 SNPs and 2293 InDels were finally filtered with PIC < 0.1, MAF < 0.05, and a call rate $\geq$ 80%. The filtered variants were analyzed for their association with AHPND tolerance. Although there were no significantly associated SNPs and InDels above the Bonferroni correction threshold, candidate variants, four SNPs and 17 InDels corresponding to $p < 0.01$, were provided for further validation of the AHPND tolerance trait. The candidate SNPs are located on an exon of the zinc finger protein 239-like gene, an intron of an uncharacterized gene, and in intergenic regions. Most of the candidate InDels are in the intergenic regions, with fewer in the intronic and exonic regions. This study provides information on SNPs and InDels for white shrimp. These markers will support the variant database of shrimp and be useful in shrimp aquaculture for breeding selection.

Keywords: *Vibrio parahaemolyticus*; acute hepatopancreatic necrosis disease; DArT sequencing; genome-wide association study; shrimp

1. Introduction

Pacific white shrimp (*Litopenaeus vannamei*) is one of the most economically important animals in several countries. However, in recent decades, acute hepatopancreatic necrosis disease (AHPND) or early mortality syndrome (EMS) has emerged in shrimp farming, leading to massive mortalities in shrimp farming and thus economic losses [1]. Vibrio parahaemolyticus was first identified as the causative agent of AHPND (Vp_{AHPND}) [2]. There have been several attempts to control AHPND in shrimp farms, such as by using antibiotic treatment, pond management, and post-larval pathogen-free control, but they have not been successful so far [1]. To achieve sustainable control of AHPND, genetic selection is one of the most promising approaches to reducing the risks of pathogen infection in aquaculture by improving genetic accuracy [3].

Molecular breeding is one of the most accurate and precise approaches for selecting animals with the desired traits using molecular markers [4]. This involves the use of marker-assisted selection (MAS) with DNA markers obtained from quantitative trait loci (QTL) and/or genome-wide association study (GWAS) analyses [5]. Single-nucleotide polymorphisms (SNPs) and insertions/deletions (InDels) are the most common DNA markers used for MAS due to their high diversity and frequency in the genome of aquatic species to facilitate breeding selection and accelerate the discovery of genes related to economic traits, such as fast growth [6,7]. In shrimp, SNPs and candidate genes were identified using GWAS for *L. vannamei* body weight [8], white spot syndrome virus (WSSV) [9], and sex phenotype [10]. However, the identification of variants in shrimp tolerant to Vp_{AHPND} has not been thoroughly studied. For example, the estimation of shrimp tolerance to Vp_{AHPND} using twelve microsatellite markers was studied [11]. The use of 508 amplicons from targeted sequencing yielded 30 candidate SNPs related to AHPND resistance in white shrimp [12].

The development of high-throughput sequencing platforms has become a more powerful tool to dissect the genetic basis underlying the traits. In shrimp, association analysis of 2b-RAD sequencing data, together with gene-based genotyping from two populations, identified a novel growth-related DNA marker in *L. vannamei* [13]. Targeted amplicon sequencing was used to analyze genotypes to provide an SNP dataset of *L. vannamei* [14]. In addition, SNPs were developed from RNA sequencing data [15] and a 50K SNP chip [9] in *L. vannamei*. DArTSeq, a similar technique to ddRAD sequencing, has the advantage of reducing complexity by digesting methylated DNA sequences using restriction enzymes to the genome in a library preparation before undergoing Illumina sequencing [16]. It has been widely used as a genotyping and sequencing approach to obtain either silico DArT and DArT SNPs or SNPs and InDels [17]. A comparison of marker-associated traits using m and GBS indicated that some QTLs are commonly found in European winter wheat [18]. In addition, it is widely used to identify variants and subsequent analysis in QTL and/or GWAS [19–21]. It was also used to identify and develop SNPs in *Penaeus monodon* [22]. To date, there is no large-scale sequencing data used to investigate an association of AHPND tolerance in shrimp. Therefore, this study aimed to identify DNA markers using DArTSeq and analyzed the association of genotypes with AHPND tolerance in *L. vannamei*. This would provide a molecular basis for the prevention of disease and accelerate the breeding of shrimp that are tolerant to AHPND.

2. Materials and Methods

2.1. Animal Samples

Post-larval (PL15–20) *L. vannamei* were obtained from three different local shrimp farms in Songkhla province, Thailand. They were acclimatized to 30 ppt seawater at 25 °C for 3 days before use. Shrimp were fed twice daily with commercial feed pellets. The ethics for the use of animals were approved by the Institutional Animal Care and Use Committee, PSU (Ref. No. 97/2021), and the protocol was performed according to the regulation regarding the Animals for Scientific Purposes Act, B.E. 2558 (A.D. 2015), Thailand.

2.2. Bacterial Culture

Vp_{AHPND} J36 was streaked and cultured on TCBS agar at 37 °C overnight. A single colony was picked and cultured overnight in tryptic soy broth (TSB + 1.5% *w*/*v* NaCl) at 37 °C, 250 rpm. After overnight cultivation, the concentration was measured as cfu/mL by spreading the serially diluted culture on TSA containing 1.5% *w*/*v* NaCl.

2.3. Mortality Determination and Sample Collection of Shrimp Infected with Vp_{AHPND}

Post-larval shrimp were reared in 1 L of 30 ppt seawater at a temperature of approximately 25 °C. The cultured seawater was changed every 4 days after the start of the experiment. Their lengths and weights were 0.94 ± 0.17 cm and 5.32 ± 2.14 mg, respectively. Approximately 150–200 individual shrimp from each population were used. They were immersed in 1×10^5 cfu/mL Vp_{AHPND} J36. Dead or moribund shrimp (D) were collected every single day as AHPND-susceptible shrimp, while surviving shrimp (S) were harvested on day 14 as AHPND-tolerant shrimp. This was performed in triplicate. Whole bodies were extracted for their DNA using a conventional phenol/chloroform extraction method. The quality and quantity of DNA samples were evaluated by agarose gel electrophoresis and Nanodrop (Thermo Scientific, Waltham, MA, USA), respectively.

2.4. Sequencing and Variant Calling Analysis

The 93 DNA samples, including 48 D and 45 S shrimp, were analyzed for their genotypes by DArTSeq using the Illumina HiSeq 2500 platform with 138 bp single-end sequencing (Australia). The quality of the raw reads was evaluated using FASTQC [23], and adaptors, low-quality sequences, and sequence lengths less than 18 nt were removed using Cutadapt v. 4.4 [24]. To identify variants, the clean reads were mapped to the *L. vannamei* genome (RefSeq Accession No. GCF_003789085) using BWA version 0.7.17 [25] with default parameters. Mark duplication and data sorting were performed using MarkDuplicateSpark before variants were called using the GATK workflow [26]. The variants were hard-filtered with qual < 30, missing values < 20%, AC < 2, Info/DP < 10, QD < 2, and MQ < 40 using bcftools [27]. Filtered variants with polymorphic information content (PIC) < 0.1 and MAF < 0.05 were removed. The final variants were annotated using the ANNOVAR program (version 2020-06-07) [28]. The GFF file was used to construct the ANNOVAR annotation file and retrieve the gene location corresponding to the variant location and the up- and downstream positions of the variants.

For population structure analysis, the final filtered SNPs were used to analyze the dissimilarity index, and factorial analysis was performed using DARwin version 6.0 [29] and the clustering method via the Bayesian approach using STRUCTURE version 2.3.4 [30]. The SNP data were calculated for dissimilarity using the simple matching method with a bootstrap of 500 in the DARwin program. Subsequently, the dissimilarity result was used for factorial analysis and hierarchical clustering with the UPGMA method. To estimate the subpopulation (K), we set the burn-in period to 5000 and MCMC to 50,000. K was set from 1–8, with 20 iterations analyzed in the STRUCTURE program. The Ln P(D) values were then analyzed for K estimation using the delta K method [31] with the STRUCTURE HARVESTER program (version 0.6.93) [32] and the K optimum of 2 was determined. Finally, the population structure was analyzed with K = 2, a burn-in period of 50,000, and an MCMC of 500,000.

2.5. Genome-Wide Association Analysis for AHPND Tolerance

The filtered SNPs and InDels were used to analyze their association with the AHPND-tolerant phenotype in shrimp using GAPIT with a multi-locus mixed model (MLMM) [33]. The number of days that the shrimp were dead after Vp_{AHPND} infection was used as phenotypic data. The variants corresponding to a *p*-value < the Bonferroni correction threshold were selected. The explanation of genetic variation (PVE) was calculated according to [34]. The identification of the gene loci corresponding to the significant SNPs and InDels was performed with the ANNOVAR program. Subsequently, the identified gene loci from the shrimp genome were searched for gene names via the NCBI database. The ANNOVAR output also provided genes upstream and downstream of the SNP and InDel locations.

3. Results

3.1. Determination of Shrimp Mortality after Vp_{AHPND} Infection

Shrimp were infected with Vp_{AHPND} J36 and a number of dead samples were collected every day. The result indicated that most shrimp were dead or moribund 2–3 days after infection. Approximately 5% of shrimp survived until day 14 (Figure 1A). Among the dead group, most shrimp samples were dead on day 2 after infection, and only a few on later days (Figure 1B). Among the three populations, 48 dead (D) and 45 surviving (S) shrimp were selected and their sequences were analyzed using DArTSeq (Table S1).

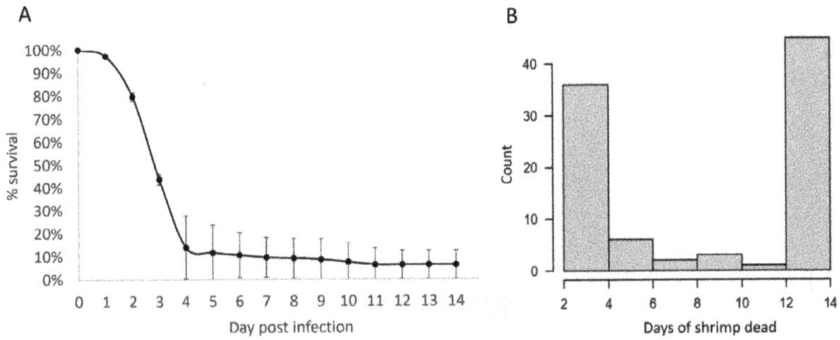

Figure 1. Shrimp mortality after Vp_{AHPND} infection. Post-larval shrimps were immersed in 1×10^5 cfu/mL Vp_{AHPND}, and the number of dead shrimps was recorded daily. The cumulative mortality of shrimp was presented in a line graph (**A**). Approximately 100–150 individuals were used in triplicate. Dots and error bars represent means and SEM, respectively. The number of dead shrimp was used to analyze their genotypes in each day, as shown (**B**).

3.2. DArTSeq and Genetic Diversity among the Three Populations of Tested Shrimp

A total of 93 DNA samples from shrimp infected with Vp_{AHPND} were analyzed using DArTseq. From an average of 3,508,523 raw reads, an average of 3,508,465 clean reads with 285,948,354 bases were obtained (Supplementary Material File S1). These raw data were deposited as a sequence read archive at NCBI (BioProject No. PRJNA1137268). After variant calling, a total of 108,983 SNPs and 17,212 InDels were obtained, and hard filtration provided 53,083 SNPs and 15,557 InDels (Table 1). After a final filtering by discarding variants with PIC < 0.1, MAF < 0.05, and a call rate < 80%, only 516 SNPs and 2292 InDels remained. The details of the filtered SNPs and InDels and their functional annotation are shown in Supplementary Material Files S2 and S3, respectively. The transition/transversion ratios of hard-filtered SNPs and final filtered SNPs were 1.533 and 1.659, respectively (Table 1 and Figure 2A). The distribution of the final filtered InDel gaps showed that most InDels had a gap size of 1 nt (~67%) or 2–4 nt (~24.8%), with fewer having larger gaps (Figure 2B). The biotype of the SNPs and InDels was identified. The results revealed that the most SNPs and InDels were found in the intergenic regions (approximately 75%), with fewer in the intronic (12–14%) and exonic (4–7%) regions (Figure 2C). Moreover, the number of synonymous SNPs was greater than that of non-synonymous SNPs, while most filtered exonic InDels caused a frameshift mutation instead of a non-frameshift mutation (Figure 2D). Variant annotation is in the Supplementary Material Files S2 and S3. The final filtered SNPs were used to construct a dendrogram (Figure 3A) and to conduct factorial analysis (Figure 3B) based on a dissimilarity index. The subpopulation (K) was estimated using the delta K method and the optimum of K was 2 (Figure S2). The result indicated that the samples from two populations (pop1 and pop2) were more closely clustered and separate from another population (pop3) (Figure 3).

Table 1. Summary of the variants analyzed from the experimental shrimp.

Description	SNPs	InDels
Total variants	108,983	17,212
Hard-filtered variants	53,083	15,557
Final filtered variants	516	2292

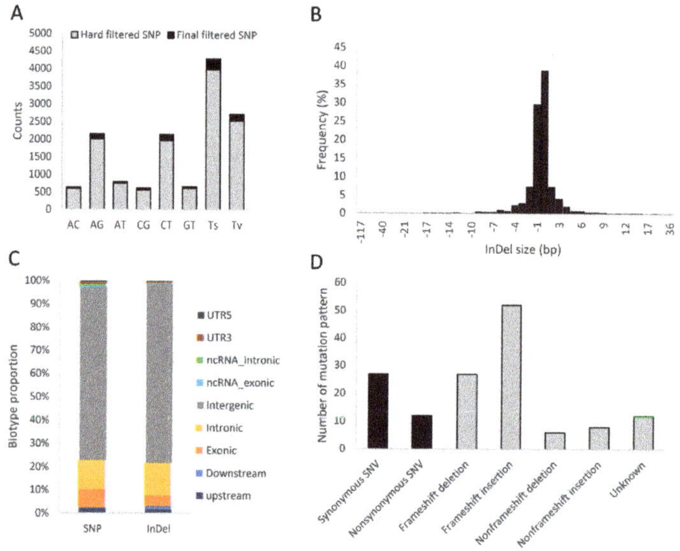

Figure 2. Summarized statistics of the filtered SNPs and InDels. The transition (Ts) and transversion (Tv) of the hard-filtered and final filtered SNPs were calculated and visualized in a bar graph (**A**). Distribution of the final filtered InDels' gaps was counted (**B**). The biotypes of the final filtered SNPs and InDels were counted as percentages (**C**). For exonic variants, the number of mutational patterns was counted (**D**).

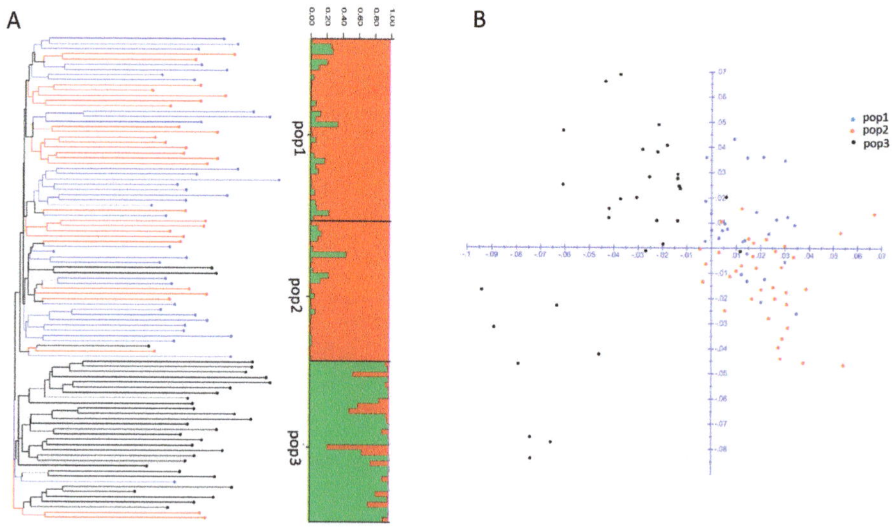

Figure 3. Clustering analysis of shrimp samples. Dendrogram of clustering analysis using the final filtered SNPs with UPGMA by DARwin and with Bayesian by STRUCTURE (**A**). The factorial analysis

was performed based on the dissimilarity index by using DARwin (**B**). The blue, red, and black lines and dots represent data from population 1, 2, and 3, respectively. The red and green bars correspond to the average estimated cluster for all individuals in the population.

3.3. Association Analysis for AHPND Tolerance

The final filtered SNPs and InDels were used to analyze the association with the AHPND-tolerant phenotype using the GAPIT program. Five models were evaluated in this study to select an appropriate model to reduce false positives. The Q-Q plots showed that the MLMM model had the least false positives compared to the other models for SNP data, while the MLMM, farmCPU, and BLINK models for InDels showed similar patterns (Figure S1). In addition, the heritability of the AHPND-tolerant trait was approximately 42% (Figure S3), indicating moderate heritability. In the association analysis, we found that there was no significant association of SNPs and InDels with the AHPND tolerance trait based on the Bonferroni correction threshold (9.6×10^{-5} and 2.14×10^{-5} for SNPs and InDels, respectively) (Figure 4). However, among the variants, four SNPs and 17 InDels had a *p*-value < 0.01 (Figure 4, Tables 2 and 3), which could be considered for further validation in future studies. Since there were too many scaffolds of the *L. vannamei* genome, we grouped the genome scaffolds into a few sequences (Tables 2 and 3).

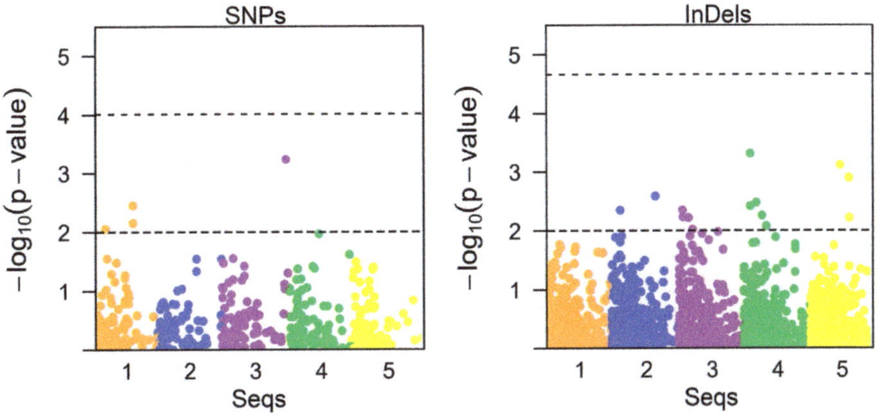

Figure 4. Manhattan plot of association analysis. The final filtered SNPs and InDels were used to perform association analysis with the AHPND-tolerant phenotype using the GAPIT program. The upper and lower dash lines represent the Bonferroni correction thresholds and a *p*-value of 0.01, respectively.

Table 2. The statistics of the associated SNPs to the AHPND-tolerant phenotype.

SNP Id	Biotype	Genome Accession No.	Position	*p*-Value	MAF	Effect	PVE (%)	Gene Name
SNP2515 C > A	Intergenic	NW_020869786.1	872,951	0.0022	0.12	−3.09	6.42%	-
SNP3827 A > C	Intergenic	NW_020870527.1	844,041	0.0033	0.19	−2.31	6.43%	-
SNP6128 A > T	Intronic	NW_020872696.1	875,105	0.0040	0.05	−3.72	9.29%	Uncharacterized (XM_027350909)
SNP2501 T > C	Exonic (Synonymous)	NW_020869780.1	528,035	0.0097	0.18	−2.14	7.07%	Zinc finger protein 239-like (XM_027364481)

Of the variants located in the intergenic regions of the *L. vannamei* genome, only SNP2501 was found on an exon of the zinc finger protein 239-like gene (ZNF239-like), while InDel3386 and InDel3811 were found on exonic regions of nascent polypeptide-associated complex subunit alpha, muscle-specific form-like (NACA), and probable phosphorylase b kinase regulatory subunit alpha (PHKA), respectively (Tables 2 and 3). The phenotypic variation explained (PVE) of the candidate variants showed moderate values in the range of 5–13% (Tables 2 and 3). In addition, extracellular protein-coding genes were mainly found near significant InDels such as vegetative cell wall protein gp1-like and proline-rich extensin-like protein EPR1, as well as some immune-responsive genes, including receptor-type guanylate cyclase Gyc76C-like and Ras GTPase-activating protein (Table 4).

Table 3. The statistics of the associated InDels to the AHPND-tolerant phenotype.

Candidate Markers	Ins/Del Gap (nt)	Biotype	Accession No.	Position	*p*-Value	MAF	Effect	PVE (%)	Gene Name
Indel0773	Ins 2	Intergenic	NW_020868718.1	1,222,723	2.4×10^{-4}	0.17	−2.53	9.88%	-
Indel3486	Del 1	Intergenic	NW_020870028.1	97,508	4.8×10^{-4}	0.31	−2.17	4.78%	-
Indel4880	Ins 2	Intergenic	NW_020870775.1	455,677	7.5×10^{-4}	0.11	−2.87	13.25%	-
Indel4460	Del 1	Intergenic	NW_020870542.1	589,428	1×10^{-3}	0.17	−2.32	4.83%	-
Indel4461	Ins 1	Intergenic	NW_020870542.1	589,433	1×10^{-3}	0.17	−2.32	5.19%	-
Indel1142	Ins 1	Intergenic	NW_020868919.1	635,311	2.5×10^{-3}	0.33	−1.72	5.26%	-
Indel3675	Ins 1	Intergenic	NW_020870115.1	185,999	3.3×10^{-3}	0.065	3.11	8.86%	-
Indel3485	Ins 1	Intergenic	NW_020870028.1	97,483	3.8×10^{-3}	0.31	−1.79	2.97%	-
Indel2533	Ins 1	Intergenic	NW_020869537.1	56,216	3.8×10^{-3}	0.20	2.09	3.83%	-
Indel3811	Ins 1	Intergenic	NW_020870177.1	120,506	4.5×10^{-3}	0.12	−2.45	10.66%	-
Indel0506	Del 1	Intronic	NW_020868577.1	271,787	5.5×10^{-3}	0.065	2.93	10.73%	probable phosphorylase b kinase regulatory subunit alpha (XM_027376870)
Indel2406	Ins 1	Intergenic	NW_020869477.1	65,359	5.9×10^{-3}	0.27	−1.73	5.06%	-
Indel3386	Del 1	Exonic (Frameshift deletion)	NW_020869981.1	594,057	5.9×10^{-3}	0.11	−2.42	14.42%	nascent polypeptide-associated complex subunit alpha, muscle-specific form-like(XM_027366427)
Indel6073	Del 1	Intergenic	NW_020871916.1	143,341	6×10^{-3}	0.65	2.95	11.85%	-
Indel6600	Ins 6	Intergenic	NW_020872438.1	335,379	8.2×10^{-3}	0.19	−1.88	5.92%	-
Indel2592	Ins 1	Intergenic	NW_020869574.1	204,883	9.5×10^{-3}	0.30	−1.68	8.24%	-
Indel2590	Del 3	Intergenic	NW_020869574.1	204,798	9.6×10^{-3}	0.26	−1.87	2.95%	-

Table 4. Neighboring gene distance to the significant variants.

Candidate Markers	Distance (kb)	Gene Name (RefSeq Accession No.)	Function
SNP2515	Up 169.9	uncharacterized LOC113812673 (XM_027364589)	-
	Down 95.7	uncharacterized LOC113812685 (XM_027364609)	-
SNP3827	Up 357.7	vegetative cell wall protein gp1-like (XM_027371696)	Extracellular region
Indel0773	Up 64.4	proline-rich extensin-like protein EPR1 (XM_027353909)	Extracellular region
	Down 70.7	uncharacterized LOC113803169 (XM_027353894)	-
Indel3486	Down114.7	Zinc finger protein 749-like (XM_027366949)	Transcription
Indel4880	Up 29.2	Mucin-2-like (XM_027374070)	Extracellular region
Indel4460	Up 128	Receptor-type guanylate cyclase Gyc76C-like (XM_027371820)	Immune response
Indel4461	Up 128	Receptor-type guanylate cyclase Gyc76C-like (XM_027371820)	Immune response
Indel1142	Up 15.5	probable Ras GTPase-activating protein (XM_027355753)	Tumorigenesis
	Down 90.5	uncharacterized LOC113804845 (XM_027355756)	-
Indel3675	Up 50.2	putative neural-cadherin 2 (XM_027367808)	Cell-cell adhesion
Indel3485	Down 114.8	zinc finger protein 749-like (XM_027366949)	Transcription
Indel2533	Up 30.5	prolyl 4-hydroxylase subunit alpha-1-like (XM_027362215)	Collagen synthesis
	Down 190.4	basic salivary proline-rich protein 2-like (XM_027362206)	Extracellular region
Indel3811	Up 70.9	nephrin-like (XM_027368425)	Secretory system
Indel2406	Down 100.5	uncharacterized histidine-rich protein DDB_G0274557-like (XM_027361657)	-
Indel6073	Up 77.9	Trnal-cag_17	Translation
	Down 10.2	cell wall protein DAN4-like (XM_027379940)	Extracellular region

Table 4. *Cont.*

Candidate Markers	Distance (kb)	Gene Name (RefSeq Accession No.)	Function
Indel6600	Up 190.5	cell surface glycoprotein 1-like (XM_027381906)	Extracellular region
Indel2592	Up 80.9	Trnaf-gaa_48	Translation
	Down 49.9	uncharacterized LOC113810924 (XM_027362580)	-
Indel2590	Up 80.9	Trnaf-gaa_48	Translation
	Down 49.9	uncharacterized LOC113810924 (XM_027362580)	-

4. Discussion

In shrimp aquaculture, AHPND is a serious disease that causes massive death within 3–7 days [1]. There have been several attempts to solve this important problem, such as by using antibiotics and/or probiotics, pond management, and post-larval stock control; however, successful control remains limited [1]. Molecular breeding is a promising tool to accelerate the development of desirable economic organisms [4]. In this study, AHPND-susceptible and -tolerant *L. vannamei* were sampled for genotyping analysis. Within the three populations, most shrimp were dead a few days after infection with Vp_{AHPND}, and only a few shrimps survived (Figure 1). This was similar to the effect of Vp_{AHPND} on shrimp mortality, including *P. monodon* [35] and *L. vannamei* [10,36,37]. These shrimps were dead within 2–3 days post-infection, indicating the virulence of Vp_{AHPND} in shrimp. In addition, among the three populations of samples analyzed, we found that two populations are closely related and different from each other.

In recent studies, the genetic variation of shrimp has been investigated from several physiological aspects, including growth performance [38], ammonia tolerance [39], and pathogen infection [10]. These findings provided genetic markers from genotyping data generated by several types of DNA markers. For example, the genetic estimation of *V. parahaemolyticus* resistance was investigated using twelve microsatellite markers in *L. vannamei* [10]. The thousands of SNPs were used to analyze the genomic selection for growth and WSSV-resistant traits [38] and growth performance [8,13] in white shrimp. In this study, we obtained hundreds of SNPs and InDels from DArTSeq (Table 1). DArTSeq is a GBS technique that has been widely used in several organisms including crustaceans. Using double digestion with the restriction enzymes cleaving methylated DNA in a library preparation step before sequencing provided silico DArT and DArT SNP data, as well as analysis for SNPs and InDels from the sequencing data [16]. For example, thousands of SNPs were obtained after being analyzed from the DArTSeq data of *P. monodon* [22,40] and *Panulirus ornatus* [41]. However, a fairly limited number of SNPs and InDels might be obtained from strict filtering steps. Likewise, approximately 7000 SNPs were obtained and used in the GWAS analysis of the WSSV resistance trait in *L. vannamei* [9]. Some studies also had similar variant filtrations. In addition, The Ts/Tv ratio of the SNPs in this study was similar to other studies. For instance, the Ts/Tv of SNPs in *L. vannamei* was 1.91–1.97 under growth performance and WSSV infection using RNA sequencing [42], and 1.49–1.53 under high- and low-fecundity populations [38]. Most InDels had short gaps (1–3 bp), similar to InDels found in other organisms, including plants and animals [43]. In addition, the proportion of SNPs and InDels was dominant in the intergenic regions and lower in the intragenic regions (Figure 2C). Similarly, the number of SNPs and InDels found in intergenic regions was greater than in intragenic regions, including introns and coding sequences in other animals such as cattle [44].

GWAS is a powerful tool for association analysis between genotypes and phenotypes without breeding information [45]. In shrimp, candidate SNPs were identified that were associated with WSSV resistance and growth performance [9]. Similarly, 18 growth-related and 11 sex-related SNPs were identified in *Macrobrachium nipponense* [46]. In addition, the estimation of Vp_{AHPND} tolerance was performed with twelve microsatellite markers in white shrimp [11]. Considering the moderate heritability of the AHPND tolerance trait [11] and our heritability effect (Figure S3), the sample size in this study appeared small. In addition, there were no significant variants associated with the AHPND tolerance phenotype in relation to the Bonferroni correction threshold, but we found that the candidate four SNPs and 17 InDels were likely involved in the AHPND-tolerant phenotype in white shrimp. This would be useful for further validation with a larger sample size and application. Although there are a few GWAS analyses regarding the AHPND resistance trait in shrimp, our DArTseq data, as large-scale data, could be further used in meta-analysis studies. The meta-analysis of GWAS in combination with publicly independent GWAS results with similar traits could lead to an increase in the sample size, statistical power, and precision of the association analysis [47]. In addition, the association analysis in this study did not include adjustment of covariates such as the length and/or weight of the tested shrimp. This might result in a lower rate of actual association with the phenotype under consideration. However, our dataset could be useful for meta-analysis with a new dataset that adjusts for covariates such as shrimp weight using the recent method [48].

Most candidate variants exhibited moderate PVEs (Tables 2 and 3). For example, some SNPs and InDels were located in the immune-responsive genes, including ZNF239-like, PHKA, and NACA, corresponding to a %PVE of 7.07%, 10.73%, and 14.42%, respectively. This suggests that polygenic involvement, including the candidate variants, has an impact on the AHPND tolerance trait. ZNF plays a role as a transcription factor that regulates several physiological processes, including cancer development [49] and the immune response [50]. PHKA is involved in the control of carbohydrate metabolism, particularly in glycan storage [51]. In shrimp, the increase in carbohydrate metabolism is triggered by pathogen infection [52]. NACA is a transcriptional cofactor that regulates AP-1 transcription, which is involved in cell differentiation [53]. It has been shown that the *EsNACA* expression was upregulated after infection with *Vibrio anguillarum* in a crab, *Eriocheir sinensis* [54]. These may indicate that the variants found on these genes have the potential to influence gene function, leading to a physiological change in AHPND tolerance in shrimp. The candidate SNPs and InDels identified in this study need to be further validated for AHPND tolerance in new shrimp populations. Expression analysis of the SNPs and InDels and their respective genes, as well as the neighboring genes located near the SNPs and InDels, would reveal crucial functions for the AHPND tolerance phenotype in shrimp. The validated ones will be used for the selection of AHPND-tolerant breeding broodstock to produce AHPND-tolerant offspring. This study provides information on genetic variants and potentially associated SNPs and InDels for the AHPND-tolerant trait in white shrimp, which could be useful for shrimp breeding programs in the future.

5. Conclusions

In this study, we conducted and collected *L. vannamei* samples from three populations that were susceptible and tolerant to Vp_{AHPND} infection. Using DArT sequencing, the polymorphic variants, SNPs and InDels, were obtained, and only four candidate SNPs and 17 InDels were selected from the AHPND-tolerant shrimp. This study provided information about variants with AHPND tolerance in white shrimp.

Supplementary Materials: The supporting information can be downloaded at: https://www.mdpi.com/article/10.3390/biology13090731/s1. File S1: Information of the DArT sequencing data; File S2: The final filtered SNPs information; File S3: The final filtered InDels information; Table S1: Number of dead shrimp selected for use in this study; Figure S1: Q-Q plots of the five models' SNPs and InDels. Figure S2: K estimation using Delta K analysis; Figure S3: Heritability of AHPND tolerance analyzed using the GAPIT program.

Author Contributions: Conceptualization, methodology, formal analysis, writing—original manuscript, S.W. and P.S. (Ponsit Sathapondecha); Investigation, P.S. (Phassorn Suksri) and A.S.; Writing—review and editing, Resources, J.T. and A.S.; Data curation, visualization, funding acquisition, P.S. (Ponsit Sathapondecha). All authors have read and agreed to the published version of the manuscript.

Funding: This study was supported by National Science, Research, and Innovation Fund (NSRF) and Prince of Songkla University (Grant No. SCI6505086S).

Institutional Review Board Statement: The animal study protocol was approved by the Institutional Animal Care and Use Committee, PSU (Protocol Ref. No. 97/2021).

Informed Consent Statement: Not applicable.

Data Availability Statement: The DArT sequencing data of 93 shrimp were deposited at the NCBI and available in BioProject No. PRJNA1137268.

Acknowledgments: We would like to thank Aupaporn Morkaew and Thi Hai Au La for the experimental assistance.

Conflicts of Interest: The authors declare no conflicts of interest.

References

1. Kumar, V.; Roy, S.; Behera, B.K.; Bossier, P.; Das, B.K. Acute hepatopancreatic necrosis disease (AHPND): Virulence, pathogenesis and mitigation strategies in shrimp aquaculture. *Toxins* **2021**, *13*, 524. [CrossRef]
2. Tran, L.; Nunan, L.; Redman, R.M.; Mohney, L.L.; Pantoja, C.R.; Fitzsimmons, K.; Lightner, D.V. Determination of the infectious nature of the agent of acute hepatopancreatic necrosis syndrome affecting penaeid shrimp. *Dis. Aquat. Org.* **2013**, *105*, 45–55. [CrossRef] [PubMed]
3. Zenger, K.R.; Khatkar, M.S.; Jones, D.B.; Khalilisamani, N.; Jerry, D.R.; Raadsma, H.W. Genomic selection in aquaculture: Application, limitations and opportunities with special reference to marine shrimp and pearl oysters. *Front. Genet.* **2019**, *9*, 693. [CrossRef] [PubMed]
4. Johnsson, M. Genomics in animal breeding from the perspectives of matrices and molecules. *Hereditas* **2023**, *160*, 20. [CrossRef] [PubMed]
5. You, X.; Shan, X.; Shi, Q. Research advances in the genomics and applications for molecular breeding of aquaculture animals. *Aquaculture* **2020**, *526*, 735357. [CrossRef]
6. Yáñez, J.M.; Barría, A.; López, M.E.; Moen, T.; Garcia, B.F.; Yoshida, G.M.; Xu, P. Genome-wide association and genomic selection in aquaculture. *Rev. Aquac.* **2023**, *15*, 645–675. [CrossRef]
7. Tsai, H.Y.; Hamilton, A.; Tinch, A.E.; Guy, D.R.; Gharbi, K.; Steer, M.J.; Matika, O.; Bishop, S.C.; Houston, R.D. Genome wide association and genomic prediction for growth traits in juvenile farmed Atlantic salmon using a high density SNP array. *BMC Genom.* **2015**, *16*, 969. [CrossRef]
8. Lyu, D.; Yu, Y.; Wang, Q.; Luo, Z.; Zhang, Q.; Zhang, X.; Xiang, J.; Li, F. Identification of growth-associated genes by genome-wide association study and their potential application in the breeding of Pacific white shrimp (*Litopenaeus vannamei*). *Front. Genet.* **2021**, *12*, 611570. [CrossRef]
9. Medrano-Mendoza, T.; García, B.F.; Caballero-Zamora, A.; Yáñez, J.M.; Montoya-Rodríguez, L.; Quintana-Casares, J.C.; Durán-Auilar, M.; Campos-Montes, G.R. Genetic diversity, population structure, linkage disequilibrium and GWAS for resistance to WSSV in Pacific white shrimp (*Litopenaeus vannamei*) using a 50K SNP chip. *Aquaculture* **2023**, *562*, 738835. [CrossRef]
10. Jones, D.B.; Nguyen, H.T.; Khatkar, M.S.; Simma, D.B.; Jerry, D.R.; Raadsma, H.W.; van der Steen, H.; Prochaska, J.; Zenger, K.R. The identification of a major sex QTL in the white-leg shrimp, *Litopenaeus vannamei*. *Aquaculture* **2020**, *529*, 735673. [CrossRef]
11. Lyu, D.; Yu, Y.; Zhang, Q.; Luo, Z.; Wang, Q.; Xiang, J.; Li, F. Estimating genetic parameters for resistance to *Vibrio parahaemolyticus* with molecular markers in Pacific white shrimp. *Aquaculture* **2020**, *527*, 735439. [CrossRef]
12. Zhang, Q.; Yu, Y.; Wang, Q.; Liu, F.; Luo, Z.; Zhang, C.; Zhang, X.; Huang, H.; Xiang, J.; Li, F. Identification of single nucleotide polymorphisms related to the resistance against acute hepatopancreatic necrosis disease in the Pacific white shrimp *Litopenaeus vannamei* by target sequencing approach. *Front. Genet.* **2019**, *10*, 700. [CrossRef] [PubMed]
13. Wang, Q.; Yu, Y.; Zhang, Q.; Zhang, X.; Yuan, J.; Huang, H.; Xiang, J.; Li, F. A novel candidate gene associated with body weight in the Pacific white shrimp *Litopenaeus vannamei*. *Front. Genet.* **2019**, *10*, 520. [CrossRef]
14. Yu, Y.; Luo, Z.; Wang, Q.; Zhang, Q.; Zhang, X.; Xiang, J.; Li, F. Development of high throughput SNP genotyping approach using target sequencing in Pacific white shrimp and its application for genetic study. *Aquaculture* **2020**, *528*, 735549. [CrossRef]
15. Huang, Y.; Zhang, L.; Ge, H.; Wang, G.; Huang, S.; Yang, Z. SNP development in *Penaeus vannamei* via next-generation sequencing and DNA pool sequencing. *Fishes* **2021**, *6*, 36. [CrossRef]
16. Sansaloni, C.; Petroli, C.; Jaccoud, D.; Carling, J.; Detering, F.; Grattapaglia, D.; Kilian, A. Diversity Arrays Technology (DArT) and next-generation sequencing combined: Genome-wide, high throughput, highly informative genotyping for molecular breeding of Eucalyptus. *BMC Proc.* **2011**, *5*, P54. [CrossRef]
17. Mujyambere, V.; Adomako, K.; Olympio, O.S. Effectiveness of DArTseq markers application in genetic diversity and population structure of indigenous chickens in Eastern Province of Rwanda. *BMC Genom.* **2024**, *25*, 193. [CrossRef]
18. Ladejobi, O.; Mackay, I.J.; Poland, J.; Praud, S.; Hibberd, J.M.; Bentley, A.R. Reference genome anchoring of high-density markers for association mapping and genomic prediction in European winter wheat. *Front. Plant Sci.* **2019**, *10*, 1278. [CrossRef] [PubMed]

19. Mogga, M.; Sibiya, J.; Shimelis, H.; Lamo, J.; Yao, N. Diversity analysis and genome-wide association studies of grain shape and eating quality traits in rice (*Oryza sativa* L.) using DArT markers. *PLoS ONE* **2018**, *13*, e0198012. [CrossRef]
20. Nowak, B.; Tomkowiak, A.; Bocianowski, J.; Sobiech, A.; Bobrowska, R.; Kowalczewski, P.Ł.; Bocianowska, M. The use of DArTseq technology to identify markers linked to genes responsible for seed germination and seed vigor in maize. *Int. J. Mol. Sci.* **2022**, *23*, 14865. [CrossRef]
21. Curtolo, M.; Cristofani-Yaly, M.; Gazaffi, R.; Takita, M.A.; Figueira, A.; Machado, M.A. QTL mapping for fruit quality in Citrus using DArTseq markers. *BMC Genom.* **2017**, *18*, 289. [CrossRef] [PubMed]
22. Guppy, J.L.; Jones, D.B.; Kjeldsen, S.R.; Le Port, A.; Khatkar, M.S.; Wade, N.M.; Sellars, M.J.; Steinig, E.J.; Raadsma, H.W.; Jerry, D.R.; et al. Development and validation of a RAD-Seq target-capture based genotyping assay for routine application in advanced black tiger shrimp (*Penaeus monodon*) breeding programs. *BMC Genom.* **2020**, *21*, 541. [CrossRef] [PubMed]
23. Andrews, S. FastQC: A Quality Control Tool for High Throughput Sequence Data [Online]. 2010. Available online: http://www.bioinformatics.babraham.ac.uk/projects/fastqc/ (accessed on 15 January 2022).
24. Martin, M. Cutadapt removes adapter sequences from high-throughput sequencing reads. *EMBnet. J.* **2011**, *17*, 10–12. [CrossRef]
25. Li, H.; Durbin, R. Fast and accurate short read alignment with Burrows-Wheeler transform. *Bioinformatics* **2009**, *25*, 1754–1760. [CrossRef]
26. Van der Auwera, G.A.; O'Connor, B.D. *Genomics in the Cloud: Using Docker, GATK, and WDL in Terra*, 1st ed.; O'Reilly Media: Newton, MA, USA, 2020.
27. Danecek, P.; Bonfield, J.K.; Liddle, J.; Marshall, J.; Ohan, V.; Pollard, M.O.; Whitwham, A.; Keane, T.; McCarthy, S.A.; Davies, R.M.; et al. Twelve years of SAMtools and BCFtools. *GigaScience* **2021**, *10*, giab008. [CrossRef]
28. Wang, K.; Li, M.; Hakonarson, H. ANNOVAR: Functional annotation of genetic variants from high-throughput sequencing data. *Nucleic Acids Res.* **2010**, *38*, e164. [CrossRef]
29. Perrier, X.; Flori, A.; Bonnot, F. Data Analysis Methods. In *Genetic Diversity of Cultivated Tropical Plants*; Science Publishers: Enfield, UK, 2003; pp. 43–76.
30. Pritchard, J.K.; Stephens, M.; Donnelly, P. Inference of population structure using multilocus genotype data. *Genetics* **2000**, *155*, 945–959. [CrossRef] [PubMed]
31. Evanno, G.; Regnaut, S.; Goudet, J. Detecting the number of clusters of individuals using the software STRUCTURE: A simulation study. *Mol. Ecol.* **2005**, *14*, 2611–2620. [CrossRef]
32. Earl, D.A.; vonHoldt, B.M. STRUCTURE HARVESTER: A website and program for visualizing STRUCTURE output and implementing the Evanno method. *Conserv. Genet. Resour.* **2012**, *4*, 359–361. [CrossRef]
33. Wang, J.; Zhang, Z. GAPIT version 3: Boosting power and accuracy for genomic association and prediction. *Genom. Proteom. Bioinform.* **2021**, *19*, 629–640. [CrossRef]
34. Shim, H.; Chasman, D.I.; Smith, J.D.; Mora, S.; Ridker, P.M.; Nickerson, D.A.; Krauss, R.M.; Stephens, M. A multivariate genome-wide association analysis of 10 LDL subfractions, and their response to statin treatment, in 1868 Caucasians. *PLoS ONE* **2015**, *10*, e0120758. [CrossRef] [PubMed]
35. Hossain, M.M.M.; Farjana, N.; Afroz, R.; Hasan-Uj-Jaman; Saha, P.K.; Roy, H.S.; Rahman, M.A.; Farid, M.A. Genes expression in *Penaeus monodon* of Bangladesh; challenged with AHPND-causing *Vibrio parahaemolyticus*. *Fish Shellfish. Immunol. Rep.* **2023**, *4*, 100092. [CrossRef] [PubMed]
36. Choi, M.; Stevens, A.M.; Smith, S.A.; Taylor, D.P.; Kuhn, D.D. Strain and dose infectivity of *Vibrio parahaemolyticus*: The causative agent of early mortality syndrome in shrimp. *Aquac. Res.* **2017**, *48*, 3719–3727. [CrossRef]
37. Soto-Rodriguez, S.A.; Gomez-Gil, B.; Lozano-Olvera, R.; Betancourt-Lozano, M.; Morales-Covarrubias, M.S. Field and experimental evidence of *Vibrio parahaemolyticus* as the causative agent of acute hepatopancreatic necrosis disease of cultured shrimp (*Litopenaeus vannamei*) in Northwestern Mexico. *Appl. Environ. Microbiol.* **2015**, *81*, 1689–1699. [CrossRef]
38. Sui, J.; Luan, S.; Cao, J.; Dai, P.; Meng, X.; Luo, K.; Chen, B.; Tan, J.; Fu, Q.; Kong, J. Genomic signatures of artificial selection in fecundity of Pacific white shrimp, *Penaeus vannamei*. *Front. Genet.* **2022**, *13*, 929889. [CrossRef]
39. Fu, S.; Liu, J. Genome-wide association study identified genes associated with ammonia nitrogen tolerance in *Litopenaeus vannamei*. *Front. Genet.* **2022**, *13*, 961009. [CrossRef]
40. Vu, N.T.T.; Zenger, K.R.; Silva, C.N.S.; Guppy, J.L.; Jerry, D.R. Population structure, genetic connectivity, and signatures of local adaptation of the giant black tiger shrimp (*Penaeus monodon*) throughout the Indo-Pacific region. *Genome Biol. Evol.* **2021**, *13*, evab214. [CrossRef]
41. Farhadi, A.; Pichlmueller, F.; Yellapu, B.; Lavery, S.; Jeffs, A. Genome-wide SNPs reveal fine-scale genetic structure in ornate spiny lobster *Panulirus ornatus* throughout Indo-West Pacific Ocean. *ICES J. Mar. Sci.* **2022**, *79*, 1931–1941. [CrossRef]
42. Santos, C.A.; Andrade, S.C.S.; Freitas, P.D. Identification of SNPs potentially related to immune responses and growth performance in *Litopenaeus vannamei* by RNA-seq analyses. *PeerJ* **2018**, *6*, e5154. [CrossRef]
43. Fan, Y.; Wang, W.; Ma, G.; Liang, L.; Shi, Q.; Tao, S. Patterns of insertion and deletion in Mammalian genomes. *Curr. Genom.* **2007**, *8*, 370–378. [CrossRef]
44. Fernandes Júnior, G.A.; de Oliveira, H.N.; Carvalheiro, R.; Cardoso, D.F.; Fonseca, L.F.S.; Ventura, R.V.; de Albuquerque, L.G. Whole-genome sequencing provides new insights into genetic mechanisms of tropical adaptation in Nellore (*Bos primigenius indicus*). *Sci. Rep.* **2020**, *10*, 9412. [CrossRef] [PubMed]

45. Uffelmann, E.; Huang, Q.Q.; Munung, N.S.; de Vries, J.; Okada, Y.; Martin, A.R.; Martin, H.C.; Lappalainen, T.; Posthuma, D. Genome-wide association studies. *Nat. Rev. Methods Primers* **2021**, *1*, 59. [CrossRef]
46. Wang, M.; Jin, S.; Liu, S.; Fu, H.; Zhao, Y.; Jiang, L. Genome-wide association study of growth and sex traits provides insight into heritable mechanisms underlying growth development of *Macrobrachium nipponense* (Oriental River prawn). *Biology* **2023**, *12*, 429. [CrossRef]
47. Evangelou, E.; Ioannidis, J. Meta-analysis methods for genome-wide association studies and beyond. *Nat. Rev. Genet.* **2013**, *14*, 379–389. [CrossRef]
48. Zhao, R.; Kundu, P.; Saha, A.; Chatterjee, N. Heterogeneous transfer learning for building high-dimensional generalized linear models with disparate datasets. *arXiv* **2023**, arXiv:2312.12786. [CrossRef]
49. Cassandri, M.; Smirnov, A.; Novelli, F.; Pitolli, C.; Agostini, M.; Malewicz, M.; Melino, G.; Raschellá, G. Zinc-finger proteins in health and disease. *Cell Death Discov.* **2017**, *3*, 17071. [CrossRef]
50. Rakhra, G.; Rakhra, G. Zinc finger proteins: Insights into the transcriptional and post transcriptional regulation of immune response. *Mol. Biol. Rep.* **2021**, *48*, 5735–5743. [CrossRef] [PubMed]
51. Almodóvar-Payá, A.; Villarreal-Salazar, M.; de Luna, N.; Real-Martínez, A.; Andreu, A.L.; Martín, M.A.; Arenas, J.; Lucia, A.; Vissing, J.; Krag, T.; et al. Preclinical research in glycogen storage diseases: A comprehensive review of current animal models. *Int. J. Mol. Sci.* **2020**, *21*, 9621. [CrossRef]
52. Kumar, R.; Huang, J.Y.; Ng, Y.S.; Chen, C.Y.; Wang, H.C. The regulation of shrimp metabolism by the white spot syndrome virus (WSSV). *Rev. Aquac.* **2022**, *14*, 1150–1169. [CrossRef]
53. Addison, W.N.; Pellicelli, M.; St-Arnaud, R. Dephosphorylation of the transcriptional cofactor NACA by the PP1A phosphatase enhances cJUN transcriptional activity and osteoblast differentiation. *J. Biol. Chem.* **2019**, *294*, 8184–8196. [CrossRef]
54. Li, S.; Peng, W.; Chen, X.; Geng, X.; Sun, J. Identification and characterization of nascent polypeptide-associated complex alpha from Chinese mitten crab (*Eriocheir sinensis*): A novel stress and immune response gene in crustaceans. *Fish Shellfish. Immunol.* **2016**, *48*, 54–61. [CrossRef] [PubMed]

biology

MDPI

Article

A Comprehensive Assessment of the Nutritional Value, Antioxidant Potential, and Genetic Diversity of *Fenneropenaeus merguiensis* from Three Different Regions in China

Yundong Li [1,2,3,4,†], Juan Chen [1,2,†], Siyao Cao [1,2], Ziyi Jiang [5], Song Jiang [1], Qibin Yang [1,3], Lishi Yang [1], Jianhua Huang [1,4], Jianzhi Shi [1], Zhenhua Ma [1,3] and Falin Zhou [1,2,3,4,*]

1 South China Sea Fisheries Research Institute, Chinese Academy of Fishery Sciences, Key Laboratory of South China Sea Fishery Resources Exploitation and Utilization, Ministry of Agriculture and Rural Affairs, Guangzhou 510300, China; liyd2019@163.com (Y.L.); chenj29725@163.com (J.C.); cao13793933909@163.com (S.C.); tojiangsong@163.com (S.J.); yangqibin1208@163.com (Q.Y.); yangls2016@163.com (L.Y.); huangjianhua@scsfri.ac.cn (J.H.); shijianzhi1989@163.com (J.S.); zhenhua.ma@scsfri.ac.cn (Z.M.)

2 College of Fisheries and Life Sciences, Dalian Ocean University, Dalian 116023, China

3 Key Laboratory of Efficient Utilization and Processing of Marine Fishery Resources of Hainan Province, Sanya Tropical Fisheries Research Institute, Sanya 572018, China

4 Shenzhen Base of South China Sea Fisheries Research Institute, Chinese Academy of Fishery Sciences, Shenzhen 518108, China

5 Keck School of Medicine, University of Southern California, 1975 Zonal Ave., Los Angeles, CA 90033, USA; ziyijian@usc.edu

* Correspondence: zhoufalin0925@163.com

† These authors contributed equally to this work.

Simple Summary: *Fenneropenaeus merguiensis*, one of the largest penaeid shrimp species, has a wide germplasm distribution along the southeastern coast of China, although its germplasm characteristics remain inadequately explored. This study conducted comprehensive analyses of basic nutritional components, amino acids, fatty acids, antioxidant indices, and genetic diversity across three *F. merguiensis* populations (FmRP, FmSZ, FmSY). The findings revealed that FmSZ had significantly higher ash, crude protein, and essential amino acid levels, indicating superior nutritional quality. FmSY showed a potential nutritional advantage due to its fatty acid profile. All populations demonstrated high levels of polyunsaturated fatty acids (DHA, EPA), along with strong antioxidant capacities. Genetic analysis showed FmSZ had the lowest inbreeding coefficient and relatively higher genetic diversity. Overall, FmSZ stands out as a high-quality candidate for aquaculture and genetic breeding, highlighting the value of *F. merguiensis* as an aquatic product with considerable potential for wider promotion and development.

Abstract: *Fenneropenaeus merguiensis* is one of the largest species of penaeid shrimp. It has a wide distribution of germplasm resources in the South China Sea and the southeastern coastal regions of China, yet its germplasm characteristics remain insufficiently understood. Therefore, we conducted analyses of basic nutritional components, amino acids, fatty acids, antioxidant indices, and genetic diversity in three *F. merguiensis* populations (FmRP, FmSZ, FmSY). The results showed a significant difference in ash content, with FmSZ having the highest ash content (1.77 g/100 g) ($p < 0.05$). A total of 17 amino acids were detected, and FmSZ exhibited higher concentrations of most amino acids, especially essential amino acids ($p < 0.05$). The analysis of fatty acid composition revealed significant differences between the populations, with FmSY potentially having a nutritional advantage overall ($p < 0.05$). In terms of antioxidant capacity and genetic diversity, there were no significant differences in total antioxidant capacity (T-AOC), catalase (CAT), or superoxide dismutase (SOD) among the populations ($p > 0.05$). However, genetic analysis indicated that FmSZ had the lowest inbreeding coefficient and relatively higher genetic diversity. In conclusion, among the three *F. merguiensis* populations, FmSZ has relatively higher nutritional quality and genetic diversity, and *F. merguiensis* is a high-quality aquatic product worth promoting. The analysis and characterization of various aspects of *F. merguiensis* from the three locations provide important information and data support for germplasm resources and genetic breeding efforts.

Keywords: *Fenneropenaeus merguiensis*; nutritional composition; germplasm resources; genetic diversity

1. Introduction

Fenneropenaeus merguiensis is widely distributed in the Southern Hemisphere, from East Africa to Australia, as well as in the Northern Hemisphere across Southeast Asia and the Indian Ocean [1–3]. In China, *F. merguiensis* is primarily distributed in the coastal areas of Guangdong, Hainan, and Guangxi. As an economically significant warm-water shrimp, it is recognized for its rapid growth, tender meat, and strong disease resistance, which make it a valuable species for commercial aquaculture [4]. Despite its high commercial value and popularity as a seafood choice, the lack of a stable, artificially bred seed supply system has hindered the development of large-scale farming. Currently, the aquaculture industry faces challenges in maintaining a consistent supply of juveniles, limiting the expansion of farming operations and the species' full potential in the market. Nevertheless, *F. merguiensis* remains a focus of research due to its promising growth characteristics and overall high-quality meat.

As the farming industry for *F. merguiensis* has developed, artificial breeding of the species has increasingly gained importance. Currently, much of the global research on shrimp germplasm resources focuses on species such as *Litopenaeus vannamei* [5] and *Penaeus monodon* [6]. *F. merguiensis* has great potential as a cultured species, but research on its germplasm resources is limited. The studies that exist mainly concentrate on its genetic diversity [1,7], growth performance [8], and disease resistance [4]. A fatty acid composition analysis of *F. merguiensis* from Pakistan identified five dominant fatty acids in its muscle: docosahexaenoic acid, eicosapentaenoic acid, oleic acid, palmitic acid, and stearic acid [9]. When assessing the impact of varying concentrations of Enteromorpha polysaccharides (EPS) on the growth of *F. merguiensis*, the results indicated that incorporating 1 g/kg of EPS into the diet notably boosted final body mass, weight gain, average daily growth rate, and specific growth rate, thereby enhancing overall growth performance. Moreover, this supplementation significantly increased the activity of various antioxidant enzymes within the hemolymph [8]. The optimal growth performance of *F. merguiensis* was observed with 35% protein content and an energy-to-protein ratio of 8.5 kcal/g, while the highest feed consumption, energy digestibility, and survival rate occurred with 40% protein and a ratio of 9.5 kcal/g. Regression analysis suggested that the ideal protein content is 34.4%, with an energy-to-protein ratio of 8.5 kcal/g [10]. Genetic diversity studies on *F. merguiensis* predominantly use microsatellite markers to characterize the genetic traits of different populations [1,11].

Currently, research on *F. merguiensis* is mainly focused on fundamental areas. Although some preliminary achievements have been made in germplasm resources and genetic diversity, studies in these fields remain relatively scarce. Especially within the context of China, comprehensive studies focusing on distinct regional demographics are particularly scarce. In order to fill the void in this area of research, we have chosen key agricultural sites across China (Raoping, Shenzhen, and Sanya) to systematically assess the germplasm resources of *F. merguiensis*, evaluate its nutritional value, and build a theoretical foundation for establishing a germplasm resource bank. Furthermore, we carried out an in-depth analysis of the genetic diversity of *F. merguiensis* across the three farming sites. This comprehensive study is designed to establish a robust basis for exploring the germplasm resources of *F. merguiensis* and to inform scientific management of shrimp resources in these agricultural regions. This will not only help promote the conservation and sustainable use of the species but also provide valuable data support for future genetic research and aquaculture practices.

2. Materials and Methods

2.1. Sample Collection

Based on the legal and regulatory provisions, as well as the diversity of *F. merguiensis* farming locations in China, we selected three representative farming sites for sample collection: Raoping, Sanya, and Shenzhen (Table 1, Figure 1). We employed calibrated equipment like pH meters, dissolved oxygen meters, and salinity meters, along with chemical test kits utilizing colorimetric techniques (Merck KGaA, Darmstadt, Germany), to track water quality indicators including salinity, ammonia, nitrite, pH levels, dissolved oxygen, and nitrate levels. The results indicated that these parameters were largely uniform across various populations, suggesting that the farming environments were broadly similar across the different regions. Specifically, the water temperature at each sampling site was maintained between 27.4 and 28.5 °C, and the pH was kept around 8.8.

Table 1. Fundamental details regarding the survey location.

Population	Body Length (mm)	Weight (g)	Category	Location
Raoping (FmRP)	179.8 ± 8.2	80.2 ± 7.9	Representative enterprise	23°66′ N, 117°00′ E
Shenzhen (FmSZ)	175.5 ± 4.3	80.1 ± 2.2	Genetic breeding center	22°49′ N, 114°59′ E
Sanya (FmSY)	172.00 ± 12.00	73.60 ± 17.82	Genetic breeding center	18°30′ N, 110°10′ E

Body length and weight are presented as mean ± standard deviation (n = 30).

Figure 1. Geographical representation of the survey location.

At each selected location, we randomly selected 30 live shrimps and their body length and weight were recorded. Subsequently, 1 g of muscle tissue was extracted from each shrimp, which was then placed into centrifuge tubes and preserved at −80 °C. For regular nutritional analysis, including amino acids, fatty acids, and enzymes, three random shrimps were chosen per parameter. The remaining samples were reserved for genetic analysis to assess germplasm traits.

2.2. General Nutrition Analysis

At every sampling location, three randomly chosen, healthy *F. merguiensis* individuals in the intermolt phase were selected to collect muscle tissue for in-depth analysis. The determination of moisture content was based on the National Food Safety Standard, specifically the "Determination of Moisture in Food" using the direct drying method as outlined in GB 5009.3-2016 [12]. The protein content was quantified using the Kjeldahl method as detailed in GB 5009.5-2016 [13], ensuring the precision of the measurement. The fat content was evaluated based on the Soxhlet extraction method described in GB/T 5009.6-2016 [14]. Furthermore, the ash content was analyzed in accordance with the Ashing method specified in GB 5006.4-2016 [15], and the total sugar content was measured using the Phenol-sulfuric acid colorimetric method specified in GB/T 15672-2009 [16], which is the standard for "Determination of Total Saccharide in Edible Mushroom." These methods ensured that our nutritional analysis was both thorough and precise.

2.3. Physiological Analysis

The composition of amino acids in the hydrolyzed samples was assessed according to the National Food Safety Standards, specifically adhering to the "Determination of Amino Acids in Food" protocol using high-performance liquid chromatography as detailed in GB 5009.124-2016 [17]. In addition, the fatty acid profile of the samples was ascertained following the "Determination of Fatty Acids in Food" using the External standard method, as per GB 5009.168-2016 [18]. For the assessment of antioxidant enzyme activities, superoxide dismutase (SOD) activity was measured using the WST-8 kit (A001-1-2) procured from Nanjing Jiancheng Bioengineering Institute. The assessment of total antioxidant capacity (T-AOC) was conducted using a suitable assay kit (A015-1), and catalase (CAT) activity was determined using a kit (A007-1-1) also supplied by Nanjing Jiancheng Bioengineering. Throughout the experimental process, strict adherence to the protocols was maintained. Samples were meticulously prepared, and reagents were accurately dispensed to initiate the enzymatic reactions. Reactions were halted at the designated time intervals to ensure the accurate detection of reaction products and the precise measurement of enzyme activities, with all findings meticulously documented.

2.4. Genetic Diversity

In this research, muscle samples from randomly selected live shrimps across various regions underwent genetic diversity analysis. The genomic DNA quality was verified, and ultrasonication was employed to fragment the DNA. Purification and end-repair processes were conducted, followed by adenylation and adapter ligation. Agarose gel electrophoresis was used to select fragments of the appropriate size, which were then amplified by PCR to generate sequencing libraries. Following quality assessment, these libraries were subjected to sequencing on the Illumina sequencing platform. Fastp software (0.23.2) processed the raw data, filtering out reads with excessive adapter sequences, low-quality bases, or high N content, resulting in clean and high-quality data. The purified data were subsequently aligned using Sentieon version 201711.03 [19] and the advanced Burrows-Wheeler Aligner (BWA) MEM algorithm developed by the Wellcome Trust Sanger Institute in Cambridge, UK. This approach not only expedited the alignment process but also ensured that the alignments maintained a high level of precision [20,21]. Genetic variants were identified using Sentieon and HaplotypeCaller in joint-genotyping mode to generate gVCF files and detect SNPs across populations. High-precision SNP detection was ensured by applying quality control based on GATK's filtering criteria. SNPs with a detection frequency over 60% were selected using GATK's SelectVariants and VariantFiltration tools. Genetic diversity metrics including SNP frequency, nucleotide diversity, inbreeding coefficient, polymorphic information content, and observed heterozygosity were determined using VCFtools and tailored scripts. SNP detection in BAM files was performed with GATK, and the F_{ST} index was used to measure genetic differentiation among populations. To improve data quality, low-quality reads were removed using SOAPnuke software (https://github.com/BGIflexlab/SOAPnuke, 2.X version, accessed by 29 October 2024) [22]. The genomic sequencing and bioinformatics analyses conducted have laid a robust foundation and provided pivotal insights into the assessment and understanding of genetic variability within the shrimp samples.

2.5. Data Analysis

In this research, a one-way ANOVA test was conducted utilizing SPSS 16.0 to assess statistical significance, with the threshold for significance defined as $p < 0.05$. Outcomes exceeding this threshold ($p > 0.05$) were deemed statistically insignificant. When the ANOVA indicated significance, Duncan's test was employed to discern specific differences among the groups. When the ANOVA indicated significance, Duncan's test was employed to discern specific differences among the groups. Data are expressed as mean ± standard deviation, ensuring the precision and reliability of the outcomes. Genetic variation was measured using various indices in our analysis, including SNP density = Number of SNPs/Genome size (Kb), PIC = $1 - \sum pi^2 - \sum\sum 2pi^2pj^2$, Ho = het/(het + hom), $\pi = n/(n - 1) \sum pipj\pi ij$, F = (Ho − He)/(n − He), F_{ST} = (HT − HS)/HT.

3. Results

3.1. Analysis of General Nutritional Composition

In the nutritional analysis of *F. merguiensis* muscle tissue, no significant differences ($p > 0.05$) were found in moisture, crude protein, crude fat, and total sugar among the three cultured populations. However, there were notable differences in ash content ($p < 0.05$, refer to Table 2). The ash content varied from 1.37 to 1.77 g/100 g, with FmSZ recording the maximum at 1.77 g/100 g. Regarding moisture, FmRP exhibited the highest content at 75.43 g/100 g, and FmSZ the lowest at 73.87 g/100 g. The crude fat content spanned from 0.90 to 1.03 g/100 g. Interestingly, FmRP showed the lowest levels in crude protein and total sugar, at 20.83 g/100 g and 0.32%, respectively. Conversely, FmSY had the highest crude protein content at 21.70 g/100 g, and FmSZ had the highest total sugar content at 0.40%.

Table 2. Nutritional composition and content in the muscle tissue across three populations of *F. merguiensis*.

Nutritional Components	FmRP	FmSZ	FmSY
Ash (g/100 g)	1.63 ± 0.06 ^{a}	1.77 ± 0.06 ^{a}	1.37 ± 0.06 ^{b}
Moisture (g/100 g)	75.43 ± 0.25 ^{a}	73.87 ± 1.02 ^{a}	74.17 ± 0.55 ^{a}
Crude fat (g/100 g)	1.00 ± 0.10 ^{a}	1.03 ± 0.15 ^{a}	0.90 ± 0.10 ^{a}
Crude protein (g/100 g)	20.83 ± 0.46 ^{a}	21.70 ± 1.71 ^{a}	21.97 ± 1.31 ^{a}
Total sugar (%)	0.32 ± 0.03 ^{a}	0.40 ± 0.08 ^{a}	0.35 ± 0.03 ^{a}

mean ± standard deviation ($n = 3$). Superscript letters (a, b) within the same row denote statistically significant differences between groups. Values sharing the same letter are not significantly different ($p > 0.05$).

3.2. Amino Acids Results of *F. merguiensis*

Following the evaluation of amino acid profiles in FmRP, FmSZ, and FmSY, we identified a total of 17 amino acids, comprising 7 essential, 2 conditionally essential, and 8 non-essential amino acids (Table 3, Figure 2). Most amino acid types exhibited significant differences ($p < 0.05$). Overall, FmRP and FmSY showed a trend of no significant differences ($p > 0.05$) in most amino acid types, while FmSZ exhibited significant differences ($p < 0.05$) compared to the other two populations. Among all detected amino acids, glutamic acid had the highest content, ranging from 2.07 to 2.33 g/100 g, while cysteine had the lowest content, ranging from 0.10 to 0.14 g/100 g. It is worth mentioning that certain amino acids, including aspartic acid (Asp), threonine (Thr), serine (Ser), glutamic acid (Glu), glycine (Gly), alanine (Ala), valine (Val), methionine (Met), isoleucine (Ile), phenylalanine (Phe), lysine (Lys), histidine (His), and arginine (Arg), were particularly abundant in the FmSZ group. Conversely, cysteine (Cys), leucine (Leu), and tyrosine (Tyr) were found to be more prevalent in the FmRP group. Overall, FmSY had lower levels of most amino acids compared to FmRP and FmSZ, while FmSZ exhibited the highest values for most amino acids. Additionally, the FmSZ samples showed the highest values in EAA (essential amino acids), SEAA (semi-essential amino acids), DAA (non-essential amino acids), TAA (total amino acids), and EAA/TAA, with significant differences among the groups.

Table 3. Amino acid profile and concentration in the muscle tissues of the three *F. merguiensis* groups.

Amino Acids (g/100 g)	FmRP	FmSZ	FmSY
Aspartic acid @	1.42 ± 0.03^{a}	1.54 ± 0.02^{b}	1.36 ± 0.04^{a}
Threonine *	0.55 ± 0.01^{a}	0.60 ± 0.01^{b}	0.55 ± 0.02^{a}
Serine	0.46 ± 0.01^{ab}	0.48 ± 0.01^{a}	0.45 ± 0.02^{b}
Glutamic acid @	2.07 ± 0.04^{a}	2.33 ± 0.03^{b}	2.10 ± 0.07^{a}
Glycine @	1.53 ± 0.03^{a}	1.09 ± 0.01^{b}	1.08 ± 0.03^{b}
Alanine @	0.87 ± 0.02^{a}	0.94 ± 0.01^{b}	0.91 ± 0.02^{b}
Cystine	0.14 ± 0.01^{a}	0.13 ± 0.02^{ab}	0.10 ± 0.01^{b}
Valine *	0.67 ± 0.01^{a}	0.74 ± 0.01^{b}	0.67 ± 0.02^{a}
Methionine *	0.36 ± 0.01^{a}	0.39 ± 0.02^{b}	0.35 ± 0.01^{a}
Isoleucine *	0.61 ± 0.01^{a}	0.67 ± 0.01^{b}	0.60 ± 0.02^{a}
Leucine *	1.33 ± 0.02^{a}	1.24 ± 0.02^{b}	1.11 ± 0.03^{a}
Tyrosine	0.29 ± 0.01^{a}	0.25 ± 0.00^{b}	0.24 ± 0.01^{c}
Phenylalanine *	0.57 ± 0.01^{a}	0.63 ± 0.01^{b}	0.56 ± 0.01^{a}
Lysine *	1.22 ± 0.03^{a}	1.33 ± 0.01^{b}	1.22 ± 0.03^{a}
Histidine &	0.27 ± 0.01^{a}	0.31 ± 0.00^{b}	0.28 ± 0.01^{a}
Arginine &	1.51 ± 0.02^{a}	1.76 ± 0.02^{b}	1.60 ± 0.05^{c}
Proline	0.67 ± 0.03^{ab}	0.62 ± 0.01^{a}	0.71 ± 0.02^{b}

Amino acid levels are presented as mean ± SD ($n = 3$). Distinct letters denote significant differences ($p < 0.05$) among treatments. @ Delicious amino acids. * Essential amino acids. & Semi-essential amino acids.

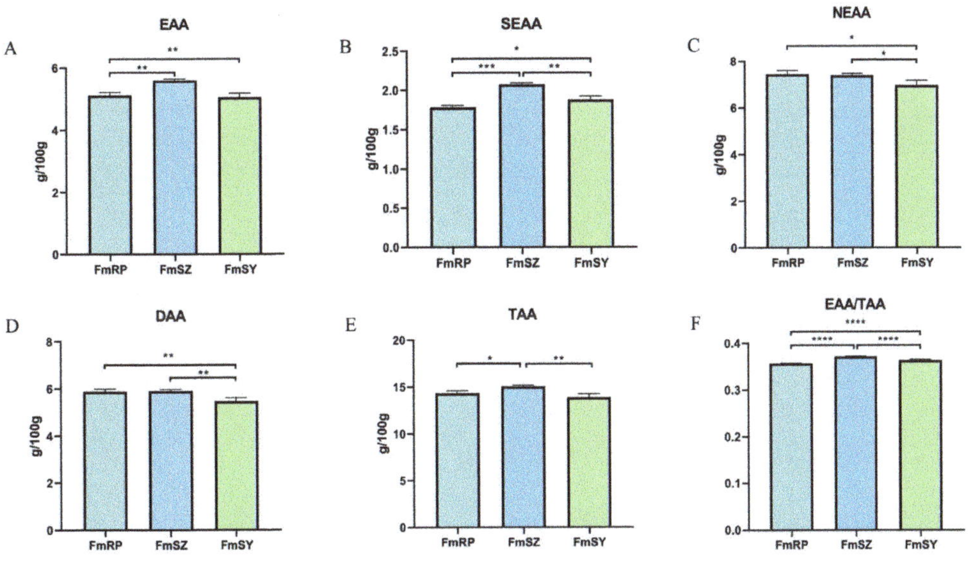

Figure 2. Muscle amino acid composition and content of three groups of *F. merguiensis*. (**A**): EAA: essential amino acids. (**B**): SEAA: total semi-essential amino acids. (**C**): NEAA: total non-essential amino acids. (**D**): DAA: delicious amino acids. (**E**): TAA: total amino acids. (**F**): EAA/TAA: essential amino acids/total amino acids. Different number of symbols between treatments indicate significant differences (* $0.01 < p < 0.05$; ** $p < 0.01$, *** $p < 0.001$, **** $p < 0.0001$).

3.3. *Results of Fatty Acids Composition*

After a detailed analysis of the fatty acid composition in the three populations (FmRP, FmSZ, and FmSY), we revealed significant differences in the content of various fatty acids ($p < 0.05$) (Table 4, Figure 3). Among the detected fatty acids, the content of C18:0 was the highest, ranging from 75.23 to 118.20 mg/100 g, while C20:2 had the lowest content, ranging from 3.70 to 5.23 mg/100 g. Notably, the highest content of C15:0 in FmSY was 8.50 mg/100 g, whereas the lowest content of C16:0 in FmSY was 66.43 mg/100 g. C16:1 and C17:0 had relatively low levels in FmSZ, while C18:1n9c was lower in FmSZ but higher in FmRP at 72.20 mg/100 g. The contents of C18:2n6c and C20:2 were relatively similar across the three samples. C22:0 had the highest content in FmRP and the lowest in FmSY, further emphasizing the differences in fatty acid composition among the populations. The polyunsaturated fatty acids C20:4n6 and C22:1n9 were higher in FmSZ and relatively lower in FmRP, which may reflect FmSZ's advantage in these health-related fatty acids. Omega-3 polyunsaturated fatty acids C20:5n3 (EPA) and C22:6n3 (DHA) were higher in FmSY and lower in FmSZ. Moreover, FmSY exhibited the highest values in the total of DHA and EPA, polyunsaturated fatty acids, monounsaturated fatty acids, saturated fatty acids, total fatty acid content, and n-6 polyunsaturated fatty acids, indicating that FmSY may have an advantage in the overall nutritional value of fatty acids. These results offer a comprehensive insight into the fatty acid profiles of the three populations, establishing a foundation for subsequent studies on nutritional value and health benefits of food.

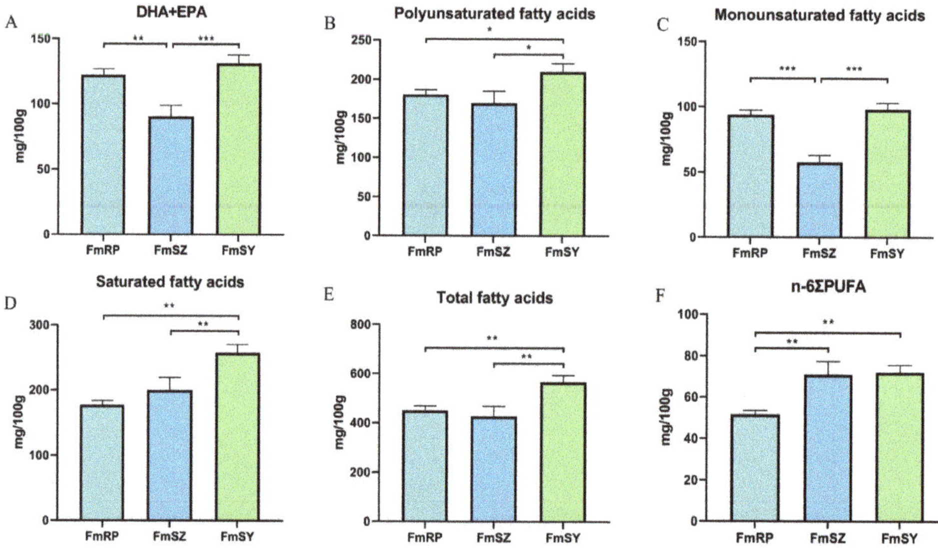

Figure 3. Muscle fatty acid composition and content of three groups of *F. merguiensis*. (**A**): the content of DHA + EPA. (**B**): the content of polyunsaturated fatty acids. (**C**): the content of monounsaturated fatty acid. (**D**): the content of saturated fatty acid. (**E**): the content of total fatty acid. (**F**): the content of n-6ΣPUFA. Different number of symbols between treatments indicate significant differences (* $0.01 < p < 0.05$; ** $p < 0.01$, *** $p < 0.001$).

Table 4. Muscle fatty acid composition and content of three groups of *F. merguiensis* (mg/100 g).

Fatty Acid	FmRP	FmSZ	FmSY
C15:0	4.37 ± 0.15 [a]	5.57 ± 0.57 [b]	8.50 ± 0.40 [c]
C16:0	69.80 ± 2.82 [a]	66.43 ± 6.44 [b]	99.97 ± 4.96 [b]
C16:1	21.60 ± 0.92 [a]	16.33 ± 1.60 [b]	31.40 ± 1.60 [c]
C17:0	12.43 ± 0.50 [a]	16.67 ± 1.55 [b]	19.97 ± 1.05 [c]
C18:0	75.23 ± 2.86 [a]	96.63 ± 9.42 [b]	118.60 ± 6.06 [c]
C18:1n9c	72.20 ± 2.72 [a]	41.13 ± 3.96 [b]	66.57 ± 3.37 [a]
C18:2n6c	6.90 ± 0.26 [a]	7.63 ± 0.78 [a]	7.57 ± 0.25 [a]
C20:2	4.43 ± 0.12 [ab]	3.70 ± 0.10 [a]	5.23 ± 0.59 [b]
C22:0	15.13 ± 0.57 [a]	14.63 ± 1.50 [a]	10.60 ± 0.50 [b]
C20:4n6	40.27 ± 1.43 [a]	59.47 ± 5.64 [b]	59.17 ± 2.86 [b]
C22:1n9	6.33 ± 0.21 [a]	8.47 ± 0.76 [b]	6.50 ± 0.40 [a]
C20:5n3 (EPA)	63.57 ± 2.51 [a]	41.77 ± 4.01 [b]	67.73 ± 3.35 [a]
C22:6n3 (DHA)	58.67 ± 2.05 [a]	48.43 ± 4.58 [b]	63.63 ± 3.01 [a]

Mean fatty acids ± standard deviation ($n = 3$). Different letters between treatments indicate significant differences ($p < 0.05$). C15:0—Pentadecanoic acid. C16:0—Palmitic acid. C16:1—Palmitoleic acid. C17:0—Heptadecanoic acid. C18:0—Stearic acid. C18:1n9c—Oleic acid. C18:2n6c—Linoleic acid. C20:2—Eicosadienoic acid. C22:0—Behenic acid. C20:4n6—Arachidonic acid. C22:1n9—Erucic acid. C20:5n3 (EPA)—Eicosapentaenoic acid. C22:6n3 (DHA)—Docosahexaenoic acid.

3.4. Oxidative Stress Indicators of *F. merguiensis*

After analyzing the physiological and biochemical indices of the three populations of *F. merguiensis* (FmRP, FmSZ, and FmSY), we observed no significant differences among the groups in total antioxidant capacity (T-AOC), catalase (CAT), and superoxide dismutase (SOD) ($p > 0.05$) (Figure 4). The values for T-AOC ranged from 2.678 to 2.689 µmol/g, CAT values ranged from 809.0 to 861.0 µmoL/(min·g), and SOD values ranged from 46.85 to 53.98 U/g. Among these, FmSZ exhibited the highest values, while FmSY had the lowest.

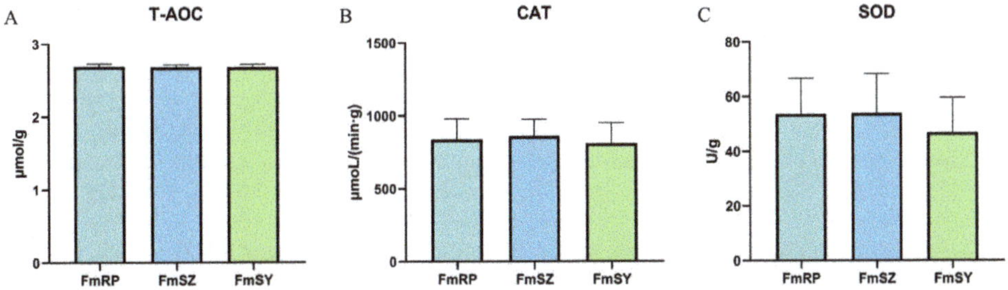

Figure 4. Muscle physiological and biochemical indicators composition and content of three groups of *F. merguiensis*. (**A**): The activity of T-AOC. (**B**): The activity of SOD. (**C**): The activity of CAT. Mean ± standard deviation ($n = 3$).

3.5. Analysis of Genetic Diversity

When evaluating the genetic variability across the three groups, the average SNP density ranged from 1.066 to 1.081, while the values for nucleotide diversity (π) were similar, approximately between 2.18×10^4 and 2.22×10^4. The inbreeding coefficient varied between 12.9% and 15.9%, with FmSZ exhibiting the lowest inbreeding coefficient, indicating the least likelihood of inbreeding. The average value of polymorphic information content (PIC) was around 0.125, reflecting the diversity of genetic markers. The average observed heterozygosity (Ho) ranged from 0.116 to 0.121, which indicates the frequency of heterozygotes in the population (Table 5). Finally, the genetic differentiation index (F_{ST}) demonstrated the genetic differences between populations (Table 6), with the lowest F_{ST} value of 0.002 between FmSZ and FmSY, suggesting minimal genetic differentiation between these two populations. In contrast, the F_{ST} value between FmRP and FmSZ was

0.004, and the value between FmRP and FmSY was also 0.004, indicating slightly greater genetic differences between these two groups.

Table 5. Statistics of genetic diversity parameters of three populations of *F. merguiensis*.

Population	SNP Density (SNP/Kb)	Nucleotide Diversity (π)	Polymorphism Information Content (PIC)	Observed Heterozygosity (Ho)	Inbreeding Coefficient (F_{HOM})
FmRP	1.066	$2.18 \times 10^{-4} \pm 3.41 \times 10^{-4}$	0.125 ± 0.114	0.116 ± 0.151	$1.59 \times 10^{-1} \pm 4.90 \times 10^{-2}$
FmSZ	1.081	$2.22 \times 10^{-4} \pm 3.47 \times 10^{-4}$	0.125 ± 0.113	0.121 ± 0.155	$1.29 \times 10^{-1} \pm 9.71 \times 10^{-1}$
FmSY	1.076	$2.20 \times 10^{-4} \pm 3.44 \times 10^{-4}$	0.124 ± 0.114	0.117 ± 0.152	$1.47 \times 10^{-1} \pm 3.06 \times 10^{-2}$

Table 6. Genetic differentiation index (F_{ST}) statistics of three populations of *F. merguiensis*.

Population-1	Population-2	F_{ST}
FmRP	FmSZ	0.004
FmRP	FmSY	0.004
FmSZ	FmSY	0.002

4. Discussion

4.1. Comparative Analysis of Basic Nutritional Components Across F. merguiensis Populations

Aquatic creatures serve as premium dietary sources, prized for their rich protein, minimal fat, and low-calorie attributes [23]. The nutritional value of aquatic animals' muscle tissue is primarily determined by the combined presence of key constituents, namely moisture, ash, crude protein, and crude fat. The muscle's water content, being its most prevalent constituent, plays a crucial role in influencing its tenderness and flavor, and greatly affects the muscle's water-holding capacity [24]. Ash, encompassing a variety of minerals, is present in small quantities but is essential for sustaining the life processes and well-being of both animals and humans. Crude protein, reflecting the muscle's total protein content in aquatic animals, is a key nutritional metric, vital for their growth, development, and tissue repair [25]. In recent decades, with the rapid improvement in socio-economic levels, the demand for high-quality aquatic food has continuously increased, and the aquaculture industry has experienced a rapid expansion. Therefore, a comprehensive assessment of these components is essential for evaluating and ensuring the nutritional value and food quality of aquatic products. When analyzing the conventional nutritional components of three populations of *F. merguiensis*, it was found that there were significant differences in the ash content among the three farmed populations ($p < 0.05$). The ash content of the three populations ranged from 1.37 to 1.77 g/100 g, with the highest ash content in FmSZ, reaching 1.77 g/100 g. Compared with *Penaeus chinensis* (1.69 g/100 g) and *L. vannamei* (1.52 g/100 g) [26], FmSZ has relatively higher ash content. The content of crude protein ranged from 20.83 to 21.97 g/100 g, which is higher than that of *L. vannamei* (15.09 g/100 g), *Penaeus japonicus* (14.60 g/100 g), *P. monodon* (13.29 g/100 g), and *Macrobrachium rosenbergii* (12.33 g/100 g) [27], among which, the crude protein of FmSZ and FmSY is slightly higher than that of FmRP. The content of crude fat is also lower than that of *p. japonicus* (2.09 g/100 g) [28]. In general, the three populations of *F. merguiensis* are a type of food with high protein and low fat, among which, FmSZ has relatively higher nutritional value.

4.2. Comparative Analysis of Amino Acids Across F. merguiensis Populations

Amino acids play a crucial role in the muscle of aquatic animals, participating in a variety of physiological functions and metabolic processes [29,30]. They are essential for the growth, development, reproduction, and tissue repair of aquatic animals [31]. In addition, amino acids are also related to the water retention, structural maintenance, and motor capacity of muscles [32]. Based on the FAO/WHO's recommended model, the

EAA/TAA ratio for high-quality proteins should ideally be 40%. Although the ratio for the three populations is slightly below this benchmark, it remains relatively close, suggesting a fairly balanced amino acid profile and classifying their protein as high-quality for human consumption. Moreover, it can be seen from the figure that the EAA of the three populations is less than 6.0 g/100 g, which is lower than that of *Haliotis japonica* (6.59 g/100 g) [33]. In addition, the composition and content of DAA can significantly influence the palatability of animal protein [34]. The results of this study found that among the three populations of *F. merguiensis*, FmSZ showed the highest values in most amino acids, and also showed the highest values in EAA, SEAA, DAA, TAA, and EAA/TAA, with significant differences between the populations. In summary, the amino acid level of *F. merguiensis* is at a medium level among shrimps, and among the three populations, FmSZ was consistently present at a notably elevated concentration, suggesting that it possesses superior nutritional benefits and a palatable flavor.

4.3. Comparative Analysis of Fatty Acids Across *F. merguiensis* Populations

Fatty acids play a vital role in aquatic animals, including providing energy, constructing cell membranes, participating in metabolic regulation, and maintaining normal growth and reproductive functions [35,36]. Furthermore, the fatty acid profile in muscle tissue is a critical determinant of both quality and taste [37]. Among the fatty acids, polyunsaturated fatty acids (PUFA), notably eicosapentaenoic acid (EPA) and docosahexaenoic acid (DHA), serve as primary precursors for the synthesis of bioactive anti-inflammatory compounds and are essential for brain development and the prevention of inflammatory cardiovascular diseases [38–40]. In this study, the total content of ΣPUFA in the three populations of *F. merguiensis* ranged from 169.5 to 209.8 mg/100 g, and ΣMUFA ranged from 57.47 to 97.97 mg/100 g, which is higher than that of *P. chinensis* (42.58 mg/100 g, 18.85 mg/100 g) [26] and *L. vannamei* (38.11–44.88 mg/100 g, 14.79–17.67 mg/100 g) [41]. All three populations of *F. merguiensis* showed a trend of ΣPUFA > ΣMUFA, which is consistent with the trend of *P. monodon* [6] and *L. vannamei* [5]. Moreover, the content of DHA + EPA in the three populations was 90.20-131.4 mg/100 g, which is higher than that of *L. vannamei* (103.30–115.10 mg/100 g) [5] and *P. japonicus* (11.58 mg/100 g) [27]. Most fatty acids showed a significant difference ($p < 0.05$), indicating differences in diet and environmental conditions among different populations. The WHO advises that the intake of saturated fats should be reduced to below 10% of total energy consumption. With the saturated fat content in *F. merguiensis* not exceeding 300 mg/100 g, it demonstrates that the saturated fat intake from *F. merguiensis* is well below this 10% benchmark. Consequently, *F. merguiensis* can be classified as an excellent source of nutrition, adhering to the healthy dietary standards recommended by the WHO. In summary, in terms of fatty acids, *F. merguiensis* has a high content of fatty acids and is of high nutritional value. Among the three populations, FmSZ has relatively higher overall values and more significant differences compared to the other two populations; thus, FmSZ has a relatively higher nutritional value.

4.4. Comparative Analysis of Physiological Indicators Across *F. merguiensis* Populations

In this study, no substantial variations were observed in the physiological indicators among the three populations of *F. merguiensis*. Antioxidant enzyme activity is a crucial indicator of an organism's antioxidant capacity and is often used to assess its response to oxidative stress [42]. Aquatic organisms rely on the antioxidant enzyme system to neutralize potentially harmful reactive oxygen species (ROS), maintaining cellular integrity and physiological functions, and maintaining the redox balance within cells [43,44]. T-AOC, serving as an encompassing indicator of antioxidant potential, mirrors the collective antioxidant profile of an organism [45]. An elevated T-AOC level generally indicates a potent antioxidant mechanism, which is beneficial for shielding cells against oxidative stress and sustaining overall well-being [46]. SOD, an essential antioxidant enzyme, neutralizes superoxide anion radicals, protecting the body against oxidative stress [43]. CAT efficiently transforms hydrogen peroxide into water and oxygen, protecting cells from oxidative stress [47]. Therefore, the higher the activity of these antioxidant enzymes, the stronger the antioxidant and immune capabilities of the organism. In our research, the T-AOC levels in the three *F. merguiensis* populations ranged from 2.68 to 2.69 μmol/g, surpassing the T-AOC activity found in *P. clarkii*, which was recorded at 2.57 U/mg protein [48]; the SOD activity in the three populations of *F. merguiensis* was between 46.85 and 53.98 U/g, which is greater than the SOD activity observed in *P. monodon*, ranging from 39.65 to 47.29 U/g [6], indicating that the three different populations of *F. merguiensis* all showed higher antioxidant enzyme activity, among which FmSZ showed higher activity in various antioxidant enzymes, thus FmSZ has higher antioxidant enzyme activity and antioxidant capacity.

4.5. Comparative Analysis of Genetic Diversity Across F. merguiensis Populations

Genetic diversity is essential for a species' ability to adapt to dynamic environments, supporting its survival and evolutionary potential. Higher genetic diversity is beneficial for enhancing a species' environmental adaptability, growth, and disease resistance [49,50]. Parameters like Polymorphic Information Content (PIC) are vital for evaluating the genetic diversity within populations [51]. The PIC value reflects the probability that a marker will exhibit different allelic genotypes when two individuals are randomly selected from a population. A PIC value greater than 0.5 is considered highly polymorphic, values between 0.25 and 0.5 are considered moderately polymorphic, and values below 0.25 are classified as low polymorphism [52]. SNP density reflects how frequently SNP sites occur across the genome, with a higher density generally associated with increased genetic diversity [53]. Nucleotide diversity evaluates the differences among nucleotide sites within a defined genetic region, serving as a vital indicator of genetic diversity among populations [54]. The F_{ST} index evaluates the level of genetic differentiation among populations, with elevated F_{ST} values suggesting substantial genetic separation [55]. In the genetic diversity analysis of the FmRP, FmSZ, and FmSY populations, it was found that their SNP density ranged from 1.066 to 1.081, nucleotide diversity (π) values were between 2.18×10^{-4} and 2.22×10^{-4}, and PIC was within the range of 0.124–0.125, indicating low genetic diversity. Additionally, the inbreeding coefficient (F_{ST}) of these three populations ranged from 0.002 to 0.004, also showing low values. These findings imply that the genetic variability within the three *F. merguiensis* populations is low, and there is no significant genetic differentiation among them. Among these populations, FmSZ exhibits a comparatively higher level of genetic diversity, whereas FmRP and the other population show a relatively greater extent of genetic differentiation.

5. Conclusions

This study analyzed the nutritional components, amino acid composition, fatty acid profile, antioxidant capacity, and genetic diversity of *F. merguiensis*. The results indicate that *F. merguiensis* is a high-quality aquatic food, characterized by high protein and low fat content. The FmSZ population stands out with significantly higher levels of ash, crude protein, and optimal amino acids, enhancing its nutritional value. All populations exhibited high levels of polyunsaturated fatty acids, particularly DHA and EPA, suggesting potential health benefits. Antioxidant assays demonstrated strong resistance to oxidative stress, with FmSZ showing superior antioxidant enzyme activity. While genetic diversity is relatively low, the overall performance of the FmSZ population underscores its significance for aquaculture and food security. Among the three *F. merguiensis* populations, FmSZ displays relatively higher nutritional quality and genetic diversity. This research positions *F. merguiensis* as a high-quality aquatic species with substantial potential for broader promotion and development. Specifically, the FmSZ population is identified as having notable potential for further advancement in the aquaculture industry, offering a nutritious and healthy option for consumers. Future research should focus on enhancing the genetic diversity of *F. merguiensis*, exploring the molecular mechanisms behind its nutritional and antioxidant properties, assessing environmental impacts, and investigating its food safety and meat quality to support its industrial applications.

Author Contributions: Z.M. and F.Z. conceived the project and supervised the work. J.C., Z.J. performed the bioinformatics analysis and prepared the manuscript, the table and figures. Y.L., J.C., S.C. and Z.J. conducted the experiment. L.Y., S.J., Q.Y., J.H. and J.S. collected the samples and performed sequencing. All authors have read and agreed to the published version of the manuscript.

Funding: This research was funded by National Key R & D Program of China (2022YFD2401900, 2022YFD2400104); Central Public-interest Scientific Institution Basal Research Fund, CAFS (2023TD34); China Agriculture Research System (CARS-48); Guangdong Basic and Applied Basic Research Foundation (2023A1515012410); Hainan Provincial Natural Science Foundation of China (323MS127); Earmarked fund for HNARS (HNARS-10-ZJ01); Project funded by China Postdoctoral Science Foundation (2023T160178); Central Public-interest Scientific Institution Basal Research Fund, South China Sea Fisheries Research Institute, CAFS (NO. 2024RC06).

Institutional Review Board Statement: The animal study was reviewed and approved by the Animal Care and Use Committee of the South China Sea Fisheries Research Institute.

Informed Consent Statement: Not applicable.

Data Availability Statement: Data are contained within the article.

Conflicts of Interest: The authors declare no conflicts of interest.

References

1. Aziz, D.; Siraj, S.S.; Arshad, A. Genetic Diversity of Banana Prawns *Fenneropenaeus merguiensis* in Malaysian Waters Using Microsatellite Markers. *J. Environ. Biol.* **2020**, *41*, 1349–1357. [CrossRef]
2. Nguyen, N.H.; Quinn, J.; Powell, D.; Elizur, A.; Thoa, N.P.; Nocillado, J.; Lamont, R.; Remilton, C.; Knibb, W. Heritability for Body Colour and Its Genetic Association with Morphometric Traits in Banana Shrimp (*Fenneropenaeus merguiensis*). *BMC Genet.* **2014**, *15*, 132. [CrossRef] [PubMed]
3. Vance, D.J.; Rothlisberg, P.C. Chapter One—The Biology and Ecology of the Banana Prawns: Penaeus Merguiensis de Man and P. Indicus H. Milne Edwards. In *Advances in Marine Biology*; Sheppard, C., Ed.; Academic Press: New York, NY, USA, 2020; Volume 86, pp. 1–139.
4. Nguyen, N.H.; Phuthaworn, C.; Knibb, W. Genomic Prediction for Disease Resistance to Hepatopancreatic Parvovirus and Growth, Carcass and Quality Traits in Banana Shrimp *Fenneropenaeus merguiensis*. *Genomics* **2020**, *112*, 2021–2027. [CrossRef] [PubMed]
5. Li, Y.; Cao, S.; Jiang, S.; Huang, J.; Yang, Q.; Jiang, S.; Yang, L.; Zhou, F. Comparative Study of Nutritional Composition, Physiological Indicators, and Genetic Diversity in Litopenaeus Vannamei from Different Aquaculture Populations. *Biology* **2024**, *13*, 722. [CrossRef]
6. Li, Y.; Chen, J.; Jiang, S.; Huang, J.; Jiang, S.; Yang, Q.; Yang, L.; Shi, J.; Zhou, F. A Comprehensive Study on Nutritional Quality, Physiological Enzyme Activity and Genetic Diversity in Six Populations of *Penaeus monodon*. *Aquacult Int.* **2024**, *32*, 10141–10157. [CrossRef]
7. Prasertlux, S.; Khamnamtong, B.; Wisuntorn, E.; Soonsan, P.; Janpoom, S.; Tang, S.; Rongmung, P.; Ratdee, O.; Ninwichian, P.; Sakamoto, T. Genetic Diversity and Population Differentiation of Wild and Domesticated Banana Shrimp *Fenneropenaeus merguiensis*: Applications for Development of Its Breeding Program. *Reg. Stud. Mar. Sci.* **2024**, *69*, 103309. [CrossRef]
8. Liu, W.-C.; Zhou, S.-H.; Balasubramanian, B.; Zeng, F.-Y.; Sun, C.-B.; Pang, H.-Y. Dietary Seaweed (Enteromorpha) Polysaccharides Improves Growth Performance Involved in Regulation of Immune Responses, Intestinal Morphology and Microbial Community in Banana Shrimp *Fenneropenaeus merguiensis*. *Fish Shellfish. Immunol.* **2020**, *104*, 202–212. [CrossRef]
9. Fatima, H.; Ayub, Z.; Siddiqui, G.; Ali, S.A. Fatty Acid Composition of Two Candidate Species of Aquaculture, *Fenneropenaeus merguiensis* and *F. penicillatus* (Crustacea: Decapoda) in Pakistan. *Pak. J. Zool.* **2012**, *44*, 969.
10. Prakoso, A.A.; Suprapto, J.; Subandiyono, D. Influence of Protein and the Level of Energy-Protein Feed Ratio on Growth of Banana Shrimp (*Fenneropenaeus merguiensis* de Man). *Int. J. Fish. Aquat. Stud.* **2020**, *8*, 280–287.
11. Gong, F.; Zhang, N.; Guo, H.; Zhu, K.; Liu, T.; Jiang, S.; Zhang, D. Development and Characterization of 23 Polymorphic Microsatellite Markers for Banana Shrimp *Fenneropenaeus merguiensis*. *Conserv. Genet. Resour.* **2016**, *8*, 9–11. [CrossRef]
12. *GB 5009.3–2016*; Determination of Moisture in Food. National Standards of the People's Republic of China: Beijing, China, 2016.
13. *GB 5009.5-2016*; National Standards for Food Safety-Determination of Proteins in Foods. National Standards of the People's Republic of China: Beijing, China, 2016.
14. *GB 5009.6–2016*; National Food Safety Standard-Determination of Crude Fat in Food. National Standards of the People's Republic of China: Beijing, China, 2016.

15. *GB 5009.4–2016;* National Food Safety Standard-Determination of Ash in Food. National Standards of the People's Republic of China: Beijing, China, 2016.
16. *GB/T 15672–2009;* Determination of Total Sugar Content in Edible Fungi. National Standards of the People's Republic of China: Beijing, China, 2009.
17. *GB 5009.124-2016;* National Food Safety Standard-Determination of Amino Acids in Foods by Hydrochloric Acid Hydrolysis. National Standards of the People's Republic of China: Beijing, China, 2016.
18. *GB 5009.168-2016;* National Food Safety Standard-Determination of Fatty Acids in Foods by Gas Chromatography. National Standards of the People's Republic of China: Beijing, China, 2016.
19. Freed, D.; Aldana, R.; Weber, J.A.; Edwards, J.S. The Sentieon Genomics Tools–A Fast and Accurate Solution to Variant Calling from next-Generation Sequence Data. *BioRxiv* **2017**. [CrossRef]
20. Li, H. Aligning Sequence Reads, Clone Sequences and Assembly Contigs with BWA-MEM. *arXiv* **2013**, arXiv:1303.3997.
21. Kendig, K.I.; Baheti, S.; Bockol, M.A.; Drucker, T.M.; Hart, S.N.; Heldenbrand, J.R.; Hernaez, M.; Hudson, M.E.; Kalmbach, M.T.; Klee, E.W. Sentieon DNASeq Variant Calling Workflow Demonstrates Strong Computational Performance and Accuracy. *Front. Genet.* **2019**, *10*, 736. [CrossRef] [PubMed]
22. Chen, Y.; Chen, Y.; Shi, C.; Huang, Z.; Zhang, Y.; Li, S.; Li, Y.; Ye, J.; Yu, C.; Li, Z. SOAPnuke: A MapReduce Acceleration-Supported Software for Integrated Quality Control and Preprocessing of High-Throughput Sequencing Data. *Gigascience* **2018**, *7*, gix120. [CrossRef]
23. Lin, Y.; Miao, L.; Sun, C.; Jiang, W.; Zhou, Q.; Liu, B.; Ge, X. New Insights on the Effects of In-Pond Raceway Aquaculture System (IRAS) with Dietary Rhubarb Extracts on the Fresh Meat Quality of *Megalobrama amblycephala. Aquaculture* **2022**, *560*, 738524. [CrossRef]
24. Lorenzo, I.; Serra-Prat, M.; Yébenes, J.C. The Role of Water Homeostasis in Muscle Function and Frailty: A Review. *Nutrients* **2019**, *11*, 1857. [CrossRef]
25. Hua, K.; Cobcroft, J.M.; Cole, A.; Condon, K.; Jerry, D.R.; Mangott, A.; Praeger, C.; Vucko, M.J.; Zeng, C.; Zenger, K. The Future of Aquatic Protein: Implications for Protein Sources in Aquaculture Diets. *One Earth* **2019**, *1*, 316–329. [CrossRef]
26. Wang, J. Comparision of Nutritional Composition in Muscle of Penaeus Chinensis, Penaeus Vannamei Boone and *Penaeus japonicuss* Bate. *Food Sci. Technol.* **2013**, *38*, 145–150.
27. Liu, Z.; Liu, Q.; Zhang, D.; Wei, S.; Sun, Q.; Xia, Q.; Shi, W.; Ji, H.; Liu, S. Comparison of the Proximate Composition and Nutritional Profile of Byproducts and Edible Parts of Five Species of Shrimp. *Foods* **2021**, *10*, 2603. [CrossRef]
28. XU XH, L.; YAN, B. Nutritional Component Analysis and Quality Evaluation of *Penaeus japonicus. Food Sci.* **2011**, *32*, 297–301.
29. Fafournoux, P.; Bruhat, A.; Jousse, C. Amino Acid Regulation of Gene Expression. *Biochem. J.* **2000**, *351*, 1–12. [CrossRef] [PubMed]
30. Hoseini, S.M.; Reverter, M. Health-Promoting Effects of Amino Acids in Fish. In *Biotechnological Advances in Aquaculture Health Management*; Gupta, S.K., Giri, S.S., Eds.; Springer Nature Singapore: Singapore, 2021; pp. 493–533, ISBN 9789811651946.
31. Duran, B.O.S.; Zanella, B.T.T.; Perez, E.S.; Mareco, E.A.; Blasco, J.; Dal-Pai-Silva, M.; Garcia de la Serrana, D. Amino Acids and IGF1 Regulation of Fish Muscle Growth Revealed by Transcriptome and microRNAome Integrative Analyses of Pacu (*Piaractus mesopotamicus*) Myotubes. *Int. J. Mol. Sci.* **2022**, *23*, 1180. [CrossRef] [PubMed]
32. Li, G.; Li, Z.; Liu, J. Amino Acids Regulating Skeletal Muscle Metabolism: Mechanisms of Action, Physical Training Dosage Recommendations and Adverse Effects. *Nutr. Metab.* **2024**, *21*, 41. [CrossRef] [PubMed]
33. Shi, L.; Hao, G.; Chen, J.; Ma, S.; Weng, W. Nutritional Evaluation of Japanese Abalone (*Haliotis Discus Hannai* Ino) Muscle: Mineral Content, Amino Acid Profile and Protein Digestibility. *Food Res. Int.* **2020**, *129*, 108876. [CrossRef] [PubMed]
34. Tan DeQing, T.D.; Wang JianWei, W.J.; Dan ShengGuo, D.S. The Ratio of Flesh to Body and Analysis on Nutritive Composition of Muscle in *Ancherythroculter nigrocauda. Acta Hydrobiol. Sin.* **2004**, *28*, 240–246.
35. Tocher, D.R. Fatty Acid Requirements in Ontogeny of Marine and Freshwater Fish. *Aquac. Res.* **2010**, *41*, 717–732. [CrossRef]
36. Sargent, J.; Bell, G.; McEvoy, L.; Tocher, D.; Estevez, A. Recent Developments in the Essential Fatty Acid Nutrition of Fish. *Aquaculture* **1999**, *177*, 191–199. [CrossRef]
37. Yuan, J.L.; Liu, M.; Ni, M.; Mi, G.Q.; Zhang, C.; Gu, Z.M. Effects of Different Culture Models on Growth Performances, Morphological Traits and Nutritional Quality in Muscles of *Micropterus salmoides. Acta Agric. Univ. Jiangxiensis* **2018**, *40*, 1276–1285.
38. Chen, J.; Liu, H. Nutritional Indices for Assessing Fatty Acids: A Mini-Review. *Int. J. Mol. Sci.* **2020**, *21*, 5695. [CrossRef]
39. Araujo, P.; Truzzi, C.; Belghit, I.; Antonucci, M. The Impact of Seawater Warming on Fatty Acid Composition and Nutritional Quality Indices of *Trematomus bernacchii* from the Antarctic Region. *Food Chem.* **2021**, *365*, 130500. [CrossRef]
40. Simonetto, M.; Infante, M.; Sacco, R.L.; Rundek, T.; Della-Morte, D. A Novel Anti-Inflammatory Role of Omega-3 PUFAs in Prevention and Treatment of Atherosclerosis and Vascular Cognitive Impairment and Dementia. *Nutrients* **2019**, *11*, 2279. [CrossRef] [PubMed]
41. Li, X.; Wang, Y.; Li, H.; Jiang, X.; Ji, L.; Liu, T.; Sun, Y. Chemical and Quality Evaluation of Pacific White Shrimp *Litopenaeus vannamei:* Influence of Strains on Flesh Nutrition. *Food Sci. Nutr.* **2021**, *9*, 5352–5360. [CrossRef] [PubMed]
42. Yaz, Y.A.; Yıldırım, N.; Yaz, Y.; Tekin, N.; İnal, M.; Şahin, F.M. Role of Oxidative Stress in Pseudoexfoliation Syndrome and Pseudoexfoliation Glaucoma. *Turk. J. Ophthalmol.* **2019**, *49*, 61.

43. Zhang LiYing, Z.L.; Zhao Meng, Z.M.; Wang YueZhi, W.Y. A Review on Superoxide Dismutases of Hydrobios. *Acta Agric. Univ. Jiangxiensis* **2012**, *34*, 800–804.
44. Attia, H.G.; El-Morshedy, S.M.; Nagy, A.M.; Ibrahim, A.M.; Aleraky, M.; Abdelrahman, S.S.; Osman, S.M.; Alasmari, S.M.; El Raey, M.A.; Abdelhameed, M.F. Citrus Clementine Peel Essential Oil Ameliorates Potassium Dichromate-Induced Lung Injury: Insights into the PI3K/AKT Pathway. *Metabolites* **2024**, *14*, 68. [CrossRef]
45. Ren, Z.; Fan, Y.; Zhang, Z.; Chen, C.; Chen, C.; Wang, X.; Deng, J.; Peng, G.; Hu, Y.; Cao, S. Sodium Selenite Inhibits Deoxynivalenol-Induced Injury in GPX1-Knockdown Porcine Splenic Lymphocytes in Culture. *Sci. Rep.* **2018**, *8*, 17676. [CrossRef]
46. Silvestrini, A.; Meucci, E.; Ricerca, B.M.; Mancini, A. Total Antioxidant Capacity: Biochemical Aspects and Clinical Significance. *Int. J. Mol. Sci.* **2023**, *24*, 10978. [CrossRef]
47. Carmo De Carvalho E Martins, M.D.; Martins; Da Silva Santos Oliveira, A.S.; Da Silva, L.A.A.; Primo, M.G.S.; De Carvalho Lira, V.B. Biological Indicators of Oxidative Stress [Malondialdehyde, Catalase, Glutathione Peroxidase, and Superoxide Dismutase] and Their Application in Nutrition. In *Biomarkers in Nutrition*; Patel, V.B., Preedy, V.R., Eds.; Biomarkers in Disease: Methods, Discoveries and Applications; Springer International Publishing: Cham, Switzerland, 2022; pp. 1–25, ISBN 978-3-030-81304-8.
48. Tan, S.H.; Deng, X.Y.; Jiang, W.M.; He, F.L. Effects of High Level Chromium on Antioxidant Enzyme System in Gill and Hepatopancreas of Procambarus Clarkii. *J. Agro-Env. Sci.* **2007**, *23*, 24–34.
49. Tang, Q.-Y.; Xie, J.-H.; Xia, Z.-L.; Cai, M.-Y.; Wu, Y.-M.; Bai, L.-H.; Du, H.-K.; Li, J.-F.; Yang, G.-L. Genetic Diversity of the Breeding Populations of Giant Freshwater Prawn *Macrobrachium rosenbergii*. *Acta Hydrobiol. Sin.* **2020**, *44*, 1097–1104.
50. O'Connell, M.; Wright, J.M. Microsatellite DNA in Fishes. *Rev. Fish Biol. Fish.* **1997**, *7*, 331–363. [CrossRef]
51. Jian-lin, L.I.; Hong-xia, L.I.; Yong-kai, T.; Ju-hua, Y.U.; Fan, Y.U. Genetic Difference Analysis of Two Genetically Improved Farmed Tilapia Populations by Using Microsatellite Marker. *J. South. Agric.* **2015**, *46*, 138.
52. Serrote, C.M.L.; Reiniger, L.R.S.; Silva, K.B.; dos Santos Rabaiolli, S.M.; Stefanel, C.M. Determining the Polymorphism Information Content of a Molecular Marker. *Gene* **2020**, *726*, 144175. [CrossRef] [PubMed]
53. Matukumalli, L.K.; Lawley, C.T.; Schnabel, R.D.; Taylor, J.F.; Allan, M.F.; Heaton, M.P.; O'Connell, J.; Moore, S.S.; Smith, T.P.; Sonstegard, T.S. Development and Characterization of a High Density SNP Genotyping Assay for Cattle. *PLoS ONE* **2009**, *4*, e5350. [CrossRef] [PubMed]
54. Luo, W.; Luo, C.; Wang, M.; Guo, L.; Chen, X.; Li, Z.; Zheng, M.; Folaniyi, B.S.; Luo, W.; Shu, D. Genome Diversity of Chinese Indigenous Chicken and the Selective Signatures in Chinese Gamecock Chicken. *Sci. Rep.* **2020**, *10*, 14532. [CrossRef] [PubMed]
55. Holsinger, K.E.; Weir, B.S. Genetics in Geographically Structured Populations: Defining, Estimating and Interpreting F ST. *Nat. Rev. Genet.* **2009**, *10*, 639–650. [CrossRef]

biology

MDPI

Article

Genome-Wide Identification and Expression of Neuropeptides and Their Expression Patterns After RNAi of CHH Genes in Pacific White Shrimp *Litopenaeus vannamei*

Long Zhang †, Lichao Sun †, Guanghao Song, Beibei Wang, Yanting Cui, Fei Liu, Yuquan Li and Zhongkai Wang *

School of Marine Science and Engineering, Qingdao Agricultural University, Qingdao 266109, China
* Correspondence: zkwang@qau.edu.cn
† These authors contributed equally to this work.

Simple Summary: Neuropeptides are small molecules that play key roles in regulating various developmental, physiological, and behavioral processes in crustaceans. In this study, we focused on the Pacific white shrimp, *Litopenaeus vannamei,* to identify and understand the genes responsible for producing these neuropeptides. Using advanced genetic analysis techniques, we discovered 125 genes that encode neuropeptides, with 54 of them being newly identified. We found that these genes are mainly active in the shrimp's nervous system, particularly in the eyestalk, brain, and nerve ganglia. Additionally, we explored how silencing two specific genes, CHH and VIH, which are important for hormone regulation, affected the expression of other neuropeptide genes. Our results showed that silencing these genes led to a significant decrease in the activity of many neuropeptide genes, affecting pathways related to metabolism and hormone production. This research provides valuable insights into the genetic mechanisms that control shrimp growth and development, which could help improve shrimp farming practices by improving growth rates and disease resistance.

Abstract: Neuropeptides are pivotal in regulating a broad spectrum of developmental, physiological, and behavioral processes throughout the life cycle of crustaceans. In this comprehensive study, we utilized a multiomics approach to characterize neuropeptide precursors and to assess the expression profiles of neuropeptide-encoding genes across various tissues and developmental stages in the Pacific white shrimp, *Litopenaeus vannamei*. Additionally, we explored the differential expression of neuropeptide genes in the eyestalk before and after the RNA interference-mediated suppression of crustacean hyperglycemic hormone (CHH) and vitellogenesis-inhibiting hormone (VIH) gene expression. Our study identified a total of 125 neuropeptide-encoding genes in *L. vannamei,* with 54 of these genes previously uncharacterized in the genome. Notably, certain neuropeptide-encoding gene families showed significant expansion, as demonstrated by the discovery of 10 adipokinetic hormone/corazonin-like peptide (ACP) genes, 55 CHH superfamily genes, and 13 pigment-dispersing hormone (PDH) genes. Alternative splicing was also found to play a crucial role in generating functionally diverse neuropeptides; for example, the agatoxin and calcitonin genes undergo alternative splicing that leads to the production of three distinct agatoxin neuropeptides and two distinct calcitonin neuropeptides, respectively. Neuropeptide genes are predominantly expressed in neuroendocrine tissues, including the eyestalk, cerebral ganglia, thoracic ganglia, and ventral ganglia. During the embryonic development of *L. vannamei,* with the exception of the molt-inhibiting hormone (MIH) gene, all monitored genes display minimal expression from the zygote stage through to the larval in membrane (Lim) stage. In contrast, the majority of these genes exhibit a steady uptick in expression from the nauplius stage onwards, culminating in the post-larval stage. Furthermore, comparative transcriptomic analysis of the eyestalk revealed that the expression of the majority of neuropeptide genes was downregulated following the suppression of CHH and VIH gene expression. This downregulation was significantly associated with the enrichment of pathways related to amino acid metabolism and hormone synthesis. The findings of this study provide valuable insights for future research aimed at elucidating the role of neuropeptides in regulating physiological functions in *L. vannamei,* potentially leading to advancements in shrimp aquaculture practices.

Keywords: neuropeptides; gene expansion; alternative splicing; CHH; VIH

1. Introduction

Neuropeptides are key signaling molecules that play an indispensable role in orchestrating a diverse array of developmental, physiological, and behavioral processes throughout the life cycle of crustaceans [1]. These molecules are directly encoded in the genome and are initially translated as large precursor molecules, called pre-prohormones, which require further post-translational processing for their biological activity and stability [2]. The mature neuropeptides are typically short peptides composed of 3–100 amino acid residues and are involved in a wide range of physiological processes, including reproduction, metabolism, growth, and locomotion [3,4]. Due to their key roles in these fundamental processes, neuropeptides are often targeted for modulating these processes to align with market demands in commercially important species.

The Pacific white shrimp, *Litopenaeus vannamei*, has gained global recognition as the most crucial species for shrimp aquaculture [5]. Its production has faced challenges due to environmental stressors and disease exposure [6–8]. Neuropeptides, which are pivotal in regulating a range of physiological and behavioral processes in crustaceans, offer substantial potential for enhancing aquaculture practices. Gaining insights into the genetic mechanisms that govern growth, development, and stress response in shrimp can lead to strategies that boost growth rates, reproductive success, and disease resistance. Consequently, advances in *L. vannamei* production could be significantly propelled by modulating its neuroendocrine system.

Neuropeptidomics aims to comprehensively identify the collection of neuropeptides in an organism, organ, tissue, or cell. The charting of the neuropeptidome is the first step towards understanding peptidergic signaling [9].With the advancements in RNA sequencing (RNA-seq), the application of bioinformatics methods in neuropeptide mining has been performed in various crustaceans, including *Macrobrachium nipponense* [10], *Portunus trituberculatus* [11], and *Cherax quadricarinatus* [12]. In the case of *L. vannamei*, computational transcriptome analysis combined with bioinformatics-based peptide prediction has also been utilized to expand the neuropeptidome [13,14]. Although these previous studies have added to the identification of known neuropeptides in shrimp, there are still numerous undiscovered neuropeptides or peptides with incomplete sequence information.

In crustaceans, the eyestalk X-organ/sinus gland (XO/SG) complex is the primary site for the synthesis and secretion of neuropeptides, which has been a focal point for neuropeptide research [15]. Among the identified neuropeptides, the crustacean hyperglycemic hormone (CHH) family is particularly significant, due to its multifunctional roles, which include the regulation of carbohydrate metabolism, molting, osmoregulation, and reproduction [16]. However, it is crucial to acknowledge that, despite the established functions of certain members within the CHH superfamily, their identities and functionalities remain largely uncharacterized. Additionally, although the XO/SG complex serves as a primary site for the synthesis and secretion of various significant neuropeptides, there is a paucity of information concerning other neuropeptides and their regulatory roles in relation to the CHH family in *L. vannamei*. RNA interference (RNAi) represents a powerful tool for the functional study of neuropeptides by specifically silencing their expression. Previous studies have employed RNAi to knockdown CHH expression in *L. vannamei*, leading to significant advancements in understanding their roles in controlling reproductive processes [17–19]. However, the overall impact of CHH silencing on the shrimp's neuropeptidome and the subsequent physiological consequences remain to be clarified.

The elucidation of the genome now facilitates a comprehensive genome-wide analysis of neuropeptides in *L. vannamei* [20]. In this study, we performed comprehensive genome and transcriptome analyses to characterize neuropeptide-encoding genes and evaluate their expression profiles across various tissues and developmental stages. Additionally, we investigated the chromosomal localization, gene structure, and phylogenetic relationships within the CHH gene family. We also examined the expression patterns of neuropeptides in the eyestalk following the silencing of two specific CHH genes. These findings provide a foundational framework for clarifying the regulatory network between CHH and other eyestalk neuropeptides and offer a valuable resource for genetic research in *L. vannamei*.

2. Materials and Methods

2.1. Experimental Animals

Adult shrimps (mean body weight: 12.1 ± 0.4 g; mean body length: 10.0 ± 0.4 cm) and juvenile shrimps (mean body weight: 5.5 ± 0.4 g; mean body length: 7.0 ± 0.4 cm) were procured from a commercial farm in Qingdao, China. Prior to the experiment, the shrimps were subjected to a one-week acclimation period in a 300 L PVC tank. The tank was continuously supplied with aerated natural seawater, maintaining dissolved oxygen levels regularly above 6.0 mg/L. The water parameters were stringently controlled, with a salinity of 30‰, a pH of 8.0, and a temperature of 28 ± 0.5 °C. The dietary regimen included administering dry pellets thrice daily at 09:00, 15:00, and 21:00 h. Daily maintenance was carefully performed, including the removal of fecal matter and uneaten feed, using a siphon tube. Water quality was rigorously monitored, with a daily exchange of 50% of the tank volume at 09:00 h. Seawater salinity was accurately measured using an YSI multi-parameter water quality monitor (YSI Incorporated, Yellow Springs, OH, USA). All experimental procedures and animal treatments adhered strictly to the ethical guidelines established by the Animal Experiment Ethics Committee of Qingdao Agricultural University.

2.2. Tissue Sampling

For the purpose of in silico transcriptome mining of neuropeptides, nine adult shrimps were randomly selected. Various tissues, including the intestine, hepatopancreas, muscle, epidermis, gill, stomach, heart, hemolymph, lymphoid organ, antennal gland, eyestalk, brain ganglia, ventral nerve, and thoracic ganglia, were meticulously dissected. The excised tissues were promptly flash-frozen in liquid nitrogen and subsequently stored at −80 °C for further analysis.

2.3. RNA Extraction and Transcriptome Sequencing

Total RNA was extracted from the collected tissue samples utilizing the TRIzol Reagent (Vazyme, Nanjing, China). The RNA quality was evaluated using the RNA Nano 6000 Assay Kit in conjunction with the Agilent Bioanalyzer 2100 system (Agilent Technologies, Santa Clara, CA, USA), and the RNA Integrity Number (RIN) was determined. Only RNA samples that met the stringent quality criteria (OD260/OD280 = 2.0 − 2.2, OD260/OD230 ≥ 2.0, RIN ≥ 8.0, and 28S:18S ≥ 1.0) were selected for library construction. For each replicate, RNA samples from three individuals were pooled in equal amounts to create a single mixed sample. Three mixed samples were prepared for each experimental group. Illumina library preparation and sequencing were conducted by Novogene Corporation (Novogene, Beijing, China) utilizing the NovaSeq 6000 platform (Illumina, San Diego, CA, USA). Initial processing of the raw reads in FASTQ format was executed using Perl scripts. Clean reads were obtained by filtering out reads containing adapters, poly-N sequences, and low-quality reads from the raw data. Furthermore, the Q20, Q30, and GC contents of the clean data were calculated.

2.4. Characterization of Neuropeptides by Bioinformatics Analysis

Trinity was utilized to assemble transcriptome data from various tissues and generate unigenes. The annotation of all unigenes was conducted based on the NCBI databases, using a cut-off E-value of 1.0×10^{-5}. To identify neuropeptides in *L. vannamei*, we performed a search in the sequence annotation file using keywords associated with known neuropeptides in other decapods. Additionally, we performed a local tblastn analysis using TBtools v2.119 software [21], employing known decapod neuropeptide precursors as query sequences [14]. Furthermore, all identified neuropeptide sequences were validated using online Blast against the genome of *L. vannamei* (GenBank number: PRJNA508983) to confirm the neuropeptide-encoding genes. The chromosomal localization and gene structure were illustrated by IBS 1.0 software [22].

The structure of the mature neuropeptides was predicted using a well-established workflow [13]. We initiated our analysis by examining the deduced precursors for the presence of signal peptides using the online tool SignalP 5.0 (https://services.healthtech.dtu.dk/services/SignalP-5.0/) (accessed on 6 November 2022). Following this, pro-hormone cleavage sites were predicted, either by adhering to the criteria established by Veenstra [23] or by aligning with recognized precursor processing schemes. Finally, the mature peptides were identified by comparing their amino acid sequences to those of known crustacean neuropeptides, ensuring a high degree of homology. Schematic diagrams of precursor structures and multiple sequence alignment of the predicted peptide sequences were generated using CLC Main Workbench v8.0 (QIAGEN, Vedbæk, Denmark). The evolutionary relationship of CHH precursors with other known orthologous precursors was analyzed, and a phylogenetic tree was constructed using MEGA X [24] through a neighbor-joining (NJ)-based approach with 1000 bootstrap trials.

2.5. Expression Profiles of the Neuropeptide Genes

In addition to the transcriptomic data collected from various tissues in this study, RNA-seq data from nine early developmental stages of *L. vannamei* were obtained from NCBI (SRA accession number: SRP094135). The expression profiles of neuropeptide-encoding genes were analyzed using the reference genome for *L. vannamei*. Clean reads were aligned to the reference genome using HISAT2 v2.0.5 and the mapped reads from each sample were subsequently assembled using StringTie v1.3.3 [25]. The expression abundance and variations of each gene were quantified by calculating the fragment per kilobase of transcript per million mapped reads (FPKM). The expression profiles of neuropeptide genes were visualized as a heatmap using TBtools v2.119 software [21]. In the heatmap, the transition from red to blue indicates a decrease in $\log_2$ (FPKM + 1) values, with red signifying high expression levels and blue denoting lower expression levels.

2.6. RNAi of *LvCHH3* and *LvVIH*

To predict the target segment containing multiple functional short-interfering RNA (siRNA) sites, we utilized the online BLOCK-iT RNAi Designer (http://rnaidesigner.thermofisher.com) (accessed on 23 June 2018) to efficiently construct double-stranded RNA (dsRNA) specific to *LvCHH3* (GenBank accession number: XM_027353902.1) and *LvVIH* (GenBank accession number: KC962398.1). Primers incorporating T7 promoter sequences were designed and synthesized based on the sequences of *LvCHH3* and *LvVIH* (Table S1). The gene fragment was subsequently amplified using these T7 promoter-conjugated primers and served as a template for in vitro transcription. The dsRNAs, specifically dsCHH3 and dsVIH, were generated using the T7 RiboMAX Express RNAi System (Promega, Madison, WI, USA) in accordance with the manufacturer's protocol. Additionally, dsGFP, targeting the green fluorescent protein (GFP), was synthesized following the same procedure.

A cohort of 45 juvenile shrimps, all in the intermolt phase, was systematically divided into 3 equal groups, each comprising 15 individuals. Each group was further subdivided into 3 replicates, with 5 shrimps per replicate. The experimental protocol involved the administration of dsCHH3, dsVIH, or dsGFP (control) via injection at a dosage of 3 μg/g body weight. After 24 h post injection, eyestalk samples were collected from a randomly selected subset of 9 shrimps per group, ensuring that 3 shrimps were taken from each replicate. The subsequent quantification of *LvCHH3* and *LvVIH* expression levels was performed using quantitative real-time PCR (qPCR) to validate the silencing of the target genes. Following this, RNA-seq was employed to examine the expression profiles of neuropeptides in the eyestalk, subsequent to CHH gene silencing.

2.7. Functional Analysis of Differentially Expressed Genes (DEGs)

Differential expression analysis of genes was performed between the RNAi group (dsVIH, dsCHH3) and control group (dsGFP) by DESeq2 R package (1.16.1) [26]. The resulting p-values were adjusted by the Benjamini–Hochberg procedure. Genes with an adjusted p-value < 0.05 and $|\log_2(\text{Fold Change})| > 1.0$ were regarded as differentially expressed. Gene Ontology (GO) and Kyoto Encyclopedia of Genes and Genomes (KEGG) enrichment analyses of DEGs were implemented by means of the clusterProfiler R package, in which gene length bias was corrected [27]. GO terms or KEGG pathways with corrected p-value less than 0.05 were considered significantly enriched in a set of DEGs. To validate the Illumina sequencing data, eight DEGs were chosen for qPCR analysis by using the same RNA samples for transcriptome sequencing.

2.8. qPCR Analysis of Target Genes

The expression profiles of the target genes were assessed utilizing qPCR. Specific primers for both the target and reference genes are detailed in Table S1. The qPCR assays were conducted in accordance with the manufacturer's protocol for the ChamQ Universal SYBR qPCR Master Mix Kit (Vazyme, Nanjing, China). Each reaction mixture had a total volume of 10 µL, comprising 2.0 µL of cDNA template (5 ng/µL), 5.0 µL of 2× ChamQ Universal SYBR qPCR Master Mix, 0.2 µL each of 10 µM forward and reverse primers, and 2.6 µL of RNase-free water. The qPCR assays were performed utilizing a CFX96 Touch Real-Time PCR Detection System (Bio-Rad, Hercules, CA, USA). The amplification protocol commenced with an initial pre-incubation phase at 95 °C for 30 s, succeeded by 40 cycles of denaturation at 95 °C for 10 s and annealing/extension at 60 °C for 30 s. To verify the specificity of the amplified qPCR products, melt-curve analysis was conducted. Each cDNA sample was subjected to analysis in triplicate technical replicates, while tissue samples from each experimental group were analyzed as triplicate biological replicates.

The relative expression levels of the target genes were normalized to the expression levels of the reference gene EF1α, which was quantified in each qPCR run. The relative gene expression levels were determined using the comparative Ct method and calculated via the $2^{-\Delta\Delta Ct}$ analysis method [28]. The qPCR data were subjected to statistical analysis using one-way ANOVA followed by Tukey's post hoc test, conducted with SPSS version 20.0 (SPSS Inc., Chicago, IL, USA). A p-value of less than 0.05 was considered indicative of a statistically significant difference.

3. Results

3.1. Identification of Neuropeptide Transcript and Peptide Prediction

Forty-two cDNA libraries were generated from 14 distinct tissues of *L. vannamei* following the prescribed methodology. The raw data have been deposited in the Genome Sequence Archive in the National Genomics Data Center, China National Center for Bioinformation (GSA: CRA019665), and are publicly accessible at https://ngdc.cncb.ac.cn/gsa (accessed on 15 October 2024). Following the removal of adaptors and quality trimming, the clean reads were employed for de novo assembly, aiming to maximize the identification of neuropeptide transcripts in *L. vannamei*. Comprehensive information regarding the quality assessment, de novo assembly statistics, and annotation summary data can be found in Tables S2 and S3. Consequently, the RNA-seq analysis provided a thorough transcriptome dataset that facilitated the accurate identification of neuropeptides in *L. vannamei*.

A total of 74 putative neuropeptide precursor transcripts were identified in the assembled transcriptome. The majority of these neuropeptides had been previously identified in other crustacean and/or insect species. The mRNA sequences of newly identified precursors have been submitted to GenBank (Table 1). Furthermore, an online BLAST analysis against the reference genome of *L. vannamei* identified 125 neuropeptide-encoding genes, of which 54 had not been previously characterized (Table 1). In addition, 27 of these neuropeptide-encoding genes lacked gene ID annotations in the genome, and 4 genes were not detected within the genomic sequence.

Table 1. Putative neuropeptide-encoding genes predicted in *L. vannamei*.

Gene ID	Gene Description	NT ID	Neuropeptides
unannotated	uncharacterized	OP797893	Adipokinetic hormone/corazonin-like peptide 1 (ACP1)
unannotated	uncharacterized	OP797894	Adipokinetic hormone/corazonin-like peptide 2 (ACP2)
unannotated	uncharacterized	OP797895	Adipokinetic hormone/corazonin-like peptide 3 (ACP3)
LOC113821010	uncharacterized	XM_027373437.1	Adipokinetic hormone/corazonin-like peptide 4 (ACP4)
LOC113826242	U8-agatoxin-Ao1a-like	OP797896 XM_027379116.1 OP797897	Agatoxin1 Agatoxin2 Agatoxin3
unannotated	uncharacterized	OP797898	Allatostatin A (AST-A)
LOC113826235	prothoracicostatic peptide-like	XM_027379108.1	Allatostatin B (AST-B)
unannotated	uncharacterized	OP797899	Allatostatin C1 (AST-C1)
LOC113815894	uncharacterized	XM_027367911.1	Allatostatin C2 (AST-C2)
LOC113806195	allatostatin-like	XM_027357317.1	Allatostatin C3 (AST-C3)
LOC113804738	bursicon-like	XM_027355629.1	Bursicon α
LOC113819040	partner of bursicon-like	XM_027371277.1	Bursicon β
LOC113820349	uncharacterized	XM_027372685.1 OP797900	Calcitonin A Calcitonin B
LOC113801805	uncharacterized	XM_027352242.1	CCHamide1
LOC113808829	uncharacterized	XM_027360325.1	CCHamide2
unannotated	uncharacterized	OP797901	CCRFamide
LOC113814975	uncharacterized	XM_027367049.1	CNMamide
LOC113826786	uncharacterized	XM_027379674.1	Corazonin1 (CRZ1)
unannotated	uncharacterized	OP797902	Corazonin2 (CRZ2)
LOC113820987	cardioactive peptide-like	XM_027373408.1	Crustacean cardioactive peptide (CCAP)
LOC113807567	uncharacterized	XM_027358856.1	Crustacean female sex hormone1a (CFSH1a)
LOC113807568	uncharacterized	XM_027358857.1	Crustacean female sex hormone1b (CFSH1b)
LOC113825772	uncharacterized	XM_027378604.1	Crustacean female sex hormone1c (CFSH1c)
LOC113812164	uncharacterized	XM_027363996.1	Crustacean female sex hormone1d (CFSH1d)
LOC113803173	crustacean hyperglycemic hormones-like	XR_003475487.1	Crustacean hyperglycemic hormone 1 (CHH1)
LOC113814049 LOC113814050 LOC113814052 LOC113814053 LOC113814056 LOC113814057 LOC113814058 LOC113814060 LOC113814061 LOC113814062 LOC113814063 LOC113814065 LOC113814068 LOC113814069	crustacean hyperglycemic hormones-like	XM_027366125.1 XM_027366126.1 XM_027366129.1 XM_027366130.1 XM_027366132.1 XM_027366133.1 XM_027366134.1 XM_027366135.1 XM_027366136.1 XM_027366137.1 XM_027366139.1 XM_027366140.1 XM_027366141.1 XM_027366142.1	Crustacean hyperglycemic hormone 2 (CHH2)
LOC113803174	crustacean hyperglycemic hormones	XM_027353902.1	Crustacean hyperglycemic hormone 3 (CHH3)
LOC113829422	crustacean hyperglycemic hormones 4	XM_027382596.1	Crustacean hyperglycemic hormone 4 (CHH4)
LOC113815763 LOC113818576	crustacean hyperglycemic hormones-like	XM_027367779.1 XM_027370778.1	Crustacean hyperglycemic hormone 5 (CHH5)

Table 1. *Cont.*

Gene ID	Gene Description	NT ID	Neuropeptides
LOC113815764	molt-inhibiting hormone-like	XM_027367781.1	Crustacean hyperglycemic hormone 6 (CHH6)
		XM_027367780.1	Crustacean hyperglycemic hormone 6_L (CHH6_L)
LOC113823331	crustacean hyperglycemic hormone 6-like	XM_027375947.1	Crustacean hyperglycemic hormone 7 (CHH7)
LOC113814054	crustacean hyperglycemic hormones-like	XM_027366131.1	Crustacean hyperglycemic hormone 8 (CHH8)
LOC113815762	crustacean hyperglycemic hormone	XM_027367778.1	Crustacean hyperglycemic hormone 9 (CHH9)
LOC113820798	crustacean hyperglycemic hormone-like	XM_027373159.1	Crustacean hyperglycemic hormone 10 (CHH10)
LOC113816918	uncharacterized	XM_027368909.1	Crustacean hyperglycemic hormone 11 (CHH11)
LOC113814051	crustacean hyperglycemic hormone-like	XM_027366128.1	Crustacean hyperglycemic hormone 12 (CHH12)
LOC113814071	crustacean hyperglycemic hormone-like	XM_027366143.1	Crustacean hyperglycemic hormone 13 (CHH13)
LOC113815761	crustacean hyperglycemic hormone 2-like	XM_027367777.1	Crustacean hyperglycemic hormone 14 (CHH14)
LOC113818816	molt-inhibiting hormone-like	XM_027371007.1	Crustacean hyperglycemic hormone 15 (CHH15)
LOC113818818	uncharacterized	XM_027371008.1	
LOC113818819	uncharacterized	XM_027371010.1	
LOC113818820	uncharacterized	XM_027371011.1	
unannotated	uncharacterized	OP797904	Crustacean hyperglycemic hormone 16 (CHH16)
LOC113815982	uncharacterized	XM_027368000.1	Diuretic hormone 31 (DH31)
LOC113803372	uncharacterized	XM_027354151.1	Diuretic hormone 45 (DH45)
LOC113828029	uncharacterized	XM_027380970.1	Ecdysis triggering hormone (ETH)
LOC113819918	eclosion hormone-like	XM_027372151.1	Eclosion hormone 1 (EH1)
LOC113819919	eclosion hormone-like	XM_027372154.1	Eclosion hormone 2 (EH2)
LOC113805280	uncharacterized	XM_027356262.1	Elevenin
LOC113807871	uncharacterized	XM_027359175.1	FLRFamide
LOC113803916	thyrostimulin alpha-2 subunit-like	XM_027354724.1	Glycoprotein-A2 (GPA2)
LOC113815740	thyrostimulin beta-5 subunit-like	XM_027367761.1	Glycoprotein-B5 (GPB5)
LOC113814627	PRQFV-amide-like	XM_027366675.1	GSEFLamide
LOC113827581	nucleolar protein dao-5-like	XM_027380478.1	Kinin
Not found	uncharacterized	AY425615.1	Molt-inhibiting hormone 1 (MIH1)
LOC113829062	molt-inhibiting hormone-like	XM_027382182.1	Molt-inhibiting hormone 2 (MIH2)
Not found	uncharacterized	MF358695.1	Molt-inhibiting hormone 3 (MIH3)
LOC113828133	molt-inhibiting hormone-like	XM_027381102.1	Molt-inhibiting hormone 4 (MIH4)
LOC113828720	molt-inhibiting hormone	XM_027381767.1	Molt-inhibiting hormone 5 (MIH5)
LOC113828964	molt-inhibiting hormone-like	XM_027382073.1	Molt-inhibiting hormone 6 (MIH6)
LOC113828814	molt-inhibiting hormone-like	XM_027381883.1	Molt-inhibiting hormone 7 (MIH7)
LOC113828891	molt-inhibiting hormone-like	XM_027381977.1	Molt-inhibiting hormone 8 (MIH8)
Not found	uncharacterized	KC962398.1	Vitellogenesis-inhibiting hormone (VIH)
LOC113829851	myosuppressin-like	XM_027383026.1	Myosuppressin
Not found	uncharacterized	OP797903	Neuroparsin 1 (NP1)
LOC113803948	neuroparsin-A-like	XM_027354760.1	Neuroparsin 2 (NP2)
LOC113824994	neuroparsin-A-like	XM_027377778.1	Neuroparsin 3 (NP3)
LOC113813942	neuropeptide F-like	XM_027366020.1	Neuropeptide F1 (NPF1)
LOC113825053	uncharacterized	XM_027377831.1	Neuropeptide F2 (NPF2)
LOC113802180	uncharacterized	XM_027352711.1	short Neuropeptide F (sNPF)
LOC113819779	orcokinin peptide type B-like	XM_027371972.1	Orcokinin
LOC113819780		XM_027371973.1	

Table 1. *Cont.*

Gene ID	Gene Description	NT ID	Neuropeptides
LOC113824694	pigment-dispersing hormone type 1-like	XM_027377462.1	
LOC113824695	pigment-dispersing hormone type 1-like	XM_027377463.1	
LOC113824696	pigment-dispersing hormone type 1-like	XM_027377464.1	
LOC113824697	pigment-dispersing hormone type 1-like	XM_027377465.1	
LOC113824698	pigment-dispersing hormone type 1-like	XM_027377466.1	
LOC113824699	pigment-dispersing hormone type 1-like	XM_027377468.1	
LOC113824700	pigment-dispersing hormone type 1-like	XM_027377469.1	Pigment-dispersing hormone (PDH)
LOC113824704	pigment-dispersing hormone type 1-like	XM_027377470.1	
LOC113824705	pigment-dispersing hormone type 1-like	XM_027377471.1	
LOC113824707	pigment-dispersing hormone type 2-like	XM_027377472.1	
LOC113824708	pigment-dispersing hormone type 2-like	XM_027377473.1	
LOC113829575	pigment-dispersing hormone type 1-like	XM_027382770.1	
LOC113829577	pigment-dispersing hormone type 1-like	XM_027382773.1	
LOC113813542	uncharacterized	XM_027365551.1	Proctolin
LOC113799929	buccalin-like	XM_027350631.1	Pyrokinin
LOC113822904	red pigment-concentrating hormone-like	XM_027375473.1	Red pigment-concentrating hormone (RPCH)
LOC113827384	red pigment-concentrating hormone-like	XM_027380264.1	
LOC113806468	uncharacterized	XM_027357611.1	RYamide
LOC113820189	FMRFamide-related neuropeptide-like	XM_027372474.1	SIFamide
LOC113801964	uncharacterized	XM_027352452.1	Sulfakinin
LOC113826877	tachykinin-like	XM_027379777.1	Tachykinin
LOC113813698	uncharacterized	XM_027365733.1	Trissin
LOC113817914	oxytocin-neurophysin 1-like	XM_027370037.1	Vasotocin–neurophysin (VT-NP)
LOC113829169	uncharacterized	XM_027382271.1	WXXXRamide

Note: Gene ID, the Genbank accession number of the gene; Gene description, the annotation of the gene in Genbank; NT ID, the Genbank accession number of the mRNA transcribed from the gene; Neuropeptides, the neuropeptides translated from the gene. Four copies each of the ACP1 and ACP2 genes, and fourteen copies of the CHH16 gene were found in the genome.

3.2. The Expansion of Neuropeptide-Encoding Genes in *L. vannamei*

3.2.1. Gonadotropin-Releasing Hormone (GnRH)-like Gene Family

In *L. vannamei*, three GnRH-like neuropeptides, namely Adipokinetic hormone/corazonin-like peptide (ACP), corazonin (CRZ), and red pigment-concentrating hormone (RPCH), were discovered. Moreover, it was observed that multiple genes were responsible for encoding the precursors of these neuropeptides (Figure 1). Specifically, four types of ACP-encoding genes were identified, with four ACP1 genes, four ACP2 genes, and one ACP3 gene arranged in tandem on one scaffold, and one ACP4 gene located on another scaffold (Figure 1A). The ACP1 and ACP2 genes were found to possess two exons, while the ACP3 and ACP4 genes exhibited three exons. The initial exon of these genes was responsible for encoding the signal peptide (SP), the mature ACP peptide of pQITFSRSWVPQamide, and a partial precursor-related peptide (PRP). Additionally, the alignment of ACP precursors demonstrated a shared consensus N-terminus, encompassing both the SP and the mature ACP peptide (Figure 1A). Two distinct CRZ precursors were encoded by two genes situated on distinct scaffolds (Figure 1B). Each gene consisted of a 5′UTR exon and three CDS exons, although the 5′UTR exon of the CRZ1 gene was located on a separate scaffold. Sequence alignment revealed that the two CRZ precursors exhibited significant similarity in their mature CRZ peptides, except for a variation at position 4 (Figure 1B). In contrast, the scenario was distinct for the RPCH precursor. Two copies of RPCH genes were identified on different scaffolds, each consisting of three exons (Figure 1C). Both genes encoded identical RPCH precursors, resulting in the production of the mature peptide QLNFSPGWamide.

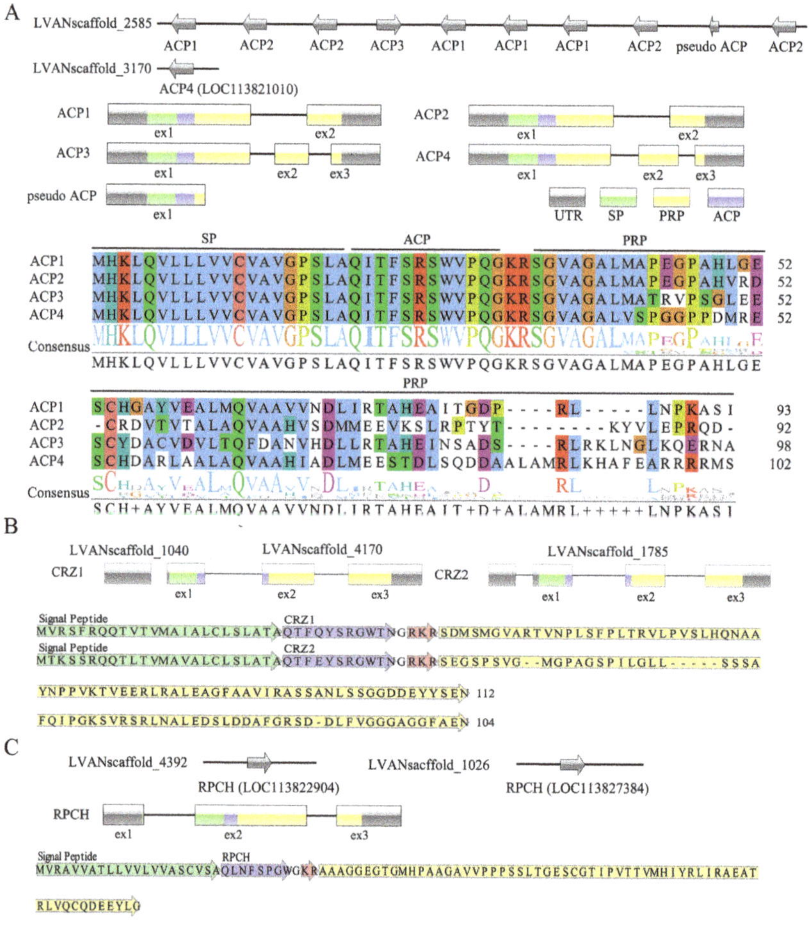

Figure 1. The genomic location and structural organization of GnRH-like genes, as well as the arrangement and alignment of neuropeptide precursors. (**A**) ACP; (**B**) CRZ; (**C**) RPCH. Arrows along the line denote genes located on a single scaffold. Gene structures are depicted with boxes for exons and lines for introns. Grey boxes indicate untranslated regions, green boxes denote sequences encoding single peptides (SPs), yellow boxes represent sequences encoding precursor-related peptides (PRPs), and blue boxes signify sequences encoding the mature peptides of ACP, CRZ, or RPCH. In the detailed sequences of CRZ and RPCH precursors, the mature peptides are indicated in blue, potential dibasic cleavage sites are marked in red, the SP is highlighted in green, and the PRP is shown in yellow.

3.2.2. CHH Gene Superfamily

This study has led to the identification of 52 genes within the *L. vannamei* genome that encode for CHH precursors, a significant finding that expands our understanding of the CHH family in this species. Some of the CHH genes may occur as the only gene copy on a scaffold, whereas most were organized in various tandem/duplication arrangements (Figure 2A–H). Four CHH genes were identified at scaffold 2640, with one gene encoding two precursors of CHH6 and CHH6_L through alternative splicing (Figure 2A). Additionally, one gene at scaffold 2640 encoded the precursor for CHH5, with a duplicated copy found at scaffold 2916 (Figure 2A). At scaffold 2490, a total of eighteen CHH genes were located, with fourteen of them being duplicated genes encoding the precursor for CHH2 (Figure 2B). In addition, the study revealed the presence of fourteen duplicated genes responsible for encoding the precursor for CHH16 across nine scaffolds, which had not been previously annotated in the genome (Figure 2F). Regarding the genes encoding MIH precursors, a cluster of six genes was observed in a tandem arrangement within scaffold 1036 (Figure 2H). Additionally, the gene encoding the CHH4 precursor was found to be incomplete in the genome. The genes encoding the precursors of MIH1, MIH3, and VIH were not detected in the genome in this study, despite their previous documentation and availability in the GenBank database.

The majority of CHH genes exhibit a tripartite exon organization (Figure 2A–H). The initial exon encodes a signal peptide, with the prevailing lengths being 18 or 15 base pairs. The subsequent exon decodes the signal peptide, the CHH precursor-related peptide (CPRP), which is absent in MIH genes, and a partial segment of the mature CHH peptide. The final exon encodes the remaining portion of the mature peptide, with a length of 106 base pairs being the most frequently observed.

The CHH genes encoded a total of 26 CHH/MIH/VIH precursors, as depicted in Figure 2I. Through multiple sequence alignment, it was observed that all mature peptides exhibited significant spatial conservation of six cysteines, which are distinctive features of the CHH superfamily. Furthermore, phylogenetic analysis classified crustacean CHH mature peptides into three subgroups (Figure 2J). Based on the results, the presence of CPRP distinguished Type-I CHH precursors, while Type-II CHH precursors were characterized by a Gly^{12} insertion in the mature peptides of MIH/VIH. Type-III CHH precursors displayed an intermediate structure, lacking both CPRP and the Gly^{12} insertion in the CHH10 and CHH11 precursors.

3.2.3. PDH Gene Family

A total of three transcripts were observed to encode putative PDH precursors, whereas a Blast analysis conducted on the *L. vannamei* genome led to the discovery of thirteen genes encoding PDH precursors (Figure 3A). Among these, eleven were found to be arranged in a tandem configuration within scaffold 3566, along with six pseudo PDH genes, while the remaining two were situated in scaffold 621. The thirteen PDH precursors were comprised of 78–82 amino acids and the C-terminal amidated PDH mature peptide was released by two dibasic cleavage sites (KR) (Figure 3B). All PDH mature peptides exhibited conserved sequence of NSELINSLLG-L/I-PKVMNDAamide (Figure 3B).

3.3. Alternative Splicing Creates Different Agatoxin and Calcitonin Precursors

Three transcripts were identified as encoding putative Agatoxin precursors, while only one gene was found in the genome. Alignment of the three Agatoxin precursors revealed that they shared the identical N-terminus and C-terminus, except for a few inserted amino acids within the precursors of Agatoxin 2 and Agatoxin 3 (Figure 4A). The Agatoxin gene consisted of four exons, with exon 4 comprising exon 4a and exon 4b. It can be inferred that the three Agatoxin precursors resulted from alternative splicing of exon 3, exon 4a, or both (Figure 4A). In a similar manner, the consensus N-terminus was shared by the precursors of Calcitonin A and Calcitonin B, while the release of two distinct types of calcitonin peptides occurred from these precursors (Figure 4B). The unique Calcitonin gene comprised four exons, and the alternative splicing of either exon 3 or exon 4 resulted in the formation of the two Calcitonin precursors. Upon sequence alignment, it was observed that both types of Calcitonin peptides possessed a predicted N-terminal disulfide bridge and a C-terminal Pro-amide sequence, except the Calcitonin B peptides in *L. vannamei*, *M. rosenbergii*, and *P. clarkii*, which were predicted to have two cysteine bridges at the N-terminus (Figure 4B).

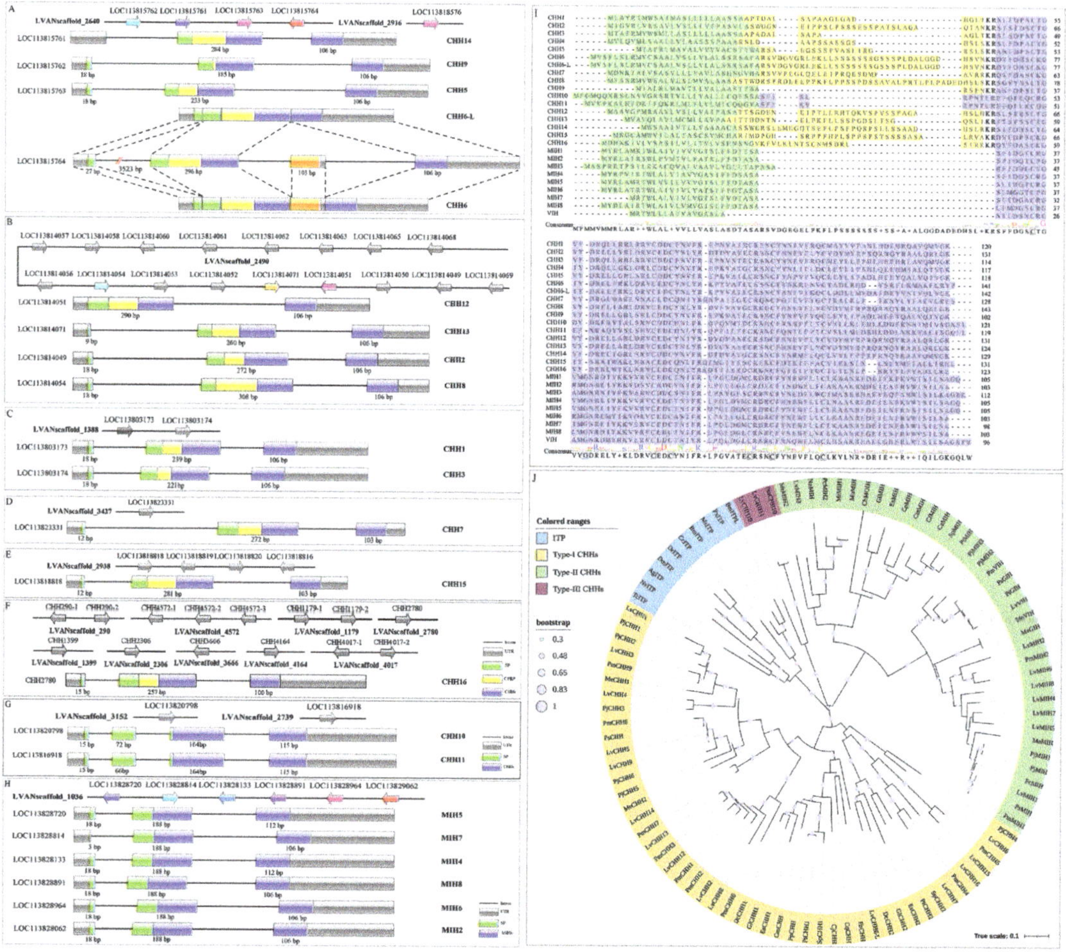

Figure 2. The genomic location and structural organization of CHH genes, the arrangement, alignment of CHH precursors, and phylogenetic analysis of crustacean CHH peptides and insect ITP. (**A–H**) Genomic location and structural organization of CHH genes; (**I**) arrangement and alignment of CHH precursors; (**J**) phylogenetic analysis of crustacean CHH peptides and insect ITP. Arrows along the line denote genes located on a single scaffold. Grey arrows represent multiple copies of the same CHH genes in a single or different scaffold. Arrows in different colors represent different CHH/MIH genes in one scaffold. Gene structures are depicted with boxes for exons and lines for introns. Grey boxes indicate untranslated regions, green boxes denote sequences encoding SP, yellow boxes represent sequences encoding CHH precursor-related peptides (CPRPs), and blue boxes signify sequences encoding the mature peptides of CHH or MIH. Specifically, the orange box is an alternative spliced exon. In the comprehensive sequences of CHH precursors, the mature peptides are denoted in blue, the SP is highlighted in green, and the CPRP is depicted in yellow. A phylogenetic tree,

constructed using the MUSCLE algorithm and the NJ method implemented in MEGA X, is based on the amino acid sequences of crustacean CHH peptides and insect ITP. The node values represent the percentage derived from 1000 bootstrap replicates. The peptide sequences are detailed in Table S4.

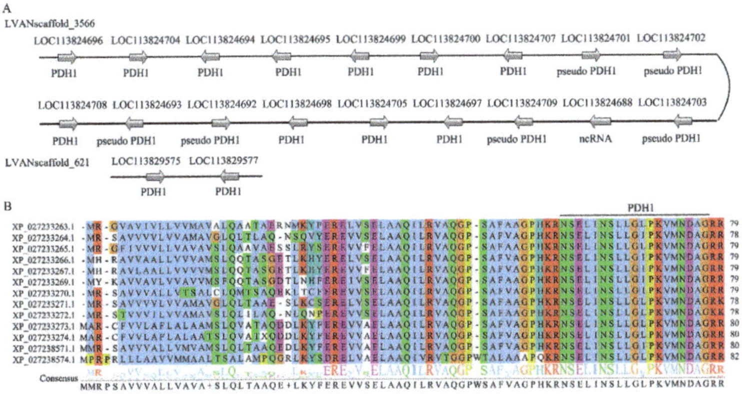

Figure 3. The genomic location of PDH genes and the alignment of PDH precursors. (**A**) Genomic location of PDH genes; (**B**) alignment of PDH precursors. Grey arrows along the line denote multiple copies of PDH genes located on a single scaffold.

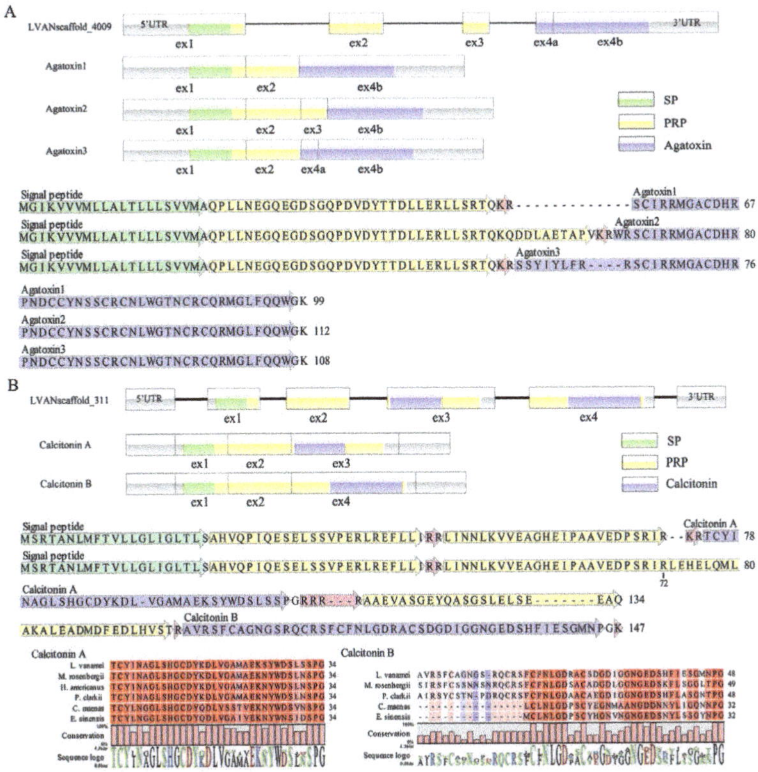

Figure 4. The structural organization and alternative splicing of Agatoxin and Calcitonin genes, as well as the arrangement and alignment of neuropeptides. (**A**) Agatoxin; (**B**) Calcitonin. Gene structures are

depicted with boxes for exons and lines for introns. Grey boxes indicate untranslated regions, green boxes denote sequences encoding SP, yellow boxes represent sequences encoding PRP, and blue boxes signify sequences encoding the mature peptides of Agatoxin or Calcitonin. In the detailed sequences of agatoxin and calcitonin precursors, the mature peptides are indicated in blue, potential dibasic cleavage sites are marked in red, the SP is highlighted in green, and the PRP is shown in yellow. Sequences alignments of predicted Calcitonin A and Calcitonin B peptides are shown. Conserved amino acids are shown in a gradient from blue to red (blue means nearly similar, dark red means exact amino acid and white is no conservation).

3.4. *The Identification of Other Neuropeptide Precursors*

The amino acid sequences of other putative neuropeptide precursors, along with their structural information including the locations of bioactive mature peptides, cleavage sites, and precursor size, are presented in Figures S1–S4. Conserved motifs were identified in several neuropeptide families, such as AST-A (XYXFGLamide, where X represents variable residues), AST-B (XWXXXXGXWamide), Orcokinin (NFDEIDRXXXGFX), Pyrokinin (XXFXFXPRLamide), and WXXXRamide (WXXRamide).

3.5. *Temporal- and Spatial-Expression Patterns of Neuropeptide Genes*

An analysis was conducted to examine the temporal- and spatial-expression patterns of neuropeptide genes using transcriptomic data obtained from various tissues and developmental stages (Figures 5 and 6). The results revealed that most of the neuropeptide genes exhibited distinct and elevated expression levels in neuroendocrine tissues such as the eyestalk, brain, thoracic ganglion, and ventral nerve (Figure 5). Significantly, a considerable proportion of these genes exhibited exclusive expression in the eyestalk. Notably, certain genes, such as sulfakinin, WXXXRamide, MIH3, and sNPF demonstrated elevated expression levels in the brain. Similarly, the genes AST-B, CCHamide, and orcokinin displayed distinct expression patterns in the thoracic ganglion, whereas bursicon α and bursicon β exhibited the highest expression levels in the ventral nerve. Intriguingly, we have also noted the presence of neuropeptide genes exhibiting significant expression in non-neuroendocrine tissues. For instance, genes including Calcitonin, CHH7, elevenin, and tachykinin displayed distinct expression patterns in the intestine. Similarly, one neuroparsin gene and one CHH15 gene showed highest expression levels in the heart, while the EH1 gene exhibited high levels in the gill and epidermis.

The expression of neuropeptide genes during early embryonic development was found to be minimal, except for the specific expression of the MIH2 gene in the blastula stage (Figure 6). Additionally, several genes exhibited initial expression at the larva in the membrane (Lim) stage. Furthermore, most of neuropeptide genes began to be expressed after hatching at the nauplius or zoea stage, and demonstrated high expression levels from the zoea stage to the post-larva stage. Notably, a significant proportion of these genes exhibited particularly high expression levels at the post-larva stage, such as the CHH3, CHH5, and CHH8 genes. Remarkably, the CHH9 and CHH15 displayed notably higher expression levels at the zoea stage, and the genes such as CFSH1b and pyrokinin exhibited higher expression levels at the mysis stage.

3.6. *RNAi Efficiency of dsCHH3 and dsVIH in the Eyestalk*

To effectively silence the expression of *LvCHH3* and *LvVIH*, an intramuscular injection of in vitro synthesized dsRNAs (dsCHH3 and dsVIH) was meticulously administered. The subsequent analysis of *LvCHH3* and *LvVIH* expression levels in the eyestalk was conducted 24 h following the injection. The results demonstrated a successful knockdown efficacy, with an impressive 80% reduction in gene expression relative to the control group, as depicted in Figure 7.

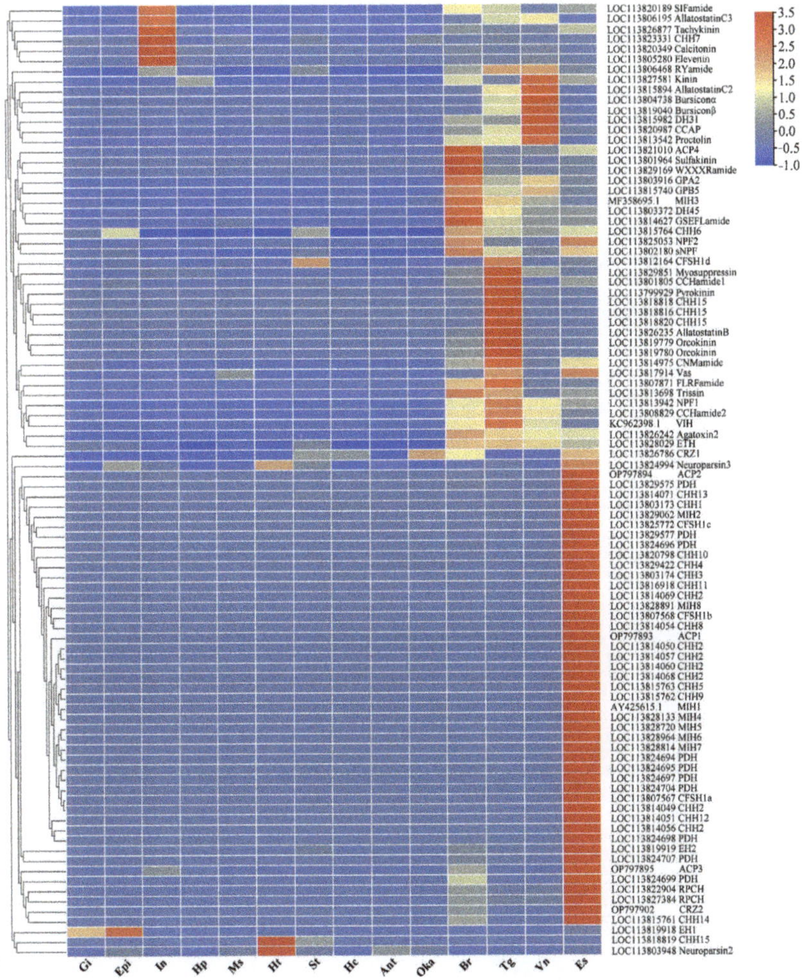

Figure 5. Hierarchical clustering of neuropeptide genes in various tissues. Colors represent relative mRNA expression as indicated in the color key. The red color shows high expression, and the blue color represents lower levels of expression. The color from red to blue represents the $\log_2$ (FPKM + 1) from large to small. Gi, gill; Epi, epidermis; In, intestine; Hp, hepatopancreas; Ms, muscle; Ht, heart; St, stomach; Hc, hemolymph; Ant, antennal gland; Oka, lymphoid organ; Br, brain ganglia; Tg, thoracic ganglia; Vn, ventral nerve; Es, eyestalk.

3.7. Functional Analysis of DEGs in the Eyestalk After RNAi

Nine high-quality RNA samples from eyestalks (with three replicates per group) were utilized for the construction of cDNA libraries, which were subsequently sequenced. Detailed information is provided in Table S4. The raw sequencing reads have been deposited in the SRA database (SRP200583). A significant number of genes were found to be up-regulated in dsCHH3 compared to dsGFP (61 genes), and in dsVIH compared to dsGFP (120 genes). Conversely, the number of down-regulated genes was 201 and 182, respectively. Moreover, a substantial number of the down-regulated genes encoded neuropeptides. The expression profiles of neuropeptide genes showed that the majority of the genes were down-regulated after the silencing of either *LvCHH3* or *LvVIH* (Figure 8). The results of the differential expression analysis were validated by qPCR, demonstrating the reliability and accuracy of the differential expression findings (Figure S5).

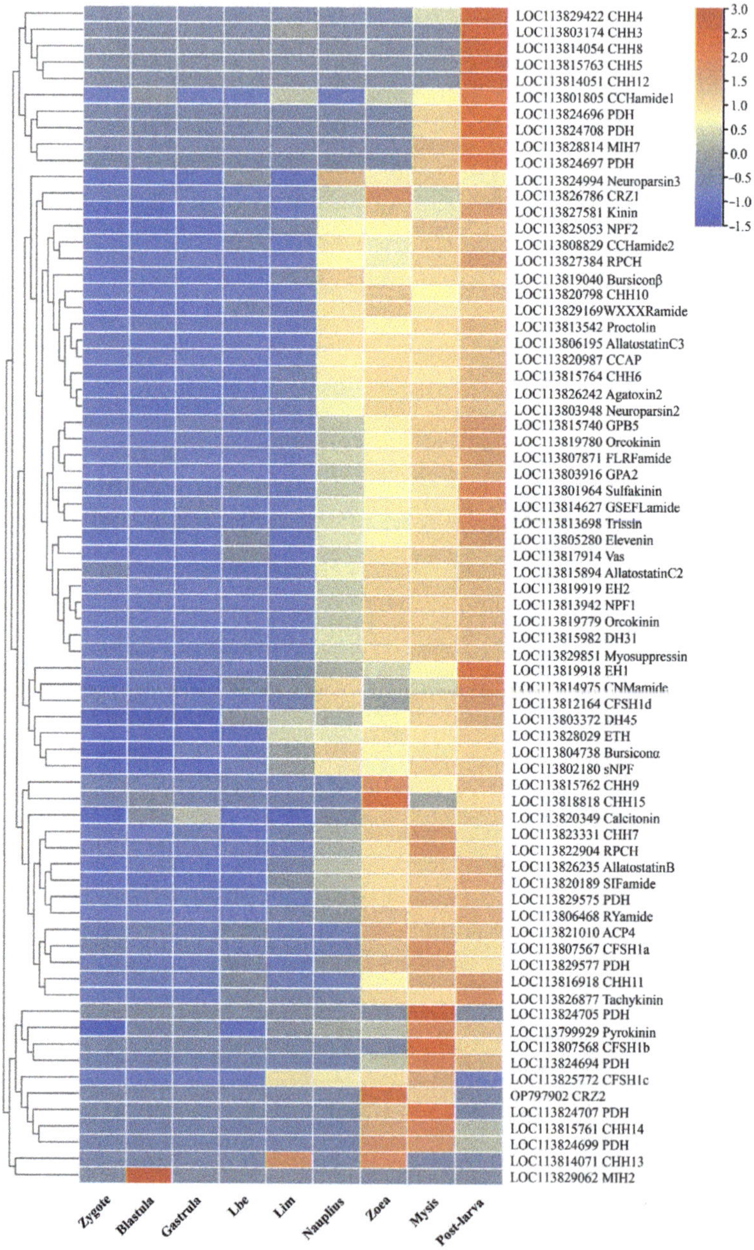

Figure 6. Hierarchical clustering of the expression of neuropeptide genes in various developmental stages. Colors represent relative mRNA expression, as indicated in the color key. The red color shows high expression, and the blue color represents lower levels of expression. The color from red to blue represents the $\log_2$ (FPKM + 1) from large to small. Lbe, limb bud embryo; Lim, larva in membrane.

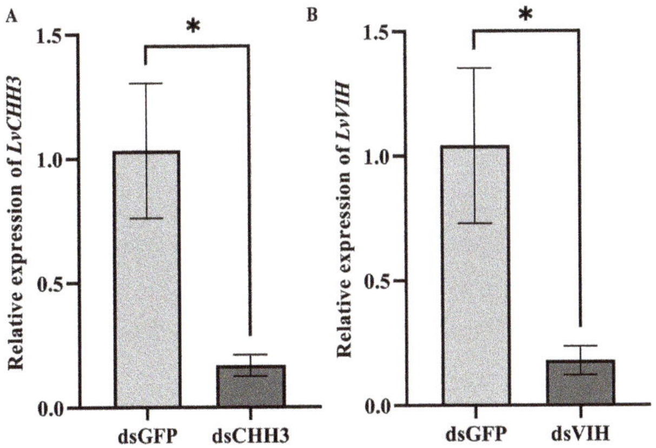

Figure 7. The RNAi efficiency of (**A**) dsCHH3 and (**B**) dsVIH. The results are presented as mean ± SEM, and asterisks indicate a significant difference ($p < 0.05$) between the dsGFP and dsCHH3 or dsVIH injection group.

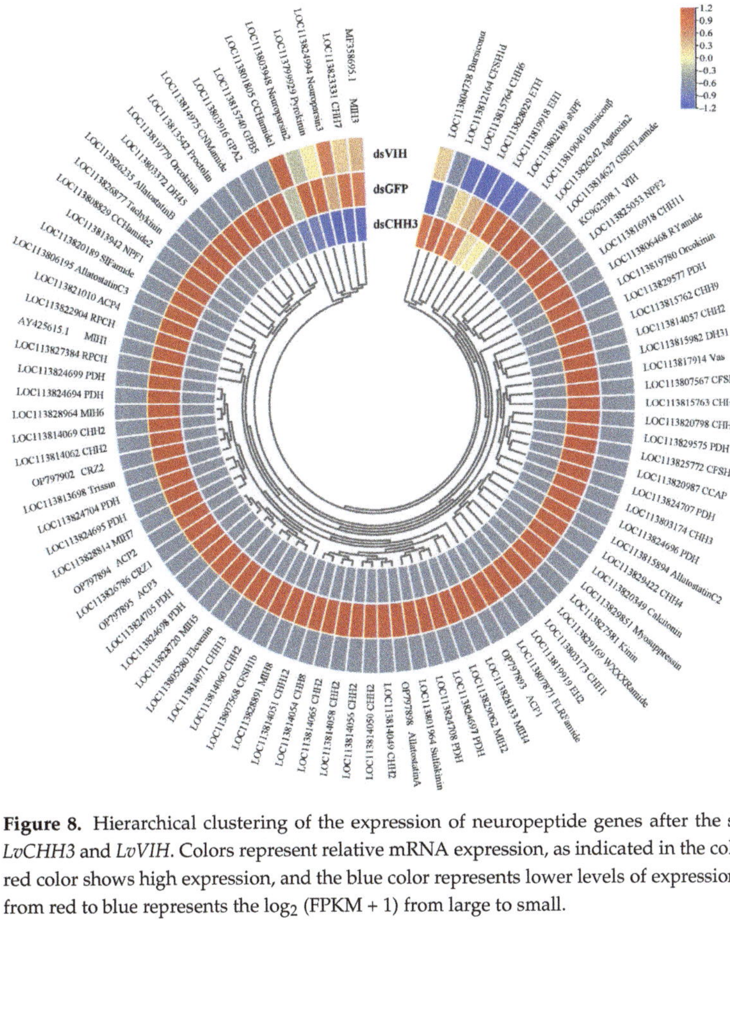

Figure 8. Hierarchical clustering of the expression of neuropeptide genes after the silencing of *LvCHH3* and *LvVIH*. Colors represent relative mRNA expression, as indicated in the color key. The red color shows high expression, and the blue color represents lower levels of expression. The color from red to blue represents the $\log_2$ (FPKM + 1) from large to small.

The top significantly enriched GO terms were consistent across the two comparisons (corrected $p < 0.05$; Table 2). These terms included the neuropeptide signaling pathway (GO: 0007218) and glucose metabolic process (GO: 0006006) within the biological processes category, the extracellular region (GO: 0005576) within the cellular component category, and neuropeptide hormone activity (GO: 0005184) and hormone activity (GO: 0005179) within the molecular function category. Among the enriched GO terms, the presence of "neuropeptide signaling pathway", "neuropeptide hormone activity", and "hormone activity" highlights the fact that the DEGs influenced by the silencing of *LvCHH3* and *LvVIH* were indeed implicated in hormonal activities. This underscored the pivotal role of these genes in processes mediated by neuropeptides.

Table 2. The significantly enriched GO terms.

GO Terms	Class	Description
GO:0007218	Biological Process	neuropeptide signaling pathway
GO:0006006	Biological Process	glucose metabolic process
GO:0051904	Biological Process	pigment granule transport
GO:0005576	Cellular Component	extracellular region
GO:0000235	Cellular Component	astral microtubule
GO:0038039	Cellular Component	G-protein coupled receptor heterodimeric complex
GO:0005184	Molecular Function	neuropeptide hormone activity
GO:0005179	Molecular Function	hormone activity

In the two comparisons, three and five KEGG pathways were significantly enriched, respectively (corrected $p < 0.05$; Table 3). The majority of these significantly enriched pathways were categorized under metabolic pathways. In the dsGFP vs. dsCHH3 comparison, the most significantly enriched pathways included those related to amino acid metabolism, such as Phenylalanine, Tyrosine, and Tryptophan biosynthesis (ko00400), as well as Phenylalanine metabolism (ko00360). Conversely, in the dsGFP vs. dsVIH comparison, the most significantly enriched pathways pertained to lipid metabolism, specifically Arachidonic acid metabolism (ko00590). Additionally, Insect hormone biosynthesis (ko00981) was also identified as a significantly enriched pathway in the dsGFP vs. dsVIH comparison.

Table 3. The significantly enriched KEGG pathways.

Comparison	Pathway	KEGG Class	Pathway	*q* Value
dsGFP vs. dsCHH3	ko00400	Amino acid metabolism	Phenylalanine, tyrosine and tryptophan biosynthesis	1.79×10^{-2}
	ko00360	Amino acid metabolism	Phenylalanine metabolism	4.32×10^{-2}
	ko04260	Circulatory system	Cardiac muscle contraction	4.32×10^{-2}
dsGFP vs. dsVIH	ko00590	Lipid metabolism	Arachidonic acid metabolism	7.94×10^{-3}
	ko00981	Metabolism of terpenoids and polyketides	Insect hormone biosynthesis	2.80×10^{-2}
	ko00910	Energy metabolism	Nitrogen metabolism	3.51×10^{-2}
	ko00591	Lipid metabolism	Linoleic acid metabolism	3.65×10^{-2}
	ko04976	Digestive system	Bile secretion	3.65×10^{-2}

4. Discussion

4.1. The Expansion of Neuropeptides in L. vannamei

In this study, we conducted in silico mining of transcriptome and genome databases, to characterize the neuropeptidome of *L. vannamei*. We identified 125 neuropeptide-encoding genes, of which 54 were previously uncharacterized or unannotated. In this study, we conducted a re-annotation of neuropeptide-encoding genes (Table 1). We acknowledge the potential for annotation errors that could misrepresent gene functions and consequently distort our comprehension of the neuropeptidome in *L. vannamei*. To mitigate this risk and ensure the precision of our gene annotations, we employed a stringent annotation pipeline using NCBI databases, setting an E-value threshold of 1.0×10^{-5} to minimize the inclusion of false positives. Furthermore, we bolstered the accuracy of our annotations by corroborating our results with known neuropeptide sequences from other decapod species. While annotation errors may present a limitation, these rigorous measures ensure that they do not significantly undermine the overall validity of this study's findings.

While the majority of these neuropeptides had been identified in other crustacean or insect species [14], several precursors are novel to *L. vannamei* in a broader context, such as the precursors of ACP2-ACP4 and CRZ2. The invertebrate neuropeptides, including ACP, CRZ, and RPCH, share intriguing evolutionary ties with GnRH, a neuropeptide that is widely acknowledged for its pivotal role in regulating reproduction in vertebrates [29,30]. However, the presence of GnRH in crustaceans has been a subject of debate, as only a limited number of studies have reported the presence of GnRH-like immunoreactivity [31,32], prompting the hypothesis that GnRH peptides might have been lost along the arthropod lineage [33]. Through our comprehensive and meticulous analysis, we have successfully identified all genes encoding proteins with characteristics similar to GnRH, thereby elucidating the expansion and diversification of genes encoding GnRH-like neuropeptides within the *L. vannamei* genome. This discovery significantly enhances our understanding of the evolutionary conservation and divergence of neuropeptide families, and paves the way for further exploration of their functional roles in crustacean physiology and reproduction.

The elucidation of the *L. vannamei* genome has indeed heralded a new era in our understanding of the CHH gene-family expansion. However, a comprehensive global perspective on this gene family has remained elusive [20]. The absence of three genes from the genomic sequence could be attributed to the challenges associated with genome assembly, a common limitation in genomic studies that can obscure the complete picture of gene-family representation. Our analysis of the identified CHH genes, including their structures and genomic locations, has unveiled a diverse genomic organization.

While some genes stand alone on their scaffolds, others are clustered, indicating a complex history of gene duplication and rearrangement. The hypothesis regarding the origin of the CHH superfamily members from a common ancestral gene has been a topic of interest in evolutionary biology, as suggested by previous research [34,35]. Our present study has yielded intriguing insights that have prompted us to formulate a more nuanced hypothesis regarding the evolutionary origins of the CHH superfamily in *L. vannamei*. We propose that the CHH6 gene emerges as a prime candidate for the ancestral gene within this species' genome. This hypothesis is bolstered by several key findings. A significant observation is the structural diversity present in the C-terminus of type-I CHH peptides. Our analysis has uncovered a divergence in the C-terminus structure, where one variant is equipped with a glycine amidation site, while another is devoid of this feature. This variation is a direct outcome of the alternative splicing events occurring within the CHH6 gene, which is capable of producing these two distinct types of CHH peptides. This discovery suggests a mechanism for functional diversification and specialization within the CHH superfamily. Moreover, the CHH6 gene's structure, which includes four exons as opposed to the three-exon configuration typical of other type-I CHH genes, lends further credence to our hypothesis. The presence of exon 3 and exon 4, both susceptible to alternative splicing, is particularly noteworthy. The coding sequences of these exons, with lengths of 103 bp and 106 bp, respectively, closely mirror the lengths of exon 3 in other type-I CHH genes (106 bp or 103 bp). This conservation in exon length between the CHH6 gene and its type-I CHH counterparts indicates a shared evolutionary heritage and provides molecular evidence of the gene duplication events that have sculpted the CHH superfamily. Additionally, the CHH6 peptide lacking the glycine amidation site has been observed for its osmoregulatory functions, akin to its homologous gene ITP found in insects [36,37]. In contrast, the type-I CHH peptide with the glycine amidation site exhibits a more pronounced role in glucose regulation, a function that appears to be unique to crustaceans [38,39]. These functional distinctions underscore the potential for the CHH6 gene to be the progenitor of the CHH superfamily in *L. vannamei*, with its alternative splicing of exon 3 potentially giving rise to a novel peptide with hyperglycemic activity. In conclusion, our study posits that the CHH6 gene could be the ancestral gene for the CHH superfamily in *L. vannamei*, highlighting the importance of alternative splicing in the generation of functionally distinct peptides.

The phenomenon of alternative splicing is not confined to the CHH superfamily, but extends to other neuropeptide precursors as well, such as agatoxin and calcitonin. Previous reports have documented the existence of three agatoxin precursors and two calcitonin precursors across various crustacean species [14]. Our study breaks new ground by unraveling the origins of these precursors for the first time, offering a window into the molecular mechanisms that have shaped the diversity of the crustacean neuropeptidome.

4.2. The Expression Profiles of Neuropeptide Genes

The present study represents the inaugural delineation of neuropeptide gene-expression patterns across various tissues in *L. vannamei*. Our findings highlight the fact that the majority of these genes are primarily expressed in the eyestalk, cerebral, thoracic, and ventral ganglia, consistent with their roles as neuroendocrine signaling molecules (Figure 4). Notably, a significant subset of these genes exhibited exclusive expression in the eyestalk, emphasizing the critical neuroendocrine function of the XO-SG complex [15]. A significant number of these genes are affiliated with the CHH superfamily, GnRH-like family, PDH family, and CFSH family, correlating with the XO-SG complex's primary role in the synthesis and secretion of hormones associated with the CHH family and pigmentary-effector molecular families [40]. However, it has also been observed that numerous neuropeptides exhibit heightened expression in the cerebral, thoracic, and ventral ganglia, suggesting that these ganglia serve as significant neuroendocrine centers warranting further investigation to elucidate the regulatory roles of neuropeptides derived from these tissues. Moreover, the study identified the fact that several neuropeptide genes demonstrate peak expression levels in non-neuroendocrine tissues, including six genes in the intestine, two in the heart, and one in the epidermis. Notably, the pronounced expression of the CHH7 gene in the intestine is associated with its osmoregulatory function in *L. vannamei* [41]. These findings suggest the potential presence of endocrine cells within these tissues, which may exert autocrine effects.

Crustaceans exhibit intricate physiological mechanisms, due to the interplay of various life processes. These encompass the developmental journey from embryos to larvae, the transformative phase of metamorphosis, and the repetitive cycle of molting [42]. Transcriptomic analysis of developing embryos revealed minimal expression of neuropeptide transcripts during embryogenesis, indicating that these neuropeptide genes are not maternally expressed. Notably, an exception is observed with the MIH2 gene, which exhibits higher expression levels during the blastula stage compared to other neuropeptide genes. Furthermore, the expression of the majority of genes demonstrated a marked upregulation from the nauplius stage to the post-larva stage, which may be associated with the enhanced neuronal proliferation observed during larval development. A study investigating tachykinin-related peptide (TRP) expression across various developmental stages of *Drosophila melanogaster* demonstrated that the number of TRP-expressing neuronal cell bodies in the brain and ventral nerve cord increases during larval development, with a notable surge occurring from the later pupal stages to adulthood [43].

4.3. The Expression of Neuropeptide Genes After the Silencing of LvCHH3 and LvVIH

A comparative transcriptome analysis was conducted to examine the effects of CHH silencing on gene expression in the eyestalk. The results revealed a downregulation of neuropeptide genes following the suppression of either *LvCHH3* or *LvVIH*. Previous studies have demonstrated that the use of dsRNA targeting the CHH gene *Liv-SGP-G* (designated as *LvCHH5* in this study) can inhibit the transcription of not only the target gene, but also closely related CHH genes *Liv-SGP-A*, *-B*, and *-C* (designated as *LvCHH1*, *2*, and *3* in this study) in *L. vannamei* [19]. The utilization of the relative dsRNA in RNA interference may enable the co-transcriptional repression of the specific *CHH* gene, as well as other homologous *CHH* genes exhibiting high sequence similarity to the dsRNA. Nevertheless, this does not clarify the observed reduction in the expression of other neuropeptide genes.

Subsequent enrichment analysis revealed that the most significantly enriched pathways were those involved in Phenylalanine, Tyrosine, and Tryptophan biosynthesis (ko00400) and Phenylalanine metabolism (ko00360) when comparing dsGFP to dsCHH3. Tryptophan, tyrosine, and phenylalanine function as biosynthetic precursors for neurotransmitters including serotonin, dopamine, and norepinephrine [44]. The regulation of hormones from the XO/SG complex appears to be partially mediated by neurotransmitters such as dopamine, serotonin, and encephalin [45,46]. It is plausible that the silencing of *LvCHH3* affected the synthesis of monoamine neurotransmitters, subsequently influencing neuropeptide synthesis. Additionally, arachidonic acid metabolism (ko00590) was identified as the most significantly enriched pathway in the comparison between dsGFP and dsVIH. The metabolism of arachidonic acid results in its conversion into prostaglandins and thromboxane. In mammals, arachidonic acid and its derivatives play a crucial role in the release of neuropeptides within the central nervous system [47]. The findings suggest that the silencing of *LvVIH* may have disrupted arachidonic acid metabolism, thereby inhibiting neuropeptide release and subsequently repressing their expression. These findings elucidate the complex regulatory mechanisms mediated by neuropeptides and provide a foundation for future investigations into the functional roles of these critical molecules in crustacean physiology and development.

The suppression of CHH and VIH genes in the eyestalk may have far-reaching consequences for the shrimp's physiological condition. The CHH superfamily, in particular, is known beyond its role in the eyestalk, for its involvement in regulating a range of critical functions such as carbohydrate metabolism, molting, osmoregulation, and reproduction [16]. The broader systemic effects of silencing these genes could disrupt these processes, thereby affecting the shrimp's overall health and growth performance. Further research is essential to investigate these systemic implications and to elucidate how the downregulation of neuropeptide genes in the eyestalk can influence other physiological systems.

4.4. Potential Roles of Neuropeptides in Aquaculture

The implications of our findings extend beyond the realm of basic biology and into the practical applications of aquaculture. Neuropeptides, as pivotal regulators of physiological processes, present genetic targets that could revolutionize shrimp production [1]. For example, genes implicated in metabolism, molting, and reproduction, particularly those within the CHH superfamily, could be strategically manipulated to mitigate molting-related stress or to boost ovarian development [16–19]. Additionally, neuropeptides linked to stress response and disease resistance might pave the way for developing shrimp strains with enhanced resilience to environmental challenges and pathogens [48–51]. Our findings on the developmental expression patterns of neuropeptide genes also suggest that targeted interventions at critical stages could significantly improve survival rates and growth efficiency. While the prospects of neuropeptide research in aquaculture are indeed promising, there are several hurdles that need to be overcome to effectively translate these findings into tangible applications. These include navigating the complexities of neuropeptide signaling pathways and addressing the ethical concerns associated with genetic manipulation. Future research should concentrate on the functional validation of these neuropeptide genes, delve into their roles in stress response, and devise targeted interventions that can be effectively implemented within aquaculture settings.

5. Conclusions

In this comprehensive study, we employed a multiomics approach to systematically identify and characterize neuropeptide-encoding genes in *L. vannamei*. Certain neuropeptide-encoding gene families demonstrated significant expansion. In addition, alternative splicing was found to play a crucial role in generating functionally diverse neuropeptides. RNAi-mediated suppression of CHH and VIH genes resulted in the downregulation of the majority of neuropeptide genes. This downregulation was significantly associated with the enrichment of pathways related to amino acid metabolism and hormone synthesis, indicating the broader regulatory impact of these hormones. Our findings provide valuable genetic targets for future research aimed at improving shrimp growth and reproduction through genetic manipulation. This could lead to advancements in shrimp aquaculture practices, enhancing productivity and sustainability.

Supplementary Materials: The following supporting information can be downloaded at: https://www.mdpi.com/article/10.3390/biology13121038/s1. Table S1: Specific primers used in the current study; Table S2: Sequencing data across the 42 libraries of various tissues from *L. vannamei*; Table S3: Functional annotation of de novo transcriptome from *L. vannamei*; Table S4: The sequences of peptides used for the phylogenetic tree in Figure 2; Table S5: Summary of sequence data generated from transcriptome sequencing; Figure S1: Complete sequences from L. vannamei of the precursors containing: (A) AST-A; (B) AST-B; (C) AST-C; Figure S2: Complete sequences from L. vannamei of the precursors containing: (A) Bursicon α; (B) Bursicon β; (C) CCRFamide; (D) CNMamide; (E) CFSH; (F) DH31; (G) DH45; (H) ETH; (I) EH; (J) Elevenin; Figure S3: Complete sequences from L. vannamei of the precursors containing: (A) FLRFamide; (B) GPA2; (C) GPA5; (D) GSEFLamide; (E) Kinin; (F) Myosuppressin; (G) Neuroparsin; (H) Neuropeptide F; (I) sNPF; (J) Orcokinin; (K) Proctolin; Figure S4: Complete sequences from L. vannamei of the precursors containing: (A) Pyrokinin; (B) RYamide; (C) SIFamide; (D) Sulfakinin; (E) Tachykinin; (F) Trissin; (G) Vasotocin-neurophysin; (H) WXXXRamide; Figure S5: qPCR validation of RNA-seq data.

Author Contributions: Conceptualization, L.Z. and Z.W.; methodology, L.Z.; software, L.S.; validation, L.Z., L.S. and G.S.; formal analysis, Y.C.; investigation, F.L.; resources, B.W.; data curation, L.S.; writing—original draft preparation, L.Z. and L.S.; writing—review and editing, Z.W.; visualization, F.L. and Z.W.; supervision, Y.L.; project administration, Z.W.; funding acquisition, Y.L., F.L., Y.C. and Z.W. All authors have read and agreed to the published version of the manuscript.

Funding: This research was funded by the National Key Research and Development Program of China (2022YFD2400203); the National Science Foundation of China (32102762); the Natural Science Foundation of Shandong Province (ZR2024MC077; ZR2024MC208); the Shrimp & Crab Innovation Team of Shandong Agriculture Research System (SDAIT-15-011); and the Major Special Project for Technological Innovation in Dongying (2023ZDJH30).

Institutional Review Board Statement: All treatments in this study were undertaken strictly in accordance with the guidelines of the Animal Experiment Ethics Committee of Qingdao Agriculture University, which also approved the protocol in May 2020 (Approval Code: 2020-026).

Informed Consent Statement: Not applicable.

Data Availability Statement: Data are contained within the article.

Conflicts of Interest: The authors declare no conflicts of interest.

Abbreviations

CHH	crustacean hyperglycemic hormone
VIH	vitellogenesis-inhibiting hormone
MIH	molt-inhibiting hormone
ACP	adipokinetic hormone/corazonin-like peptide
CRZ	corazonin
RPCH	red pigment-concentrating hormone
PDH	pigment-dispersing hormone
AST-A	allatostatin A
AST-B	allatostatin B
AST-C	allatostatin C
CCAP	crustacean cardioactive peptide
DH	diuretic hormone

ETH	ecdysis triggering hormone
EH	eclosion hormone
GPA2	glycoprotein-A2
GPB5	glycoprotein-B5
NP	neuroparsin
NPF	neuropeptide F
sNPF	short Neuropeptide F
VT-NP	aasotocin–neurophysin

References

1. Gäde, G.; Hoffmann, K.-H. Neuropeptides Regulating Development and Reproduction in Insects. *Physiol. Entomol.* **2005**, *30*, 103–121. [CrossRef]
2. De Haes, W.; Van Sinay, E.; Detienne, G.; Temmerman, L.; Schoofs, L.; Boonen, K. Functional Neuropeptidomics in Invertebrates. *BBA Proteins Proteom.* **2015**, *1854*, 812–826. [CrossRef] [PubMed]
3. Hartenstein, V. The Neuroendocrine System of Invertebrates: A Developmental and Evolutionary Perspective. *J. Endocrinol.* **2006**, *190*, 555–570. [CrossRef] [PubMed]
4. Salio, C.; Lossi, L.; Ferrini, F.; Merighi, A. Neuropeptides as Synaptic Transmitters. *Cell Tissue Res.* **2006**, *326*, 583–598. [CrossRef]
5. Yuan, J.; Yu, Y.; Zhang, X.; Li, S.; Xiang, J.; Li, F. Recent Advances in Crustacean Genomics and Their Potential Application in Aquaculture. *Rev. Aquacult.* **2023**, *15*, 1501–1521. [CrossRef]
6. Farhadi, A.; Liu, Y.; Xu, C.; Han, T.; Wang, X.; Li, E. Evidence from Transcriptome Analysis Unravelled the Roles of Eyestalk in Salinity Adaptation in Pacific White Shrimp (*Litopenaeus vannamei*). *Gen. Comp. Endocrinol.* **2022**, *329*, 114120. [CrossRef]
7. Duan, Y.; Xiong, D.; Wang, Y.; Li, H.; Dong, H.; Zhang, J. Toxic Effects of Ammonia and Thermal Stress on the Intestinal Microbiota and Transcriptomic and Metabolomic Responses of *Litopenaeus vannamei*. *Sci. Total Environ.* **2021**, *754*, 141867. [CrossRef]
8. Yuan, J.; Yu, Y.; Li, S.; Zhang, X.; Zhang, C.; Li, R.; Hu, J.; Si, S.; Zhang, C.; Xiang, J.; et al. Shrimp Shapes a Resistance Trait against Vibriosis by Memorizing the Colonization Resistance of Intestinal Microbiota. *PLoS Pathog.* **2024**, *20*, e1012321. [CrossRef]
9. Christie, A.E. Identification of Putative Neuropeptidergic Signaling Systems in the Spiny Lobster, *Panulirus Argus*. *Invert. Neurosci.* **2020**, *20*, 2. [CrossRef]
10. Sun, S.; Zhu, M.; Pan, F.; Feng, J.; Li, J. Identifying Neuropeptide and G Protein-Coupled Receptors of Juvenile Oriental River Prawn (*Macrobrachium nipponense*) in Response to Salinity Acclimation. *Front. Endocrinol.* **2020**, *11*, 623. [CrossRef]
11. Wang, D.; Liu, X.; Zhang, J.; Gao, B.; Liu, P.; Li, J.; Meng, X. Identification of Neuropeptides Using Long-Read RNA-Seq in the Swimming Crab *Portunus Trituberculatus*, and Their Expression Profile Under Acute Ammonia Stress. *Front. Physiol.* **2022**, *13*, 910585. [CrossRef] [PubMed]
12. Nguyen, T.V.; Cummins, S.F.; Elizur, A.; Ventura, T. Transcriptomic Characterization and Curation of Candidate Neuropeptides Regulating Reproduction in the Eyestalk Ganglia of the Australian Crayfish, *Cherax Quadricarinatus*. *Sci. Rep.* **2016**, *6*, 38658. [CrossRef] [PubMed]
13. Christie, A.E. Expansion of the *Litopenaeus Vannamei* and *Penaeus Monodon* Peptidomes Using Transcriptome Shotgun Assembly Sequence Data. *Gen. Comp. Endocrinol.* **2014**, *206*, 235–254. [CrossRef] [PubMed]
14. Veenstra, J.A. Similarities between Decapod and Insect Neuropeptidomes. *PeerJ* **2016**, *4*, e2043. [CrossRef] [PubMed]
15. Simões, L.A.R.; Normann, R.S.; Chung, J.S.; Vinagre, A.S. A Brief and Updated Introduction to the Neuroendocrine System of Crustaceans. *Mol. Cell Endocrinol.* **2024**, *590*, 112265. [CrossRef] [PubMed]
16. Chen, H.-Y.; Toullec, J.-Y.; Lee, C.-Y. The Crustacean Hyperglycemic Hormone Superfamily: Progress Made in the Past Decade. *Front. Endocrinol.* **2020**, *11*, 578958. [CrossRef]
17. Feijó, R.G.; Braga, A.L.; Lanes, C.F.C.; Figueiredo, M.A.; Romano, L.A.; Klosterhoff, M.C.; Nery, L.E.M.; Maggioni, R.; Wasielesky, W.; Marins, L.F. Silencing of Gonad-Inhibiting Hormone Transcripts in *Litopenaeus Vannamei* Females by Use of the RNA Interference Technology. *Mar. Biotechnol.* **2016**, *18*, 117–123. [CrossRef]
18. Kang, B.J.; Bae, S.-H.; Suzuki, T.; Niitsu, S.; Wilder, M.N. Transcriptional Silencing of Vitellogenesis-Inhibiting Hormone (VIH) Subtype-I in the Whiteleg Shrimp, *Litopenaeus Vannamei*. *Aquaculture* **2019**, *506*, 119–126. [CrossRef]
19. Kang, B.J.; Sultana, Z.; Wilder, M.N. Assessment of the Effects of Double-Stranded RNAs Corresponding to Multiple Vitellogenesis-Inhibiting Hormone Subtype I Peptides in Subadult Female Whiteleg Shrimp, *Litopenaeus Vannamei*. *Front. Endocrinol.* **2021**, *12*, 594001. [CrossRef]
20. Zhang, X.; Yuan, J.; Sun, Y.; Li, S.; Gao, Y.; Yu, Y.; Liu, C.; Wang, Q.; Lv, X.; Zhang, X.; et al. Penaeid Shrimp Genome Provides Insights into Benthic Adaptation and Frequent Molting. *Nat. Commun.* **2019**, *10*, 356. [CrossRef]
21. Chen, C.; Wu, Y.; Li, J.; Wang, X.; Zeng, Z.; Xu, J.; Liu, Y.; Feng, J.; Chen, H.; He, Y.; et al. TBtools-II: A "one for all, all for one" bioinformatics platform for biological big-data mining. *Mol. Plant.* **2023**, *16*, 1733–1742. [CrossRef] [PubMed]
22. Liu, W.; Xie, Y.; Ma, J.; Luo, X.; Nie, P.; Zuo, Z.; Lahrmann, U.; Zhao, Q.; Zheng, Y.; Zhao, Y.; et al. IBS: An Illustrator for the Presentation and Visualization of Biological Sequences. *Bioinformatics* **2015**, *31*, 3359–3361. [CrossRef] [PubMed]
23. Veenstra, J.A. Mono- and Dibasic Proteolytic Cleavage Sites in Insect Neuroendocrine Peptide Precursors. *Arch. Insect Biochem. Physiol.* **2000**, *43*, 49–63. [CrossRef]

24. Kumar, S.; Stecher, G.; Li, M.; Knyaz, C.; Tamura, K. MEGA X: Molecular Evolutionary Genetics Analysis across Computing Platforms. *Mol. Biol. Evol.* **2018**, *35*, 1547–1549. [CrossRef]
25. Pertea, M.; Kim, D.; Pertea, G.M.; Leek, J.T.; Salzberg, S.L. Transcript-Level Expression Analysis of RNA-Seq Experiments with HISAT, StringTie and Ballgown. *Nat. Protoc.* **2016**, *11*, 1650–1667. [CrossRef]
26. Love, M.I.; Huber, W.; Anders, S. Moderated Estimation of Fold Change and Dispersion for RNA-Seq Data with DESeq2. *Genome Biol.* **2014**, *15*, 550. [CrossRef]
27. Yu, G.; Wang, L.-G.; Han, Y.; He, Q.-Y. clusterProfiler: An R Package for Comparing Biological Themes among Gene Clusters. *Omics* **2012**, *16*, 284–287. [CrossRef]
28. Livak, K.J.; Schmittgen, T.D. Analysis of Relative Gene Expression Data Using Real-Time Quantitative PCR and the $2^{-\Delta\Delta CT}$ Method. *Methods* **2001**, *25*, 402–408. [CrossRef]
29. Hauser, F.; Grimmelikhuijzen, C.J.P. Evolution of the AKH/Corazonin/ACP/GnRH Receptor Superfamily and Their Ligands in the Protostomia. *Gen. Comp. Endocrinol.* **2014**, *209*, 35–49. [CrossRef]
30. Millar, R.P. GnRHs and GnRH Receptors. *Anim. Reprod. Sci.* **2005**, *88*, 5–28. [CrossRef]
31. Ngernsoungnern, P.; Ngernsoungnern, A.; Kavanaugh, S.; Sobhon, P.; Sower, S.A.; Sretarugsa, P. The Presence and Distribution of Gonadotropin-Releasing Hormone-Liked Factor in the Central Nervous System of the Black Tiger Shrimp, *Penaeus Monodon. Gen. Comp. Endocrinol.* **2008**, *155*, 613–622. [CrossRef] [PubMed]
32. Tinikul, Y.; Poljaroen, J.; Nuurai, P.; Anuracpreeda, P.; Chotwiwatthanakun, C.; Phoungpetchara, I.; Kornthong, N.; Poomtong, T.; Hanna, P.J.; Sobhon, P. Existence and Distribution of Gonadotropin-Releasing Hormone-like Peptides in the Central Nervous System and Ovary of the Pacific White Shrimp, *Litopenaeus Vannamei. Cell Tissue Res.* **2011**, *343*, 579–593. [CrossRef] [PubMed]
33. Roch, G.J.; Busby, E.R.; Sherwood, N.M. GnRH Receptors and Peptides: Skating Backward. *Gen. Comp. Endocrinol.* **2014**, *209*, 118–134. [CrossRef]
34. Chan, S.-M.; Gu, P.-L.; Chu, K.-H.; Tobe, S.S. Crustacean Neuropeptide Genes of the CHH/MIH/GIH Family: Implications from Molecular Studies. *Gen. Comp. Endocrinol.* **2003**, *134*, 214–219. [CrossRef]
35. Montagné, N.; Desdevises, Y.; Soyez, D.; Toullec, J.-Y. Molecular Evolution of the Crustacean Hyperglycemic Hormone Family in Ecdysozoans. *BMC Evol. Biol.* **2010**, *10*, 62. [CrossRef]
36. Lago-Lestón, A.; Ponce, E.; Muñoz, M.E. Cloning and Expression of Hyperglycemic (CHH) and Molt-Inhibiting (MIH) Hormones mRNAs from the Eyestalk of Shrimps of *Litopenaeus Vannamei* Grown in Different Temperature and Salinity Conditions. *Aquaculture* **2007**, *270*, 343–357. [CrossRef]
37. Tiu, S.H.K.; He, J.-G.; Chan, S.-M. The LvCHH-ITP Gene of the Shrimp (*Litopenaeus Vannamei*) Produces a Widely Expressed Putative Ion Transport Peptide (LvITP) for Osmo-Regulation. *Gene* **2007**, *396*, 226–235. [CrossRef]
38. Camacho-Jiménez, L.; Sánchez-Castrejón, E.; Díaz, F.; Aguilar, M.B.; Muñoz-Márquez, M.E.; Ponce-Rivas, E. Cloning and Expression of the Recombinant Crustacean Hyperglycemic Hormone Isoform B2 (rCHH-B2) and Its Effects on the Metabolism and Osmoregulation of the Pacific White Shrimp *Litopenaeus Vannamei. Gen. Comp. Endocrinol.* **2017**, *253*, 33–43. [CrossRef]
39. Katayama, H.; Ohira, T.; Aida, K.; Nagasawa, H. Significance of a Carboxyl-Terminal Amide Moiety in the Folding and Biological Activity of Crustacean Hyperglycemic Hormone. *Peptides* **2002**, *23*, 1537–1546. [CrossRef]
40. Hopkins, P.M. The Eyes Have It: A Brief History of Crustacean Neuroendocrinology. *Gen. Comp. Endocrinol.* **2012**, *175*, 357–366. [CrossRef]
41. Ventura-López, C.; Gómez-Anduro, G.; Arcos, F.G.; Llera-Herrera, R.; Racotta, I.S.; Ibarra, A.M. A Novel CHH Gene from the Pacific White Shrimp *Litopenaeus Vannamei* Was Characterized and Found Highly Expressed in Gut and Less in Eyestalk and Other Extra-Eyestalk Tissues. *Gene* **2016**, *582*, 148–160. [CrossRef] [PubMed]
42. Mykles, D.L.; Chang, E.S. Hormonal Control of the Crustacean Molting Gland: Insights from Transcriptomics and Proteomics. *Gen. Comp. Endocrinol.* **2020**, *294*, 113493. [CrossRef] [PubMed]
43. Winther, A.M.E.; Siviter, R.J.; Isaac, R.E.; Predel, R.; Nässel, D.R. Neuronal Expression of Tachykinin-Related Peptides and Gene Transcript during Postembryonic Development of *Drosophila. J. Comp. Neurol.* **2003**, *464*, 180–196. [CrossRef] [PubMed]
44. Fernstrom, J.D. Dietary Amino Acids and Brain Function. *J. Am. Diet. Assoc.* **1994**, *94*, 71–77. [CrossRef]
45. Ollivaux, C.; Dircksen, H.; Toullec, J.-Y.; Soyez, D. Enkephalinergic Control of the Secretory Activity of Neurons Producing Stereoisomers of Crustacean Hyperglycemic Hormone in the Eyestalk of the Crayfish *Orconectes Limosus. J. Comp. Neurol.* **2002**, *444*, 1–9. [CrossRef]
46. Sarojini, R.; Nagabhushanam, R.; Fingerman, M. Dopaminergic and Enkephalinergic Involvement in the Regulation of Blood Glucose in the Red Swamp Crayfish, *Procambarus Clarkii. Gen. Comp. Endocrinol.* **1995**, *97*, 160–170. [CrossRef]
47. Ojeda, S.R.; Urbanski, H.F.; Junier, M.P.; Capdevila, J. The Role of Arachidonic Acid and Its Metabolites in the Release of Neuropeptides. *Ann. N. Y. Acad. Sci.* **1989**, *559*, 192–207. [CrossRef]
48. Tong, R.; Pan, L.; Zhang, X.; Li, Y. Neuroendocrine-Immune Regulation Mechanism in Crustaceans: A Review. *Rev. Aquacult.* **2022**, *14*, 378–398. [CrossRef]
49. Wang, Y.; Liu, A.; Huang, Y.; Lu, L.; Guo, S.; Ye, H. Role of Crustacean Female Sex Hormone in Regulating Immune Response in the Mud Crab, *Scylla Paramamosain. Fish Shellfish Immunol.* **2023**, *142*, 109094. [CrossRef]

50. Xu, Z.; Wei, Y.; Guo, S.; Lin, D.; Ye, H. Short Neuropeptide F Enhances the Immune Response in the Hepatopancreas of Mud Crab (*Scylla Paramamosain*). *Fish Shellfish Immunol.* **2020**, *101*, 244–251. [CrossRef]
51. Wei, Y.; Lin, D.; Xu, Z.; Gao, X.; Zeng, C.; Ye, H. A Possible Role of Crustacean Cardioactive Peptide in Regulating Immune Response in Hepatopancreas of Mud Crab. *Front. Immunol.* **2020**, *11*, 711. [CrossRef]

biology

MDPI

Article

Investigation of the Influence of Hypercapnia on the Physiology of Ovigerous West Coast Rock Lobsters, *Jasus lalandii*, and Their Embryonic Development

Annika Ritter [1], Christopher R. Bridges [1] and Lutz Auerswald [2,3,*]

[1] Institute of Metabolic Physiology/Ecophysiology, Heinrich-Heine University, D-40225 Düsseldorf, Germany; annika.c.ritter@gmail.com (A.R.); bridges@tunatech.de (C.R.B.)
[2] Branch Fisheries Management, Department of Forestry Fisheries and the Environment (DFFE), Roggebaai, Cape Town 8012, South Africa
[3] Department of Animal Sciences, Stellenbosch University, Stellenbosch 7602, South Africa
* Correspondence: lutz.auerswald@gmail.com

Simple Summary: The West Coast rock lobster (WCRL), *Jasus lalandii*, inhabits the dynamic Benguela Current upwelling system along the west coast of southern Africa and supports an important fishery. WCRLs have to respond to frequent and rapid changes in pH and other environmental impacts that are predicted to become worse. Responses to such conditions are known for mature male and juvenile lobsters, but not for mature females or embryonic development. We therefore analysed the sensitivity of ovigerous ("berried") female WCRLs and their eggs/embryos to acute and chronic hypercapnia (high pCO_2, low pH). We applied extracellular acid–base analysis, stereomicroscopic examination of egg growth and development, and SEM of female exoskeleton structure and egg membranes. Berried females efficiently respond to acute and chronic hypercapnia by means of increasing bicarbonate concentrations in the haemolymph. Embryo growth and development are not impacted by chronic hypercapnia, but growth shows geographical differences. We conclude that females and embryos of *J. lalandii* are as resilient to hypercapnia as previously shown for males and juveniles of the same species.

Abstract: The West Coast rock lobster, *Jasus lalandii*, is a key ecological species and provides an important fishery resource in South Africa and Namibia. It is found along the west coast of southern Africa in the dynamic Benguela Current upwelling system. The low seawater pH of this system is expected to decline further due to ocean acidification and increase in upwelling in terms of frequency and severity. The lobster has therefore to respond to frequent and rapid changes in pH and other environmental impacts that are predicted to become more adverse in future. Although responses to such conditions are known for mature male and juvenile lobsters, there is a lack of information on mature females and later embryonic development. We addressed this by analysing the sensitivity of ovigerous ("berried") female WCRLs and their eggs/embryos to hypercapnia (high pCO_2, low pH) and formulated the following research questions: (1) Can berried female WCRLs respond swiftly to large changes in pH? (2) What physiological mechanisms facilitate a potential response to a rapidly declining pH, i.e., acute hypercapnia? (3) Does a potential response persist during prolonged hypercapnia? (4) Are eggs/embryos impacted by hypercapnia? To investigate this, we exposed berried WCRLs to acute (pH 7.5) and chronic (up to 60 days at pH 7.5 and 7.8) hypercapnia. We applied extracellular acid–base analysis, microscopic examination of egg growth and development, and SEM of female exoskeleton structure and egg membranes. The results revealed that berried females efficiently respond to acute and chronic hypercapnia by means of increasing bicarbonate concentrations in the haemolymph. Moreover, embryo growth and development are not impacted by chronic

 https://doi.org/10.3390/biology14020132

hypercapnia, but growth shows geographical area-specific differences. We conclude that females and embryos of *J. lalandii* are as resilient to hypercapnia as previously shown for males and juveniles.

Keywords: hypercapnia; ocean acidification; upwelling; acid–base regulation; embryonic development; ovigerous; physiology; maternal care

1. Introduction

The West Coast rock lobster (WCRL), *Jasus lalandii*, is a slow-growing palinurid species from the southern African Atlantic coast that inhabits cold to temperate waters. The fishery of the WCRL is one of the most important in South Africa due to its high economic value and directly and indirectly employs about 4300 people in marine- and land-based employment [1]. The Branch: Fisheries Management of the respective national governmental department (currently Department of Forestry, Fisheries and the Environment—DFFE) manages the West Coast rock lobster (WCRL) resource per zone (Figure 1) by means of total allowable catch (TAC), by additional effort control, a minimum legal size, and a ban on retaining ovigerous ("berried") females [2].

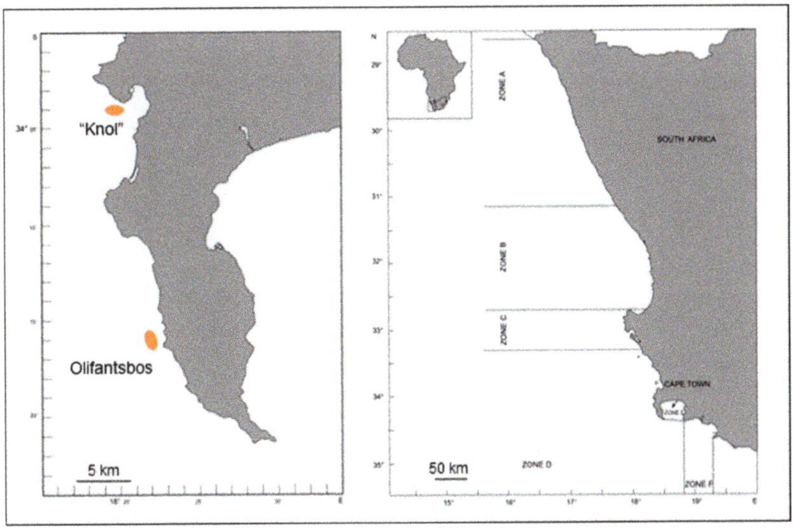

Figure 1. West Coast rock lobster capture area, showing the two sampling sites "Knol" in Hout Bay and Olifantsbos in fishing Zone D.

The WCRL is found mainly along the west coast of southern Africa in the Benguela Current System (BCLME), which is characterised by frequent upwelling events that result in hypercapnia in the range of pH 7.4–7.6 compared with normocapnic conditions of around pH 7.9–8.1 [3–5]. In addition, low-oxygen events caused by the decay of algal blooms and bacterial respiration can in some areas cause the water pH to drop to levels as low as 6.6 for several days [4]. In the long term, a further decline in pH is expected due to ocean acidification and increase in upwelling in terms of frequency and severity [6]. The rock lobster therefore has to respond to frequent and often rapid changes in pH and other environmental impacts such as rapid temperature changes and variations in oxygen levels. Responding to rapid environmental changes is not uncommon for crustaceans, however, which populate a variety of habitats and are therefore naturally exposed to a wide range of environmental parameters, resulting in a certain degree of physiological plasticity [7,8].

Previous research has revealed that *J. lalandii* is well adapted to this highly changeable habitat: the species is capable of efficiently maintaining its acid–base balance when exposed to acute and chronic hypercapnia [9,10]. This is achieved by a rapid increase in the bicarbonate levels in the haemolymph and is reversible. Bicarbonate buffering protects the pH-sensitive oxygen transport capacity of haemocyanin under hypercapnic conditions and provides optimum pH conditions for oxygen binding in the presence of a strong Bohr effect. Moreover, the oxygen affinity of haemocyanin is enhanced by an intrinsic modification of its molecular structure [9]. The immune response of this species is also very efficient and not impacted following chronic hypercapnia [11].

Mature female WCRLs moult in late austral autumn to early winter and mate shortly thereafter [12], when the female exoskeleton is still relatively soft and pleopodal setae are fresh for egg attachment [13,14]. In contrast to males, which use 99% of energy reserves accumulated during intermoult for growth, females use 95% for reproduction [15]. As a result, females grow slower than males. The fecundity of female *J. lalandii* varies along the Cape West Coast, with greater fecundity (more eggs) produced in areas where somatic growth is rapid [16]. Between the two sites investigated here, fecundity differs from approximately 104,000 to 137,000 eggs per lobster in the 75–79 mm carapace length class, a difference of 32% [16]. In times of nutritional stress, *J. lalandii* females sacrifice growth to optimize egg production [17]. Embryonic development occurs externally with the eggs attached to the ovigerous female's pleopods ("berry" stage), which exposes them to ambient conditions.

Crustacean eggs rely exclusively on the maternal provision of biomaterials and metabolites (such as vitellogenin, lipoproteins, carotenoids, carbohydrates, amino acids, nucleotides, and nucleic acids), rendering the eggs "self-sufficient capsules" [18]. This is also true for *J. lalandii*, where the developing embryo relies exclusively on its yolk for energy and structural material until hatching as naupliosoma larvae after about 70–90 days [19,20]. Lipids and proteins accumulate concomitantly in adult crustacean ovaries during maturation and growth, and the GSI (gonadosomatic index) increases [18]. This too has been shown for WCRL ovaries [21,22]. Female lipid and protein accumulation success therefore results in optimal ovary development and translates directly into a successful maternal export to the larvae [14,18].

Early life stages of marine invertebrates are assumed to be exceptionally sensitive to environmental stress and may represent a bottleneck in species success [23]. The hypothesis of embryo sensitivity comes from the assumption that hypercapnic stress may lead to a higher metabolic demand (for example, for acid–base regulation) and depletion of the finite energy stores (see above) prior to hatching [24,25]. This in turn would compromise the hatching process and species fitness. However, although crustacean embryos and larvae are potentially susceptible to climate change stressors, information about the vulnerability to stressors in early ontogenetic stages is still lacking for many species [7,26,27].

Changes in environmental conditions in the habitat due to climate change can impact sexual reproduction. Identification of the most sensitive stages in reproductive cycles will therefore enable better prediction of a species' response to environmental change [28]. However, in this regard, only mature male and juvenile WCRLs have previously been investigated (see above). Impacts of hypercapnia on females and sensitivity of embryonic development are currently a knowledge gap for *J. lalandii* and palinurids in general.

We addressed this shortfall in the present study by analysing the sensitivity of ovigerous female WCRLs and their eggs to hypercapnia. Accordingly, we formulated the following research questions: (1) Can berried female WCRLs respond swiftly to large changes in pH? (2) What physiological mechanisms facilitate a potential response to a rapidly declining pH, i.e., acute hypercapnia? (3) Does a potential response persist during prolonged hypercapnia? (4) Are eggs/embryos impacted by hypercapnia?

2. Materials and Methods

2.1. Experimental Animals

All female West Coast rock lobsters (WCRLs), *Jasus lalandii*, were caught at two sites on the Atlantic coast of South Africa (Figure 1) by large, baited lobster traps at depths of approximately 80–200 m during the spawning season in the austral winter (August). The Knol station (34°04.50″ S 18°19.00″ E) is located inside an area where fast somatic growth has been observed, whereas the Olifantsbos station (34°13.00″ S 18°19.00″ E) is within a slow somatic growth area. A total of 63 berried WCRLs were captured, of which 42 of a mean carapace length (CL) of 69.5 ± 2.2 mm were caught at the Knol station and 21 lobsters of a CL of 64.7 ± 2.0 mm at the Olifantsbos station. Aboard the ship, berried females (i.e., carrying eggs) were transferred to a flow-through system until docking at the harbour of Cape Town (after approximately two days). Lobsters were transported cool and moist to the research aquarium of the Department of Forestry, Fisheries and the Environment (DFFE) within 30 min. Here, they were kept in 1500 L tanks with a flow-through system (flow rate ~430 L h^{-1}) provided with aeration for one week until experimentation started (ambient seawater temperature (T_A) ranged from 9 to17 °C, salinity 34.5–35.0‰). They were fed weekly with pieces of frozen sardines (*Sardinops sagas*) *ad libitum*.

2.2. Acute Response to Hypercapnia

Nine berried female lobsters of similar size from the Olifantsbos area were selected from the holding tanks and transferred to a 1000 L tank, where they were kept for 24 h prior to experimentation at water conditions given in Table 1 to acclimatise to experimental temperatures. Following these 24 h, a 0.5 mL haemolymph sample was taken from each lobster (see below), and four of the lobsters (CL: 69.5 ± 1.2 mm) were transferred to individual compartments inside two glass tanks (141 L) with normocapnic conditions (pH ~8.0), and the remaining five (CL: 69.6 ± 3.3 mm) were transferred to another two glass tanks with the same dimensions and compartments, but filled with hypercapnic seawater (pH ~7.5). The normocapnic pH represented that of the incoming seawater on this day and is close to the level during non-upwelling periods in the subtidal zone. In order to minimize the time difference in exposure to the set seawater parameters of the respective treatments, consecutive lobsters were placed in alternating treatments after the initial haemolymph sample (0 h) was withdrawn (i.e., first hypercapnia, second normocapnia etc.). After 1.5, 3, 5, 8, and 24 h, 0.5 mL pre-branchial haemolymph was withdrawn through the arthrodial membrane at the base of the fifth pair of pereiopods by syringe with a hypodermic needle (Neomedic 1 mL, 29 G), avoiding tail flips (i.e., accumulation of L-lactate) by securing the abdomen with a firm grip. Circulation in the tanks was maintained by JVP-202 12,000 L h^{-1} propellers (JVP, Ningbo, China) and air was provided from the aquarium's compressed air system. The pH of the hypercapnic tank was set using a pH controller and the CO_2 storage tank connected to a solenoid valve and a pH electrode (TUNZE, Penzberg, Germany) whereby CO_2 was bubbled into the seawater as described previously [10]. Seawater pCO_2 and [HCO_3^- + CO_3^{2-}] were calculated using measured pH, salinity, and T_A and A_T [29] as constants in CO2SYS_v2.1. software [30]. Oxygen concentration was determined using a Multi 350i meter set (WTW, Weilheim, Germany). Water quality was monitored by measuring NH_3 concentration (ammonia test kit, Sera, Heinsberg, Germany) and never exceeded 0.3 mg L^{-1}.

Table 1. Physicochemical seawater conditions recorded during exposure of ovigerous *J. lalandii* to normocapnic and hypercapnic conditions.

Treatment	T_A °C	pH	A_T µmol kg^{-1}	O_2 %	Salinity ‰	Ca^{2+} mmol L^{-1}	Mg^{2+} mmol L^{-1}	pCO_2 Torr	pCO_2 (µatm)	HCO_3^- mmol L^{-1}	CO_3^{2-} mmol L^{-1}
Acclimation	14.2 ± 0.0	7.93 ± 0.02	2001 ± 21	97.1 ± 0.8	34.9 ± 00	10.1 ± 0.3	52.0 ± 1.9	0.3 ± 0.0	(460 ± 26)	1.7 ± 0.0	0.1 ± 0.0
Acute exposure											
Normocapnia	14.0 ± 0.2	7.95 ± 0.04	1997 ± 24	96.0 ± 1.4	34.9 ± 0.1	10.2 ± 0.6	50.5 ± 2.7	0.3 ± 0.0	(443 ± 37)	1.7 ± 0.0	0.1 ± 0.0
Hypercapnia 7.5	14.1 ± 0.3	7.47 ± 0.07	2003 ± 15	95.4 ± 0.8	34.9 ± 0.2	10.0 ± 0.2	51.2 ± 3.8	1.1 ±0.2	(1463 ± 236)	1.9 ± 0.0	0.0 ± 0.0
Chronic exposure											
Normocapnia	17.1 ± 1.4	7.99 ± 0.10	2034 ± 40	93.6 ± 1.5	35.0 ± 0.0	10.3 ± 0.6	52.8 ± 1.9	0.3 ± 0.1	(410 ± 130)	1.7 ± 0.1	0.1 ± 0.0
Hypercapnia 7.8	17.2 ± 1.2	7.80 ± 0.01	2029 ± 42	92.4 ± 1.4	35.0 ± 0.0	10.2 ± 0.5	52.4 ± 1.5	0.5 ± 0.0	(665 ± 25)	1.8 ± 0.0	0.1 ± 0.0
Hypercapnia 7.5	16.9 ± 1.1	7.52 ± 0.04	2032± 42	94.6 ± 1.8	35.0 ± 0.0	10.2 ± 0.3	52.6 ± 1.3	1.0 ± 0.1	(1338 ± 138)	1.9 ± 0.0	0.1 ± 0.0

2.3. Chronic Response to Hypercapnia

A systematic experimental design of a flow-through system according to guidelines by Riebesell et al. [31] (Figure 6.7–A-3) and Cornwall and Hurd [32] (Figure 3d) was essentially used: seawater from a subtidal area of Table Bay in front of the facility was continuously pumped into the aquarium facility and after filtration stored in the main water storage tank of the aquarium. From here, seawater flowed to replicate mixing tanks, where pH was adjusted. Seawater subsequently flowed to culture tanks (= sub-replicates). The normocapnic treatment used incoming pH conditions (ca. pH 8, Table 1), and the hypercapnic treatments were controlled in such a way that pH 7.8 and 7.5, respectively, were maintained in the culture tanks. The pH control systems as described above were used to reduce pH to ~7.8 and ~7.5 respectively. Specimens were separated into two groups for taking exclusively haemolymph or egg samples. Ovigerous females from both areas were individually incubated in four replicates for each of three treatments per sampling location to investigate embryonic development. Lobsters were from the Knol area (n = 12, CL = 69.5 ± 2.3 mm, w = 194 ± 6 g) and from the Olifantsbos area (n = 12, CL = 64.7 ± 2.1 mm, w = 180 ± 6 g). Only lobsters (n = 30) from the Knol area (CL = 72.5 ± 2.0 mm, w = 202 ± 6 g) were chosen for analysis of haemolymph chemistry after 60 days of incubation. In the respective two hypercapnic treatments, lobsters were acclimatized by gradually reducing the pH within three days. Lobsters were kept in three replicates per treatment (4, 3 and 3 WCRLs) and fed *ad libitum* weekly with black mussels (*Chromomytilus meridionalis, Mytilus galloprovincialis*). Haemolymph sampling after 60 days of exposure was carried out as described above. To minimise tank/placement effects, replicate tanks and culture tanks were routinely alternated. No mortalities occurred throughout the experiment.

2.4. Haemolymph Acid–Base Balance

Haemolymph pH was measured within 20 s after sampling using an Orion 3 star pH meter equipped with an Orion 8220 BNWP micro pH electrode (Thermo Scientific, Waltham MA, USA). Calibration was performed with NBS precision buffers (AppliChem, Darmstadt, Germany) at the same temperature as that of ambient seawater of the lobster tanks. To measure total CO2 (cCO_2), a haemolymph subsample (100 μL) was immediately injected into a degassing (magnetic stirrer) chamber containing 200 μL of 100 mM H_2SO_4 and the liberated total CO_2 (cCO_2) determined by CO_2 analyser (SBA4, PP Systems, Amesbury, MA, USA) using CO_2-free N_2 (technical) as carrier gas (50 mL min−1) calibrated against freshly made Na_2HCO_3 standards (1–10 mM). From these measured parameters (pH and cCO_2), pCO_2, and $[HCO_3^- + CO_3^{2-}]$ were calculated using derivatives of the Henderson–Hasselbalch equation, using constants derived from Truchot [33]. Concentrations of Ca_2^+, Mg_2^+, and L-lactate were determined using commercial kits (Diaglobal, Berlin, Germany; Roche, Penzberg, Germany) on small subsamples from each lobster. Haemocyanin (at 335 nm) and protein (at 280 nm) concentrations were measured spectrophotometrically in 1:50-diluted haemolymph vs. *J. lalandii* Ringer solution (0.52 M NaCl, 0.015 M $MgSO_4$, 0.013 M $CaSO_4$, 0.005 M KCl, 0.005 M $NaHCO_3$, pH 7.8) [9].

2.5. Embryonic Development

For examination of developmental stages of embryos, two ovigerous setae were separated from the fourth pair of pleopods of female lobsters at each sampling time point. Except for the initial sampling (45–60 eggs), from each of these setae, 25 to 30 eggs were gently removed, starting from the apical tip, because egg development was not homogenous along a seta, but apical egg development is more rapid (Figure 2). This procedure was carried out at the start of the incubation (day 0) and repeated every 10 days until the embryos hatched (day 30 for Knol eggs, day 50 for Olifantsbos eggs). Embryonic development stage (Figure 3) was determined according to Silberbauer [19] using a light microscope (SMZI 1500, Nikon, Tokyo, Japan) equipped with a digital camera (DSFI 1, Nikon, Japan). From the same pictures, egg size (diameter) was analysed by NIS software Ver5.22.00 for Windows (basic research package BR, Nikon Imaging Systems, Japan). Since no females were caught with eggs at stage 1, embryonic development was observed using females whose eggs were at stage 2 of the scale (Figure 3). Berried setae were also analysed for fungus-infected or decaying eggs.

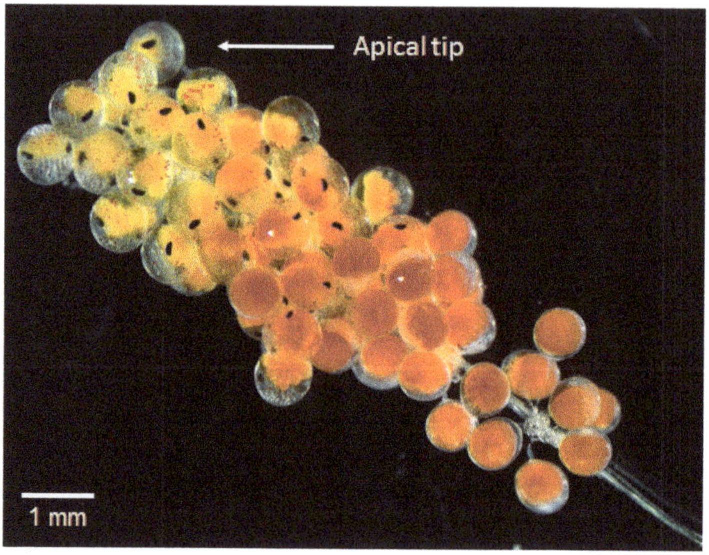

Figure 2. Dissection microscope picture of an ovigerous seta. Note the difference in developmental stage along the seta from R to L.

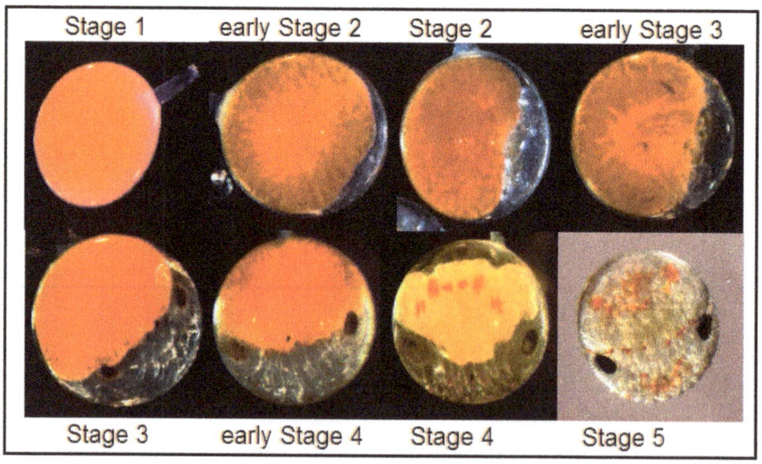

Figure 3. Staging index used for determination of developmental stages of fertilised eggs (own reference pictures staged according to Silberbauer [19].

2.6. Hepatosomatic Index (HSI) and Gonadosomatic Index (GSI)

Females used for analysing embryonic development were incubated for four more weeks after embryos had hatched. After their total body mass was measured, they were anaesthetised and then euthanised in a mixture of seawater and ice and dissected to measure the mass of gonads and the hepatopancreas. Values for GSI and HSI were calculated as described in Munian et al. [22]. Macroscopic analyses of gonad development were conducted according to Heydorn [12].

2.7. Scanning Electron Microscopy (SEM) Analysis

Samples were kept in formalin to analyse the surface of the egg membrane (chorion) using SEM. One seta of each female was later fixed in 10% formalin with a phosphate buffer and stored at room temperature. Dehydration of egg samples was accomplished through a graded acetone series from 35% to 100% according to Gil-Turnes and Fenical [34] and critical point drying (FL-9496, Blazer Union) using CO_2 as an intermediate fluid was then carried out. A defined area of the 4th right pereiopod (*merus*), as well as a defined area (a triangle cut from the posterior end of the carapace), was cut from three animals of each incubation group, fixed in 10% formalin with a phosphate buffer, and stored at room temperature. To dehydrate samples, formalin was washed out using double-deionised water. Samples were then dried in ambient air to avoid artefacts through dehydration in a drying oven. Embryo and exoskeleton samples were sputter-coated with gold (Agar Sputter Coater) at 1×180 s and observed on a Leo 1430 VP scanning electron microscope.

2.8. Statistical Analysis

All data were tested for normality using the Shapiro–Wilk test and for homogeneity of variance using the Brown–Forsythe test. Statistics were determined using Sigma Plot version 14.0. The significance level chosen throughout was $p < 0.05$. Results of acute exposure were analysed by two-way ANOVA, followed by post hoc Dunnett's test to test for differences from initial value (t_0) within treatments and between time pairs of treatments. GSI, HSI, acid–base parameters, and haemolymph data of the chronic exposure were analysed by one-way ANOVA followed by post hoc Tukey test. Egg growth rates during chronic exposure were analysed by two-way ANOVA followed by post hoc Holm–Šidák test.

3. Results

3.1. Acute Experiments—Acid Base Regulation

Throughout the acute exposure, haemolymph parameters showed different patterns of variation under normocapnic and hypercapnic conditions: In the normocapnic group, haemolymph pH decreased marginally by 0.06 units within the first hour. From then on, pH increased steadily during the next 3 to 8 h and reached a pH of 7.85 after 24 h (Figure 4A, Table 2). Extracellular total CO_2 (cCO_2) levels declined steadily to a low of 1.9 mmol L^{-1} after 3 to 5 h of exposure. Subsequently, total CO_2 concentration increased to reach 3.6 mmol L^{-1} after 24 h (Table 2). The calculated bicarbonate + dissolved CO_2 concentration, $[HCO_3^- + CO_3^{2-}]$, hereafter called "bicarbonate" for simplicity, declined steadily to a low of 1.9 mmol L^{-1} after 3 to 5 h of exposure. Subsequently, concentration increased to 3.5 mmol L^{-1} after 24 h (Figure 4B, Table 2). A similar course of changes was detected for pCO_2 values (Table 2). The Henderson–Hasselbalch diagram of the normocapnic treatment showed very little change in bicarbonate levels, pCO_2, and pH (Figure 5A).

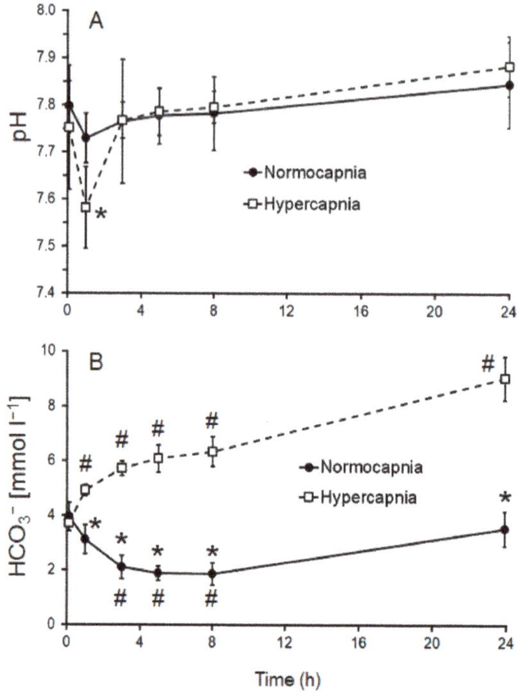

Figure 4. Time course of acid–base parameters in haemolymph of berried *J. lalandii* from the Olifantsbos area: (**A**) measured pH and (**B**) calculated $[HCO_3^- + CO_3^{2-}]$ during acute exposure to normocapnic seawater and hypercapnia for 24 h. Values are means ± S.D. # Significant difference from initial value (t_0) within the treatment; * significant difference from the same respective sampling time of the normocapnic treatment (repeated-measure ANOVA; $p < 0.05$).

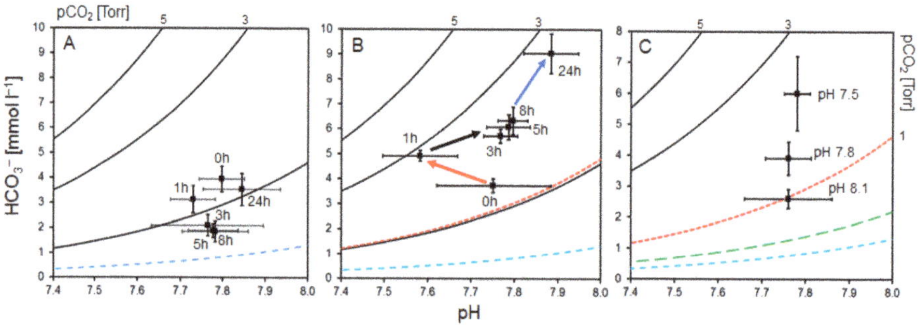

Figure 5. Henderson–Hasselbalch (pH–bicarbonate) diagrams for haemolymph of berried *J. lalandii* from the Knol area constructed from the time course of values during acute exposure presented in Table 2 and after experimental incubation for 60 days: (**A**) during 24 h normocapnia, (**B**) during 24 h hypercapnia (pH = 7.5), and (**C**) during incubation for 60 days (data from Table 3). pCO_2 isopleths were derived from the Henderson–Hasselbalch equation [9]. Appropriate values for the first dissociation constant (pK′1) and solubility coefficient (α) were derived from Truchot [33]. Values are means ± S.D. Blue dashed line = normocapnic seawater isopleth, red dashed line = hypercapnic seawater isopleth (pH = 7.5), green dashed line = hypercapnic seawater isopleth (pH = 7.8). Arrows indicate in (**B**) course of bicarbonate buffering from start to 24 h hypercapnic incubation (red = acidosis, blue = alkalosis).

Table 2. In vivo haemolymph parameters of ovigerous *J. lalandii* from the Olifantsbos area after acute exposure to normocapnic and hypercapnic seawater conditions.

Exposure Time (h)	pH	cCO_2 mmol L^{-1}	pCO_2 Torr	pCO_2 µatm	$[HCO_3^- + CO_3^{2-}]$ mmol L^{-1}	Ca^{2+} mmol L^{-1}	Mg^{2+} mmol L^{-1}	Protein mg mL^{-1}	Haemocyanin mg mL^{-1}	L-lactate mmol L^{-1}
Normocapia										
0	7.79 ± 0.05	4.0 ± 0.5	1.5 ± 0.4	2632 ± 395	4.0 ± 0.5	15.6 ± 1.7	7.9 ± 0.2	28.9 ± 3.9	20.0 ± 3.9	0.2 ± 0.1
1	7.73 ± 0.05	3.2 ± 0.5 #	1.3 ± 0.1	2105 ± 394	3.1 ± 0.5	15.7 ± 2.5	8.1 ± 1.5	29.0 ± 5.8	19.7 ± 5.2	0.3 ± 0.1
3	7.77 ± 0.13	2.1 ± 0.4 #	0.9 ± 0.2	1447 ± 263	2.1 ± 0.4 #	15.4 ± 2.3	8.9 ± 0.8	27.7 ± 5.4	18.0 ± 5.4	0.2 ± 0.1
5	7.78 ± 0.06	1.9 ± 0.3 #	0.7 ± 0.1 #	1316 ± 132	1.9 ± 0.3 #	14.4 ± 1.4	8.0 ± 1.4	23.3 ± 2.1	13.2 ± 1.1	0.2 ± 0.1
8	7.78 ± 0.08	1.9 ± 0.4 #	0.7 ± 0.0	1184 ± 264	1.9 ± 0.4 #	13.0 ± 2.6	9.7 ± 0.6	20.9 ± 4.4	9.8 ± 4.0	0.1 ± 0.1
24	7.85 ± 0.09	3.6 ± 0.7 #	1.2 ± 0.5	2368 ± 262	3.5 ± 0.6	16.1 ± 1.1	7.7 ± 0.8	22.3 ± 2.8	10.7 ± 2.6	0.2 ± 0.1
Hypercapnia										
0	7.76 ± 0.13	3.8 ± 0.3	1.6 ± 0.2	2634 ± 260	3.7 ± 0.3	12.9 ± 0.8	9.1 ± 1.2	29.6 ± 7.0	18.5 ± 7.7	0.4 ± 0.2
1	7.58 ± 0.09 *	5.1 ± 0.2 *	2.5 ± 0.1	3289 ± 132	4.9 ± 0.2 #*	11.6 ± 0.9	9.6 ± 1.0	30.2 ± 8.1	19.2 ± 9.0	0.4 ± 0.1
3	7.77 ± 0.04	5.8 ± 0.3 #*	2.9 ± 0.1 #*	3816 ± 136	5.7 ± 0.3 #*	12.7 ± 3.0	10.3 ± 0.6	28.7 ± 4.9	16.7 ± 4.2	0.3 ± 0.1
5	7.79 ± 0.05	6.2 ± 0.5 #*	3.0 ± 0.3 #*	3947 ± 398	6.1 ± 0.5 #*	10.5 ± 2.2	10.6 ± 0.9	25.3 ± 4.3	15.1 ± 6.2	0.3 ± 0.2
8	7.80 ± 0.03	6.4 ± 0.5 #*	3.1 ± 0.3 #*	4079 ± 391	6.3 ± 0.5 #*	12.0 ± 2.4	11.8 ± 0.7	23.6 ± 2.6	12.3 ± 2.5	0.3 ± 0.2
24	7.88 ± 0.06	9.1 ± 0.8 #*	4.6 ± 0.3 #*	6053 ± 393	9.0 ± 0.8 #*	11.7 ± 3.3	9.2 ± 1.2	23.1 ± 3.1	12.0 ± 2.7	0.2 ± 0.2

Values are means ± S.D. (n = 4–5). # Significant difference from the initial value (t_0) within treatment; * significant difference from the same respective sampling time of the normocapnic treatment (repeated-measure ANOVA; $p < 0.05$).

Table 3. In vivo haemolymph parameters of ovigerous *J. lalandii* from the Knol area after chronic exposure to normocapnic and hypercapnic seawater conditions for 60 days.

Exposure Time	pH	cCO_2 mmol L^{-1}	pCO_2 Torr	pCO_2 µatm	$[HCO_3^- + CO_3^{2-}]$ mmol L^{-1}	Ca^{2+} mmol L^{-1}	Mg^{2+} mmol·L^{-1}	Protein mg mL^{-1}	Haemocyanin mg mL^{-1}	L-lactate mmol L^{-1}
Normocapnia	7.76 ± 0.10	2.7 ± 0.3	1.0 ± 0.3	1316 ± 395	2.6 ± 0.3	12.9 ± 2.1	9.5 ± 1.4	23.5 ± 3.6	11.9 ± 4.1	0.2 ± 0.1
Hypercapnia pH 7.8	7.76 ± 0.05	4.0 ± 0.5 *	1.5 ± 0.2 *	1974 ± 263	3.9 ± 0.5 *	12.1 ± 1.7	10.2 ± 0.9	23.6 ± 3.9	14.6 ± 4.2	0.2 ± 0.1
Hypercapnia pH 7.5	7.78 ± 0.02	6.1 ± 1.2 *	2.2 ± 0.5 *	2632 ± 658	6.1 ± 1.1 *	14.8 ± 3.2	9.9 ± 1.2	24.2 ± 3.8	16.5 ± 2.9 *	0.1 ± 0.1 *

Values are means ± S.D. (n = 10). * Significant difference from the normocapnia treatment (one-way ANOVA; $p < 0.05$).

In contrast, following an ambient pH decrease of 0.5 units, a substantial decrease in extracellular pH by 0.17 units was measured after 1 h of exposure, followed by a steep increase to pH 7.77 after 3 h. Subsequently, there was a gradual and continuous increase to reach a final pH of 7.88 after 24 h (Table 2, Figure 4A). This is an overcompensation of pH by approximately 0.12 pH units or reduction in [H^+] by 5 nM (28%) compared with initial levels. Extracellular total CO_2 (cCO_2) levels increased steadily throughout 24 h of exposure (Table 2). Calculated bicarbonate levels also steadily increased throughout the entire exposure (Table 2, Figure 4B). pCO_2 values followed a similar trend (Table 2). Data were used to construct a Henderson–Hasselbalch diagram (Figure 5B). During the first hour of exposure to hypercapnia, a respiratory acidosis is apparent, indicated by a shift to the left (red arrow). Simultaneously, bicarbonate starts to increase, buffering the extracellular acidosis and reversing the acidosis after 3 h of exposure, indicated by a shift back to the right (black arrow). Subsequently, the continued bicarbonate increase leads to an alkalosis when compared to the initial pH. This is indicated by a shift further to the right (blue arrow). Levels of molecular modulators of haemocyanin oxygen affinity, Ca_2^+, Mg_2^+, and L-lactate did not change during the course of the experiment and did not differ between treatments (Table 2). Protein concentration dropped steadily in both treatments until the end of experimentation (Table 2). Simultaneously, protein and haemocyanin concentration steadily declined in both treatments to reach nearly half the initial level after 24 h (Table 2), probably due to the serial sampling of the animals.

3.2. Chronic Incubation—Acid–Base Regulation

After 60 days of chronic hypercapnic exposure at pH = 7.8 and pH = 7.5, haemolymph chemistry was analysed and compared with normocapnic exposure of pH = 8.1. Extracellular (haemolymph) pH was very similar in all treatments at a level of pH = 7.76 to 7.78 (Table 3, Figure 5C). The level of extracellular total CO_2 (cCO_2) was lowest in the normocapnic treatment at 2.7 ± 0.3 mmol L^{-1}, whereas 4.0 ± 0.5 mmol L^{-1} was measured at pH = 7.8 treatment and higher in the pH = 7.5 treatment at 6.1 ± 1.2 mmol L^{-1}. Calculated bicarbonate concentration was lowest in normocapnia at 2.6 ± 0.3 mmol L^{-1}, whereas 3.9 ± 0.5 mmol L^{-1} was measured in pH = 7.8 hypercapnia and higher in pH = 7.5 hypercapnia at 6.1 ± 1.1 mmol L^{-1}. Levels of pCO_2 changed similarly to bicarbonate levels (Table 3). The Henderson–Hasselbalch diagram revealed that at similar pH in lobsters from all treatments, elevation of pCO_2 and bicarbonate caused vertical shifts in the haemolymph buffer lines in lobsters incubated at two levels of hypercapnia, with those from the pH = 7.5 treatment highest (Figure 5C). This indicated bicarbonate buffering that resulted in alkalosis in the hypercapnic groups. Both pCO_2 and bicarbonate were significantly higher compared to the normocapnic group as well as within the two hypercapnic treatments, where the pH 7.5 treatment yielded a 55% higher bicarbonate level than the pH 7.8 treatment. At the same time, there were no changes in absolute protein concentration throughout the 60-day incubation for any treatment, but there were increases in the haemocyanin concentration from normocapnic levels of 11.9 mg ml L^{-1} to 14.6 and 16.5 mg L^{-1} at pH 7.5 and 7.8, respectively.

3.3. Chronic Exposure—Embryonic Development

Embryonic development was monitored by staging and size measurement of eggs. Initial embryo stage [19] was from 2.0 to 2.2 for eggs from females from the Knol area and 2.0 for eggs from females from Olifantsbos (Figure 6). At the end of the exposure, developmental stages were between 4.3 and 4.8 in the treatments of eggs from the Knol area (30 days of incubation), whereas they were between 3.9 and 4.2 in the treatments from

Olifantsbos females after 50 days (Figure 6). Due to the non-linearity of the results and uncertainties described (see comments in Discussion), a statistical analysis was omitted.

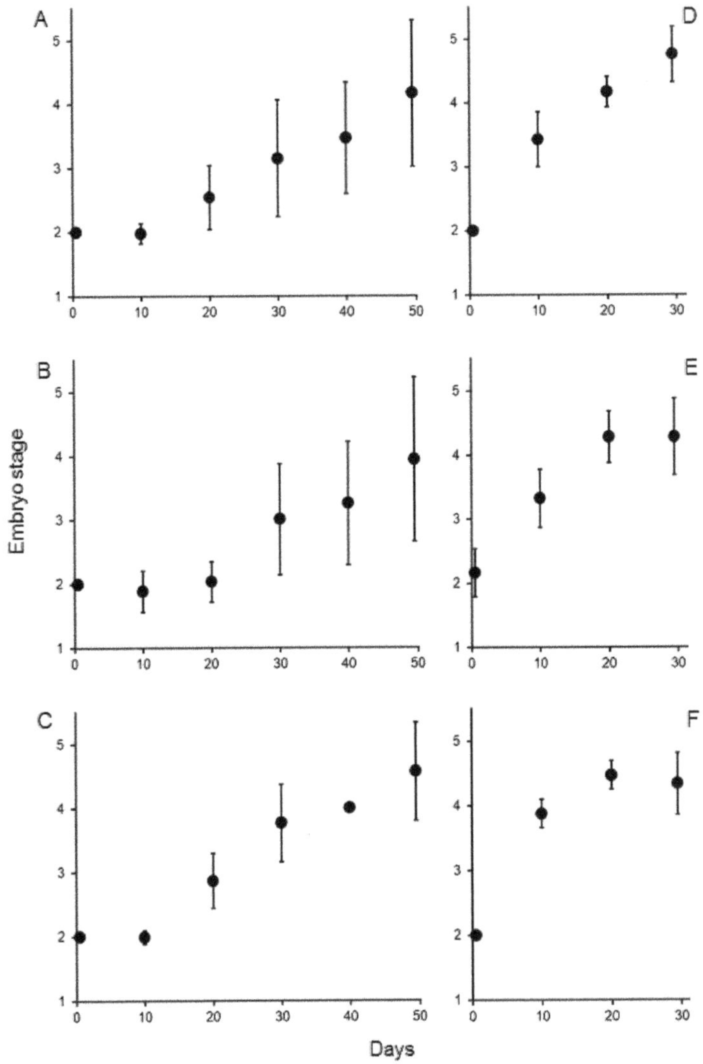

Figure 6. Mean (± S.D.) developmental stage of embryos [18] from berried lobsters from the Olifantsbos (n = 676; left column of figures) and Knol areas (n = 680; right column of figures) during 30 or 50 days of incubation. (**A**,**D**) Normocapnia, (**B**,**E**) pH 7.8, (**C**,**F**) pH 7.5. Note: x-axes of left and right panels are of the same scale to ensure comparability.

Initial egg diameters were approximately 637 ± 21 μm (n = 680) for eggs from Knol females, whereas they were 612 ± 25 μm (n = 676) for those from Olifantsbos females (Figure 7). After incubation of 30 days, diameters ranged from 724 to 746 μm in treatments of Knol embryos and from 712 to 720 μm after 51 days of incubation from treatments of Olifantsbos embryos (Figure 7). Growth in egg size was linear in all treatments of eggs from both sampling areas (Figure 7). Growth rates ranged from 3.2 to 3.6 μm d^{-1} in treatments of embryos from the Knol area and from 2.1 to 2.2 μm d^{-1} in treatments of those from the

Olifantsbos area (Figure 8). There was no significant difference between treatments within the respective areas. However, growth rates of Knol eggs from all treatments were 47–64% higher than those from all treatments of the Olifantsbos area (Figure 8). No fungal infected or decayed eggs/embryos were observed during sampling.

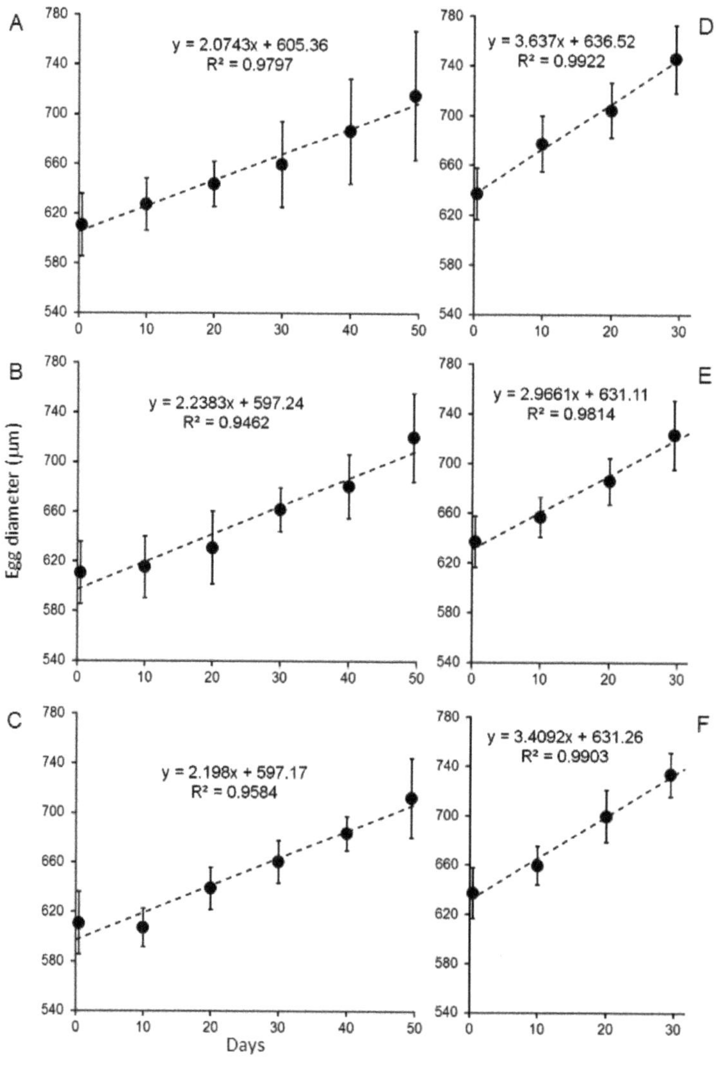

Figure 7. Egg growth, expressed as mean (± S.D.) diameter, in berried lobsters from the Olifantsbos (n = 676; left column of figures) and Knol areas (n = 680; right column of figures) during 50 and 30 days of incubation, respectively. **(A,D)** Normocapnia, **(B,E)** pH 7.8, **(C,F)** pH 7.5. Note: x-axes of left and right panels are of the same scale to ensure comparability.

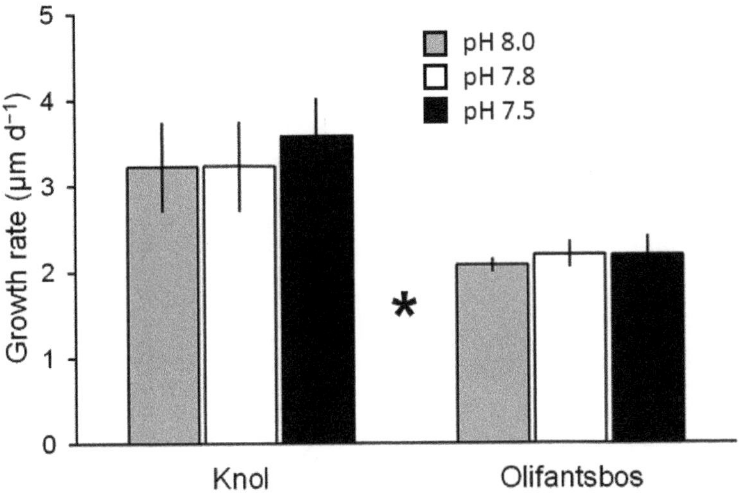

Figure 8. Daily growth rate (µm d^{-1}) of embryos/eggs from berried rock lobsters from the Knol and Olifantsbos areas under three pH treatments (see Table 1) calculated from correlation slopes of egg growth from individual females (see Figure 7). * Significant difference between locations ($p < 0.001$, two-way ANOVA and post hoc Holm–Šidák test).

3.4. Chronic Exposure—Gonadosomatic and Hepatosomatic Index

Females from the Olifantsbos area had lower GSIs than females from the Knol area. The GSI of the Olifantsbos normocapnic group was calculated at 0.7 ± 0.2% (mean ± S.D., n = 12). A significantly lower GSI was estimated for the pH 7.8 treatment—0.3% ± 0.1%. The pH 7.5 treatment was not significantly different from hypercapnia, but was slightly lower, at 0.4 ± 0.2%. Hypercapnic lobsters from the Knol area had a GSI of 0.85% ± 0.6%. The pH 7.8 and 7.5 treatments, were similar with 0.6 ± 0.1% and 0.5 ± 0.1%, respectively, both lower than the normocapnic treatment. Normocapnic treatments had higher GSIs compared with the respective hypercapnic treatments. The HSI values of normocapnic (pH = 8.0) females from the Olifantsbos area were 1.9 ± 0.3% (mean ± S.D., n = 12), whereas those in the hypercapnic treatments were 2.1 ± 0.5% and 1.9 ± 0.3%, respectively. Normocapnic females from the Knol area had a lower HSI at 2.2 ± 0.1% than both hypercapnic treatments at 2.4 ± 0.4%. However, this difference was not significant.

3.5. Chronic Exposure—Structural SEM Analysis of the Cuticle and Eggs

Scanning electron microscopic (SEM) analysis of transversally fractured/cross-fractured cuticle from *merus* samples revealed no obvious abnormalities or structural damages that could have derived from chronic hypercapnic treatments, such as decalcification of the cuticle layers (Figure 9). Similarly, carapace samples had no obvious damage or abnormalities (Figure 10). Hair-like structures in some panels of Figures 8 and 9 depict setae that are rooted in the cuticle and partially cover the exoskeleton. SEM analysis of the embryonic cuticle (chorion) of eggs from all treatments at the end of experimentation did not reveal structural differences or damage to the surface, such as cracks or dissolution of the egg membrane itself (Figure 11). Folding and dents or depressions are artefacts of the sample treatment process for SEM determinations.

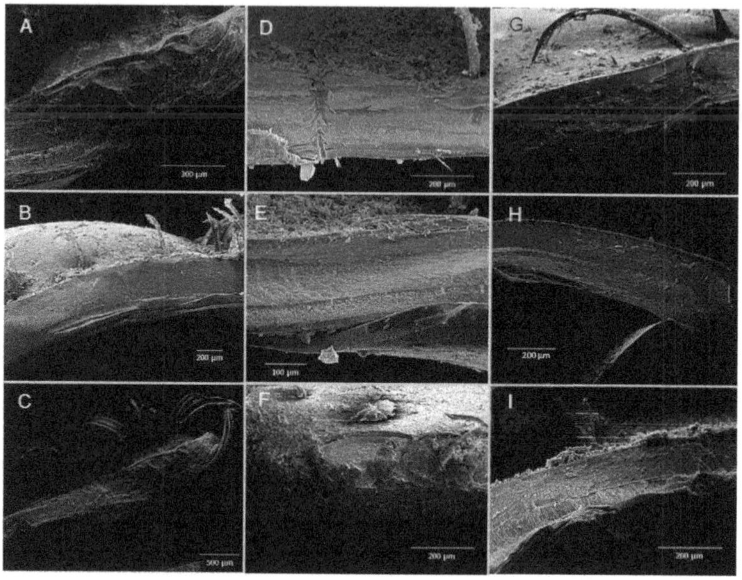

Figure 9. SEM micrographs of transversally fractured/cross-fractured cuticle of a defined *merus* sample each from individual berried *J. lalandii* from the Olifantsbos area after experimental exposure of 50 days. (**A–C**) Normocapnia, (**D–F**) pH 7.8, (**G–I**) pH 7.5. Scale bars are given in individual panels.

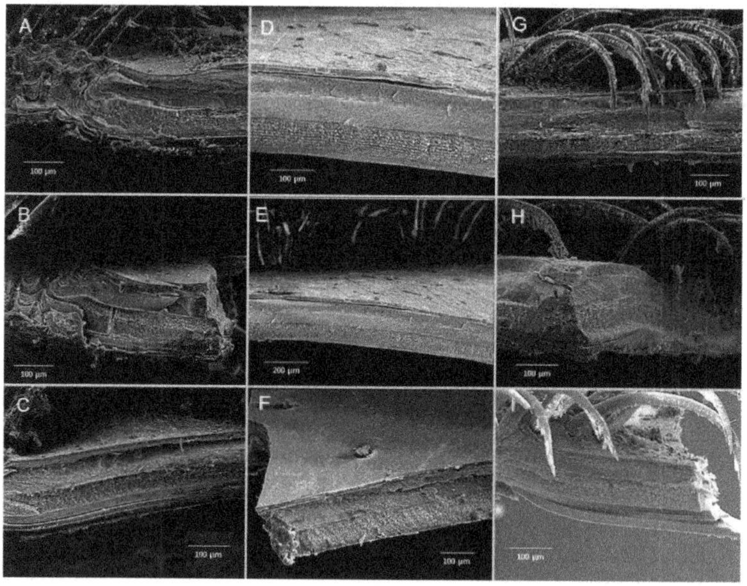

Figure 10. SEM micrographs of a transversally fractured/cross-fractured cuticle from a defined carapace area from individual berried *J. lalandii* from the Olifantsbos area after experimental exposure of 50 days. (**A–C**) Normocapnia, (**D–F**) pH 7.8, (**G–I**) pH 7.5. Scale bars are given in individual panels.

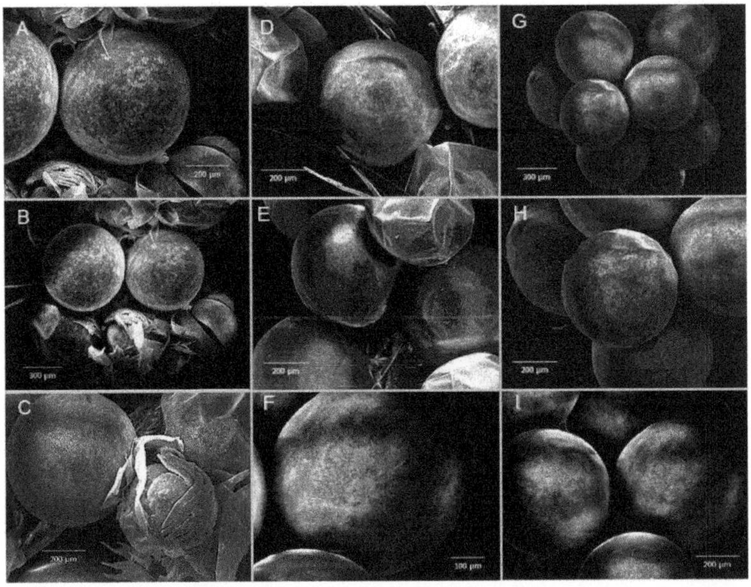

Figure 11. SEM micrographs of egg membranes (chorion) from individual berried *J. lalandii* from the Olifantsbos area after experimental exposure of 50 days. (**A–C**) Normocapnia, (**D–F**) pH 7.8, (**G–I**) pH 7.5. Scale bars are given in individual panels.

4. Discussion

We previously reported that mature males and juveniles of the WCRL are resilient to acute and chronic hypercapnia [9–11]. Our present results extend this observation for the first time to ovigerous females and their embryos according to comparable criteria.

This is apparent in the regulation of acid–base balance during acute hypercapnia: as in mature males of similar size [10], berried females adjust their acid–base balance rapidly by means of bicarbonate, which is elevated by a net 5.2 mmol L^{-1} (+133%) after 24 h. This increase in haemolymph bicarbonate levels in response to a pH = 7.5 is even more pronounced than that of adult male WCRLs after acute hypercapnia at a lower pH of 7.3 [10]. Bicarbonate buffering is a common response in crustaceans (mainly crabs) to respiratory or extracellular acidosis [27,35–39]. It is also required to compensate for disturbance of acid–base balance to achieve resting pH levels [38] or close to original values [35–37], i.e., within the physiological range. When exposed to an ambient pH of 7.5, ovigerous WCRLs fully compensated for the decreased environmental pH within 3 h post-exposure (Figure 4A, Table 2). This compensation is rapid and complete when compared with other crustacean species, but similar to male WCRLs [10]. A pronounced Bohr effect was measured in juvenile [9] and mature male WCRLs [10]. Therefore, oxygen affinity needs to be safeguarded to ensure oxygen loading of haemocyanin at the gills and offloading at the tissues. This is accomplished by the above-mentioned bicarbonate buffering of their haemolymph to provide optimum pH conditions. Bicarbonate buffering also guarantees an outward gradient for metabolic CO_2 removal by maintaining the extracellular pCO_2 above environmental pCO_2. Here, the gradient after 24 h hypercapnic incubation is 3.5 Torr, whereas it is only 0.9 Torr under normocapnia.

The initial prebranchial extracellular pH for ovigerous *J. lalandii* throughout experiments was approximately 7.8 (T_A range: 14–17 °C, Tables 2 and 3), i.e., approximately 0.2 pH units below ambient pH. This is similar to adult male WCRLs [10]. During acute exposure to a pH of 7.5, berried lobsters reached a minimum extracellular pH after 1 h, which was 44% lower (acidosis) than the initial value (Figure 4A, Table 2). During previous work on mature male *J. lalandii*, however, a strong respiratory acidosis set in during the first 25 min of hypercapnia at pH = 7.4 before bicarbonate buffering took effect [10]. This is indicative of a transient acidosis. At the 1 h sampling point, this acidosis was most likely already partially compensated by buffering in berried females. Additionally, it may have been masked by an alkalosis caused by hyperventilation [40]. This effect was considered small, however, as judged by the stable lactate concentration (Table 2).

The Henderson–Hasselbalch diagrams provide insight into the interaction of extracellular pH, calculated haemolymph bicarbonate, and pCO_2: In the WCRL group under normocapnic conditions, there is very little change compared with hypercapnic exposure (Figure 5A,B). The small changes observed, however, were caused by the changes in extracellular pCO_2, pH, and bicarbonate (Table 2, Figure 5A), possibly as a result of handling stress. After exposure to hypercapnic conditions, there is a strong transitional acidosis within 1 h after which compensation by bicarbonate becomes effective (Figure 5B). After 3 h of exposure, compensation by increased bicarbonate levels (along a constant pCO_2 isopleth) becomes apparent and the pH starts to increase (shift to the right). Subsequently, further increasing bicarbonate levels cause a move into an alkalosis and overcompensation (Figure 5B). This cause of events resembles that found in male adult WCRLs during hypercapnia [10].

The above acid–base regulation is still present after 60 days of chronic hypercapnia: berried WCRLs are still capable of bicarbonate buffering of their haemolymph to provide optimum pH conditions for oxygen binding in the presence of a strong Bohr effect (Figure 5C). This ensures functioning of respiration during prolonged hypercapnia and is similar to the response previously observed in juvenile WCRLs after chronic hypercapnia [9] and in other crustaceans [27,38,39]. After 24 h of acute hypercapnia, the extracellular pH was 0.2 units higher than ambient pH (Tables 1 and 2). This differential was maintained during chronic hypercapnia (Tables 1 and 3). However, the high outward pCO_2 gradient found after acute hypercapnia (see above) was reduced to 1.0–1.2 Torr, not much above the respective normocapnic gradient of 0.7 Torr (Tables 1 and 3). This suggests that the high gradient in the acute experiment was because of overcompensation for the sudden increase in environmental pCO_2. This overcompensation, causing alkalosis in the acute experiment, is therefore not maintained (Figure 5C). Acid–base regulation is associated with high energetic costs [25,41,42]. It is therefore possible that other metabolic functions of the animal are reduced to make this additional energy available [25]. One of those functions is maternal care. This means aerating the eggs by lifting the abdomen, fanning with the pleopods, and grooming the eggs with a special appendage on the fifth pair of pereiopods [19,43,44]. The latter specifically prevents fouling of the eggs in ovigerous *Jasus* females by removing fungus infected or decaying eggs and therefore improves general survival [45]. Fanning and lifting of the abdomen provide oxygenation of eggs, especially those that are located away from the apical tip of setae. Oxygen supply of eggs is critical for their development and survivorship, as shown for crabs [46]. Also, in crab species (*Ovalipes trimaculatus*, *Paralithodes camtschaticus*), brood care was shown to increase energetic cost to the brooding female by more than 100% of the measured oxygen consumption [47,48]. This is in addition to the cost of acid–base regulation during hypercapnia (see above). Successful growth and development of embryos as well as hatching of larvae from all treatments suggests that essential maternal care was not impaired by chronic hypercapnia (Figures 6–8). Moreover, no fungus-infected, fouled, or decaying eggs were observed during egg sampling, probably because the females may have removed them. Therefore, successful growth, development, and hatching of embryos and a constant extracellular pH suggest that the females can balance the sum of costs for maternal care and acid–base regulation.

It is generally assumed that early life stages of marine invertebrates are the most vulnerable life stage to ocean acidification [23]. However, there are extremely limited data to support this statement for larvae and embryos of decapods, mostly due to a lack of research [7,25,27]. A recent meta-analysis [27] and the few publications available demonstrate low or no impact on early life stages from other decapod crustaceans. Examples are crab eggs (*Leptuca thayeri*) that were resilient to very low pH during very short development (and incubation) time of 10 days [26]. Embryos of the snow crab (*Chionoecetes opilio*) were unaffected by hypercapnia of pH 7.8 and 7.5 during development/incubation periods of up to two years at very low temperatures (~2 °C) [8]. Larvae of another crab species, the red king crab (*Paralithodes camtschaticus*), are resilient in terms of survival and development to simulated ocean acidification of almost two years in length [48]. Moreover, *Nephrops norvegicus* embryos also were found to be insensitive to low pH and elevated temperatures [49]. Such insensitivity of embryos to low pH might be explained by adaptation to a pH-reduced external habitat and/or internal hypercapnia during incubation [49]. In contrast, there are several studies that do reveal sensitivities of decapod embryos and larvae to hypercapnia. For example, low pH levels decreased larval development rates in crustacean species, such as the barnacle *Semibalanus balanoides*, the crabs *Hyas araneus* and *Chionoecetes bairdi*, as well as the lobster *Homarus americanus* [50–54]. Larvae hatched from embryos of the red king crab *Paralithodes camtschaticus* that were incubated in hypercapnia (pH = 7.7) had a reduced survival rate when starved [55]. Research results from the present study place the WCRLs into the hypercapnia-resistant group of crustaceans: they revealed a relative resilience of egg development and growth against different levels of hypercapnia.

Previously, there were no data available on the impact of hypercapnia on palinurid embryos, whereas temperature effects were well studied, showing that embryo development accelerates with increasing temperature (for example, in *J. edwardsii*; [56,57]). We could close this knowledge gap somewhat by showing that development and growth of eggs/embryos are insensitive to hypercapnia: data from egg size measurements show that growth in diameter is linear for eggs from both collection sites and for all treatments (Figure 6). Because of this linearity and less subjectivity than egg development staging (see below), growth rates were used for further statistical analysis. The latter revealed no differences between the treatments for either location at which berried females were caught (Figure 8), indicated by similar slopes within charts for a respective catch location (Figure 7). Moreover, SEM analysis of eggs revealed no surface damage or differences between treatments (Figure 11). In contrast to measured egg diameters, results from developmental stages must be evaluated with caution. This is due to the more subjective, and potentially biased, nature of staging and non-linearity of results (Figure 6). In addition, the seasonal biological cycle of adult WCRLs (moulting, mating, spawning) commences first at northern latitudes and gradually moves south. As a result, egg development in the more northern Knol area is slightly ahead of Olifantsbos, indicated by the more advanced stage 2 eggs at Knol. The consequence was a shorter incubation period of 30 vs. 50 days at the aquarium, since Knol embryos were closer to hatching. The faster development and growth (see below) rates of Knol eggs also contributed to the shorter exposure period. Egg stage data are therefore difficult to statistically analyse and deemed not useful to conclude on the potential impact of hypercapnia on embryonic development.

Our study revealed an additional, interesting aspect of the developing eggs: independently of treatment, egg growth rates differed significantly between the two sampling sites. Rates at Knol were about 1.0–1.4 $\mu m\ d^{-1}$ higher than those from Olifantsbos, independently of treatment (Figure 8). This is a 47–64% faster growth rate. Temperature was shown to affect the duration of embryonic development in the closely related *J. edwardsii* [56,57]. However, incubation temperature (and all other conditions) was identical for eggs from both sampling sites in the present study and can be ruled out as factor for the growth differential. Even after the Knol embryos had already hatched, growth of the Olifantsbos embryos remained linear under those conditions. The reason for the differential growth rate could therefore rather reflect the exposure of the mother in its environment before capture. The growth advantage of Knol embryos therefore carries over the favourable conditions experienced by Knol females: The larger diameter of eggs from Knol lobsters at the time of catch are most likely due to the above-mentioned advanced progress into stage 2 compared with those from Olifantsbos females. However, there may also be larger eggs for females from that area due to better growth conditions for adult lobsters. The difference in male somatic growth rates and female fecundity is well known for the two areas [16]. These differences have been shown to persist in recent years in terms of male somatic growth rates [58] and in female GSI [22]. Higher fecundity is concluded from the larger number of eggs. However, eggs may also be larger in high-fecundity areas. Consequently, larger eggs in the Knol area could be the result of favourable ambient conditions for parent lobsters. Since physicochemical conditions in the Knol and Olifantsbos areas are very similar, it is assumed that food availability and composition cause such favourable conditions [22]. Egg size differences at spawning, however, are so far not reported and will need future analysis.

In decapods, the response of exoskeleton calcification to hypercapnia is species-specific and depends on duration and severity of hypercapnia [27,59,60]. In addition, it can be specific to the body region [61]. Therefore, samples from two different areas of the WCRL exoskeleton were analysed by SEM in the present study. Neither surfaces nor structures of merus and carapace pieces revealed any visible differences between treatment groups (Figures 9 and 10). This was different from a similar hypercapnic experiment where the surface of different exoskeleton parts of the *Cancer pagurus* crab became porous and small cavities developed during hypercapnia. In addition, cracks developed between exocuticle and endocuticle [62]. It must be noted, however, that although the incubation period was similarly short (60 and 56 days, respectively), the crabs were exposed to a lower hypercapnia at pH 7.35 than the WCRLs (pH 7.5 and 7.8, respectively). In both cases, the experimental period did not include a moult. A developing new exoskeleton would probably be differently affected than an existing old and calcified one [25]. This relative resilience of the exoskeleton is not surprising, since crustaceans are well capable of maintaining or increasing the calcification rate during exposure to acidified water [25,27,38]. Moreover, in the decapod exoskeleton, calcium carbonate is embedded in its amorphous form into a matrix of chitin–protein fibres [61,63]. It is therefore not as exposed to ambient conditions as, for example, mollusc shells and sea urchin tests are. However, maintenance of this structure may have consequences for energy expenditure and/or diversion from other key biological aspects if extended for longer periods [25,64–67]. The present study was not designed to investigate this aspect.

Reproductive organ size was analysed at the end of the incubation for a few females in the present study. Both HSI and GSI values were higher in Knol females. This is similar to a previous study in which many more female WCRLs were analysed [22]. Moreover, at ovary stage 5, ovaries from Knol females were relatively larger than those from Olifantsbos area [21]. The growth differences of eggs reported here (Figure 8) are in agreement with this, and may point to better maternal transfer from females to eggs in the Knol area. Therefore, larger ovaries of lobsters from Hout Bay may have resulted in larger eggs with larger reserves, leading to a development advantage over those from Olifantsbos females. This hypothesis is difficult to prove in the field and needs further analysis.

5. Conclusions

Our study has shown that ovigerous female WCRLs regulate acid–base balance similarly to males and juveniles in response to acute and chronic hypercapnia. Moreover, embryonic development was resilient to prolonged hypercapnia. Differences in development occurred, however, between the two sampling areas of the respective ovigerous females. This is in line with historic differences in male somatic growth and female fecundity between these two distinct geographical areas, which are assumed to be caused by differences in food availability and composition.

Author Contributions: A.R., C.R.B. and L.A. contributed to the study conception, validation, and experimental design. A.R. and L.A. performed animal rearing, investigation, data collection, curation, and formal analysis. C.R.B. and L.A. provided resources. A.R. and L.A. wrote the original draft. A.R., C.R.B. and L.A. reviewed and edited the original draft. A.R., C.R.B. and L.A. read and approved the final manuscript. A.R., C.R.B. and L.A. agree to be accountable for all aspects of the work. All authors have read and agreed to the published version of the manuscript.

Funding: This research received no external funding.

Institutional Review Board Statement: Handling, transport, and experiments were conducted in accordance with the standard operating procedures established by the Aquaculture Animal Ethics Committee of the Department of Forestry, Fisheries and the Environment of South Africa.

Informed Consent Statement: Not applicable.

Data Availability Statement: Relevant information is included in the article. Raw data supporting the conclusions are available from the author, L.A., upon request.

Acknowledgments: We are grateful to A. Busby and team from the Seapoint research aquarium and staff from the Rock Lobster group for their support.

Conflicts of Interest: The authors declare no conflicts of interest.

References

1. Eggers, J.M.; Cochrane, K.L.; Sauer, W.H.H. Estimating the economic income and social contributions derived from the South African west coast rock lobster fishery. *Afr. J. Mar. Sci.* **2022**, *44*, 255–269. [CrossRef]
2. DFFE (Department of Forestry, Fisheries and the Environment). *Status of the South African Marine Fishery Resources*; DFFE: Cape Town, South Africa, 2023. [CrossRef]
3. Pitcher, G.C.; Probyn, T.A. Red tides and anoxia: An example from the southern Benguela Current System. In Proceedings of the 14th International Conference on Harmful Algae. International Society for the Study of Harmful Algae and Intergovernmental Oceanographic Commission of UNESCO 2013, Creete, Greece, 1–5 November 2010; Pagou, P., Hallegraeff, G., Eds.; pp. 159–161.
4. Pitcher, G.C.; Figueiras, F.G.; Hickey, B.M.; Moita, M.T. The physical oceanography of upwelling systems and the development of harmful algal blooms. *Prog. Oceanogr.* **2010**, *85*, 5–32. [CrossRef] [PubMed]
5. Pitcher, G.C.; Probyn, T.A.; du Randt, A.; Lucas, A.J.; Bernard, S.; Evers-King, H.; Lamont, T.; Hutchings, L. Dynamics of oxygen depletion in the nearshore of a coastal embayment of the southern Benguela upwelling system. *J. Geophys. Res. Oceans* **2014**, *119*, 2183–2200. [CrossRef]
6. Cooley, S.; Schoeman, D.; Bopp, L.; Boyd, P.; Donner, S.; Ghebrehiwet, D.Y.; Ito, S.-I.; Kiessling, W.; Martinetto, P.E.; Ojea, E.; et al. Oceans and Coastal Ecosystems and Their Services. In *Climate Change 2022: Impacts, Adaptation and Vulnerability. Contribution of Working Group II to the Sixth Assessment Report of the Intergovernmental Panel on Climate Change*; Pörtner, H.-O., Roberts, D.C., Tignor, M., Poloczanska, E.S., Mintenbeck, K., Alegría, A., Craig, M., Langsdorf, S., Löschke, S., Möller, V., et al., Eds.; Cambridge University Press: Cambridge, UK; New York, NY, USA, 2022; pp. 379–550. [CrossRef]
7. Bednaršek, N.; Ambrose, R.; Calosi, P.; Childers, R.K.; Feely, R.A.; Litvin, S.Y.; Long, W.C.; Spicer, J.I.; Štrus, J.; Taylor, J.; et al. Synthesis of thresholds of ocean acidification impacts on decapods. *Front. Mar. Sci.* **2021**, *8*, 651102. [CrossRef]
8. Long, W.C.; Swiney, K.M.; Foy, R.J. Direct, carryover, and maternal effects of ocean acidification on snow crab embryos and larvae. *PLoS ONE* **2023**, *18*, e0276360. [CrossRef]
9. Knapp, J.L.; Bridges, C.R.; Krohn, J.; Hoffman, L.C.; Auerswald, L. Acid–base balance and changes in haemolymph properties of the South African rock lobsters, *Jasus lalandii*, a palinurid decapod, during chronic hypercapnia. *Biochem. Biophys. Res. Commun.* **2015**, *461*, 475–480. [CrossRef]

10. Knapp, J.L.; Bridges, C.R.; Krohn, J.; Hoffman, L.C.; Auerswald, L. The effects of hypercapnia on the West Coast rock lobster (*Jasus lalandii*) through acute exposure to decreased seawater pH—Physiological and biochemical responses. *J. Exp. Mar. Biol. Ecol.* **2016**, *476*, 58–64. [CrossRef]
11. Knapp, J.L.; Hoffman, L.C.; Auerswald, L.; Macey, B.M. Effects of chronic hypercapnia and elevated temperature on the immune response of the spiny lobster, *Jasus lalandii. Fish. Shellf. Immunol.* **2019**, *93*, 752–762. [CrossRef]
12. Heydorn, A.E.F. The rock lobster of the South African West Coast *Jasus lalandii* (H. Milne Edwards):—Part 2; Population studies, behavior, reproduction, molting, growth and migration. *Investig. Rep. Div. Sea Fish. S. Afr.* **1969**, *71*, 1–52.
13. Jeffs, A.G.; Gardner, C.; Cockcroft, A. Jasus and *Sagamariasus* species. In *Lobsters: Biology, Management, Aquaculture and Fisheries*; Phillips, B.F., Ed.; John Wiley & Sons, Ltd.: Oxford, UK, 2013; pp. 259–288.
14. Munian, A.; Cockcroft, A.; Hoffman, L.; Auerswald, L. Depot lipids in mature palinurid decapods (Crustacea). *Invertebr. Reprod. Dev.* **2021**, *65*, 1–11. [CrossRef]
15. Zoutendyk, P. Gonad output in terms of carbon and nitrogen by the Cape rock lobster *Jasus lalandii* (H. Milne-Edwards 1837) (Decapoda, Palinuridae). *Crustaceana* **1990**, *59*, 180–191. [CrossRef]
16. Beyers, B.; Goosen, P. Variations in fecundity and size at sexual maturity of female rock lobster *Jasus lalandii* in the Benguela ecosystem. *S. Afr. J. Mar. Sci.* **1987**, *5*, 513–521. [CrossRef]
17. Cockcroft, A.C. Biochemical composition as growth predictor of the West Coast rock lobster *Jasus lalandii. Mar. Freshw. Res.* **1997**, *48*, 845–856. [CrossRef]
18. Harrison, K.E. The role of nutrition in maturation, reproduction and embryonic development of decapod crustacean: A review. *J. Shellfish Res.* **1990**, *9*, 1–28.
19. Silberbauer, B.I. The biology of the South African rock lobster *Jasus lalandii* (H. Milne-Edwards) 1. Development. *Investl. Rep. Div. Sea Fish. S. Afr.* **1971**, *92*, 1–70.
20. Kittaka, J. Culture of the palinurid *Jasus lalandii* from egg stage to puerulus. *Nippon. Suisan Gakk.* **1988**, *54*, 87–93. [CrossRef]
21. van Rooy, C. Aspects of the Breeding Cycle and Fecundity of the Female West Coast Rock Lobster, *Jasus lalandii.* Master's Thesis, University of the Western Cape, Cape Town, South Africa, 1998.
22. Munian, A. The Search for Biomarkers in the Growth Prediction of the West Coast Rock Lobster, *Jasus lalandii.* Ph.D. Thesis, Stellenbosch University, Stellenbosch, South Africa, 2022.
23. Byrne, M. Impact of ocean warming and ocean acidication on marine invertebrate life history stages: Vulnerabilities and potential for persistence in a changing ocean. *Oceanogr. Mar. Biol. Annu. Rev.* **2011**, *49*, 1–42.
24. Pörtner, H.-O.; Langenbuch, M.; Reipschläger, A. Biological impact of elevated ocean CO_2 concentrations: Lessons from animal physiology and earth history. *J. Oceanogr.* **2004**, *60*, 705–718. [CrossRef]
25. Whiteley, N.M. Physiological and ecological responses of crustaceans to ocean acidification. *Mar. Ecol. Prog. Ser.* **2011**, *430*, 257–271. [CrossRef]
26. Pardo, C.F.; Costa, T.M. Multiple-stressor effects of warming and acidification on the embryonic development of an estuarine fiddler crab. *Estuar. Coast. Shelf Sci.* **2021**, *254*, 107296. [CrossRef]
27. Ocampo, E.H.; Nunez, J.D.; Ribeiro, P.D.; Perez García, M.; Bas, C.C.; Luppi, T.A. Disparate response of decapods to low pH: A meta-analysis of life history, physiology and behavior traits across life stages and environments. *Mar. Pollut. Bull.* **2024**, *202*, 116293. [CrossRef] [PubMed]
28. Padilla-Gamiño, J.L.; Alma, L.; Spencer, L.H.; Venkataraman, Y.R.; Wessler, L. Ocean acidification does not overlook sex: Review of understudied effects and implications of low pH on marine invertebrate sexual reproduction. *Front. Mar. Sci.* **2022**, *9*, 977754. [CrossRef]
29. Sarazin, G.; Michard, G.; Prevot, F. A rapid and accurate spectroscopic method for alkalinity measurements in sea water samples. *Water Res.* **1999**, *33*, 290–294. [CrossRef]
30. Pierrot, D.E.; Lewis, E.; Wallace, D.W.R. MS excel program developed for CO_2 system calculations. ORNL/CDIAC-105a. In *Carbon Dioxide Information Analysis Center*; Oak Ridge National Laroratory, US Department of Energy: Oak Ridge, TN, USA, 2006.
31. Riebesell, U.; Fabry, V.J.; Hansson, L.; Gattuso, J.-P. *Guide to Best Practices for Ocean Acidification Research and Data Reporting. [Reprinted Edition Including Erratum]*; EUR 24872 EN; Publications Office of the European Union: Luxembourg, 2011; 258p. [CrossRef]
32. Cornwall, C.E.; Hurd, C.L. Experimental design in ocean acidification research: Problems and solutions. *ICES J. Mar. Sci.* **2015**, *73*, 572–581. [CrossRef]
33. Truchot, J.P. Carbon dioxide combining properties of the blood of the shore crab *Carcinus maenas* (L.): Carbon dioxide solubility coefficient and carbonic acid dissociation constants. *J. Exp. Biol.* **1976**, *64*, 45–57. [CrossRef]
34. Gil-Turnes, M.S.; Fenical, W. Embryos of *Homarus americanus* are protected by epibiotic bacteria. *Biol. Bull.* **1992**, *182*, 105–108. [CrossRef]
35. Cameron, J.N. Effects of hypercapnia on blood acid–base status, NaCl fluxes, and trans-gill potential in freshwater blue crabs, *Callinectes sapidus. J. Comp. Physiol. B* **1978**, *123*, 137–141. [CrossRef]

36. Henry, R.P.; Kormanik, G.A.; Smatresk, N.J.; Cameron, J.N. The role of $CaCO_3$ dissolution as a source of HCO_3^- for the buffering of hypercapnic acidosis in aquatic and terrestrial decapod crustaceans. *J. Exp. Biol.* **1981**, *94*, 269–274. [CrossRef]
37. Pane, E.F.; Barry, J.P. Extracellular acid–base regulation during short-term hypercapnia is effective in a shallow-water crab, but ineffective in a deep-sea crab. *Mar. Ecol. Prog. Ser.* **2007**, *334*, 1–9. [CrossRef]
38. Spicer, J.I.; Raffo, A.; Widdicombe, S. Influence of CO_2-related seawater acidification on extracellular acid–base balance in the velvet swimming crab *Necora puber*. *Mar. Biol.* **2007**, *151*, 1117–1125. [CrossRef]
39. Small, D.P.; Calosi, P.; Rastrick, S.P.S.; Turner, L.M.; Widdicombe, S.; Spicer, J.I. The effects of elevated temperature and pCO_2 on the energetics and haemolymph pH homeostasis of juveniles of the European lobster, *Homarus gammarus*. *J. Exp. Biol.* **2020**, *223*, jeb209221. [CrossRef] [PubMed]
40. Burnett, L.E.; Johansen, K. The role of branchial ventilation in hemolymph acid–base changes in the shore crab *Carcinus maenas* during hypoxia. *J. Comp. Physiol. B* **1981**, *141*, 489–494. [CrossRef]
41. Reipschläger, A.; Pörtner, H.-O. Metabolic depression during environmental stress: The role of extracellular versus intracellular pH in *Sipunculus nudus*. *J. Exp. Biol.* **1996**, *199*, 1801–1807. [CrossRef]
42. Pörtner, H.-O.; Bock, C.; Reipschläger, A. Modulation of the cost of pH_i regulation during metabolic depression: A ^{31}P-NMR study in invertebrate (*Sipunculus nudus*) isolated muscle. *J. Exp. Biol.* **2000**, *203*, 2417–2428. [CrossRef]
43. von Bonde, C.; Marchand, J.M. The natural history and utilisation of the Cape crawfish, kreef or spiny lobsters, *Jasus (Palinurus) lalandii* (Milne Edwards) Ortmann. *Fish. Bull.* **1935**, *1*, 1–55.
44. Ansell, A.D.; Robb, L. The spiny lobster *Palinurus elephas* in Scottish waters. *Mar. Biol.* **1977**, *43*, 63–70. [CrossRef]
45. Key, M.M.; Smith, A.M.; Hanns, B.; Kane-Sanderson, P. Rare report of bryozoan fouling of rock lobsters (*Jasus edwardsii*: Decapoda: Palinuridae) from the North Island of New Zealand. *N. Z. J. Mar. Freshw. Res.* **2021**, *57*, 229–241. [CrossRef]
46. Fernández, M.; Bock, C.; Pörtner, H.-O. The cost of being a caring mother: The ignored factor in the reproduction of marine invertebrates. *Ecol. Lett.* **2020**, *3*, 487–494. [CrossRef]
47. Fernandez, M.; Brante, A. Brood care in Brachyuran crabs: The effect of oxygen provision on reproductive costs. *Rev. Chil. Hist. Nat.* **2003**, *76*, 157–168. [CrossRef]
48. Long, W.C.; Conrad, A.L.; Gardner, J.L.; Foy, R.J. Red king crab larval survival and development are resilient to ocean acidification. *J. Exp. Mar. Biol. Ecol.* **2024**, *577*, 152028. [CrossRef]
49. Styf, H.K.; Nilsson Sköld, H.; Eriksson, S.P. Embryonic response to long-term exposure of the marine crustacean *Nephrops norvegicus* to ocean acidification and elevated temperature. *Ecol. Evol.* **2013**, *3*, 5055–5065. [CrossRef] [PubMed]
50. Findlay, H.S.; Kendall, M.A.; Spicer, J.I.; Widdicombe, S. Future high CO_2 in the intertidal may compromise adult barnacle *Semibalanus balanoides* survival and embryonic development rate. *Mar. Ecol. Prog. Ser.* **2009**, *389*, 193–202. [CrossRef]
51. Walther, K.; Anger, K.; Pörtner, H.-O. Effects of ocean acidification and warming on the larval development of the spider crab *Hyas araneus* from different latitudes (54° vs. 79° N). *Mar. Ecol. Prog. Ser.* **2010**, *417*, 159–170. [CrossRef]
52. Keppel, E.A.; Scrosati, R.A.; Courtenay, S.C. Ocean acidification decreases growth and development in American lobster (*Homarus americanus*) larvae. *J. Northw. Atl. Fish. Sci.* **2012**, *44*, 61–66. [CrossRef]
53. Swiney, K.M.; Long, W.C.; Foy, R.J. Effects of high pCO_2 on tanner crab reproduction and early life history—Part I: Long-term exposure reduces hatching success and female calcification, and alters embryonic development. *ICES J. Mar. Sci.* **2016**, *73*, 825–835. [CrossRef]
54. Long, W.C.; Swiney, K.M.; Foy, R.J. Effects of high pCO_2 on Tanner crab reproduction and early life history, part II: Carryover effects on larvae from oogenesis and embryogenesis are stronger than direct effects. *ICES J. Mar. Sci.* **2016**, *73*, 836–848. [CrossRef]
55. Long, W.C.; Swiney, K.M.; Foy, R.J. Effects of ocean acidification on the embryos and larvae of red king crab, *Paralithodes camtschaticus*. *Mar. Pollut. Bull.* **2013**, *69*, 38–47. [CrossRef]
56. Tong, L.J.; Moss, G.A.; Pickering, T.D.; Paewai, M.P. Temperature effects on embryo and early larval development of the spiny lobster *Jasus edwardsii*, and description of a method to predict larval hatch times. *Mar. Freshw. Res.* **2000**, *51*, 243–248. [CrossRef]
57. Smith, G.G.; Ritar, A.J.; Thompson, P.A.; Dunstan, G.A.; Brown, M.R. The effect of embryo incubation temperature on indicators of larval viability in Stage I phyllosoma of the spiny lobster, *Jasus edwardsii*. *Aquaculture* **2002**, *209*, 157–167. [CrossRef]
58. Munian, A.; Cockcroft, A.; Hoffman, L.; Auerswald, L. Analysis of biological and biochemical parameters of adult male spiny lobsters *Jasus lalandii* for identification of possible growth predictors. *Fish. Res.* **2021**, *243*, 106061. [CrossRef]
59. Siegel, K.R.; Kaur, M.; Grigal, A.C.; Metzler, R.A.; Dickinson, G.H. Meta-analysis suggests negative, but pCO_2-specific, effects of ocean acidification on the structural and functional properties of crustacean biomaterials. *Ecol. Evol.* **2022**, *12*, e8922. [CrossRef] [PubMed]
60. Algayer, T.; Mahmoud, A.; Saksena, S.; Long, W.C.; Swiney, K.M.; Foy, R.J.; Steffel, B.V.; Smith, K.E.; Aronson, R.B.; Dickinson, G.H. Adult snow crab, *Chionoecetes opilio*, display body-wide exoskeletal resistance to the effects of long-term ocean acidification. *Mar. Biol.* **2023**, *170*, 63. [CrossRef]

61. Dickinson, G.H.; Bejerano, S.; Salvador, T.; Makdisi, C.; Patel, S.; Long, W.C.; Swiney, K.M.; Foy, R.J.; Steffel, B.V.; Smith, K.E.; et al. Ocean acidification alters properties of the exoskeleton in adult Tanner crabs, *Chionoecetes bairdi*. *J. Exp. Biol.* **2021**, *224*, jeb232819. [CrossRef] [PubMed]
62. Klumpen, E. Einfluss der Acidifizierung der Meere auf die Physiologie des Taschenkrebses *Cancer pagurus*. Master's Thesis, Düsseldorf University, Düsseldorf, Germany, 2010.
63. Al-Sawalmih, A.; Li, C.; Siegel, S.; Fabritius, H.; Yi, S.; Raabe, D.; Fratzl, P.; Paris, O. Microtexture and chitin/calcite orientation relationship in the mineralized exoskeleton of the American lobster. *Adv. Funct. Mater.* **2008**, *18*, 3307–3314. [CrossRef]
64. Wood, H.L.; Spicer, J.I.; Widdicombe, S. Ocean acidification may increase calcification rates, but at a cost. *Proc. R. Soc. B* **2008**, *275*, 1767–1773. [CrossRef]
65. Stumpp, M.; Wren, J.; Melzner, F.; Thorndyke, M.C.; Dupont, S.T. CO_2 induced seawater acidification impacts sea urchin larval development I: Elevated metabolic rates decrease scope for growth and induce developmental delay. *Comp. Biochem. Physiol. A* **2011**, *160*, 331–340. [CrossRef]
66. Figuerola, B.; Hancock, A.M.; Bax, N.; Cummings, V.J.; Downey, R.; Griffiths, H.J.; Smith, J.; Stark, J.S. A review and meta-analysis of potential impacts of ocean acidification on marine calcifiers from the Southern Ocean. *Front. Mar. Sci.* **2021**, *8*, 584445. [CrossRef]
67. Feely, R.A.; Carter, B.R.; Alin, S.R.; Greeley, D.; Bednaršek, N. The combined effects of ocean acidification and respiration on habitat suitability for marine calcifiers along the west coast of North America. *J. Geophys. Res. Oceans* **2024**, *129*, e2023JC019892. [CrossRef]

biology

MDPI

Article

Life History Traits and Fishery Dynamics of Speckled Shrimp, *Metapenaeus monoceros* (Fabricius, 1798), Along the Saudi Arabian Red Sea Coast

Sheeja Gireesh [1], Eyüp Mümtaz Tıraşın [1,2,*], Goutham Bharathi Muthu Palani [1], Santhosh Kumar Charles [1], Sirajudheen Thayyil Kadengal [1], Ronald Grech Santucci [1], Ricardo Clapis Garla [1], Zahra Okba [1], Adel M. S. Adam [1] and Mark Dimech [1]

1 KAUST Beacon Development Department, National Transformation Institute, King Abdullah University of Science and Technology, Thuwal 23955-6900, Saudi Arabia; sheeja.gireesh@kaust.edu.sa (S.G.); goutham.muthupalani@kaust.edu.sa (G.B.M.P.); santhosh.charles@kaust.edu.sa (S.K.C.); sirajudheen.kadengal@kaust.edu.sa (S.T.K.); ronald.grechsantucci@kaust.edu.sa (R.G.S.); ricardo.garla@kaust.edu.sa (R.C.G.); zahra.okba@kaust.edu.sa (Z.O.); adel.adam@kaust.edu.sa (A.M.S.A.); mark.dimech@kaust.edu.sa (M.D.)

2 Institute of Marine Sciences and Technology, Dokuz Eylül University, İnciraltı, İzmir 35340, Türkiye

* Correspondence: eyupmumtaz.tirasin@kaust.edu.sa

Simple Summary: Marine shrimp fisheries primarily target penaeid species, which are in high demand in the global market. The considerable ecological and commercial importance of these species highlights the need for effective stock assessments to support appropriate management strategies. This study examines key population parameters, including growth, mortality, maturity, and morphometric relationships, to assess the status of the speckled shrimp *Metapenaeus monoceros* stock along the Saudi Arabian Red Sea coast. Growth parameter estimates indicate that females attain a larger asymptotic carapace length than males but grow at a slower rate. The exploitation rate for both sexes exceeds the optimum level, raising concerns regarding the sustainability of the fishery. The high proportion of immature individuals in the catch, combined with elevated fishing mortality and exploitation rates, strongly suggests that overfishing is occurring in the region. To ensure the long-term sustainability of this commercially valuable species, continued monitoring and the implementation of effective management measures are recommended.

Abstract: *Metapenaeus monoceros* is the second most abundant commercially important shrimp species caught along the Saudi Arabian Red Sea coast. However, knowledge of its life history and population dynamics in the region remains limited, posing challenges for sustainable management. This study assessed the growth, maturity, and mortality of *M. monoceros* based on fishery-independent samples collected from shrimp trawling grounds off Al Qunfudhah and Jizan (October 2022–September 2023). Altogether, 10,859 specimens were examined, with a carapace length (CL) ranging from 7.98 to 49.10 mm and a total weight (W) in the range of 0.41 to 41.28 g. The observed sex ratio was 1.56:1 in favor of females. The CL–W relationship was $W = 0.0024 \cdot CL^{2.58}$. The seasonalized von Bertalanffy growth parameters were $CL_\infty = 50.66$ mm, $K = 0.35$ year^{-1}, $t_{anchor} = 0.43$ years, $C = 0.64$, and $t_s = 0.61$ years for females, and $CL_\infty = 38.97$ mm, $K = 0.55$ year^{-1}, $t_{anchor} = 0.51$ years, $C = 0.84$, and $t_s = 0.39$ years for males. Females reached their first sexual maturity at 21.12 mm CL. The exploitation rates (0.60 for both sexes) indicated overfishing. Spawning potential ratio analysis showed that the current fishing pressure reduces the spawning stock biomass to 23% of unexploited levels. A 50% reduction in fishing effort, extended seasonal closures, and selective gear modifications are recommended to rebuild the stock and ensure sustainability.

https://doi.org/10.3390/biology14040406

Keywords: crustacean fisheries; shrimp trawling; Penaeidae; growth; mortality; maturity; stock assessment; spawning potential ratio; overfishing

1. Introduction

The speckled shrimp *Metapenaeus monoceros* (Fabricius, 1798) is a widely distributed penaeid species with a remarkable geographical range spanning the Indo-West Pacific region. Its native distribution encompasses a vast area, including the east coast of Africa, Madagascar, Tanzania, the Red Sea, the entire Indian coastal belt, Sri Lanka, Pakistan, Malaysia, the Straits of Malacca, Indonesia, Australia, and Japan [1–3]. Notably, *M. monoceros* has successfully migrated through the Suez Canal, establishing itself as a Lessepsian migrant in the eastern Mediterranean Sea [4]. This expansion has resulted in established populations along the coasts of Egypt, Israel, Syria, Lebanon, Turkey, Greece, and Tunisia [5–8]. Economically, the species plays a crucial role in regional fisheries, particularly in Bangladesh, where it constitutes more than 50% of the total shrimp landings [9]. This commercial importance underscores the species' significance not only from an ecological perspective but also as a critical resource for local fishing communities and national seafood industries.

M. monoceros inhabits marine and brackish water environments, typically ranging from 10 to 30 m in depth, with occasional observations extending to a maximum recorded depth of 170 m [3,10]. The species demonstrates a strong preference for sandy and muddy habitats, which provide essential ecological niches for feeding, reproduction, and shelter. The species exhibits remarkable sexual dimorphism, with males attaining a maximum total length (*TL*) of 199 mm and females reaching 226 mm [11]. The species displays a relatively short life cycle, typically spanning 2–3 years [12,13]. Growth patterns also differ between sexes; males are heavier than females up to 77 mm *TL*, after which females exhibit greater weight due to maturation processes [14]. Furthermore, females are generally more abundant than males in the population [11,15–17]. Reproduction in *M. monoceros* is characterized by continuous spawning throughout the year, punctuated by seasonal peaks [18–22].

The fishery biology and stock characteristics of *M. monoceros* have been extensively studied in India [13,15,23–25]. Notably, trawl fisheries have led to overfishing of this species in both the Arabian Sea and the Bay of Bengal. Specifically, overfishing occurred along the Karwar, Veraval, and Mumbai coasts in the Arabian Sea [13,25]. Moreover, Nandakumar and Srinath [15] reported overfishing off the Kerala coast in the same region. Similarly, trawling has resulted in the overfishing of *M. monoceros* along the Kakinada and Visakhapatnam coasts in the Bay of Bengal [13,25]. The species is also targeted by trawling in the eastern Mediterranean, according to records for the southern coast of Turkey, and the continental shelf off Palestine, Egypt, and Tunisia [5,6,26–28]. Furthermore, overfishing has been documented in the industrial fishing zone of Bangladesh [29].

The shrimp fishery along the southeastern Red Sea coast of Saudi Arabia is predominantly based on *Penaeus semisulcatus* and *M. monoceros* [30]. Statistics from the Saudi Arabian Ministry of Environment, Water, and Agriculture (MEWA) indicate that annual shrimp landings along the Saudi Arabian Red Sea coast between 2017 and 2023 (reported collectively for all commercial shrimp species) ranged from 500 to 835 tonnes, with an average of 620 tonnes [31]. Despite the commercial and ecological importance of these species, scientific literature documenting their stock assessments along the Saudi Arabian Red Sea coast remains remarkably limited, creating a substantial knowledge gap in regards to regional marine resource management. Understanding the life history traits, growth and mortality parameters, and reproductive capacity of a species under current fishing pressure is essential for evaluating its stock status and ensuring sustainable management. This study aims to address this critical knowledge gap by investigating the growth, mortality, and reproduction of *M. monoceros* and conducting a preliminary assessment of stock status along the southeastern Red Sea coast. The findings of this study are expected to provide essential information for the sustainable management of this commercially important species.

2. Materials and Methods

2.1. Study Area

Al Qunfudhah and Jizan are the only ports along the Red Sea coast of Saudi Arabia where shrimp trawl fisheries are legally permitted (Figure 1). Shrimp trawling in these areas is conducted exclusively by vessels registered as industrial vessels. These vessels are typically constructed from wood (often reinforced with fiberglass) or sheet metal [32]. According to national regulations [33], industrial vessels must be equipped with modern instrumentation, including echosounders, GPS, and communication systems, as well as safety equipment, refrigeration facilities, and winches. The length overall and main engine power of these vessels must not exceed 20 m and 250 HP, respectively. The MEWA retains the authority to issue or renew licenses, thereby regulating the number of operational fishing vessels and determining ownership, including for investor fishers who may not be directly involved in fishing activities. Additional input controls imposed by Saudi Arabia on the shrimp fishery include a seasonal closure between April and August, restricting trawling to designated fishing grounds, and prohibiting fishing within two nautical miles of the coastline or islands. The only technical measure currently in place is a regulation that prohibits the use of codend mesh sizes smaller than 38 mm [33].

Jizan, situated near the Farasan Islands, is the largest and most important fishing port along the Red Sea coast in terms of fleet size and fish production (Figure 1). The industrial fishing fleet registered in Jizan comprises 151 vessels, whereas only 24 vessels are registered in Al Qunfudhah [32,34]. In Jizan, trawling is typically conducted at an acute angle to the coastline, with fishing depths ranging from 10 to 40 m. In contrast, in Al Qunfudhah, trawling operations are primarily conducted parallel to the coastline, with fishing grounds ranging in depth from 10 to 30 m. The trawling grounds in both areas are primarily sandy, interspersed with patches of muddy substrate [34,35]. In addition to trawling, industrial vessels may also engage in purse seining, depending on the seasonal closure of the shrimp trawl fishery [32].

2.2. Biological Data Collection

Monthly biological sampling was carried out off the coasts of Al Qunfudhah and Jizan, along the southeastern Red Sea (Figure 1), from October 2022 to September 2023. For experimental trawl surveys, commercial shrimp trawlers registered at these ports were chartered. The trawl hauls in these surveys closely replicated commercial fishing operations, maintaining consistency in trawl net specifications, fishing locations, haul duration, towing speed, and other operational parameters. All hauls were conducted at night at an average towing speed of three knots, with each tow lasting approximately three hours. The number of hauls varied from month to month due to weather conditions, but an average of nine hauls per month were conducted at each location. Over the course of the study, a total of 110 hauls were completed in Al Qunfudhah and 108 in Jizan. Every haul utilized the standard 40 mm diamond mesh codend, which is commonly used in commercial shrimp trawling operations in the region [32,35]. A special permit was obtained from the MEWA to continue experimental trawl surveys during the closed season between April and August.

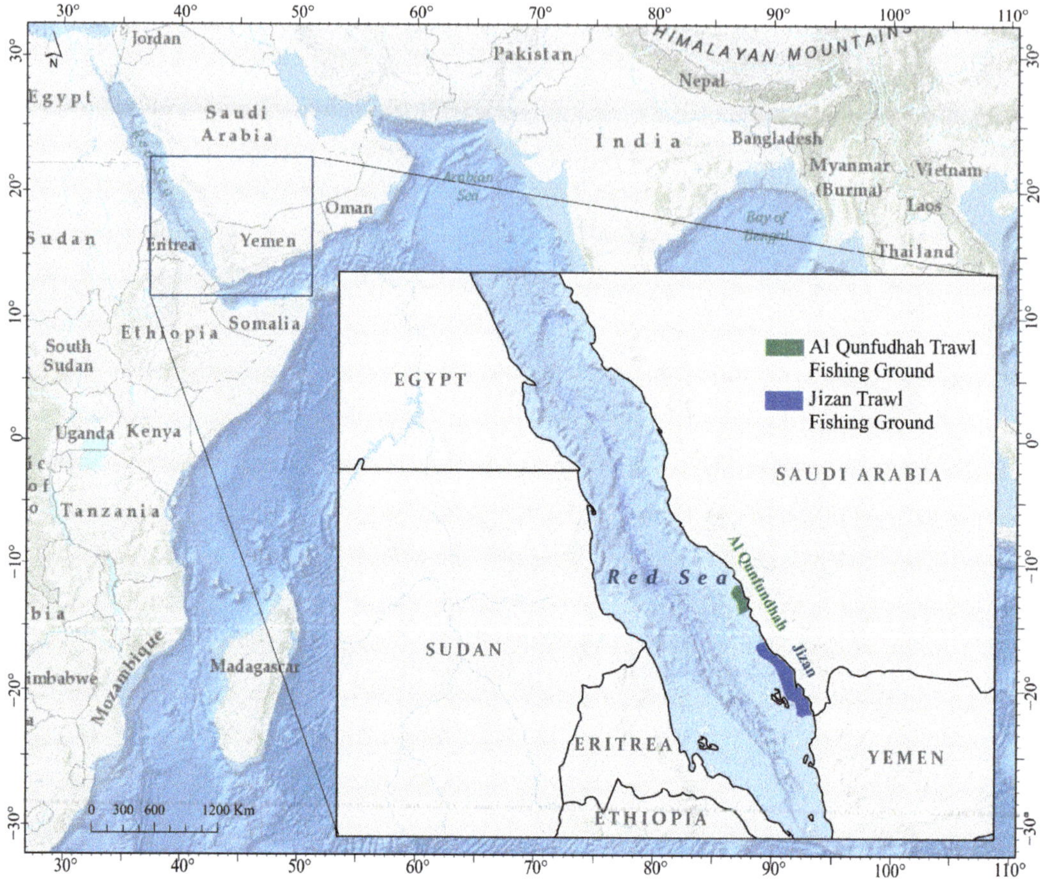

Figure 1. Locations of Al Qunfudhah and Jizan trawl fishing grounds along the southeastern Red Sea coast of Saudi Arabia.

The primary target species for shrimp trawlers operating within Saudi Arabian territorial waters in the Red Sea include *P. semisulcatus*, *M. monoceros*, *P. pulchricaudatus*, *P. indicus*, and *P. hathor* [30,34]. After the trawl was brought onboard, all target shrimp species were separated from the rest of the catch. Shrimp from each haul, regardless of species, were stored separately on ice in separate containers and transported to the laboratory for detailed processing. In the laboratory, specimens were first sorted by species. All *M. monoceros* individuals were processed, and various biological variables were registered. Carapace length (*CL*) was measured to the nearest 0.01 mm using a digital caliper, and total weight (*W*) was recorded at the nearest 0.01 g for each specimen. Sex was determined macroscopically based on the presence of petasma in males or thelycum in females [36].

Gonadal maturity stages were determined only in female shrimp using a modified methodology described by Amanat and Qureshi [37] and Abdallah et al. [27]. Ovarian maturation stages were classified into four distinct stages, based on coloration and size of the ovaries—(Stage I) Immature or undeveloped: ovaries are thin, translucent, and underdeveloped, with poorly defined anterior and lateral lobes; (Stage II) Developing: ovaries appear creamy white to yellowish and are larger in size, and in more advanced cases within this stage, they occupy most of the dorsal portion and range from yellow to light green in color; (Stage III) Mature: ovaries are fully developed, occupying the entire dorsal region, are dark green in color, and are visible through the exoskeleton; (Stage IV) Spent: ovaries resemble those in Stage II in color but are more enlarged, flaccid in texture, and reduced in turgidity, indicating post-spawning condition (Figure S1).

2.3. Length–Weight Relationship

Estimated mean values of *CL* and *W* measurements by sex were compared using independent *t*-tests. Prior to these analyses, the assumptions of normality and homoscedasticity were evaluated using *F*-tests, Shapiro–Wilk normality tests, and normal quantile-quantile plots [38]. If the normality assumption was met but variances were heteroscedastic, Welch's approximate *t*-test was applied. For data that violated both assumptions, a log transformation was attempted. If this transformation did not resolve the issues, the non-parametric Mann–Whitney test was employed [38]. The relationship between *CL* and *W* was examined using linear regression analysis after log-transforming all data pairs, based on the assumption of a multiplicative error structure [39].

$$W = a \cdot CL^b \quad (1)$$

In this nonlinear relationship, *a* represents the coefficient for the ratio between *W* and *CL*, while *b* is the exponent describing how the body mass (*W*) of a shrimp changes as its size (*CL*) increases. An analysis of covariance (ANCOVA) [38] was conducted to determine whether the pairs of *a* and *b* values differed between female and male shrimp. Departures from isometric growth, i.e., whether the estimated *b* significantly deviated from the hypothesized value of 3, were assessed by determining whether the confidence intervals for each *b* comprised 3 or not [39,40].

2.4. Size at First Maturity

The median size at first sexual maturity, i.e., the *CL* at which 50% of speckled shrimp in the population reach sexual maturity (CL_{50}), was estimated exclusively for females using logistic regression analysis [41], as described by Aydın and Tıraşın [42]. The maturity status of each female shrimp was classified as binary data (Stage I was deemed "immature = 0", while Stage II and above were designated as "mature = 1"). The logistic model was fitted to these binomial maturity data. To evaluate the uncertainty of the CL_{50} estimate, the bootstrap method [42,43] was applied, generating 5000 bootstrap samples. A nonparametric 95% confidence interval for CL_{50} was then estimated using the bias-corrected and accelerated (BCa) method [43].

2.5. Growth

The growth of *M. monoceros* in the southeastern Red Sea was described by means of the seasonally oscillating version of the von Bertalanffy growth model [44], as follows:

$$CL_t = CL_\infty \cdot \left(1 - e^{-K\cdot(t-t_0)-\left(\frac{C\cdot K}{2\pi}\right)\cdot(sin(2\pi\cdot(t-t_s))-sin(2\pi\cdot(t_0-t_s)))}\right) \quad (2)$$

where CL_t is the expected *CL* at age *t* years, CL_∞ is the asymptotic *CL*, *K* is the growth coefficient or curvature parameter indicating the rate at which shrimp grow towards their CL_∞, t_0 is the theoretical age at 0 mm *CL*, *C* is a constant which shows the amplitude of the growth oscillation, and t_s is the fraction of a year defining the beginning of the sinusoidal growth oscillation with respect to t_0, i.e., relative to the age of recruitment [44,45]. When *C* equals 0, then there is no seasonal oscillation, and the above model reduces to the traditional parameterization of the von Bertalanffy growth model created by Beverton and Holt [46].

The recently developed TropFishR package (version 1.6.4) [47,48], which incorporates the electronic length frequency analysis (ELEFAN) method [45,49], was utilized to estimate the parameters of the seasonally oscillating von Bertalanffy growth model for female and male shrimp separately, except for t_0. The von Bertalanffy growth model requires actual age information (t) as an independent variable; however, length frequency samples only provide the time interval between samplings, making it impossible to estimate t_0 solely with length frequency data [45]. Instead, TropFishR provides a related parameter, "t_{anchor}", which represents the fraction of the year ranging between 0 and 1, where the growth curve crosses 0 mm *CL* for a given cohort [47]. A powerful new optimization procedure in TropFishR, the ELEFAN_GA function, which simultaneously searches across all parameters, was employed using a moving average (MA) over five *CL* intervals. Using the ELEFAN_GA function offers a significant advantage over the basic ELEFAN, as it efficiently models the seasonally oscillating version of the von Bertalanffy growth model. Monthly *CL* distributions, grouped into 2 mm class intervals for each sex, were used as input data. A comparison of growth performances between females and males was conducted using the growth performance index (φ') [50], as follows:

$$\varphi' = \log_{10}(K) + 2 \cdot \log_{10}(CL_\infty) \tag{3}$$

2.6. Mortality

The total mortality rate (Z) for each sex was estimated from the linearized length-converted catch curve analysis [51]. In accordance with the work of Quinn and Deriso [39], Cope and Hamel [52], and Maunder et al. [53], the natural mortality rate (M) was estimated for each sex using three different methods recommended by Maunder et al. [53] to enhance methodological robustness and reduce potential bias from relying on a single approach [54,55].

$$M = \frac{5.4}{t_{\max}} \tag{4}$$

$$M = 1.55 \cdot K \tag{5}$$

$$M = 4.118 \cdot K^{0.73} \cdot CL_\infty^{-0.33} \tag{6}$$

$t_{\max}$ in Equation (4) represents the maximum age, or longevity, of a species. Based on previous studies [12,13], t_{max} was assumed to be 3 years for female shrimp and 2 years for males. CL_∞ and K are the parameters of the von Bertalanffy growth model, as described earlier. The M value ultimately used for further analysis was determined by averaging the three independent estimates derived from the abovementioned methods. Since Z consists of both M and the fishing mortality rate (F), once Z and M were obtained, F was estimated using the following relationship:

$$F = Z - M \tag{7}$$

2.7. Stock Status Evaluation

A preliminary evaluation of the stock status of *M. monoceros* in the southeastern Red Sea was conducted by estimating the current levels of the exploitation rate (E) and the spawning potential ratio (SPR). The E was calculated separately for each sex as the ratio of F to Z, as follows:

$$E = \frac{F}{Z} \tag{8}$$

SPR is a key metric used to assess the reproductive potential of a stock under current fishing pressure, with a particular focus on females, as they are responsible for egg production, which drives recruitment. Goodyear [56] defined SPR as the equilibrium spawning stock biomass per recruit (SSBpR) at a given level of fishing mortality, divided by the equilibrium SSBpR in the absence of fishing (i.e., when only natural mortality affects the stock). SPR analysis is an extension of the prediction models, such as those developed by Thompson and Bell [51] or Beverton and Holt [46], which are traditionally used to estimate yield per recruit (YpR) and biomass per recruit (BpR). BpR represents the average biomass contributed by each recruit in the stock. The calculation of SSBpR for a fished stock is analogous to the computation of BpR; however, instead of considering the entire life span of a recruiting cohort, the focus is on the mature females within the cohort that are capable of spawning [56,57]. Like BpR, SPR has a maximum value of 1 when no fishing occurs ($F = 0$), and it declines towards 0 as F increases. SPR may also be expressed as a percentage, representing the proportion of the unfished SSBpR that remains under a given F value (e.g., at $F = 0$, %SPR is at 100%).

A length-based Thompson and Bell YpR model [51] was applied to estimate the SPR of female speckled shrimp. This analysis focused exclusively on females due to their direct contribution to egg production and consequently, their critical role in determining population reproductive capacity. The model required sex-specific growth and mortality parameters, including CL_{∞}, K, and M, as well as the number of shrimp in each 2 mm CL-class (using the same size intervals and counts employed for the von Bertalanffy growth model parameter estimation). Additional model inputs included the mean weights by CL-class and the CL at first capture (CL_c), which was set at 17 mm, based on findings from a recent study by Santucci et al. [35] conducted with commercial trawl boats and gear off Jizan. The proportion of mature females in each CL-class was predicted using logistic regression analysis. The YpR, BpR, and SSBpR were estimated as a function of varying F levels and expressed in mass (g).

Biological reference points (BRPs) derived from YpR models are expressed as F rates, which serve as indicators of stock status [39]. From the YpR model, two key BRPs are often derived: F_{max} and $F_{0.1}$. F_{max}; the F that maximizes YpR is considered a limit reference point (LRP) [58]. $F_{0.1}$ is the F at which the slope of the YpR curve is reduced to 10% of its slope at the origin [40]. This reference point is often regarded as a precautionary target reference point (TRP) for sustainable fishery management [59]. The LRP defines a critical threshold that should not be breached to prevent serious harm to the stock, indicating an unsustainable biomass or fishing pressure level. In contrast, the TRP represents the desired level of F or stock biomass, aimed at achieving long-term sustainability and optimal yield while maintaining the stock's health and productivity [58–60].

YpR analysis focuses primarily on assessing growth overfishing and does not account for the impact of fishing on the reproductive capacity of the stock [39,61,62]. Alternatively, SPR analysis offers critical information on recruitment overfishing by evaluating the effects of fishing on the stock's ability to replenish itself through adequate SSB [39,63,64]. The minimum SPR value widely recommended as a TRP to safeguard against stock depletion and prevent recruitment overfishing is 0.4 [56,61,62]. The F associated with this TRP is commonly referred to as $F_{40\%}$. Although an SPR value as low as 0.2 has been proposed as an LRP [56,63], Walters and Martell [64] emphasized that historical stock-recruitment data show a considerably higher risk of recruitment overfishing when SPR falls below 0.3. Accordingly, an SPR of 0.3 is adopted as the LRP in this evaluation.

Additionally, in this study, an E value of 0.5 has been employed as an LRP [58,65], representing a threshold of sustainable fishing pressure that should ideally be avoided, to evaluate the status of speckled shrimp in the region.

All statistical tests were conducted with R software version 4.4.1 [66] with a significance level of 5%.

3. Results

3.1. Distribution of the Samples

A total of 103,758 individuals from six commercially valuable shrimp species were sampled from the fishing grounds in the southeastern Red Sea during the study period.

The catch composition was dominated by *P. semisulcatus* (83%) and *M. monoceros* (10%). *P. pulchricaudatus* and *P. indicus* each contributed approximately 3%, while *P. hathor* and *P. monodon* made only minor contributions, accounting for 1% and 0.02% of the total number of individuals, respectively. From the 10,859 specimens of *M. monoceros* examined, 6617 were females and 4242 were males. Between the two sampling locations, Jizan accounted for the majority of *M. monoceros* specimens (6098; 56%), while Al Qunfudhah contributed 4761 specimens (44%).

3.2. Carapace Length Frequency Distribution

For all *M. monoceros* samples studied, *CL* ranged from 7.98 to 49.10 mm, while *W* varied between 0.41 and 41.28 g. Summary statistics, including the range, mean, standard deviation (SD), median, and interquartile range (IQR) of *CL* and *W* measurements by sex and combined data, are presented in Table 1. Females exhibited a wider observed range for both *CL* and *W* compared to males (Table 1). Additionally, the variability, as indicated by the SD of both *CL* and *W*, was greater in females, reflecting higher dispersion in these measurements (*F*-test for *CL*, $F = 3.87$, df = 6616, 4241, $p < 0.001$; for *W*, $F = 5.55$, df = 6616, 4241, $p < 0.001$). Females also displayed significantly larger mean values for *CL* and *W* compared to those for males (Table 1, Welch's approximate *t*-test for *CL*, $t = 36.26$, df = 10,391, $p < 0.001$ and Mann–Whitney test for *W*, $U = 9{,}625{,}587$, $p < 0.001$). The overall *CL* frequency distribution of *M. monoceros* samples from the southeastern Red Sea, stratified by sex and grouped into 2 mm class intervals, is illustrated in Figure 2. One female shrimp with a *CL* of 7.98 mm was included in the 8–10 mm class, while another female with a CL of 49.10 mm was grouped into the 44–46 mm class. In addition, the monthly *CL* frequency distributions of *M. monoceros*, separated by sex, are presented in Figure 3. Notable differences in *CL* distributions between males and females were observed across the months, with females generally exhibiting a broader size range and a higher proportion of larger individuals compared to the results for males (Figure 3).

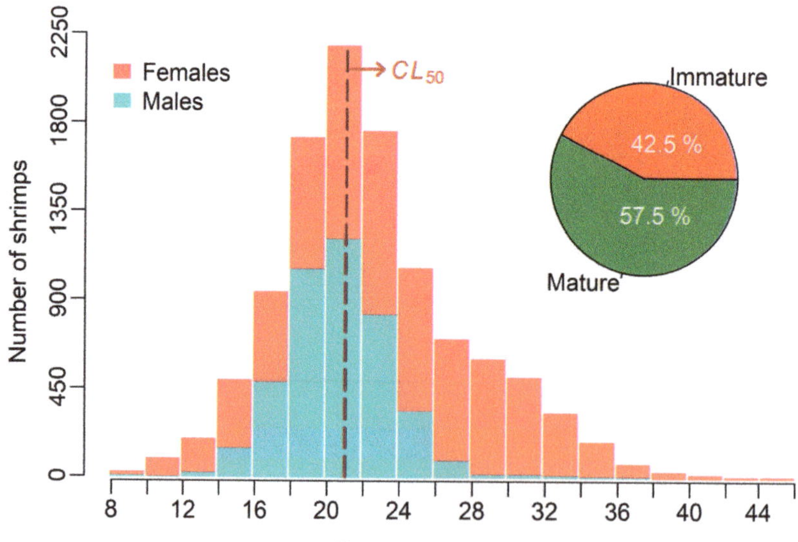

Figure 2. *CL* frequency distribution (in 2 mm intervals) of *M. monoceros* samples from the southeastern Red Sea, stratified by sex. Of 10,859 total sampled specimens, 6617 were females and 4242 were males. CL_{50} denotes the *CL* at first sexual maturity for female shrimp, estimated as 21.12 mm. Proportions of immature shrimp ($CL < CL_{50}$) and mature shrimp ($CL \geq CL_{50}$) in the overall catch are also shown.

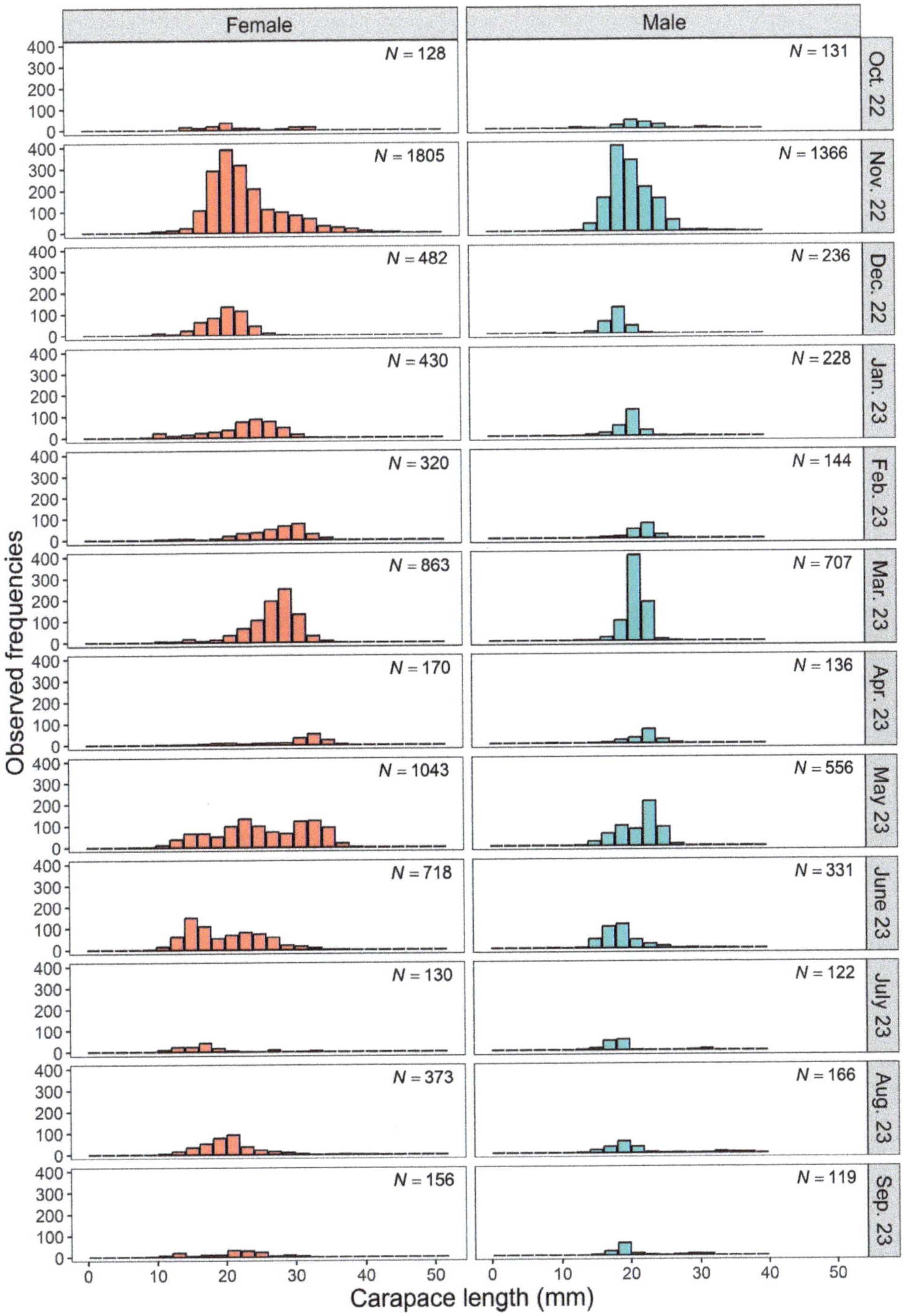

Figure 3. Monthly *CL* frequency distributions of *M. monoceros* by sex.

Table 1. Descriptive statistics for carapace length (*CL*) and weight (*W*) of *M. monoceros* collected from the southeastern Red Sea, presented for the total sample and separated by sex. *N*, SD, and IQR denote sample size, standard deviation, and interquartile range, respectively.

Sex	*N*	Carapace Length (mm)				Weight (g)			
		Range	Mean ± SD	Median	IQR	Range	Mean ± SD	Median	IQR
Female	6617	7.98–49.10	23.76 ± 5.83	23.32	19.70–28.00	0.41–41.28	9.52 ± 5.71	8.35	5.04–13.44
Male	4242	8.40–37.56	20.68 ± 2.96	20.60	18.70–22.42	0.46–25.88	6.21 ± 2.42	6.08	4.60–7.50
Total	10,859	7.98–49.10	22.56 ± 5.14	21.75	19.22–25.30	0.41–41.28	8.29 ± 5.02	7.00	4.78–10.52

3.3. Sex Ratio and Maturity

The overall sex ratio indicated a predominance of females, at 1.56:1 (Figure 2). The exact binomial test confirmed that this observed sex ratio deviated significantly from unity. The monthly percentage distributions of female speckled shrimp maturity stages are presented in Figure 4. Mature females at Stage III were observed in nearly all sampling months, except for July 2023. They were particularly abundant between February 2023 and May 2023, with the highest proportion (52.3%) was recorded in April 2023 (Figure 4). The smallest mature female at Stage III measured 14.70 mm *CL*, while the mean *CL* of females at this stage was 28.48 ± 4.13 mm.

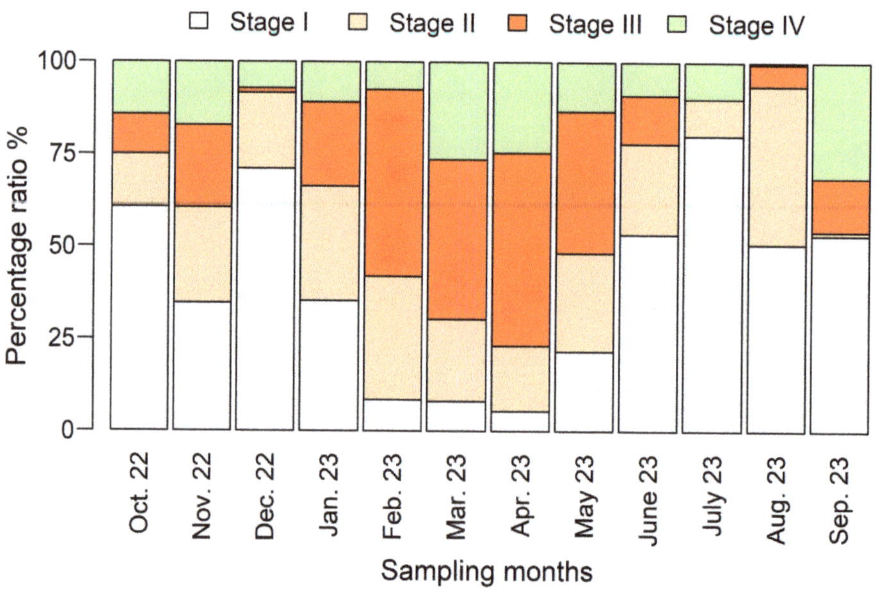

Figure 4. Monthly percentage distribution of female *M. monoceros* maturity stages in the southeastern Red Sea.

The CL_{50}, estimated by logistic regression analysis, was 21.12 mm for females, with a nonparametric 95% confidence interval of 20.91–21.29 mm (Figure 5). When examining the overall *CL* distribution of the samples, immature speckled shrimp ($CL < CL_{50}$) accounted for approximately 44% of all specimens caught during the study period (Figure 2). In contrast, mature individuals ($CL \geq CL_{50}$) comprised only slightly more than 56% of the total samples.

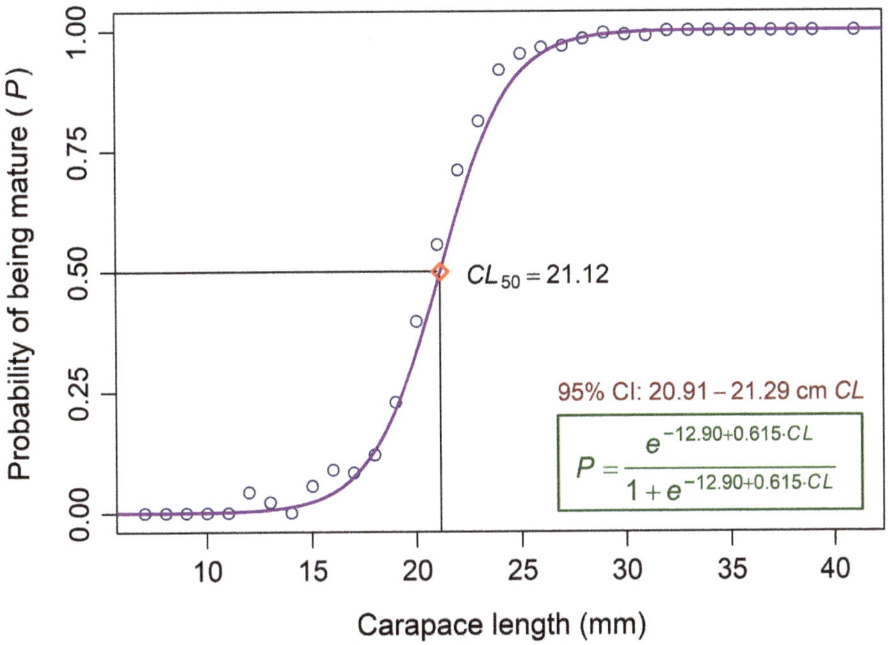

Figure 5. The estimated *CL* at first sexual maturity (CL_{50}) of female *M. monoceros*. The curve shows the fitted logistic model to the binomial maturity data. Circles represent the observed proportions of mature females relative to the total sampled female shrimp. CI denotes confidence intervals.

3.4. Length and Weight Relationship

The ANCOVA of *CL* and *W* data revealed no significant differences in the *a* and *b* values between female and male shrimp (ANCOVA, $F = 0.32$, df = 1, 10,855, $p = 0.58$). When all data were combined, the estimated *a* and *b* values (with 95% confidence intervals) for the *CL–W* relationship were 0.00236 (0.00229–0.00244) and 2.585 (2.575–2.595), respectively, with a determination coefficient (r^2) of 0.97. The estimated *b* value, significantly lower than 3, indicates a negative allometric growth pattern in this species.

3.5. Growth and Mortality

The estimated parameters of the seasonally oscillating von Bertalanffy growth model for females were $CL_\infty = 50.66$ mm, $K = 0.35$ year^{-1}, $t_{anchor} = 0.43$ year, $C = 0.64$, and $t_s = 0.61$ year (Figure S2); for males, they were $CL_\infty = 38.97$ mm, $K = 0.55$ year^{-1}, $t_{anchor} = 0.51$ year, $C = 0.84$, and $t_s = 0.39$ year (Figure S3). The resulting growth performance indices (φ') were 2.96 for females and 2.92 for males. According to the growth parameter estimates, female shrimp grew more slowly than males but reached a larger asymptotic size. Despite the differences in growth characteristics, the overall growth performance remained consistent across sexes.

The *Z* values (with 95% confidence intervals), estimated from the linearized length-converted catch curve analysis, were 2.38 (2.03–2.72) year^{-1} for females and 3.65 (2.66–4.64) year^{-1} for males (Figure S4). The *M* estimates derived from three different methods (in the order previously described) were 1.8, 0.55, and 0.53 year^{-1} for females and 2.7, 0.85, and 0.79 year^{-1} for males. The average *M* values calculated across these methods were 0.96 year^{-1} for females and 1.45 year^{-1} for males. The *F* estimates, based on sex-specific *Z* and *M* values, were 1.42 year^{-1} for females and 2.20 year^{-1} for males.

3.6. Stock Evaluation

The E values derived from the sex-specific mortality estimates were 0.60 for both female (0.597) and male (0.604) speckled shrimp, exceeding the LRP of 0.5, hence indicating unsustainable fishing pressure.

The BRPs derived from the YpR and SPR analysis for female shrimp are presented in Figure 6. The left y-axis represents SPR, while the right y-axis shows the YpR (g) for a single female shrimp recruiting to the stock. The figure illustrates the decline in SSBpR relative to the unfished level and changes in YpR as a function of F. The estimated F_{max} was 1.20 year^{-1}, corresponding to a maximum YpR of 1.66 g. Beyond this F level, YpR decreased consistently. The more conservative reference point, $F_{0.1}$, was 0.66 year^{-1}. The current F for female shrimp, estimated at 1.42 year^{-1}, substantially exceeded both the F_{max} and $F_{0.1}$. At this F value, the SSB was approximately 23% of its unexploited level. The $F_{40\%}$ value associated with the TRP for SPR analysis was 0.69 year^{-1}, and as shown in Figure 6, the $F_{40\%}$ and $F_{0.1}$ estimates were very close.

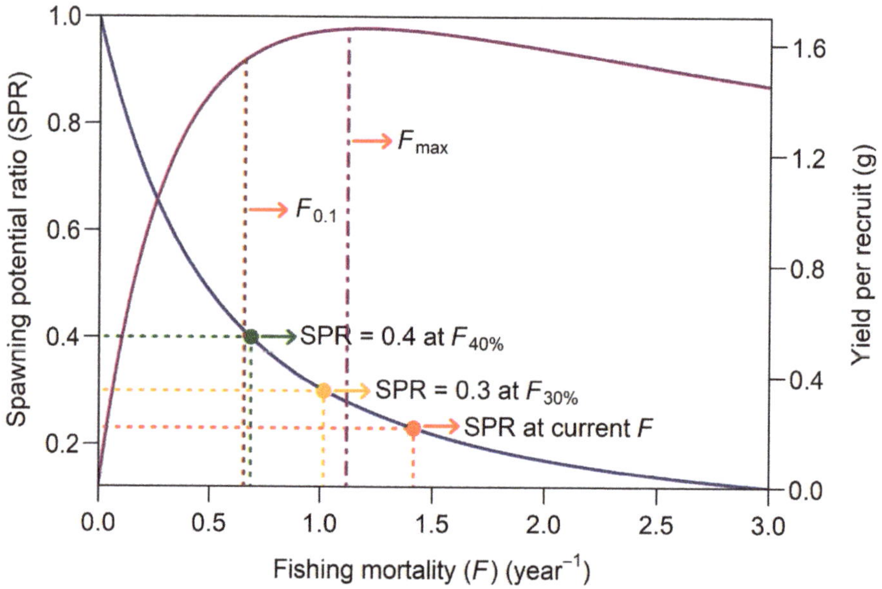

Figure 6. Relationship between SSBpR (dark blue line, corresponding to the left SPR axis) and total YpR (dark red line, corresponding to the right y-axis) as a function of F. The SPR and YpR levels were estimated using the Thomson and Bell YpR analysis. The figure also illustrates the SPR values corresponding to the current F and various BPRs, including F_{max}, $F_{0.1}$, $F_{40\%}$, and $F_{30\%}$.

4. Discussion

Marine fishery resources have been declining globally, with developing countries experiencing particularly severe declines due to a combination of overfishing, pollution, increasing market demand, and inadequate management policies. The depletion of these resources poses a significant threat to marine biodiversity, ecosystem stability, and the livelihoods of communities dependent on fisheries. Therefore, it is crucial to implement effective strategies for stock recovery to ensure long-term ecological and economic sustainability. Stock assessment and science-based resource management are essential tools for the conservation and sustainable utilization of living marine resources [67–71]. This study provides new and comprehensive insights into the population dynamics, stock status, and biological characteristics of *M. monoceros* in the southeastern Red Sea. By presenting key population parameters and stock assessment findings, comparisons with similar studies conducted in other regions are facilitated.

The experimental trawl surveys confirmed that *M. monoceros* was the second most abundant shrimp species after *P. semisulcatus*, contributing 10% of the total shrimp catch by number in the southeastern Red Sea. A similar trend was observed by Ghamrawy [30], who reported that this species was also the second most abundant in Jizan, contributing almost 32% of the total shrimp catch. The observed difference in relative abundance between studies may be attributed to temporal variations, including fluctuations in environmental conditions, increased fishing pressure, and changes in ecosystem dynamics over time.

The *CL* frequency distribution for combined sex in the present study ranged from 7.98 mm to 49.1 mm. In comparison, Barua et al. [72] reported a *CL* frequency distribution range of 18–74 mm in Bangladesh. Similarly, Abdallah et al. [28] found that males of *M. monoceros* were predominant within the *TL* range of 70–100 mm, while females became predominant beyond this size range.

The estimated *b* value of 2.58 in the present study indicates a negative allometric growth pattern, meaning that *M. monoceros* grows in *CL* at a faster rate than in *W* for both sexes. This finding aligns with previous studies reporting similar negative allometric growth patterns for this species in different regions. For instance, Sukumaran et al. [13] reported a *b* value of 2.76 along the Indian coast, while Manaşırlı [73] recorded a *b* value of 2.81 in Iskenderun Bay in the northeastern Mediterranean (Turkey). Comparable results were also observed by Lalita Devi [12] and George [18] from India, Ghobashi et al. [74] from the Egyptian Mediterranean, and Mustafa et al. [75] from Bangladesh.

The overall sex ratio in the present study indicated a predominance of females (1.56:1). This pattern is consistent with findings from various regions, where several studies, including those by CMFRI [11], Nandakumar and Srinath [15], Nandakumar [16], Abdallah et al. [17], Abdallah et al. [27,28], Manaşırlı [73], Bayhan et al. [76], and Yılmaz et al. [8] (Table 2), have also reported female-biased sex ratios in *M. monoceros*.

Table 2. Summary of maximum size, sex ratio, size at first maturity, and von Bertalanffy growth parameters estimated for *M. monoceros* in different regions of the world. M and F denote males and females, respectively. When a value is reported for combined sexes, it is presented in between the M and F columns. Measurements based on *CL* rather than *TL* are marked with an asterisk (*).

Location	Max Size (mm)		Sex Ratio (F:M)	TL_{50} (mm)		TL_∞ (mm)		K (Year^{-1})		W_∞ (g)		References
	M	F		M	F	M	F	M	F	M	F	
Tunisia			1.5:1	76.5	122.3							Abdallah et al. [27]
			1.7:1			169.5	193.3	2.06	1.36			Abdallah et al. [28]
	160	178		76.7	117.6							Hadj Hamida [77]
India	199	226	1.25:1	105.4	121.3	208.9	237.3	1.6	1.5			CMFRI [11]
						208.4	216.2	1.0	1.0			Lalitha Devi [12]
				96	115.5	180	210	1.8	1.8	37	71	Sukumaran et al. [13]
	163	187				178.4	207.3	1.68	1.62	36	68	Rao and Krishnamoorthi [23]
				135.5		190	225					George et al. [78]
				118								Nalini [79]
Egypt				96								Abdel Razek et al. [80]

Table 2. *Cont.*

Location	Max Size (mm)		Sex Ratio (F:M)	TL_{50} (mm)		TL_∞ (mm)		K (Year^{-1})		W_∞ (g)		References
	M	F		M	F	M	F	M	F	M	F	
Turkey	145	165	2.8:1	75	115							Yılmaz et al. [8]
	150	155	1.4:1			162.8	178.4	1.39	1.51			Manaşırlı [73]
	155	190	1.3:1			157.5	199.5	0.71	0.52	26	40	Bayhan et al. [76]
Bangladesh	74 *			33.3 *		85.4 *						Barua et al. [72]
Red Sea, Saudi Arabia	37.6 *	49.1 *	1.56:1		21.12 *	38.97 *	50.66 *	0.55	0.35	31	60	Present study

The estimated CL_{50} of female speckled shrimp in the present study was 21.12 mm. Ghamrawy [30] reported that *M. monoceros* off the Jizan coast (Saudi Arabia) reached maturity at smaller sizes, with the minimum *CL* to attain Stage III and Stage IV of ovarian maturation being 21 mm and 24 mm, respectively. According to Rao [81], the *TL* at first sexual maturity (TL_{50}) of *M. monoceros* in the Kakinada area (India) was 96 mm in males and 116 mm in females. Nalini [79] estimated a very close TL_{50} value of 118 mm for females off Cochin (southwestern India). George et al. [78] reported that the species reaches maturity at approximately 135 mm *TL* along the Karwar coast (India). Abdel Razek et al. [80] documented the TL_{50} for females as 96 mm from the Egyptian Mediterranean coast (Table 2).

Several studies [18–22] have noted that reproduction in *M. monoceros* is characterized by continuous spawning throughout the year, with distinct seasonal peaks. Based on the maturity stages of female gonads observed in the present study (Figure 4), the peak spawning season for this species occurs between February and May in the southern Red Sea. This is further supported by the increased presence of young recruiting individuals, particularly between June and September (Figure 3). Similar findings have been reported in other regions, particularly from the Indian waters. Srivatsa [82] observed that the major spawning season of *M. monoceros* in the Gulf of Kutch occurs from February to April. Sukumaran et al. [13] reported comparable results from the Indian coast, identifying January to April as the peak spawning period. In contrast, Yılmaz et al. [8], who studied the spawning season of speckled shrimp in Antalya Bay (Turkey), reported a different pattern. Contrary to the results of the present study, they found that spawning occurs between December and January. Several other studies have confirmed year-round reproduction in *M. monoceros*. Abdallah et al. [27] and Nalini [79] documented continuous reproduction throughout the year, while Nandakumar [16] reported the presence of advanced mature females year-round along the Kerala coast (India). Additionally, Rao [81] noted that the species follows a bi-monthly spawning cycle, spawning once every two months.

The CL_∞ of female speckled shrimp was estimated at 50.66 mm, while that of males was 38.97 mm in the present study. In contrast, Barua et al. [72] reported a much larger CL_∞ of 85.4 mm for both sexes in Bangladesh. In Karwar, India, Sukumaran et al. [13] estimated the asymptotic *TL* (TL_∞) at 210 mm for females and 180 mm for males, whereas Abdallah et al. [28] reported TL_∞ estimates of 193.3 mm for females and 169.5 mm for males from the Gulf of Gabès in the central Mediterranean Sea (Tunisia). This pattern of sexual dimorphism in size, where females attain a larger asymptotic length than males, has been documented in this species by various researchers across different regions [12,13,23,28,76,78] (Table 2). The *K* estimate in the present study was lower in females than in males, indicating that females reach their CL_∞ at a slower rate compared to males. Despite these differences in growth characteristics, the overall growth performance index remained consistent across sexes.

The Z was estimated at 2.38 year^{-1} for females and 3.35 year^{-1} for males in the present study. In comparison, much higher Z values have been reported for *M. monoceros* in several other regions, with males frequently exhibiting higher mortality rates than females. For instance, in the Cochin waters of Kerala, the Z of *M. monoceros* was estimated at 7.06 year^{-1} for females and 7.75 year^{-1} for males [15]. Similarly, Sukumaran et al. [13] observed higher Z values for males across three different regions along both the eastern and western coasts of India, reporting Z estimates of 8.05, 5.85, and 4.28 year^{-1} for males, compared to 6.33, 4.68, and 4.17 year^{-1} for females, respectively. Lalita Devi [12], Rao [24], and George et al. [78] also recorded higher Z values for male shrimp in Indian waters. In the northeastern Mediterranean Sea, Manaşırlı [73] reported a Z value of 5.34 year^{-1} for males, slightly higher than the 5.11 year^{-1} estimated for females. In contrast, findings by Pradana et al. [83] from Indonesia deviated from this pattern, as their study reported a higher Z value for females (3.29 year^{-1}) than for males (1.9 year^{-1}). Overall, the observed sex-based differences in Z are likely driven by the combined effects of life history strategies in male penaeid shrimp. Males typically exhibit shorter lifespans than females, a pattern consistent with their faster growth rates and higher natural mortality [36].

The M values estimated in the present study were 0.96 year^{-1} for females and 1.45 year^{-1} for males, while the corresponding F estimates were 1.42 year^{-1} and 2.20 year^{-1}, respectively. Comparatively, Hadj Hamida [77] estimated higher M values for *M. monoceros* in the Gulf of Gabès, with 1.73 year^{-1} for females and 2.74 year^{-1} for males. Similarly, Manaşırlı [73] reported higher M and F values for females in the northeastern Mediterranean Sea, with M = 1.65 year^{-1} and F = 1.64 year^{-1}. Nandakumar and Srinath [15] calculated even higher F values, reaching 4.71 year^{-1} for females and 5.40 year^{-1} for males in the Cochin waters.

The current E, based on the mortality parameters estimated in the present study, was 0.60 for both sexes, indicating a similar level of fishing pressure on male and female shrimp. This estimate exceeds the LRP of 0.5, clearly signaling a state of overfishing for the *M. monoceros* stock in the southeastern Red Sea. Similar overexploitation trends have been reported for other stocks of this species in different regions. Mwakosya et al. [84] estimated an E value of 0.83 in Tanzanian waters, suggesting intense fishing pressure in that region. High E values, ranging from 0.60 to 0.85 for males and 0.57 to 0.77 for females, have also been recorded by several authors in Indian waters [12,13,15,78]. In Iskenderun Bay, Manaşırlı [73] reported E values of 0.74 for male speckled shrimp and 0.56 for females, further supporting concerns about the overexploitation of this species.

Based on the YpR estimates derived from the Thompson and Bell model analysis for female shrimp (Figure 6), the current F of 1.42 year^{-1} substantially exceeds both $F_{0.1}$ and F_{max}, indicating a state of growth overfishing. This suggests that female speckled shrimp are being harvested at a rate that prevents them from reaching their maximum growth potential, thereby reducing YpR and potentially compromising the long-term productivity of the stock. Further evidence of growth overfishing, beyond the YpR analysis, is apparent from the overall CL distribution of the sampled shrimp. Nearly half of the catch, a remarkably high proportion, consisted of immature speckled shrimp ($CL < CL_{50}$) during the study period (Figure 2). If growth overfishing persists, further increases in fishing pressure will likely lead to recruitment overfishing, where the reproductive capacity of the population is impaired. The results of the SPR analysis provide evidence that recruitment overfishing may already be occurring, as the current F level reduces the SSB to approximately 23% of its unexploited level, which is below both the TRP and the LRP (Figure 6).

It should be noted, however, that the Thompson and Bell YpR model operates under the assumption of a steady state in the fishery, characterized by a constant fishing pattern over time. This implies that F and other input parameters are assumed to have remained relatively stable for a sufficient period, allowing the stock to reach an equilibrium condition. In this study, the absence of a long-term, species-specific time series and reliance on only one year of data present important limitations, particularly for short-lived species like *M. monoceros*. These limitations may introduce biases or uncertainty in the assessment of stock status derived from the YpR model. Therefore, it is essential to continue to expand systematic data collection efforts to enable long-term monitoring and enhance the accuracy of stock assessment.

Even though the YpR and SPR analyses focused solely on females due to their direct contribution to egg production and their critical role in determining the stock's reproductive potential, the assessment can reasonably be extended to the male component of the population. Males are subject to the same fishing pressure and exploitation level as females, and their removal from the population could indirectly affect reproductive success by altering the operational sex ratio and mating dynamics. An overall evaluation of the current E and F levels relative to the selected BRPs strongly indicates that the *M. monoceros* stock in the southeastern Red Sea is overexploited. The fishing pressure on both sexes exceeds sustainable levels, posing a significant risk to the long-term viability of the stock and highlighting the urgent need for immediate management interventions. A substantial reduction in F to the levels corresponding to either $F_{40\%}$ or $F_{0.1}$ is highly recommended, which would require at least a 50% reduction in fishing effort. This could be achieved by implementing measures such as limiting the number of fishing days, reducing the number of active trawlers, or extending the current five-month seasonal closure for shrimp trawling in both Al Qunfudhah and Jizan to provide additional protection during critical spawning and recruitment periods.

Adjustments to gear selectivity could further contribute to the sustainability of the fishery. Using Bayesian inference, Doll et al. [85] demonstrated that when the length at capture substantially exceeds the length at maturity, both yield and SSB can be maintained at high levels, even under high fishing pressure. The selectivity of the commercial trawl gear presently used in the region needs to be improved, as nearly half of the sampled speckled shrimp in this study consisted of immature individuals (Figure 2). In a recent selectivity study conducted in the Jizan fishing grounds, Santucci et al. [35] showed that square mesh codends were significantly more effective than currently used diamond mesh codends in allowing small, immature *M. monoceros* to escape. Transitioning from diamond to square mesh codends could increase the current CL_c of 17 mm, thereby reducing the proportion of immature individuals in the catch and enhancing the reproductive capacity of the stock.

5. Conclusions

This study provides the first comprehensive assessment of the *M. monoceros* fishery in the southeastern Red Sea, encompassing key population parameters and a stock status evaluation. This species was identified as the second most predominant shrimp species, accounting for 10% of the total shrimp catch in the area under investigation. The preliminary stock assessment suggests that current fishing activity in the Saudi Arabian Red Sea exceeds the optimum level and is being conducted at an unsustainable rate, leading to overexploitation. In addition to stock assessment indicators signaling a state of overfishing, nearly half of the total catch of the species consisted of immature individuals, highlighting low gear selectivity as a contributing factor. This underscores the urgent need for immediate management actions to mitigate overfishing and improve sustainability.

Given the economic importance of the shrimp fishery industry, effective management strategies are essential to safeguard its long-term viability. To enhance sustainability, the use of more selective square mesh codends should be implemented to reduce the capture of immature shrimp. Additional measures, such as limiting the number of fishing days, reducing the number of active trawlers, introducing catch quotas, as well as extending the current five-month seasonal closure for trawling, should be considered to provide further protection during critical spawning and recruitment periods.

It is imperative to maintain robust data collection efforts for continuous monitoring and assessment. Additionally, integrating environmental variables into the modeling framework will be essential for a comprehensive understanding of population dynamics. Given the potential impacts of climate change on marine living resources, future research should incorporate climate-related factors into management strategies to ensure the long-term sustainability of the *M. monoceros* fishery.

All these recommended measures and monitoring efforts should be embedded within a multiannual, ecosystem-based management plan developed in collaboration with relevant stakeholders. Such a plan would provide a coordinated framework for implementing regulatory actions, enhancing data collection, and supporting adaptive management through continuous stock assessment and environmental monitoring.

Supplementary Materials: The following supporting information can be downloaded at: https://www.mdpi.com/article/10.3390/biology14040406/s1, Figure S1: Ovarian maturation stages in *M. monoceros*: (a) Stage I—Immature or undeveloped; (b,c) Stage II—Developing; (d,e) Stage III—Mature; (f) Stage IV—Spent; Figure S2: (a) Monthly carapace length frequency distributions of female *M. monoceros* from the southeastern Red Sea, using a bin size of 2 mm; (b) restructured length frequency data with a moving average applied over five carapace length classes (bins), along with the seasonally oscillating von Bertalanffy growth curves (dashed dark red curves). Both plots were generated using the TropFishR package in R; Figure S3: (a) Monthly carapace length frequency distributions of male *M. monoceros* from the southeastern Red Sea, using a bin size of 2 mm; (b) restructured length frequency data with a moving average applied over five carapace length classes (bins), along with the seasonally oscillating von Bertalanffy growth curves (dashed dark red curves). Both plots were generated using the TropFishR package in R; Figure S4: Linearized length-converted catch curve analysis for estimating the total mortality rate (Z) of *M. monoceros*, based on simple linear regression analysis: (a) females and (b) males. CI denotes the confidence interval of Z. The shaded grey area around each regression line represents the 95% confidence bands.

Author Contributions: Conceptualization, E.M.T. and M.D.; methodology, E.M.T. and M.D.; formal analysis, S.G., E.M.T., and Z.O.; investigation, S.G., G.B.M.P., S.K.C., S.T.K., R.G.S., R.C.G., and A.M.S.A.; resources, M.D.; data curation, S.G., G.B.M.P., S.K.C., S.T.K., R.G.S., and Z.O.; writing—original draft preparation, S.G. and E.M.T.; writing—review and editing, G.B.M.P., S.K.C., S.T.K., R.G.S., R.C.G., Z.O., A.M.S.A., and M.D.; visualization, E.M.T.; supervision, E.M.T. and A.M.S.A.; project administration, M.D.; funding acquisition, M.D. All authors have read and agreed to the published version of the manuscript.

Funding: This study was funded by the Ministry of Environment, Water and Agriculture of Saudi Arabia, under Grant number: 83092.

Institutional Review Board Statement: This study did not involve the use of live animals in a laboratory setting. Sampling was conducted onboard commercial shrimp trawlers, and only specimens captured by commercial fishing gear were studied. All sampling procedures complied with local and international regulations governing fishing practices, ensuring that the species were collected exclusively from legally designated fishing areas.

Informed Consent Statement: Not applicable.

Data Availability Statement: The data presented in this study are available on request from the corresponding author (The data are not publicly available due to privacy restrictions set by the funding institution).

Acknowledgments: The authors express their gratitude to the Ministry of Environment, Water, and Agriculture of Saudi Arabia for funding the research project on the assessment of major shrimp stocks in the Red Sea Waters of the Kingdom. They also extend their appreciation to their colleagues at the KAUST KBD Fisheries Program for their valuable assistance and support during the field and lab work.

Conflicts of Interest: The authors declare that the research was conducted in the absence of any commercial or financial relationships that could be construed as potential conflicts of interest.

References

1. Crosnier, A. Les Crevettes Penaeides du Plateau Continental Malgache. *Cah. O.R.S.T.O.M. Ser. Océanogr.* **1965**, 3, 158. Available online: https://horizon.documentation.ird.fr/exl-doc/pleins_textes/cahiers/oceanographie/19483.pdf (accessed on 15 January 2025).
2. Racek, A.A.; Dall, W. Littoral *Penaeinae (Crustacea Decapoda)* from Northern Australia, New Guinea and adjacent waters. *Verh. K. Ned. Akad. Wet. II* **1965**, *56*, 1–119.
3. Fischer, W.; Bianchi, G. *FAO Species Identification Sheets for Fishery Purposes. Western Indian Ocean (Fishing Area 51). Prepared and Printed with the Support of the Danish International Development Agency (DANIDA)*; FAO: Rome, Italy, 1984; Volume 5, Available online: https://openknowledge.fao.org/handle/20.500.14283/ad468e (accessed on 15 January 2025).
4. Balss, H. Bericht über die Crustacea Decapoda (Natantia und Anomura). Zoological results of the Cambridge Expedition to the Suez Canal, 1924. XIV. *Trans. Zool. Soc. London* **1927**, *22*, 221–227. [CrossRef]
5. Holthuis, L.B. Invertébrés Marins: *Les Crevettes*. In *Fiches FAO D'identification des Espèces Pour les Besoins de la Pêche. Zone 37. Révision*; Fisher, W., Bauchot, M.L., Schneider, M., Eds.; FAO: Rome, Italy, 1987; Volume 1, pp. 189–292. Available online: https://www.fao.org/4/x0169f/x0169f25.pdf (accessed on 15 January 2025).
6. Galil, B.; Froglia, C.; Nöel, P. CIESM Atlas of Exotic Species in the Mediterranean. In *Crustaceans: Decapods and Stomatopods*; Briand, F., Ed.; CIESM Publishers: Monaco City, Monaco, 2002; Volume 2, p. 192. Available online: http://hdl.handle.net/10261/2400 (accessed on 15 January 2025).
7. Can, M.F.; Mazlum, Y.; Demirci, A.; Aktaş, M. The catch composition and catch per unit of swept area (CPUE) of penaeid shrimps in the bottom trawls from Iskenderun Bay, Turkey. *Turk. J. Fish. Aquat. Sci.* **2004**, *4*, 87–91. Available online: https://www.trjfas.org/abstract.php?id=235 (accessed on 15 January 2025).
8. Yılmaz, S.; Özvarol, Z.A.B.; Özvarol, Y. Fisheries and Shrimp Economy, Some Biological Properties of the Shrimp *Metapenaeus monoceros* (Fabricius, 1798) in the Gulf of Antalya (Turkey). *J. Anim. Vet. Adv.* **2009**, *8*, 2530–2536.
9. FAO. Tropical Shrimp Fisheries and their Impact on Living Resources. Shrimp Fisheries in Asia: Bangladesh, Indonesia and the Philippines; in the Near East: Bahrain and Iran; in Africa: Cameroon, Nigeria and the United Republic of Tanzania; in Latin America: Colombia, Costa Rica, Cuba, Trinidad and Tobago, and Venezuela, Fishery Technology Service, Fishery Industries Division. FAO Fisheries Department: Rome, Italy, 2001. Available online: https://www.fao.org/4/y2859e/y2859e01.htm (accessed on 15 February 2025).
10. Holthuis, L.B. FAO Species Catalogue. Vol. 1. Shrimps and prawns of the world. An annotated catalogue of species of interest to fisheries. *Crustacea FAO Fish. Synop.* **1980**, *125*, 271. Available online: https://www.fao.org/4/ac477e/ac477e00.htm (accessed on 15 January 2025).
11. CMFRI. *Annual Report 2016-17*; Central Marine Fisheries Research Institute: Kochi, India, 2017; p. 292. Available online: http://eprints.cmfri.org.in/id/eprint/11964 (accessed on 20 January 2025).
12. Lalita Devi, S. Growth and Population Dynamics of the three Penaeid Prawns in the Trawling Grounds off Kakinada. *Indian J. Fish.* **1987**, *34*, 245–263. Available online: https://eprints.cmfri.org.in/350/1/Article_04.pdf (accessed on 20 February 2025).
13. Sukumaran, K.K.; Deshmukh, V.D.; Rao, G.S.; Alagaraja, K.; Sathianandan, T.V. Stock assessment of the penaeid prawn *Metapenaeus monoceros* Fabricius along the Indian Coast. *Indian J. Fish.* **1993**, *40*, 20–34. Available online: http://eprints.cmfri.org.in/id/eprint/248 (accessed on 20 January 2025).
14. Rao, G.S. Studies on the feeding biology of *Metapenaeus monoceros* (Fabricius) along the Kakinada coast. *J. Mar. Biol. Ass. India* **1988**, *30*, 171–181. Available online: http://eprints.cmfri.org.in/id/eprint/1091 (accessed on 20 January 2025).
15. Nandakumar, G.; Srinath, M. Stock assessment of *Metapenaeus monoceros* (Fabricius) from Cochin Waters, Kerala. *Indian J. Fish.* **1999**, *46*, 221–226. Available online: http://eprints.cmfri.org.in/id/eprint/162 (accessed on 20 January 2025).
16. Nandakumar, G. Reproductive biology of the speckled shrimp *Metapenaeus monoceros* (Fabricius). *Indian J. Fish.* **2001**, *48*, 1–8. Available online: http://eprints.cmfri.org.in/id/eprint/124 (accessed on 20 January 2025).
17. Abdallah, O.B.H.H.B.; Jarboui, O.; Missaoui, H.; Hadj Hamida, N.B. Croissance relative, sex-ratio et exploitation de la crevette blanche *Metapenaeus monoceros* (Fabricius, 1798) du golfe de Gabès (Tunisie). *Bull. De L'institut Natl. Sci. Technol. Mer Salammbô* **2003**, *30*, 49–54.

18. George, M.J. On the breeding of penaeids and the recruitment of their post-larvae into the backwaters of Cochin. *Indian J. Fish.* **1962**, *9*, 110–116. Available online: http://eprints.cmfri.org.in/id/eprint/1993 (accessed on 21 January 2025).
19. Gopalakrishnan, V. The biology of the Hooghly Matalah estuarine system (West Bengal, India) with special reference to its fisheries. *J. Mar. Biol. Ass. India* **1971**, *13*, 182–194. Available online: https://www.mbai.org.in/uploads/manuscripts/Article%2039%20(406-410)795054613.pdf (accessed on 20 January 2025).
20. Subrahmanyam, M.; Ganapati, P.N. Observations on the post-larval prawns from the Godavari estuarine systems (Andhra Pradesh, India) with notes on their role in capture and culture fisheries. *J. Mar. Biol. Ass. India* **1971**, *13*, 195–202. Available online: https://www.mbai.org.in/jmbai-details/NjA2/NDM= (accessed on 15 January 2025).
21. Thomas, M.M. Reproduction, fecundity and sex ratio of the green tiger prawn, *Penaeus semisulcatus* de Haan. *Indian J. Fish.* **1974**, *21*, 153–163. Available online: http://eprints.cmfri.org.in/id/eprint/1145 (accessed on 20 January 2025).
22. Ghamrawy, M.S. Studies on the Ecology and Biology of Penaeid Shrimp in the Region of Jeddah, Saudi Arabia. Ph.D. Thesis, University of Wales, Cardiff, UK, 1982; p. 203. Available online: https://drepo.sdl.edu.sa/handle/20.500.14154/48233 (accessed on 15 January 2025).
23. Rao, G.S.; Krishnamoorthi, B. Age and growth of *Metapenaeus monoceros* along the Kakinada Coast. *J. Mar. Biol. Ass. India* **1990**, *32*, 154–161. Available online: http://eprints.cmfri.org.in/id/eprint/1122 (accessed on 21 January 2025).
24. Rao, G.S. Mortality rates and stock assessment of *Metapenaeus monoceros* along the Kakinada coast. *J. Mar. Biol. Ass. India* **1994**, *36*, 1–18. Available online: http://eprints.cmfri.org.in/id/eprint/684 (accessed on 21 January 2025).
25. Dineshbabu, A.P.; Zacharia, P.U.; Thomas, S.; Kizhakudan, S.J.; Rajesh, K.M.; Vivekanandan, E.; Pillai, S.L.; Sivadas, M.; Ghosh, S.; Ganga, U.; et al. Assessment of stock vulnerability of Indian marine fishes to past changes in climate and options for adaptation. *Clim. Res.* **2020**, *79*, 175–192. Available online: https://www.jstor.org/stable/26926470 (accessed on 25 January 2025). [CrossRef]
26. Missaoui, H.; Zaouali, J. Apparition de Nouveaux Crustacés Dans les pêches Crevettières du Golfe de Gabès, Tunisie. *Mar. Life* **1995**, *5*, 27–34. Available online: https://api.semanticscholar.org/CorpusID:132046377 (accessed on 25 January 2025).
27. Abdallah, O.B.H.H.B.; Hamida, N.B.H.; Jarboui, O.; Fiorentino, F.; Missaoui, H. Reproductive biology of the speckled shrimp *Metapenaeus monoceros* (Fabricius, 1798) (Decapoda: Penaeidae) in the Gulf of Gabès (Southern Tunisia, Eastern Mediterranean). *Cah. Biol. Mar.* **2009**, *50*, 231–240.
28. Abdallah, O.B.H.H.B.; Hamida, N.B.H.; Jarboui, O.; Missaoui, H. Age and growth of the speckled shrimp *Metapenaeus monoceros* (Fabricius, 1798) in the Gulf of Gabès (Southern Tunisia, Central Mediterranean). *Cah. Biol. Mar.* **2010**, *51*, 265–274.
29. Alam, M.S.; Liu, Q.; Schneider, P.; Mozumder, M.M.H.; Uddin, M.M.; Monwar, M.M.; Hoque, M.E.; Barua, S. Stock Assessment and Rebuilding of Two Major Shrimp Fisheries (*Penaeus monodon* and *Metapenaeus monoceros*) from the Industrial Fishing Zone of Bangladesh. *J. Mar. Sci. Eng.* **2022**, *10*, 201. [CrossRef]
30. Ghamrawy, M.S. Notes on the Biology of Penaeid Shrimp at Gizan, Red Sea. *J.K.A.U. Mar. Sci.* **1990**, *1*, 67–76. Available online: https://www.kau.edu.sa/Files/320/Researches/51760_21893.pdf (accessed on 15 January 2025). [CrossRef]
31. MEWA. The Statistical Book. 2023; pp. 184–203. Available online: https://www.mewa.gov.sa/ar/InformationCenter/Researchs/Reports/GeneralReports/%D8%A7%D9%84%D9%83%D8%AA%D8%A7%D8%A8%20%D8%A7%D9%84%D8%A5%D8%AD%D8%B5%D8%A7%D8%A6%D9%8A%202023%D9%85.pdf (accessed on 6 March 2025).
32. Kadengal, S.T.; Ceyhan, T.; Tosunoğlu, Z.; Gireesh, S.; Charles, S.K.; Santucci, R.G.; Adam, A.M.S.; Tıraşın, E.M.; Ünal, V.; Dimech, M. Toward Sustainable Fisheries: Assessing Catch per Unit Effort, Retained Bycatch, and Discard Ratios in the Red Sea Shrimp Trawl Fishery of the Kingdom of Saudi Arabia. *Sustainability* **2024**, *16*, 10285. [CrossRef]
33. MEWA. Implementing Regulation of the Agriculture Law Issued by Royal Decree No (M/64) Dated 10/8/1442 Hijri. 23 March 2021. Available online: https://www.mewa.gov.sa/en/InformationCenter/DocsCenter/RulesLibrary/Docs/Implementing%20Regulations%20of%20The%20Agriculture%20Law.pdf (accessed on 2 March 2025).
34. KAUST Beacon Development (KBD). *Assessment and Management of Shrimp Trawl Fisheries in the Red Sea Waters of the Kingdom. Report for the Ministry of Environment, Water and Agriculture (MEWA)*; KBD: Thuwal, Saudi Arabia, 2023.
35. Santucci, R.G.; Tosunoğlu, Z.; Cilbiz, M.; Charles, S.K.; Gireesh, S.; Kadengal, S.T.; Adam, A.M.S.; Tıraşın, E.M.; Ünal, V.; Dimech, M. Assessing Codend Mesh Selectivity: Comparing Diamond and Square Mesh Codend in the Red Sea Shrimp Trawl Fishery of Saudi Arabia. *J. Mar. Sci. Eng.* **2024**, *12*, 1848. [CrossRef]
36. Bauer, R.T. *Shrimps: Their Diversity, Intriguing Adaptations and Varied Lifestyles*; Springer: Cham, Switzerland, 2023; p. 720. [CrossRef]
37. Amanat, Z.; Qureshi, N.A. Ovarian Maturation Stages and Size at Sexual Maturity of *Penaeus indicus* (H. Milne Edwards, 1937) in the Lagoon Water of Sonmiani Bay, Balochistan. *Pakistan J. Zool.* **2011**, *43*, 447–459.
38. Sokal, R.R.; Rohlf, F.J. *Biometry. The Principles and Practice of Statistics in Biological Research*, 4th ed.; W. H. Freeman: New York, NY, USA, 2012; p. 937.
39. Quinn, T.J.; Deriso, R.B. Quantitative Fish Dynamics. Oxford University Press: New York, NY, USA, 1999; p. 560.
40. Tıraşın, E.M. Balık populasyonlarının büyüme parametrelerinin araştırılması. *Turkish J. Zool.* **1993**, *17*, 29–82.

41. Hosmer, D.W.; Lemeshow, S.; Sturdivant, R.X. *Applied Logistic Regression*, 3rd ed.; John Wiley and Sons: Hoboken, NJ, USA, 2013; p. 510. [CrossRef]
42. Aydın, C.M.; Tıraşın, E.M. Information on the deep-water giant red shrimp *Aristaeomorpha foliacea* (Risso, 1827) (Crustacea, Decapoda, Aristeidae) population in Antalya Bay (Eastern Mediterranean Sea, South of Turkey) based on the MEDITS protocol. *Reg. Stud. Mar. Sci.* **2023**, *60*, 102885. [CrossRef]
43. Efron, B.; Tibshirani, R.J. *An Introduction to the Bootstrap*; Chapman and Hall: New York, NY, USA, 1993; p. 436.
44. Somers, I.F. On a seasonally oscillating growth function. *Fishbyte* **1988**, *6*, 8–11. Available online: https://EconPapers.repec.org/RePEc:wfi:wfbyte:39518 (accessed on 20 January 2025).
45. Pauly, D. A Review of the ELEFAN System for Analysis of Length-Frequency Data in Fish and Aquatic Invertebrates. In *Length-Based Methods in Fisheries Research*; Pauly, D., Morgan, G.R., Eds.; Conference Proceedings No. 13; International Center for Living Aquatic Resources Management: Manila, Philippines, 1987; Volume 13, pp. 7–34. Available online: https://hdl.handle.net/20.500.12348/3366 (accessed on 21 January 2025).
46. Beverton, R.J.H.; Holt, S.J. *On the Dynamics of Exploited Fish Populations*; Fisheries Investigations Series 2: London, UK, 1957; Volume 14, pp. 31–35. [CrossRef]
47. Mildenberger, T.K.; Taylor, M.H.; Wolff, M. TropFishR: An R package for fisheries analysis with length-frequency data. *Methods Ecol. Evol.* **2017**, *8*, 1520–1527. [CrossRef]
48. Taylor, M.H.; Mildenberger, T.K. Extending electronic length frequency analysis in R. *Fish. Manag. Ecol.* **2017**, *24*, 330–338. [CrossRef]
49. Pauly, D.; David, N. ELEFAN I, a BASIC Program for the Objective Extraction of Growth Parameters from Length-Frequency Data. *Meeresforsch* **1981**, *28*, 205–211. Available online: http://legacy.seaaroundus.s3.amazonaws.com/doc/Researcher+Publications/dpauly/PDF/1981/Journal+Articles/ELEFAN_I_BasicProgramObjectiveExtractionGrowthParametersLeng.pdf (accessed on 21 January 2025).
50. Pauly, D.; Munro, J.L. Once More on the Comparison of Growth in Fish and Invertebrates. *Fishbyte* **1984**, *2*, 1. Available online: https://www.seaaroundus.org/doc/ResearcherPublications/dpauly/PDF/1984/Other/OnceMoreOnGrowthComparison.pdf (accessed on 20 January 2025).
51. Sparre, P.; Venema, S.C. *Introduction to Tropical Fish Stock Assessment. FAO Fisheries Technical Papers, No. 306/1, Rev. 2 (Part 1)*; FAO: Rome, Italy, 1998; Available online: https://openknowledge.fao.org/handle/20.500.14283/w5449e (accessed on 20 January 2025).
52. Cope, J.M.; Hamel, O.S. Upgrading from M version 0.2: An application-based method for practical estimation, evaluation and uncertainty characterization of natural mortality. *Fish. Res.* **2022**, *256*, 106493. [CrossRef]
53. Maunder, M.N.; Hamel, O.S.; Lee, H.H.; Piner, K.R.; Cope, J.M.; Punt, A.E.; Ianelli, J.N.; Castillo-Jordán, C.; Kapur, M.S.; Methot, R.D. A review of estimation methods for natural mortality and their performance in the context of fishery stock assessment. *Fish. Res.* **2023**, *261*, 106489. [CrossRef]
54. Hamel, O.S.; Cope, J.M. Development and considerations for application of a longevity-based prior for the natural mortality rate. *Fish. Res.* **2022**, *256*, 106477. [CrossRef]
55. Then, A.Y.; Hoenig, J.M.; Hall, N.G.; Hewitt, D.A. Evaluating the predictive performance of empirical estimators of natural mortality rate using information on over 200 fish species. *ICES J. Mar. Sci.* **2015**, *72*, 82–92. [CrossRef]
56. Goodyear, C.P. Spawning stock biomass per recruit in fisheries management: Foundation and current use. *Can. Spec. Publ. Fish. Aquat. Sci.* **1993**, *120*, 67–81.
57. Gabriel, W.L.; Sissenwine, M.P.; Overholtz, W.J. Analysis of spawning stock biomass per recruit: An example for Georges Bank haddock. *N. Am. J. Fish. Manag.* **1989**, *9*, 383–391. [CrossRef]
58. Caddy, J.F.; Mahon, R. *Reference Points for Fisheries Management; FAO Fisheries Technical Paper 347*; FAO: Rome, Italy, 1995; Available online: http://www.fao.org/3/a-v8400e.pdf (accessed on 20 January 2025).
59. Caddy, J.F.; McGarvey, R. Targets or limits for management of fisheries? *N. Am. J. Fish. Manag.* **1996**, *16*, 479–487. [CrossRef]
60. Prager, M.H.; Porch, C.E.; Shertzer, K.W.; Caddy, J.F. Targets and limits for management of fisheries: A simple probability-based approach. *N. Am. J. Fish. Manag.* **2003**, *23*, 349–361. [CrossRef]
61. Clark, W.G. $F_{35\%}$ revisited ten years later. *N. Am. J. Fish. Manag.* **2002**, *22*, 251–257. [CrossRef]
62. Mace, P.M. Relationships between common biological reference points used as thresholds and targets of fisheries management strategies. *Can. J. Fish. Aquat. Sci.* **1994**, *51*, 110–122. [CrossRef]
63. Mace, P.M.; Sissenwine, M.P. How Much Spawning per Recruit is Enough? In *Canadian Special Publication of Fisheries and Aquatic Sciences*; National Research Council of Canada: Ottawa, OT, Canada, 1993; Volume 120, pp. 110–118. Available online: https://fish.gov.au/Archived-Reports/2014/Documents/2014_refs/12.%20Mace%20and%20Sissenwine%201993_in%20Smith%20et%20al%20(eds)%201993.pdf (accessed on 20 January 2025).
64. Walters, C.; Martell, S.J.D. *Fisheries Ecology and Management*; Princeton University Press: Princeton, NJ, USA, 2004; p. 399. Available online: https://api.pageplace.de/preview/DT0400.9780691214634_A39698941/preview-9780691214634_A39698941.pdf (accessed on 20 January 2025).

65. Froese, R.; Winker, H.; Gascuel, D.; Sumaila, U.R.; Pauly, D. Minimizing the impact of fishing. *Fish Fish.* **2016**, *17*, 785–802. [CrossRef]
66. R Core Team. *R: A Language and Environment for Statistical Computing. R Foundation for Statistical Computing*, Version 4.4.1.; R Core Team: Vienna, Austria, 2024. Available online: https://www.R-project.org/ (accessed on 20 January 2025).
67. Butterworth, D.S.; Johnston, S.J.; Brandao, A. Pretesting the Likely Efficacy of Suggested Management Approaches to Data-Poor Fisheries. *Mar. Coast. Fish.* **2010**, *2*, 131–145. Available online: http://hdl.handle.net/11427/17907 (accessed on 21 January 2025). [CrossRef]
68. Watson, R.A.; Cheung, W.W.L.; Anticamara, J.A.; Sumaila, R.U.; Zeller, D.; Pauly, D. Global marine yield halved as fishing intensity redoubles. *Fish Fish.* **2013**, *14*, 493–503. [CrossRef]
69. Pauly, D.; Zeller, D. Catch reconstructions reveal that global marine fisheries catch are higher than reported and declining. *Nat. Commun.* **2016**, *7*, 10244. [CrossRef]
70. Benson, A.J.; Stephenson, R.L. Options for integrating ecological, economic, and social objectives in evaluation and management of fisheries. *Fish Fish.* **2018**, *19*, 40–56. [CrossRef]
71. Punt, A.E. Spatial stock assessment methods: A viewpoint on current issues and assumptions. *Fish. Res.* **2019**, *213*, 132–143. [CrossRef]
72. Barua, S.; Liu, Q.; Alam, M.S.; Schneider, P.; Chowdhury, S.K.; Mozumder, M.M.H. Assessment of Three Major Shrimp Stocks in Bangladesh Marine Waters Using Both Length-Based and Catch-Based Approaches. *Sustainability* **2023**, *15*, 12835. [CrossRef]
73. Manaşırlı, M. Population Dynamics of the *Metapenaeus monoceros* (Fabricius, 1798) in North-eastern Mediterranean Sea. *Iran. J. Fish. Sci.* **2014**, *13*, 955–966. Available online: http://jifro.ir/article-1-1780-en.html (accessed on 18 January 2025).
74. Ghobashi, A.F.A.; Abdel Razek, F.A.; Khafage, A.R.; Taha, S.M. Biological studies of Red shrimp *Metapenaeus monoceros* (Fabricius) in Egyptian Mediterranean waters 1- Morphometric characteristics. *Egypt. J. Aquat. Res.* **2009**, *35*, 527–535.
75. Mustafa, M.G.; Ali, M.S.; Azadi, M.A. Some Aspect of Population Dynamics of Three Penaeid Shrimps (*Penaeus monodon, Penaeus semisulcatus* and *Metapenaeus monoceros*) from the Bay of Bengal, Bangladesh. *Chittagong Univ. J. Sci.* **2006**, *30*, 97–102. Available online: https://hdl.handle.net/20.500.12348/1773 (accessed on 2 March 2025).
76. Bayhan, Y.K.; Ünlüer, T.; Özdöl, M.; Ataoğlu, H.; Çiçek, E. Growth, reproduction, and mortality rates of *Metapenaeus monoceros* (Fabricius, 1798) in 1981-1982 from the north-eastern Mediterranean coasts of Turkey. *Cah. Biol. Mar.* **2023**, *64*, 115–124. [CrossRef]
77. Hadj Hamida, O.B.A.B. Etude biologique et dynamique de la crevette mouchetée *Metapenaeus monoceros* (Fabricius, 1798) exploitée dans le golfe de Gabès, Tunisia. Ph.D. Thesis, National Agronomic Institute of Tunisa, Tunis, Tunisa, 2012; p. 281.
78. George, M.J.; Alagaraja, K.; Sukumaran, K.K.; Nandakumar, G.; Ramamurthy, S.; Telang, K.Y. The present status of shrimp trawling and its impact on shrimp stocks of Karnataka Coast. In *Seminar Proceedings Problems and Prospects of Marine Fishing and Fish Processing in Karnataka*; College of Fisheries: Mangalore, India, 1988; pp. 1–14.
79. Nalini, C. Observations on the Maturity and Spawning of *Metapenaeus monoceros* (Fabricius) at Cochin. *Indian J. Fish.* **1976**, *23*, 23–30. Available online: http://eprints.cmfri.org.in/id/eprint/949 (accessed on 7 March 2025).
80. Abdel Razek, F.A.; Ghobashi, A.F.A.; Khafage, A.R.; Taha, S.M. Biological studies of Red shrimp *Metapenaeus monoceros* (Fabricius) in Egyptian Mediterranean waters 2-Maturation and Spawning. *Egypt. J. Aquat. Res.* **2009**, *35*, 537–548.
81. Rao, G.S. Studies on the reproductive biology of the brown prawn *Metapenaeus monoceros* (Fabricius: 1798) along the Kakinada coast. *Indian J. Fish.* **1989**, *36*, 107–123. Available online: http://eprints.cmfri.org.in/id/eprint/305 (accessed on 7 March 2025).
82. Srivatsa, K.R. *A Survey and Comparative Analysis of the Prawn (Shrimp) Fishery of the Gulf of Kutch in Saurashtra, India*; Department of Industries and Supplies, Government of Saurashtra: Gujarat, India, 1953.
83. Pradana, R.T.; Syahrir, R.M.; Abdunnur, A. Population Dynamics of Yellow Prawn (*Metapenaeus monoceros*) Captured at Night in the Waters of Samboja Kuala, Kutai Kartanegara Regency. *Nusant. Trop. Fish. Sci. (Ilmu Perikan. Trop. Nusant.)* **2023**, *2*, 113–121. [CrossRef]
84. Mwakosya, C.A.; Kuguru, B.; Senkondo, E.; Kulekana, J.J.; Ulotu, E.; Sululu, J.; Mhitu, H.; Matola, H.; Budeba, Y.L.; Ngatunga, B.P.; et al. The Assessment of the Effectiveness of Closure of Commercial Prawn Trawling in Tanzania Coastal Waters. Report for MACEMP Project Submitted to Ministry of Livestock and Fisheries Development. Tanzania Fisheries Research Institute (TAFIRI): Dar es Salam, Tanzania, 2010; p. 95. Available online: https://www.researchgate.net/publication/265133904 (accessed on 13 February 2025).
85. Doll, J.C.; Lauer, T.E.; Clark-Kolaks, S. Yield-per-recruit modeling of two piscivores in a Midwestern reservoir: A Bayesian approach. *Fish. Res.* **2017**, *191*, 200–210. [CrossRef]

biology

MDPI

Article

Cryopreservation Protocol Optimization for *Penaeus monodon* Sperm: Reagent Screening and Parameter Refinement

Dewei Kong [1,†], Song Jiang [1,2,†], Jianzhi Shi [1], Qibin Yang [1], Jianhua Huang [1], Yundong Li [1], Yangyang Ding [1], Jieyi Wang [1], Xinyu Qi [1], Tianmi Liu [3,*] and Falin Zhou [1,2,*]

1 South China Sea Fisheries Research Institute, Chinese Academy of Fishery Sciences, Key Laboratory of South China Sea Fishery Resources Exploitation and Utilization, Ministry of Agriculture and Rural Affairs, Guangzhou 510300, China; kdw0204@163.com (D.K.); tojiangsong@163.com (S.J.); shijianzhi1989@163.com (J.S.); yangqibin1208@163.com (Q.Y.); hjh210440@sina.com.cn (J.H.); liyd2019@163.com (Y.L.); gaoguimao@gmail.com (Y.D.); 15020485966@163.com (J.W.); qjgqxy@163.com (X.Q.)

2 State Key Laboratory of Mariculture Biobreeding and Sustainable Goods, Qingdao 266071, China

3 Fisheries Technology Extension Center of Hainan Province, Haikou 570311, China

* Correspondence: ltm600@163.com (T.L.); zhoufalin0925@163.com (F.Z.)

† These authors contributed equally to this work.

Simple Summary: This study addresses critical issues in the cryopreservation protocol for *Penaeus monodon* shrimp sperm by proposing an optimized long-term preservation strategy. While previous studies validated the efficacy of 5% DMSO in sperm cryopreservation and established corresponding protocols, they lacked systematic analysis of ultrastructural integrity, enzymatic activity, and long-term preservation outcomes. To bridge these gaps, we present three innovations: First, the cryoprotectant formulation was optimized through the integration of 10% DMSO with natural seawater as a biocompatible carrier. Second, programmable controlled-rate freezing technology was implemented to minimize cryodamage during the freezing process. Third, post-thaw sperm quality was comprehensively evaluated via ultrastructural microscopy (SEM/TEM) coupled with metabolic activity profiling. After 15 days of cryopreservation, our protocol achieved 85% sperm viability with only minor acrosomal membrane damage, outperforming conventional methods in functional preservation. These advancements not only extend the effective cryopreservation window but also elucidate mechanisms underlying cryodamage mitigation, addressing critical gaps in crustacean reproductive biotechnology. By enabling reliable preservation of genetic resources, this work provides a technical foundation for sustainable aquaculture and shrimp germplasm bank development.

Abstract: *Penaeus monodon* (black tiger shrimp) is one of the important shrimp species in aquaculture. Cryopreserving its sperm not only provides technical support for breeding but also effectively prevents the decline of genetic resources, promoting the sustainable development of its aquaculture industry. This study screened different types of diluents, cryoprotectants, and concentrations and explored equilibration time, cooling protocols, and thawing conditions, ultimately determining the optimal cryopreservation protocol for *P. monodon* sperm. The results showed that the optimal cryopreservation protocol involved using natural seawater as the diluent with 10% dimethyl sulfoxide (DMSO) as the cryoprotectant, in which the sperm suspension and cryoprotectant were mixed at a 1:1 (v/v) ratio and equilibrated at 4 °C for 30 min. Subsequently, cooling was performed using a programmable controlled-rate freezer: the temperature was reduced to −20 °C at −5 °C/min and held for 5 min; then cooled to −80 °C at −10 °C/min and held for 5 min; finally, the temperature was reduced to −180 °C at −20 °C/min. After cooling, the sperm samples were transferred to liquid nitrogen for long-term storage. The results demonstrated that thawing in a 37 °C water bath achieved the highest sperm motility compared to conditions at 27 °C, 32 °C, 42 °C, and 60 °C. After 15 days of liquid nitrogen storage, the sperm survival rate was 53.33 ± 9.18%. Scanning electron microscopy (SEM) and transmission electron microscopy (TEM) observations revealed that the sperm structure was intact before freezing, with a rounded head, a distinct acrosomal spike anterior to the head, a concentrated nucleus in the head, dense chromatin, and a smooth cell membrane surface. However, after freezing and thawing, the acrosomal spikes of some sperm were fractured, and the membrane structure was damaged. Enzyme activity analysis showed that during liquid nitrogen storage from 0 to 15 days, the enzyme activity of alkaline phosphatase (AKP) and acid phosphatase (ACP) in sperm gradually increased with significant differences observed compared to day 0 ($p < 0.05$). The activity of malondialdehyde (MDA) showed a gradual increase at 0, 5, and 10 days, but then decreased at day 15. The enzyme activity of catalase (CAT) showed no significant changes from 0 to 10 days ($p > 0.05$) but significantly increased on day 15 ($p < 0.05$). The activity of total superoxide dismutase (T-SOD) showed no significant changes from 0 to 5 days ($p > 0.05$) but significantly increased from days 10 to 15 ($p < 0.05$). These findings provide valuable insights into the cryopreservation of *P. monodon* sperm and will guide the optimization of cryoprotectant combinations and freezing protocols aimed at improving sperm survival rates.

Keywords: *Penaeus monodon*; cryopreservation; sperm; ultrastructure; cryoinjury mechanisms

1. Introduction

Penaeus monodon (black tiger shrimp) holds significant economic importance in China's mariculture sector [1]. Remarkable advancements have been achieved and implemented in the full artificial breeding technology of *P. monodon*, both in production practices and research endeavors. The widespread promotion of new, artificially bred varieties of *P. monodon* has catalyzed the rapid growth of the industry. However, in practical production and research scenarios, the effective utilization period of male shrimp is relatively brief, generally confined to the reproductive season. Consequently, the long-term preservation of reproductive cells from high-quality germplasm resources has emerged as a pressing scientific challenge within the *P. monodon* industry. Furthermore, for the conservation of germplasm resources and the development of novel germplasm, it is imperative to ensure the prolonged preservation of reproductive cells to optimize resource utilization.

Cryopreservation technology, as a reliable method, plays an important role in aquaculture [2]. Sperm cryopreservation technology not only allows long-term storage of sperm, preserving its desirable traits, but also helps avoid issues related to in vivo preservation caused by factors such as gene loss, generation gaps, or genetic drift [3]. This technology significantly reduces breeding costs, addresses issues of geographic isolation and asynchronous development of male and female gametes, and provides convenience for genetic breeding research. In addition, cryopreservation technology can also minimize the impact of external environmental factors on germplasm resources, prevent the spread of animal-derived diseases, provide high-quality materials for artificial breeding and the creation of new germplasm, expand the selection of excellent trait populations, and open up new ways to establish a germplasm resource bank [4,5].

To date, research on fish sperm cryopreservation has matured, with successful cryopreservation achieved in at least 200 species of fish [6,7], providing significant value for the preservation of fish germplasm resources and genetic breeding research. Furthermore, a reliable procedure for cryopreservation of sperms has been documented in decapod crustaceans.

Examples include *Sicyonia ingentis* [8], *Macrobrachium rosenbergii* [9–11], *Fenneropenaeus indicus* [12], *Penaeus chinensis* [13], *Penaeus monodon* [14,15], and *Litopenaeus vannamei* [16].

The cryopreservation technology for *P. monodon* sperm has gradually developed into a defined research framework. Early studies primarily focused on screening cryoprotectant types and concentrations. Vuthiphandchai et al. [15] first established a cryopreservation protocol using 5% dimethyl sulfoxide (DMSO), demonstrating its efficacy in maintaining sperm viability for up to 60 days. However, this study did not explore cryoinjury mechanisms or long-term preservation effects (>90 days). Subsequent research highlighted distinctions between spermatophore-based and individual sperm cryopreservation. Bart et al. [14] validated the effectiveness of 5% DMSO in spermatophore cryopreservation, proposing that the structural integrity of spermatophores may enhance sperm survival by providing physical protection. Nevertheless, the advantages of spermatophore preservation may obscure unique challenges inherent to individual sperm cryopreservation, such as variability in osmotic tolerance and increased membrane fragility. Consequently, protocol optimization for single-sperm cryopreservation requires independent investigation to address these specific cryobiological constraints.

Recent advancements have expanded cryoprotectant selection. Rosmiati et al. [17] systematically compared the cryoprotective efficacy of DMSO, methanol, and glycerol on *P. monodon* sperm, identifying 10% glycerol as superior for maintaining sperm viability in liquid nitrogen storage. However, this evaluation was limited to short-term effects within 30 days. Grace and Natarajan [18] reported enhanced protection of sperm motility using 10% DMSO compared to lower concentrations, yet failed to assess potential structural damage caused by elevated cryoprotectant concentrations. These findings underscore a trade-off between the cryoprotectant concentration and cryoinjury severity. Current research remains constrained by short-term evaluations of single cryoprotectants, lacking comprehensive analyses of combined cryoprotectants, freeze–thaw protocols, and cellular damage mechanisms.

Although existing studies have identified critical parameters for protocol optimization (e.g., cryoprotectant selection, extender composition, cooling rates), technical translation remains hindered by systemic limitations. As noted by Aquino et al. [19] in their review, key challenges in crustacean sperm cryopreservation include membrane lipid sensitivity to hypothermic stress and the absence of standardized damage assessment metrics. Current methodologies predominantly rely on superficial indicators, such as survival rates, while quantitative analyses of mitochondrial function, membrane integrity, and enzymatic activities (e.g., ATPase, antioxidant enzymes) remain underdeveloped. This knowledge gap impedes the mechanistic understanding of cryoinjury and consequently restricts precision optimization of cryoprotectant formulations.

Although previous studies have investigated cryoprotectants and their concentrations for sperm cryopreservation in *P. monodon*, further optimization remains necessary. This study aimed to improve sperm survival rates and biological activity during long-term cryopreservation by refining cryoprotectant formulations, diluent compositions, and freeze–thaw protocols. Ultrastructural analysis and enzyme activity assays were integrated to elucidate cellular damage mechanisms during cryopreservation, thereby providing a theoretical foundation for protocol development. The findings are expected to advance reproductive regulation, selective breeding of superior varieties, and sustainable aquaculture practices for *P. monodon*.

2. Materials and Methods

2.1. Broodstock Management

The *P. monodon* used in the experiment was sourced from the Shenzhen Experimental Base of the South China Sea Fisheries Research Institute, Chinese Academy of Fishery Sciences. Prior to the experiment, the shrimp were cultured in filtered, aerated seawater at 28 ± 1 °C, with a pH of 7.8–8.2 and a salinity of 30–33 ppt, and were fed commercial shrimp formula feed (produced by Dongteng Feed Company, Zhanjiang, China). The main nutritional components of this feed include: crude protein 41.0%, crude fat 4.0%, crude fiber 5.0%, crude ash 16.0%, moisture 12.0%, calcium 2.20%, phosphorus 4.00%, and it is rich in various vitamins, such as vitamin A, D3, E, B1, B2, and B6. They were fed three times per day, with each feeding amounting to 3–5% of body weight. The sperm bundle collection of *P. monodon* was carried out using the squeezing method [20]. Thirty sexually mature male shrimp were selected, and all tools (collection tubes, pipettes, etc.) were sterilized at high temperatures or treated with 70% ethanol before the experiment. The entire operation was completed in a clean laboratory, with work surfaces and equipment disinfected using ultraviolet light, and the operators wore sterile gloves and masks. The specific procedure was as follows: the shrimp body was rinsed with sterile seawater, and the development of the testes at the base of the fifth pereiopod was visually inspected (white opaque masses were the determining indicator). A 5% iodine solution or potassium permanganate solution was used to disinfect the genital pore surface. The sperm bundle waas expelled by applying directional pressure with the thumb and index finger on the base of the fifth pereiopod, and the sample was immediately collected with sterile tweezers to avoid contamination.

2.2. Sperm Collection

Fresh sperm bundles of *P. monodon* were dried with sterile filter paper to remove surface moisture (the processing time was strictly controlled to within 2 min to prevent sperm vitality decline due to exposure to room temperature) and then weighed and transferred to a 400-mesh gauze screen. The gauze screen was immersed in a beaker containing 4 °C pre-treated seawater (filtered through sand and disinfected by UV), and sperm were quickly squeezed out of the sperm bundles using sterile tweezers. The entire process was carried out under ice bath conditions (with sperm bundle sample temperature maintained at 4 °C to ensure cell viability, and the operation time not exceeding 3 min). The resulting homogenate was transferred to a 2 mL centrifuge tube and purified by gradient centrifugation: the first stage was centrifugation at 500× g for 10 min to remove seminal plasma proteins and cell debris; the second stage was centrifugation at 2000× g for 8 min to enrich intact sperm cells. The centrifugation parameters were optimized based on the cell sedimentation curve analysis from pre-experiments.

2.3. Sperm Viability Assessment

Sperm vitality was assessed using the trypan blue staining method [7,12,21,22] to quantitatively measure the survival rate (n = 30), using a hemocytometer (25 × 16 small squares, 0.1 mm^3) for detection. Before the experiment, the sperm suspension was diluted to 10^7 sperm/mL and mixed with trypan blue staining solution at a 1:1 ratio, followed by a 3-min staining period before microscopic observation. Viable sperm remained unstained due to their intact structure, while dead sperm appeared blue. For each sample, 500 sperm were randomly counted, and only samples with a survival rate ≥85% were used for further study. All experiments were repeated independently three times and performed using the trypan blue staining cell viability detection kit (Beyotime, C0011, Hainan Luhengyuan Biological Technology Co., Ltd., Haikou, China) to ensure data reliability and experimental stability.

2.4. Screening of Cryopreservation Conditions for *P. monodon*

2.4.1. Determination of Suitable Diluents

Based on preliminary screening, this study selected natural seawater (collected from the coastal waters of Dapeng Peninsula, Shenzhen, China, filtered through a 0.45 μm membrane and sterilized at 121 °C under high pressure), Ringer's solution, Hanks' balanced salt solution (HBSS), and D-Hanks balanced salt solution (all purchased from Shantou Xilong Scientific Co., Ltd., Shantou, China, analytical grade with purity $\geq$ 99%) as the basic diluents. All solutions were independently tested three times for osmolarity using an osmometer (Gonotec Osmomat 3000): natural seawater had an osmolarity of 1020 $\pm$ 15 mOsm/L, Ringer's solution was 305 $\pm$ 3 mOsm/L (containing Ca^{2+} 2.4 mM, K^{+} 5.0 mM), Hanks' balanced salt solution was 290 $\pm$ 2 mOsm/L, and D-Hanks balanced salt solution was 295 $\pm$ 4 mOsm/L. Before use, all solutions were calibrated with a pH meter (Mettler Toledo FE28) to pH 7.8 $\pm$ 0.1 and stored in the dark at 4 °C for no more than 72 h to ensure the stability of their physicochemical properties. Dimethyl sulfoxide (DMSO, from the same brand) was selected as the cryoprotectant. DMSO plays a key protective role in sperm cryopreservation by inhibiting ice crystal formation and balancing the osmotic pressure inside and outside the cells. The cryoprotectant solution was prepared by mixing DMSO with the four diluents at final concentration gradients of 5%, 10%, and 15% (for example, the 5% concentration group was mixed at a DMSO: diluent = 1:19 volume ratio). The sperm suspension (500 μL) was mixed with an equal volume of cryoprotectant (1:1 ratio) on ice, and the mixture was left to stand for 30 min to achieve osmotic equilibrium. A three-phase cooling program was then applied: first, cooling at a rate of -5 °C/min to -20 °C and holding for 5 min; second, cooling at -10 °C/min to -80 °C and holding for 5 min; finally, cooling at -20 °C/min to -180 °C before immersion in liquid nitrogen for long-term storage. During thawing, the samples were rapidly rewarmed in a 37 °C water bath for 2 min, and three biological replicates were set for each experimental group.

2.4.2. Determination of Suitable Cryoprotectants

Methanol, DMSO, ethylene glycol, 1,2-propanediol, and glycerol were chosen as cryoprotectants, and concentrations of 5%, 10%, and 15% were prepared. Natural seawater served as the diluent, and the sperm was mixed with the cryoprotectant in a 1:1 ratio. The cooling process was carried out as previously described, and the samples were then stored in liquid nitrogen. After one hour, the samples were thawed, and sperm viability was measured. This experiment was repeated three times for accuracy.

2.4.3. Different Equilibration Times at 4 °C

Using natural seawater as the diluent, sperm was mixed with a 10% DMSO cryoprotectant in a 1:1 ratio and then allowed to stand at 4 °C for various durations: 0, 10, 20, 30, and 40 min. Following the cooling protocol (P-2), the samples were stored in liquid nitrogen. After one hour, the samples were thawed, and sperm viability was measured. This experiment was repeated three times to ensure accuracy.

2.4.4. Development of Cryopreservation Protocol

Using natural seawater as the diluent, sperm was mixed with a 10% DMSO cryoprotectant in a 1:1 ratio and allowed to stand at 4 °C for 30 min. Various cooling protocols were then applied to the samples, which were subsequently stored in liquid nitrogen for one hour. After this period, the samples were thawed in a 37 °C water bath, and sperm viability was measured. This experiment was repeated three times to ensure accuracy.

The experiment utilized a programmed cooling device (Model CJL37, produced by Chengdu Haier Biomedical Science Co., Ltd., Chengdu, China) and a liquid nitrogen tank (Model YDZ-100, also produced by Chengdu Haier Biomedical Science Co., Ltd., Chengdu, China) for programmed cooling of experimental samples, configuring four different freezing cooling protocols (Table 1). The chamber of the cooling device was dried for 30 min to ensure it remained dry, and the cooling program was set to maintain equilibration at 4 °C before placing the sperm samples from the four treatment groups inside. After cooling the sperm samples to the set temperature using the programmed cooling method, they were quickly transferred to the liquid nitrogen tank. The samples were thawed in a 37 °C water bath, and three parallel samples from the four treatment groups were placed in the water bath and gently stirred for uniform thawing for 2 min before measuring the survival rate. Three biological replicates were performed for each experimental group.

Table 1. Cooling protocols for cryopreservation of *Penaeus monodon* sperm using a programmable cryocooler.

Serial Number	Cooling Program
P-1	−5 °C/min to −80 °C; hold at −80 °C for 5 min; −20 °C/min to −180 °C; stored in liquid nitrogen
P-2	−5 °C/min to −20 °C; hold at −20 °C for 5 min; −10 °C/min to −80 °C; hold at −80 °C for 5 min; −20 °C/min to −180 °C; stored in liquid nitrogen
P-3	−5 °C/min to −20 °C; hold at −20 °C for 5 min; −20 °C/min to −80 °C; hold at −80 °C for 5 min; stored in liquid nitrogen
P-4	−5 °C/min to −80 °C; stored in liquid nitrogen

To systematically investigate the effects of different cooling pathways on sperm cryopreservation efficacy, this study established a multi-stage gradient cooling protocol from 4 °C to −180 °C, comprising four distinct programs (P-1 to P-4). Compared with the conventional two-step cooling method (slow cooling to a specific temperature followed by direct immersion in liquid nitrogen), this protocol introduces additional cooling phases and rate variations to optimize cellular physiological adaptation during phase transitions and mitigate cryoinjury.

Program P-1 served as the baseline reference, employing a single cooling rate (−5 °C/min) to −80 °C before deep cryopreservation. Programs P-2 and P-3 incorporated a 5-min equilibrium plateau at −20 °C, followed by subsequent cooling at −10 °C/min and −20 °C/min respectively, enabling comparative analysis of intermediate cooling rates on cell membrane stability and ice crystal regulation. Program P-4 represented the extreme control, implementing direct cooling from ambient temperature to −80 °C at −5 °C/min, followed by immediate immersion in liquid nitrogen (−196 °C). All programs included a standardized 5-min equilibration at −80 °C, a parameter derived from preliminary experiments on cryoprotectant permeation kinetics, designed to mitigate osmotic stress from thermal shocks and ensure metabolic arrest.

Notably, all protocols concluded with a rapid cooling phase (−20 °C/min) to traverse the ice crystal formation-sensitive zone (−60 °C to −120 °C), thereby suppressing large ice crystal growth and reducing structural damage. This multi-stage design constitutes a systematic optimization based on prior ice morphology observations, membrane fluidity assessments, and thermodynamic simulations. By fine-tuning gradient nodes, equilibrium durations, and cooling rate combinations, the protocol aims to precisely regulate intracellular/extracellular water transport and ice nucleation processes, ultimately enhancing post-thaw cell viability and integrity. The four-phase cooling strategy seeks to overcome limitations of conventional two-step methods through refined thermal control tactics, theoretically better aligning with cellular physiological demands during dynamic cooling processes.

2.4.5. Determination of Sperm Thawing Temperature

Using natural seawater as the diluent, sperm was mixed with a 10% DMSO cryoprotectant solution in a 1:1 ratio and allowed to stand at 4 °C for 30 min. Following the standard cooling protocol, the samples were stored in liquid nitrogen for 1 h. After this, the samples were thawed in water baths at temperatures of 27 °C, 32 °C, 37 °C, 42 °C, and 60 °C, respectively, and then sperm viability was measured. This entire experiment was repeated three times to ensure accuracy.

2.4.6. Cryopreservation Duration in Liquid Nitrogen

Using natural seawater as the diluent, sperm was mixed with a 10% DMSO cryoprotectant solution in a 1:1 ratio and allowed to stand at 4 °C for 30 min. This cooling protocol adhered to the general procedure, and subsequently, the samples were stored in liquid nitrogen for 15 days. Every 3 days, samples were taken, thawed in a 37 °C water bath, and then sperm viability was measured. This entire experiment was repeated three times to ensure accuracy.

2.5. *Sperm Morphology and Ultrastructure Analysis*

Sperm morphology and ultrastructure analyses were conducted based on the descriptions provided by Ulloa-Rodríguez et al. [23] and Quagio-Grassiottoand Oliveira [24], with some necessary modifications. After the sperm samples frozen in liquid nitrogen were rapidly rewarmed in a 37 °C water bath for 2 min, they were immediately centrifuged at 300× *g* for 5 min to remove the cryoprotectant. Subsequently, the samples were washed three times with pre-chilled 0.1 mol/L PBS (pH = 7.4) at 4 °C to maintain membrane stability, following the same pretreatment procedure as for fresh samples. For scanning electron microscopy (SEM), the sperm-containing supernatant was carefully placed in a clean centrifuge tube and centrifuged at 300× *g* for 5 min to collect the pellet. The sperm were then resuspended in electron microscope fixative (Servicebio, G1102, Wuhan Saiwei Biotechnology Co., Ltd, Wuhan, China), fixed at room temperature for 2 h, and stored at 4 °C. Subsequently, the sperm samples were blocked in 1% OsO_4 in 0.1 M PBS (pH = 7.4) at room temperature for 1–2 h. After washing three times with PBS (15 min each), the samples were gradually dehydrated in a series of ethanol solutions with increasing concentrations (30%, 50%, 70%, 80%, 90%, 95%, and 100%, twice; 15 min each). The samples were then treated with isoamyl acetate (Sinopharm Chemical Reagent Co., 10003128, Beijing, China) for 15 min and dried using a critical point dryer (Quorum, K850, Quorum Technologies, Newhaven, UK). The dried samples were securely attached to metal stubs using carbon adhesive tabs and coated with gold for 30 s using an ion sputter coater (Hitachi, MC1000, Hitachi High-Technologies Corporation, Tokyo, Japan). Finally, the samples were imaged using a scanning electron microscope (Hitachi, SU8100, Hitachi High-Technologies Corporation, Tokyo, Japan) at Wuhan Saiwei Biotechnology Co., Ltd.

For transmission electron microscopy (TEM), the sperm suspension samples were fixed in electron microscope fixative at room temperature for 1 h and then transferred to fresh TEM fixative in Eppendorf (EP) tubes for further fixation. The samples were washed three times with 0.1 M PBS (pH 7.4) for 15 min each. The sperm samples were then fixed with 1% OsO_4 in PBS at room temperature for 2 h and washed again with PBS three times (15 min each). The samples were gradually dehydrated in a series of ethanol solutions with increasing concentrations (30%, 50%, 70%, 80%, 95%, and 100%, twice; 20 min each) and in acetone (Sinopharm Chemical Reagent Co., 10000418, Beijing, China) for 15 min. The samples were then infiltrated and embedded in resin using the following steps: a mixture of acetone ands 812 resin (SPI, 90529-77-4, SPI Supplies, West Chester, PA, USA) in a 1:1 ratio was soaked at 37 °C for 2–4 h; a mixture of acetone and 812 resin in a 1:2 ratio was soaked at 37 °C overnight; and pure EMBed 812 was reacted at 37 °C for 5–8 h. After that, pure EMBed 812 was poured into embedding molds, and the sperm were carefully inserted into the resin and stored overnight at 37 °C. The embedding molds with resin and samples were then transferred to a 65 °C oven for polymerization for more than 48 h. The resin blocks were subsequently removed from the molds and stored at room temperature for later use. The resin blocks were sectioned into ultrathin slices (60–80 nm) using an ultramicrotome (Leica, Leica UC7, Leica Microsystems, Wetzlar, Germany), and the sections were floated onto 150-mesh copper grids using Formvar films. The copper grids were stained in the dark with a 2% uranyl acetate saturated ethanol solution for 8 min, washed three times with 70% ethanol, and rinsed three times with ultrapure water. The grids were then stained with 2.6% lead citrate in the dark for 8 min and washed three times with ultrapure water. After drying with filter paper, the copper grids were placed on grid racks and left to dry at room temperature overnight. The grids were observed using a transmission electron microscope (Hitachi, HT7800, Hitachi High-Technologies Corporation, Tokyo, Japan), and the images were captured by Wuhan Sair Biotechnology Co., Ltd., Wuhan, China By comparing the integrity of the plasma membrane, organelle structure, and ice crystal damage characteristics before and after freezing, the study systematically evaluates the effects of different freezing protocols on the ultrastructure of sperm.

2.6. Measurement of Antioxidant Enzyme Activity

2.6.1. Sperm Cryopreservation and Pretreatment

Based on the optimal cryopreservation protocol for sperm survival rate obtained through preliminary screening, this study applied the protocol to evaluate the long-term preservation of *P. monodon* sperm samples. The specific procedure was as follows: the sperm suspension was mixed with 10% DMSO cryoprotectant at a 1:1 volume ratio and subjected to programmed gradient cooling. First, the temperature was reduced from 0 °C to −20 °C at a rate of −5 °C/min and held for 5 min to promote cryoprotectant permeation and cell dehydration. Then, the temperature was reduced from −20 °C to −80 °C at a rate of −10 °C/min and held for 5 min. Finally, the temperature was rapidly lowered to −180 °C at −20 °C/min, and the samples were transferred to liquid nitrogen for long-term storage. To evaluate preservation efficacy, thawing tests were performed at 5, 10, and 15 days of storage. The samples were quickly removed from liquid nitrogen and placed in a 37 °C water bath for 2 min of shaking to thaw. Immediately after, they were centrifuged at 300× g for 5 min to remove the DMSO-containing supernatant. Then, 0.5 mL of pre-chilled PBS (4 °C, pH 7.4) was added to the sperm pellet and gently resuspended to ensure a cell density $\geq 10^6$ cells/mL. To fully release intracellular components, the suspension was transferred to a pre-chilled 2 mL glass homogenizer and manually homogenized in an ice–water mixture for 3 min using an intermittent cycle (10 s of homogenization/5 s of pause), until >95% of the cells were broken, as confirmed under the microscope. The homogenate was filtered through a 0.22 μm filter membrane, then aliquoted. Each experimental group was independently repeated 6 times to control operational deviations.

2.6.2. Enzyme Activity Measurement

The activities of several enzymes, including catalase (CAT), total superoxide dismutase (T-SOD), alkaline phosphatase (AKP), and acid phosphatase (ACP), as well as the level of malondialdehyde (MDA), were determined in both fresh and frozen-thawed *P. monodon* sperm samples using enzyme activity assay kits provided by the Nanjing Jiancheng Bioengineering Institute. All procedures were carried out in strict accordance with the manufacturer's instructions, and each assay was repeated three times to ensure consistency and accuracy. Additionally, each experimental group was replicated six times to further enhance the reliability of the results.

2.7. Data Analysis

Statistical analysis was performed using SPSS 21.0 (SPSS Inc., Michigan Avenue, Chicago, IL, USA). A one-way analysis of variance (ANOVA) was used to test for effects. When statistically significant differences were found ($p < 0.05$), Duncan's multiple range test was used for further comparison of group means. The results are expressed as the mean ± SD ($n = 3$).

3. Results

3.1. Suitable Diluents

Two-way ANOVA results showed significant differences in the survival rate of *P. monodon* in different types and concentrations of diluents ($p < 0.05$). Specifically, the type of diluent (natural seawater, Ringer's solution, D-Hanks solution) and DMSO concentration (5%, 10%, 15%) had a significant effect on the sperm survival rate. Additionally, the interaction between concentration and diluent was also significant ($p < 0.05$). Under the same DMSO concentration, the group with natural seawater as the diluent performed better than the other diluent groups. Particularly, at a 10% DMSO concentration, the natural seawater group had the highest sperm survival rate, reaching 73.08 ± 7.31%, significantly higher than the other diluent groups ($p < 0.05$). The sperm survival rate in the Ringer's solution (10% DMSO) group was 56.91 ± 6.83%, which performed relatively well. However, the sperm survival rate in the D-Hanks solution group with 15% DMSO was only 16.13 ± 2.12%, the poorest result (Figure 1A). Therefore, natural seawater is considered the most suitable diluent for freezing and preserving *P. monodon* sperm.

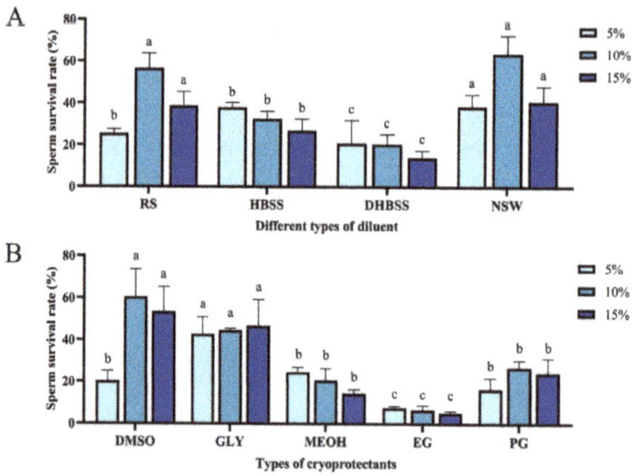

Figure 1. (**A**,**B**) both use two-way ANOVA to show the effects of different DMSO concentrations (5%, 10%, 15%), diluents (RS, HBSS, DHBSS, NSW), and cryoprotectants (glycerol, DMSO, methanol, ethylene glycol, propylene glycol) on the sperm survival rate of *Penaeus monodon* after freezing. The analysis results show that both concentration and the type of diluent/cryoprotectant significantly affect sperm survival rate ($p < 0.05$), and there is a significant interaction between concentration and diluent/cryoprotectant ($p < 0.05$). The letters (a, b, c) in the figure indicate significant differences between different groups ($p < 0.05$). The sample size for all experimental groups was n = 30. The diluents include RS (Ringer's solution), HBSS (Hanks' balanced salt solution), DHBSS (DHanks balanced salt solution), and NSW (natural seawater); cryoprotectants include glycerol (Gly), DMSO, methanol (MeOH), ethylene glycol (EG), and propylene glycol (PG).

3.2. Suitable Cryoprotectants

The effect of different cryoprotectants (glycerol, dimethyl sulfoxide DMSO, ethylene glycol, etc.) at various concentrations (5%, 10%, 15%, 20%) was analyzed by two-way ANOVA, which showed that both cryoprotectant type and concentration significantly affected the sperm survival rate ($p < 0.05$). At a concentration of 5%, glycerol showed the best cryoprotectant effect. At 10% and 15% concentrations, dimethyl sulfoxide (DMSO) was more effective than other cryoprotectants. Specifically, at a 10% DMSO concentration, the sperm survival rate reached 60.49 ± 13.14%, significantly higher than that of other cryoprotectants ($p < 0.05$). In contrast, ethylene glycol at 15% concentration performed the worst, with a sperm survival rate of only 5.31 ± 0.84% (Figure 1B). Therefore, dimethyl sulfoxide (DMSO) is considered the most suitable cryoprotectant for freezing and preserving *P. monodon* sperm, while glycerol is the second best choice.

3.3. The Effect of Different Equilibration Times at 4 °C

The post-cryopreservation survival rate of *P. monodon* sperm varied significantly depending on the equilibration time. As shown in Figure 2A, the best results were obtained with an equilibration time of 30 min at 4 °C, yielding a significantly higher sperm survival rate (44.49 ± 2.83%) compared to both the 10-min and 40-min groups ($p < 0.05$). Conversely, the experimental group that was directly cryopreserved in liquid nitrogen after only 10 min of equilibration showed the worst results, with a sperm survival rate of just 19.71 ± 4.97% (Figure 2A). These findings suggest that both excessively short and excessively long equilibration times can negatively impact sperm survival rates, with 30 min being the optimal equilibration time.

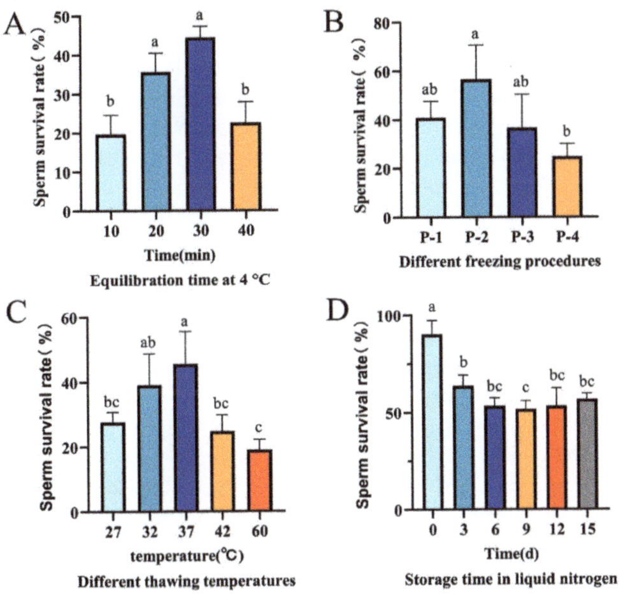

Figure 2. Post-cryopreservation sperm survival rates of *P. monodon* at different equilibration times (**A**), under different freezing protocols (**B**), thawing conditions (**C**), and cryopreservation durations (**D**). Different letters indicate significant differences between groups ($p < 0.05$).

3.4. Suitable Cooling Programs on Cryopreservation

The selection of the cryopreservation protocol played a crucial role in determining the survival rate of *P. monodon* sperm. The results indicated that protocol P-2 achieved the best outcomes, with a sperm survival rate of 56.65 ± 13.95% (Figure 2B). On the other hand, protocol P-4 performed the worst, yielding a survival rate of only 24.83 ± 5.21%. Notably, the P-2 group had a significantly higher survival rate compared to the P-4 group ($p < 0.05$). Therefore, it can be concluded that protocol P-2 is the most suitable method for the cryopreservation of *P. monodon* sperm.

3.5. Suitable Thawing Temperatures

As shown in Figure 2C, different thawing temperatures had a significant effect on the survival rate of frozen sperm ($p < 0.05$). The experimental results showed that when thawing at 37 °C in a water bath for 2 min, the sperm survival rate peaked (45.59 ± 9.99%), significantly higher than the other temperature groups ($p < 0.05$). The survival rate for thawing at 32 °C in a water bath for 2.5 min was 39.11 ± 9.67%, with no significant difference compared to the 37 °C group ($p > 0.05$) but significantly higher than the low and high temperature groups. The survival rates for thawing at 27 °C (4 min) and 42 °C (1.5 min) in a water bath were 27.65 ± 3.16% and 24.75 ± 5.08%, respectively. Thawing at 60 °C in a water bath (45 s) resulted in the lowest survival rate (18.98 ± 3.21%) due to heat shock causing more severe cell structure damage. The thawing times mentioned above were determined through preliminary experiments, representing the shortest time required for the complete thawing of liquid nitrogen-preserved samples (with the disappearance of ice crystals as the endpoint). Considering both efficiency and cell protection, 37 °C water bath thawing for 2 min was selected as the standard condition. This parameter balances the dual needs of rapid rewarming to reduce recrystallization damage and preventing high-temperature-induced membrane protein denaturation, providing a reliable basis for subsequent experiments.

3.6. Suitable Freezing Durations on Cryopreservation

Under identical cryopreservation conditions, the sperm survival rate was assessed over a 15-day storage period. The results revealed a notable decrease in the sperm survival rate within the first 0–3 days after cryopreservation. However, from days 3 to 15, the survival rate stabilized at around 50%, with no significant difference observed ($p > 0.05$) (Figure 2D).

3.7. Results of Sperm Morphology and Ultrastructure Analysis

3.7.1. Scanning Electron Microscopy

Scanning electron microscopy observations revealed minor morphological differences in *P. monodon* sperm before and after freezing. Before freezing, the ultrastructure of the sperm cells appeared sound: the body of the sperm cells, acrosomal processes, and the outer layer of the cells had a slightly irregular surface but were undamaged (Figure 3A,B). After freezing, the overall morphology of the sperm cells remained largely unaffected, and the acrosomal processes were well-preserved. Although there was a slight increase in surface roughness, no obvious structural damage was observed (Figure 3C,D). Quantitative data showed that, based on scanning electron microscopy (SEM) measurements, the sperm spike length before freezing was 4.481 ± 0.03 μm, and the sperm width was 4.315 ± 0.09 μm. After freezing, the sperm spike length was 2.718 ± 0.08 μm, and the sperm width was 4.320 ± 0.05 μm. These findings suggest that the freezing treatment has minimal influence on sperm cell morphology and can effectively preserve its form.

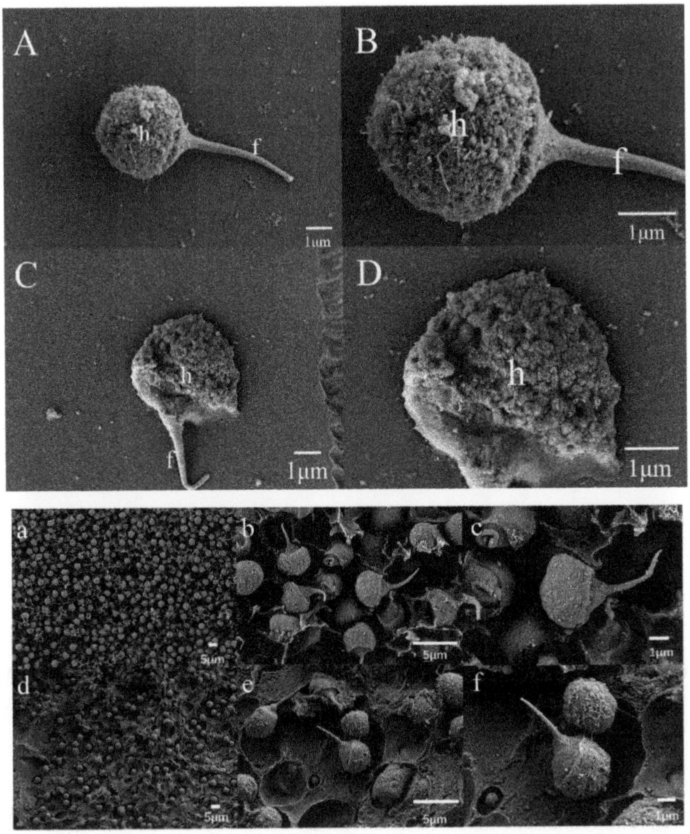

Figure 3. Scanning electron micrographs (SEM) of *P. monodon* sperm. (**A**,**B**) show sperm before cryopreservation, and (**C**,**D**) show sperm after cryopreservation. For quantitative analysis, images (**a**–**c**) depict sperm before cryopreservation, and (**d**–**f**) depict sperm after cryopreservation. h: head; f: flagellum.

3.7.2. Transmission Electron Microscopy

Transmission electron microscopy (TEM) images revealed distinct differences in the ultrastructure of *P. monodon* sperm between pre-freezing and post-freezing conditions. In the pre-freezing specimens (Figure 4), the acrosomal processes exhibited complete structural integrity with well-defined boundaries. The nuclear–subacrosomal interface was clearly demarcated, and no significant structural abnormalities were observed. The cytoplasm displayed uniform granularity without aggregation, and all organelles maintained their characteristic morphology, indicating that the sperm retained optimal physiological status at this stage. In contrast, post-freezing sperm cells (Figure 4) showed marked ultrastructural alterations. The acrosomal processes underwent partial detachment accompanied by morphological deformation, while the acrosome appeared compressed with reduced electron density compared to pre-freezing conditions. Freezing-induced damage manifested as blurring of the nuclear–subacrosomal boundary. Numerous irregular vesicles were frequently observed within the cytoplasm, suggesting cryo-injury-induced redistribution or degradation of intracellular components. Quantitative analysis demonstrated significant dimensional changes: sperm spike length decreased from 2.368 $\pm$ 0.04 μm (pre-freezing) to 2.074 $\pm$ 0.07 μm (post-freezing), and sperm width was reduced from 4.180 $\pm$ 0.06 μm to

3.786 ± 0.02 μm. These findings collectively indicated that cryopreservation compromised sperm structural integrity through multiple mechanisms.

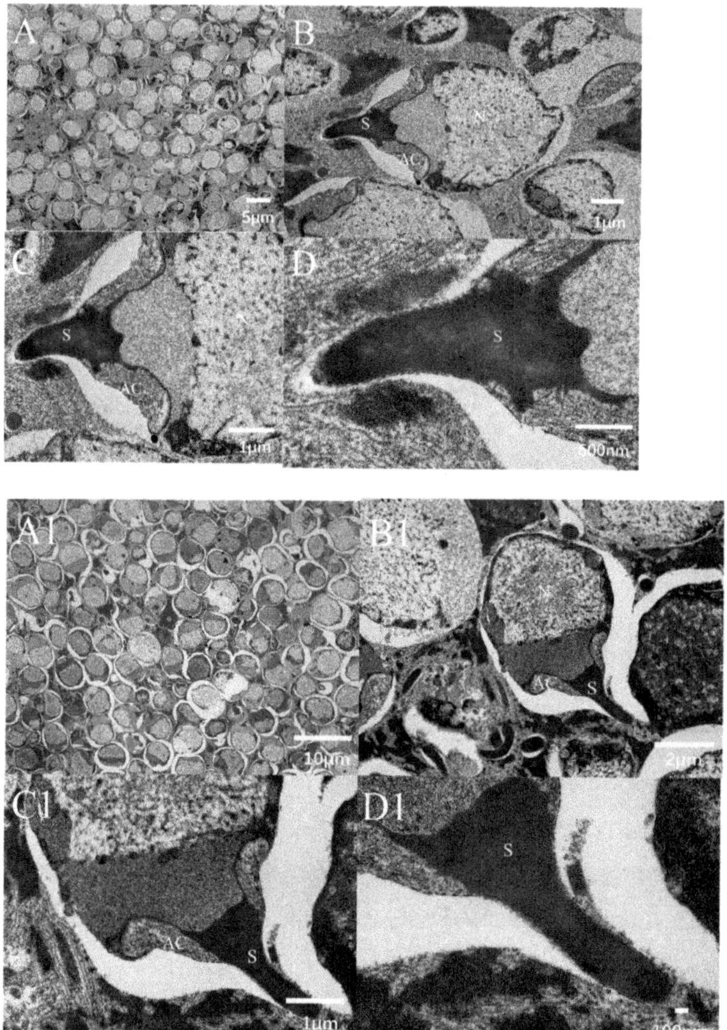

Figure 4. Transmission electron micrographs (TEM) of *P. monodon* sperm. The top row (**A–D**) depicts the TEM structure of sperm before freezing, while the bottom row (**A1–D1**) illustrates the TEM structure of sperm after freezing and thawing. (**A**,**A1**) show the general morphology of sperm; (**B**,**B1**) highlight the acrosome (AC), spike (S), and nucleus (N); (**C**,**C1**) zoom in to detail the acrosome and spike; and (**D**,**D1**) provide close-up views of the spike.S: spike; AC: acrosome; N: nucleus.3.8. Antioxidant Enzymes.

The enzyme activities of AKP and ACP in frozen sperm of *P. monodon* gradually increased as freezing time extended, showing significant differences ($p < 0.05$). Initially, the activity of MDA rose, but then it decreased with prolonged freezing time. As for CAT enzyme activity, there were no significant differences from 0 to 10 days ($p > 0.05$); however, a noticeable upward trend emerged at 15 days ($p < 0.05$). Similarly, T-SOD did not show significant differences from 0 to 5 days ($p > 0.05$), but it exhibited an upward trend from 10 to 15 days ($p < 0.05$) (Figure 5).

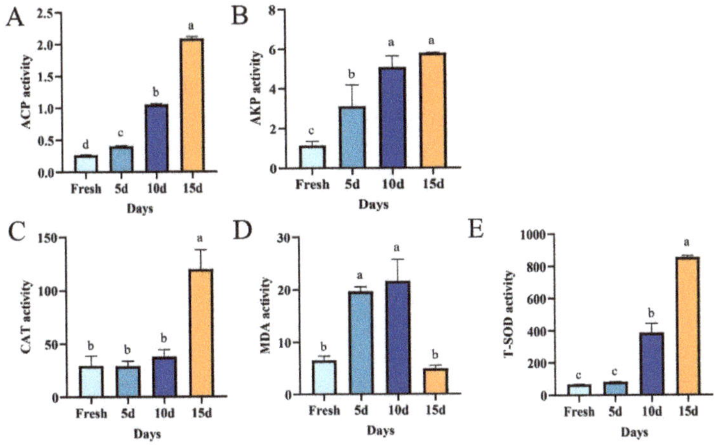

Figure 5. Effects of different freezing durations on antioxidant enzyme activities of *P. monodon* sperm. Panel (**A**) shows the ACP activity, panel (**B**) depicts the AKP activity, panel (**C**) shows the CAT activity, panel (**D**) shows MDA activity, and panel (**E**) shows T-SOD activity. Different letters represent significant differences between groups ($p < 0.05$).

4. Discussion

4.1. Effects of Dilutors on Sperm Survival Rate

In crustacean sperm cryopreservation, selecting an appropriate diluent is essential for maintaining sperm viability and function. Diluent solutions are critical in regulating osmotic pressure, pH, and ionic composition. Natural seawater and calcium-free artificial seawater have been widely used in crustacean sperm cryopreservation studies [8,25] Bray and Lawrence found that both seawater and calcium-free artificial seawater can serve as effective buffers for sperm morphological evaluation [26]. In studies on the acrosome reaction of *Portunus trituberculatus* sperm, Wang et al. found that Ca^{2+}-FSW was the optimal sperm buffer [27].

In this study, we evaluated the cryopreservation effectiveness of four different diluents—NSW, RS, HBSS, and D-Hanks—on *P. monodon* sperm. The osmolarities of each diluent were as follows: NSW was 1020 $\pm$ 15 mOsm/L, RS was 305 $\pm$ 3 mOsm/L, HBSS was 290 $\pm$ 2 mOsm/L, and D-Hanks was 295 $\pm$ 4 mOsm/L. Osmotic pressure and ionic composition are key factors affecting sperm viability. Our experimental results demonstrate that natural seawater (NSW) provides the best protection during cryopreservation, with post-thaw sperm viability significantly higher than that observed with other diluents (Figure 1). This phenomenon may be attributed to its osmotic pressure and ionic composition, which closely mimic the physiological state of prawn sperm in their natural environment. The salinity, pH, and abundance of trace elements (such as calcium and magnesium) in natural seawater likely act synergistically to maintain sperm membrane integrity and metabolic activity. These findings are consistent with previous studies. This result is consistent with the study by Chen et al. [28], who showed that Na^+ and K^+ ions play a crucial role in maintaining sperm vitality. Similarly, in their cryopreservation study of sea cucumber (*Apostichopus japonicus*) sperm, Xu et al. [29] observed that filtered natural seawater combined with an optimized cryoprotectant significantly improved post-thaw sperm viability (exceeding 65%). They suggested that the complex ionic environment of natural seawater helps reduce stress-induced damage during the freezing process. In contrast, although Ringer's solution differs in ionic composition from natural seawater, its balanced design in terms of osmotic pressure and key ions (e.g., Na^+, K^+, Ca^{2+}) results in relatively good preservation effects, ranking only slightly behind natural seawater in our experiments. However, HBSS and D-Hanks performed poorly, with significantly lower post-thaw sperm viability. This may be due to their ionic composition and osmotic pressure not fully matching the cryopreservation requirements for prawn sperm. Le et al. [30] also found in their study on the sperm of striped Jewfish (*Stereolepis doederleini*) that osmotic pressures below or above 35 PSU lead to reduced sperm viability, indicating that deviations from the natural osmotic state may cause irreversible damage to sperm. Additionally, Chang et al. [31] demonstrated in their cryopreservation study of *Portunus trituberculatus* sperm that the use of Ca^{2+}-free artificial seawater as a diluent significantly reduced both sperm viability and enzyme activity, further supporting the importance of ionic components (such as Ca^{2+}) in maintaining sperm functionality. The efficacy of different diluents has also been confirmed in other aquaculture studies. For example, studies have shown that using calcium-free artificial seawater with 20% glycerol significantly improves the sperm survival rate of *P. trituberculatus*, similar to optimized natural seawater and calcium balance [31]. Likewise, research on *Sebastes schlegelii* sperm cryopreservation demonstrated that even under optimized diluent conditions, the freezing process still affects sperm DNA integrity and mitochondrial activity, underscoring the cellular damage caused by cryopreservation. Transcriptome and methylome analyses in this study revealed the impact of cryopreservation on the expression of multiple genes, further emphasizing the need to optimize diluent composition to minimize cellular damage [32]. These differences in cryoprotective effects highlight the significant impact of diluent composition, including calcium, sodium, and potassium ion concentrations, on sperm viability and survival rates.

Our study demonstrates that osmotic pressure and ionic equilibrium of the diluent should be prioritized as critical factors when selecting extenders for sperm cryopreservation in *P. monodon*. Natural seawater, owing to its chemical properties that closely resemble the species' natural habitat, has demonstrated the most superior performance in maintaining sperm motility and cryopreservation efficacy. Nevertheless, considering the regional variability in composition of natural seawater, future investigations should focus on optimizing artificial seawater formulations (e.g., through adjustment of Na^+, K^+, and Ca^{2+} concentrations or supplementation with trace elements) to achieve the dual objectives of standardization and efficient cryopreservation.

4.2. Effects of Cryoprotectants on Sperm Survival Rate

Due to its excellent permeability and cryoprotective properties, dimethyl sulfoxide (DMSO) has been extensively employed in cryopreservation protocols for crustacean spermatozoa, with applications typically utilizing the minimum effective concentration to mitigate its inherent cytotoxic effects, including *Sicyonia ingentis* [8], *Scylla serrata* [22], *Penaeus monodon* [14,15], and *Litopenaeus vannamei* [16].

Previous studies have demonstrated that 10% DMSO as a standard cryoprotectant concentration achieves sperm survival rates of 38–56% in the cryopreservation of *P. monodon* sperm [14,17]. In the present study, a significantly enhanced survival rate of 60.49% was observed when using 10% DMSO, with this improvement potentially attributable to three critical technical modifications: (1) the implementation of natural seawater (30ppt salinity) as the diluent instead of the artificial seawater conventionally utilized in prior literature; (2) the development of a stepwise cooling protocol (4 °C → −20 °C → −80 °C → liquid nitrogen with 5 min holds at each transitional temperature stage), contrasting with the conventional practice of direct liquid nitrogen immersion; and (3) the optimization of pre-cooling equilibration duration to 30 min at 4 °C prior to freezing initiation, effectively balancing osmotic efficiency with reduced cryoprotectant toxicity exposure.

The synergistic interaction between the cooling rate and the composition of the carrier solution may represent a critical determinant. Szurek et al. [33] further conducted a systematic comparison of the toxicity of DMSO, ethylene glycol, and propylene glycol, noting that DMSO has low toxicity at commonly used concentrations (such as 5–10%), with relatively minimal negative effects on cell survival rate, fertilization rate, and development, whereas propylene glycol showed higher toxicity. This property makes DMSO particularly suitable for cell types that are sensitive to toxicity during cryopreservation, such as sperm and oocytes. Therefore, DMSO has significant advantages as the preferred cryoprotectant for sperm cryopreservation. Claudet et al. [11] found that combining 10% DMSO and 10% PG significantly enhanced sperm viability in *Macrobrachium rosenbergii*, surpassing the effects of single cryoprotectants like DMSO, PG, and glycerol, suggesting that combined cryoprotectants may further improve sperm survival rates.

Furthermore, the optimal concentration of DMSO exhibits a close correlation with sperm morphological characteristics. Anchordoguy et al. [8] observed that 5% DMSO was optimal for *Sicyonia ingentis*, while Ke and Cai [13] recommended a combination of 10% DMSO and 5–10% glycerol for *P. chinensis*, indicating a high degree of species dependency in cryoprotectant choice and concentration among crustaceans. Future studies could further explore the combined use of DMSO with other cryoprotectants to optimize cryopreservation outcomes.

4.3. The Effect of Equilibration Time on Sperm Survival Rate

In sperm cryopreservation, equilibration time is crucial for the penetration of cryoprotectants and for sperm adaptation to low temperatures. This study shows that the optimal equilibration time for *P. monodon* sperm is 30 min, which provides the best cryoprotective effect; both shorter and longer equilibration times lead to reduced survival rates. Regarding the optimization effect of equilibration time on cryoprotectant permeation, multiple pieces of evidence have been accumulated in related crustacean studies. For example, in the study of Japanese marsh shrimp (*Macrobrachium nipponense*), sperm resting at 4 °C for 30 min significantly improved survival rates, which the researchers attributed to the improvement in cryoprotectant permeation efficiency [34]. Similarly, Liu et al. [35] observed in their experiment on Japanese crab (*Charybdis japonica*) that a 20-min equilibration time achieved a balance between cryoprotectant absorption and toxicity damage. These studies collectively suggest that the specific equilibration time must be adjusted according to the biological characteristics of the species, but the core mechanism points to optimizing the interaction between the cryoprotectant and the cells through time control.

Similarly, Liu et al. showed that the optimal equilibration time for *C. japonica* sperm was 20 min, with both longer and shorter times reducing sperm viability, possibly due to metabolic depletion or the accumulation of cryoprotectant toxicity [35]. Ke and Cai found that an equilibration time of 20–30 min at 0–4 °C effectively prevented the acrosome reaction in *P. chinensis* sperm, further validating the importance of equilibration time in cryoprotection [13].

Other studies have shown that equilibration time should be adjusted according to the type of cryoprotectant and species-specific characteristics. For example, Nimrat et al. found that a shorter equilibration time (e.g., 20 min) with low-concentration DMSO significantly reduced cryodamage in *Fenneropenaeus merguiensis* sperm, thereby improving sperm survival rates [36]. This suggests that a shorter equilibration time helps prevent the accumulation of cryoprotectant toxicity, thus preserving sperm viability. In contrast, Vuthiphandchai et al. observed that extending the equilibration time to 60 min or more significantly decreased sperm viability, indicating that prolonged equilibration may lead to the buildup of cryoprotectant toxicity, detrimental to sperm survival [15]. This observation is consistent with our experiment, where longer equilibration times led to decreased survival rates in *P. monodon* sperm, further emphasizing the critical balance required in equilibration time to ensure cryoprotectants provide protection without excessive toxicity.

Additionally, Chang et al. found that 15 min was the optimal equilibration time for *P. trituberculatus*, suggesting that a short equilibration time allows effective cryoprotectant penetration while avoiding oxidative stress or metabolic consumption associated with longer times [31]. This is consistent with Gwo and Lin's finding that prolonged equilibration time led to decreased embryo survival rates in *Penaeus japonicus*. Prolonged equilibration may cause sperm to metabolize continuously at low temperatures, thereby increasing oxidative stress and leading to cellular damage [37]. Therefore, while cryoprotectants provide effective protection over short equilibration times, extended equilibration may induce additional stress responses, negatively impacting sperm viability.

4.4. Effects of Cooling Protocol and Thawing Conditions on Sperm Survival Rate

In sperm cryopreservation, the cooling rate and thawing temperature significantly impact cell survival. Using a programmable cooling device, we employed a slow freezing method, cooling *P. monodon* sperm at a rate of −5 °C/min down to −20 °C, equilibrating, and then gradually lowering the temperature to −180 °C, where it was stored in liquid nitrogen, achieving optimal cryoprotection. This stepwise cooling process facilitated the gradual formation of extracellular ice crystals, effectively reducing the risk of intracellular ice crystal formation and minimizing cell dehydration and morphological deformation due to osmotic pressure changes. This is consistent with the theory proposed by Woods et al., which states that slow freezing improves cell survival rate [38,39].

In contrast, Vuthiphandchai et al. found that a freezing rate of −2 °C/min achieved higher survival rates for *P. monodon* spermatophores, whereas faster rates led to complete sperm death [15]. This difference may be due to the gel matrix layer in the spermatophore structure, which affects cryoprotectant penetration and thus creates different cryopreservation requirements between spermatophores and individual sperm cells.

The choice of thawing temperature is equally crucial. Studies have shown that sperm frozen at optimal cooling rates are susceptible to damage during thawing if an appropriate thawing method is not used [5]. Chang et al. reported that thawing *P. trituberculatus* sperm in a 37 °C water bath yielded the best results, consistent with Yang et al.'s findings on *Litopenaeus vannamei* [31,40]. This may be because the 37 °C temperature helped maintain cell membrane integrity during rapid thawing, avoiding further damage from higher temperatures. Studies indicate that the optimal thawing temperature for *P. chinensis* sperm is 35–40 °C [13], while for *P. trituberculatus* sperm, it is 55 °C [5], highlighting that different species have specific thawing temperature requirements. This suggests that coordinating the cooling and thawing processes in cryopreservation is essential to minimize cell damage and improve survival rates. Future research could further investigate the interactions between cooling and thawing methods to optimize cryopreservation techniques for sperm and other sensitive cells.

4.5. Effects of Cryopreservation on the Ultrastructure of Sperm in P. monodon

Sperm morphological characteristics hold significant taxonomic value in crustacean species identification, particularly within the Penaeidae family, where the evolutionary specialization of sperm structures reflects adaptive differentiation in reproductive strategies. The sperm of *P. monodon* exhibits a characteristic acrosomal spike complex, comprising three distinct regions: the head region (nuclear compartment), the acrosomal vesicle (enzyme storage domain), and a non-motile acrosomal spike. In this study, we employed scanning electron microscopy (SEM) and transmission electron microscopy (TEM) to examine the ultrastructure of *P. monodon* sperm, revealing the structural changes that occur before and after cryopreservation.

Prior to freezing, *P. monodon* sperm exhibited a characteristic non-motile acrosomal spike complex. The sperm head appeared robust, with a clearly visible and structurally intact acrosomal spike showing no signs of damage. Transmission electron microscopy (TEM) revealed a well-defined boundary between the nucleus and subacrosomal region, evenly distributed cytoplasm, and well-preserved organelles, collectively indicating optimal physiological integrity at this stage. Following cryopreservation, scanning electron microscopy (SEM) observations showed that while the overall external morphology of the sperm remained relatively intact, the surface of the acrosomal spike appeared slightly roughened. Notably, no significant rupture or structural damage was detected, and the acrosomal spike remained largely intact without apparent fragmentation or detachment.

However, TEM analysis revealed internal structural changes after freezing, including vacuolation and granulation in the cytoplasm. This could be attributed to incomplete elimination of intracellular water during the freezing process, leading to ice crystal formation. This also resulted in a loss of distinction between the nucleus and subacrosome, indicating that mechanical stress during freezing had an irreversible impact on the internal sperm structure.

Similar effects of cryopreservation on sperm ultrastructure have been observed in other species. For example, studies on rainbow trout (*Oncorhynchus mykiss)* sperm [41] reported membrane rupture, flagellar breakage, and organelle damage due to freezing, which are consistent with the damage patterns seen in *P. monodon* sperm in this study. In African turquoise killifish (*Nothobranchius furzeri*) sperm cryopreserved under certain conditions [42] it was found that after freeze–thaw, the sperm motility was poor, although the structural shape remained fairly good.

When DMSO was used as a cryoprotectant with appropriate freezing and thawing rates, the sperm mitochondria and flagellum of *P. monodon* remained largely intact, similar to the mitochondrial stability observed in other species' cryopreserved sperm. This indicates that sperm tolerance to cryopreservation varies significantly among species. For instance, research on blue mussel (*Mytilus galloprovincialis*) [43] showed that different cooling rates significantly affected sperm activation rate and ultrastructural damage, which aligns with the damage patterns observed in *P. monodon* sperm in this study. However, it appears that the currently selected preservation temperatures do not prevent ultrastructural sperm damage, as internal damage, such as vacuole formation and loss of mitochondria, still occurs during cryopreservation.

Furthermore, the structural responses of sperm to cryopreservation differ across species. The external morphological damage to cryopreserved *P. monodon* sperm was relatively minor, with the flagellum and acrosomal spike on the head remaining intact. Compared to some teleost fish, such as ayu, *P. monodon* sperm demonstrates greater tolerance to freezing, particularly in protecting external structures.

It is noteworthy that partial acrosomal spikes exhibited characteristic curling after cryopreservation, with morphological alterations closely resembling the early-stage acrosome reaction patterns reported by Pongtippatee et al. [44] in *P. monodon*. This structural observation, coupled with the relatively high viability rate (82.3 $\pm$ 3.1%) demonstrated by trypan blue exclusion assay, suggests potential retention of partial functional activity in post-thaw spermatozoa. However, given the absence of direct functional assessments and in Vitro acrosome reaction induction experiments in the current investigation, it remains undetermined whether these ultrastructural modifications represent biologically significant changes or merely cryo-induced physical deformations. Subsequent studies should incorporate functional assays to elucidate the actual impacts of cryopreservation on acrosomal reactivity and comprehensive functional competence of sperm cells.

4.6. *Effects of Cryopreservation on the Antioxidant Function of Sperm in P. monodon*

T-SOD and CAT are crucial antioxidant enzymes in sperm, playing pivotal roles in neutralizing reactive oxygen species (ROS) and shielding cells from oxidative damage. Their activities are closely tied to the overall antioxidant enzyme activity of organisms or cells, and fluctuations in this activity can serve as an indicator of stress or damage inflicted by external environmental factors [45,46].

In our study, all five sperm enzymes exhibited an increase, with the exception of MDA. These results are consistent with the majority of published studies on the impact of cryopreservation on sperm enzyme activity [45]. This rise may be attributed to mechanical damage during the freezing and thawing process, which compromises the cell membrane structure, leading to the leakage of enzymes within the sperm [46].

Notably, the activities of T-SOD, AKP, and ACP in sperm significantly increased after cryopreservation, while CAT activity remained relatively stable from 0 to 10 days. On one hand, this suggests that the cryopreservation technique used in this study effectively protected *P. monodon* sperm, as the internal enzyme contents were well-preserved without significant leakage. On the other hand, it also indicates that sperm cells experienced environmental stress during cryopreservation, which may have impacted energy metabolism and ROS scavenging processes essential for surviving adverse conditions.

Interestingly, Kankofer et al. reported a decrease in SOD and CAT activities in cryopreserved horse (*Equus caballus*) sperm [47], which contrasts with our findings. This discrepancy may be attributed to various factors affecting antioxidant enzyme activity during the sperm preservation process, such as freezing stress, membrane damage, oxidative stress-related responses, and cell-native adaptation mechanisms. These changes underscore that cells undergo numerous physiological and biochemical stressors during cryopreservation, and different species may respond differently to these stressors.

5. Conclusions

In this study, we refined the cryopreservation protocol for *P. monodon* sperm. We utilized natural seawater as a diluent and incorporated 10% dimethyl sulfoxide (DMSO) as a cryoprotectant. The optimal conditions for standing, cooling, and thawing were meticulously determined: a 30-min standing period at 4 °C, followed by a cooling program of −5 °C/min to −20 °C with a 5-min hold, then −10 °C/min to −80 °C with another 5-min hold, and ultimately −20 °C/min to −180 °C for storage in liquid nitrogen. Sperm samples that underwent this cryopreservation process and were stored in liquid nitrogen for 15 days exhibited high vitality. Scanning electron microscopy (SEM) confirmed that the sperm cells sustained minimal damage both before and after freezing, thanks to the optimized cryopreservation procedure, which had minimal impact on the overall cell structure. However, transmission electron microscopy (TEM) observations revealed some internal damage to the sperm cells post-freezing. This indicates that while cell viability remained high, changes in the ultrastructure still warrant attention during the cryopreservation process. Furthermore, enzyme activity assays suggested that the frozen sperm possessed the high biological activity necessary for fertilization. These results provide valuable insights for the conservation of *P. monodon* genetic resources, streamline the breeding of parent stocks in aquaculture, and address the challenge of asynchronous sexual maturation between male and female parents.

In subsequent studies, we will investigate the use of two or more cryoprotectants to further enhance the cryopreservation formula and, consequently, improve sperm survival rates.

Author Contributions: Conceptualization, D.K., S.J. and F.Z.; methodology, Y.L., J.S., Y.D. and Q.Y.; data curation, J.W., X.Q. and J.H.; writing—original draft preparation, D.K.; writing—review and editing, T.L. and F.Z. All authors have read and agreed to the published version of the manuscript.

Funding: This study was supported by the National Key R & D Program of China (2022YFD2400104), the earmarked fund for CARS-48, Central Public-interest Scientific Institution Basal Research Fund, CAFS (No. 2023TD34), the earmarked fund for HNARS (HNARS-10-ZJ01), and the Innovative Team Building Project of Guangdong Modern Agricultural Industrial Technology System (2024CXTD25).

Institutional Review Board Statement: All experimental protocols and methods in this study were approved by the Animal Care and Use Ethics Committee in the South China Sea Fisheries Research Institute (SCSFRI96-255, 15 December 2021).

Informed Consent Statement: Not applicable.

Data Availability Statement: The raw data used to support the findings of this study are available from the corresponding author upon request: T.L. and F.Z., zhoufalin0925@163.com, ltm600@163.com.

Conflicts of Interest: The authors declare no conflicts of interest.

References

1. Wang, Y.; Duan, Y.; Huang, J.; Wang, J.; Zhou, C.; Jiang, S.; Lin, H.; Zhang, Z. Characterization and Functional Study of Nuclear Factor Erythroid 2-Related Factor 2 (Nrf2) in Black Tiger Shrimp (*Penaeus monodon*). *Fish Shellfish Immunol.* **2021**, *119*, 289–299. [CrossRef] [PubMed]
2. Tanaka, S.; Zhang, H.; Horie, N.; Yamada, Y.; Okamura, A.; Utoh, T.; Mikawa, N.; Oka, H.P.; Kurokura, H. Long-term Cryopreservation of Sperm of Japanese Eel. *J. Fish Biol.* **2002**, *60*, 139–146. [CrossRef]
3. Okutsu, T.; Yano, A.; Nagasawa, K.; Shikina, S.; Kobayashi, T.; Takeuchi, Y.; Yoshizaki, G. Manipulation of Fish Germ Cell: Visualization, Cryopreservation and Transplantation. *J. Reprod. Dev.* **2006**, *52*, 685–693. [CrossRef]
4. Asturiano, J.F.; Cabrita, E.; Horváth, Á. Progress, Challenges and Perspectives on Fish Gamete Cryopreservation: A Mini-Review. *Gen. Comp. Endocrinol.* **2017**, *245*, 69–76. [CrossRef]
5. Cabrita, E.; Sarasquete, C.; Martínez-Páramo, S.; Robles, V.; Beirão, J.; Pérez-Cerezales, S.; Herráez, M.P. Cryopreservation of Fish Sperm: Applications and Perspectives: Cryopreservation of Fish Sperm. *J. Appl. Ichthyol.* **2010**, *26*, 623–635. [CrossRef]
6. Liu, Y.; Li, X.; Robinson, N.; Qin, J. Sperm Cryopreservation in Marine Mollusk: A Review. *Aquac. Int.* **2015**, *23*, 1505–1524. [CrossRef]
7. Li, J.C.; Ke, W.J.; Yang, H.; Yu, H.; Yang, Y. Advances in cryopreservation technology applied to fish sperm preservation. *Jiangsu Agric. Sci.* **2019**, *47*, 66–69. [CrossRef]
8. Anchordoguy, T.; Crowe, J.H.; Griffin, F.J.; Clark, W.H. Cryopreservation of Sperm from the Marine Shrimp *Sicyonia ingentis*. *Cryobiology* **1988**, *25*, 238–243. [CrossRef]
9. Akarasanon, K.; Damrongphol, P.; Poolsanguan, W. Long-Term Cryopreservation of Spermatophore of the Giant Freshwater Prawn, *Macrobrachium rosenbergii* (de Man). *Aquac Res.* **2004**, *35*, 1415–1420. [CrossRef]
10. Chow, S.; Tam, Y.; Ogasawara, Y. Cryopreservation of Spermatophore of the Fresh Water Shrimp, *Macrobrachium rosenbergii*. *Biol. Bull.* **1985**, *168*, 471–475. [CrossRef]
11. Valentina claudet, P.; Narasimman, S.; Natesan, M. Effect of Cryoprotectants and Cooling Rates on Fertility Potential of Sperm in the Giant Freshwater Prawn, *Macrobrachium rosenbergii* (De Man). *Anim. Reprod. Sci.* **2016**, *171*, 49–57. [CrossRef] [PubMed]
12. Narasimman, S.; Krishnamoorthy, D.; Buelah, C.D.; Natesan, M. Retrieval and Cryopreservation of Sperm in Spermatophores from Cadaveric Indian White Shrimp, *Fenneropenaeus indicus* (H. Milne Edwards, 1837). *Anim. Reprod. Sci.* **2018**, *192*, 185–192. [CrossRef] [PubMed]
13. Ke, Y.F.; Cai, N.E. Study on sperm cryopreservation of *Penaeus chinensis*. *Oceanol. Limnol. Sin.* **1996**, *27*, 187–193. [CrossRef]
14. Bart, A.N.; Choosuk, S.; Thakur, D.P. Spermatophore Cryopreservation and Artificial Insemination of Black Tiger Shrimp, *Penaeus monodon* (Fabricius). *Aquac. Res.* **2006**, *37*, 523–528. [CrossRef]
15. Vuthiphandchai, V.; Nimrat, S.; Kotcharat, S.; Bart, A.N. Development of a Cryopreservation Protocol for Long-Term Storage of Black Tiger Shrimp (*Penaeus monodon*) Spermatophores. *Theriogenology* **2007**, *68*, 1192–1199. [CrossRef]
16. Lezcano, M.; Granja, C.; Salazar, M. The Use of Flow Cytometry in the Evaluation of Cell Viability of Cryopreserved Sperm of the Marine Shrimp (*Litopenaeus vannamei*). *Cryobiology* **2004**, *48*, 349–356. [CrossRef]

17. Rosmiati, R.; Trismawanti, I.; Lante, S. Effect of Various Cryoprotectants on Preservation of Black Tiger *Penaeus monodon* Shrimp Spermatozoa. *Indones. Aquac. J.* **2021**, *15*, 77. [CrossRef]
18. Grace, B.L.; Natarajan, P. Cryopreservation of *Penaeus monodon* (Fabricius, 1798) spermatozoa. *Indian J. Fish.* **2003**, *50*, 415–419. Available online: https://www.academia.edu/67386861/Cryopreservation_of_Penaeus_monodon_Fabricius_1798_spermatozoa (accessed on 11 March 2025).
19. Aquino, J.I.L.; Elliott, L.; Zeng, C.; Paris, D.B.B.P. Recent Developments in Male Fertility Evaluation, Sperm Cryopreservation and Artificial Fertilisation, and Their Potential Application to Decapod Crustacean Aquaculture. *Rev. Aquac.* **2022**, *14*, 848–889. [CrossRef]
20. Hong, X. Application of artificial spermatophore transplantation technology in the breeding of *Penaeus monodon*. *Fujian Fish.* **2012**, *34*, 331–334. [CrossRef]
21. Leelatanawit, R.; Uawisetwathana, U.; Khudet, J.; Klanchui, A.; Phomklad, S.; Wongtripop, S.; Angthoung, P.; Jiravanichpaisal, P.; Karoonuthaisiri, N. Effects of Polychaetes (*Perinereis nuntia*) on Sperm Performance of the Domesticated Black Tiger Shrimp (*Penaeus monodon*). *Aquaculture* **2014**, *433*, 266–275. [CrossRef]
22. Leung-Trujillo, J.R.; Lawrence, A.L. Observations on the Decline in Sperm Quality of *Penaeus setiferus* under Laboratory Conditions. *Aquaculture* **1987**, *65*, 363–370. [CrossRef]
23. Ulloa-Rodríguez, P.; Contreras, P.; Dumorné, K.; Figueroa, E.; Risopatrón, J.; Valdebenito, I.; Farías, J.G. Sperm Morphology and Ultrastructure of Patagonian Blenny (*Eleginops maclovinus*). *Tissue Cell* **2019**, *57*, 66–69. [CrossRef]
24. Quagio-Grassiotto, I.; Oliveira, C. Sperm Ultrastructure and a New Type of Spermiogenesis in Two Species of Pimelodidae, with a Comparative Review of Sperm Ultrastructure in Siluriformes (Teleostei: Ostariophysi). *Zool. Anz. A J. Comp. Zool.* **2008**, *247*, 55–66. [CrossRef]
25. Bhavanishankar, S.; Subramoniam, T. Cryopreservation of Spermatozoa of the Edible Mud crab *Scylla serrata* (Forskal). *J. Exp. Zool.* **1997**, *277*, 326–336. [CrossRef]
26. Bray, W.A.; Lawrence, A.L. Male Viability Determinations in *Penaeus vannamei*: Evaluation of Short-Term Storage of Spermatophores up to 36 h and Comparison of Ca-Free Saline and Seawater as Sperm Homogenate Media. *Aquaculture* **1998**, *160*, 63–67. [CrossRef]
27. Wang, Y.L.; Zhang, Z.P.; Xie, F.J.; Li, S.J. Study on the acrosome reaction of *Scylla paramamosain* sperm. *Acta Zool. Sin.* **2001**, *47*, 310–316+365–366. [CrossRef]
28. Chen, D.H.; Zhou, Z.L.; Fan, L.J.; Qu, D.; Zhao, Y.L.; Wang, Q. Effects of preservation solution and conditions on sperm viability of *Eriocheir sinensis*. *J. East China Norm. Univ. (Nat. Sci. Ed.)* **2007**, *4*, 86–94. [CrossRef]
29. Xu, S.; Liu, S.; Sun, J.; Zhang, L.; Lin, C.; Sun, L.; Yang, H. Optimizing cryopreservation of sea cucumber (*Apostichopus japonicus*) sperm using a programmable freezer and computer-assisted sperm analysis. *Front. Mar. Sci.* **2022**, *9*, 917045. [CrossRef]
30. Le, M.H.; Van Ngo, M.; Dinh, K.V.; Lim, H.K. Effects of environmental factors on the activity of chilled storage and cryopreservation of striped Jewfish *Stereolepis doederleini* sperm. *Reg. Stud. Mar. Sci.* **2024**, *81*, 104006. [CrossRef]
31. Chang, L.; Lu, C.; Zhu, J.; Chen, Y.; Wang, C.; Mu, C.; Hou, C. Optimization of Sperm Cryopreservation Formulation in *Portunus trituberculatus*. *Int. J. Mol. Sci.* **2023**, *24*, 4358. [CrossRef] [PubMed]
32. Niu, J.; Wang, X.; Liu, P.; Liu, H.; Li, R.; Li, Z.; He, Y.; Qi, J. Effects of Cryopreservation on Sperm with Cryodiluent in Viviparous Black Rockfish (*Sebastes schlegelii*). *Int. J. Mol. Sci.* **2022**, *23*, 3392. [CrossRef] [PubMed]
33. Szurek, E.A.; Eroglu, A. Comparison and Avoidance of Toxicity of Penetrating Cryoprotectants. *PLoS ONE* **2011**, *6*, e27604. [CrossRef]
34. Liao, X.; Ge, J.C.; Ding, S.Y.; Chen, C.J.; Lu, Q.P.; Huang, Y.H. Research on cryopreservation technology of *Macrobrachium nipponense* sperm. *J. Nanjing Univ. (Nat. Sci. Ed.)* **2008**, *44*, 421–426. [CrossRef]
35. Liu, C.L.; Zou, Y.; Wu, Y.Y.; Wang, Y.J.; Liu, T.; Song, A.H.; Tang, X.X.; Liu, H.J. Study on sperm collection and cryopreservation technology for *Portunus trituberculatus*. *J. Ocean Univ. China (Nat. Sci. Ed.)* **2020**, *50*, 87–93. [CrossRef]
36. Nimrat, S.; Noppakun, S.; Sripuak, K.; Boonthai, T.; Vuthiphandchai, V. Cryopreservation of Banana Shrimp (*Fenneropenaeus merguiensis*) Spermatophores with Supplementation of Medicinal Plant Extracts: Development of a Programmable Controlled-Rate Method and a Practical Method. *Aquaculture* **2020**, *515*, 734537. [CrossRef]
37. Gwo, J.-C. Preliminary Experiments on the Cryopreservatioh of Penaeid Shrimp (*Penaeus japonicus*) Embryos, Nauplii and Zoea. *Theriogenology* **1998**, *49*, 1289–1299. [CrossRef] [PubMed]
38. Chen, S.L. Advances and prospects in the cryopreservation of fish gametes and embryos. *J. Fish. China* **2002**, *26*, 161–168. [CrossRef]
39. Woods, E.J.; Benson, J.D.; Agca, Y.; Critser, J.K. Fundamental Cryobiology of Reproductive Cells and Tissues. *Cryobiology* **2004**, *48*, 146–156. [CrossRef]
40. Yang, C.L.; Zhao, Y.Z.; Chen, X.L.; Li, Y.M.; Peng, M.; Yang, Y.H.; He, P.P.; Chen, X.H. Study on sperm cryopreservation technology of *Litopenaeus vannamei*. *J. South. Agric.* **2013**, *44*, 1382–1389. [CrossRef]

41. Gu, W.; Zhang, Y.Q.; Hu, G.; Xu, G.F.; Wang, B.Q. Observation of Oncorhynchus mykiss sperm under scanning and transmission electron microscopy. *J. Fish. Sci.* **2014**, *27*, 24–27. [CrossRef]
42. Dolfi, L.; Suen, T.K.; Ripa, R.; Antebi, A. Sperm Cryopreservation and in Vitro Fertilization Techniques for the African Turquoise Killifish Nothobranchius Furzeri. *Sci. Rep.* **2021**, *11*, 17145. [CrossRef] [PubMed]
43. Liu, Y.; Liu, S.; Liu, B.; Qin, J.; Xu, T.; Li, X. Cryopreservation of Strip Spawned Sperm Using Programmable Freezing Technique in the Blue Mussel Mytilus Galloprovincialis. *J. Oceanol. Limnol.* **2018**, *36*, 2351–2357. [CrossRef]
44. Pongtippatee, P.; Vanichviriyakit, R.; Chavadej, J.; Plodpai, P.; Pratoomchart, B.; Sobhon, P.; Withyachumnarnkul, B. Acrosome Reaction in the Sperm of the Black Tiger Shrimp *Penaeus monodon* (Decapoda, Penaeidae). *Aquac. Res.* **2007**, *38*, 1635–1644. [CrossRef]
45. Ding, X.Y.; Li, Z.T.; Duan, P.F.; Qiu, Y.S.; Wang, X.Y.; Li, L.L.; Wang, L.N.; Liu, Y.; Li, W.S.; Wang, Q.B.; et al. Effects of long-term cryopreservation on sperm ultrastructure and enzyme activity of Epinephelus lanceolatus. *J. Fish. China* **2023**, *47*, 50–66. [CrossRef]
46. Shi, Y.X.; Cheng, S.; Zhu, J.Q.; Wu, X.F. Cryopreservation and enzyme activity detection of *Lateolabrax japonicus* sperm. *Acta Hydrobiol. Sin.* **2015**, *39*, 1241–1247. [CrossRef]
47. Kankofer, M.; Kolm, G.; Aurich, J.; Aurich, C. Activity of Glutathione Peroxidase, Superoxide Dismutase and Catalase and Lipid Peroxidation Intensity in Stallion Semen during Storage at 5 °C. *Theriogenology* **2005**, *63*, 1354–1365. [CrossRef]

biology

MDPI

Article

Insights into the Regulatory Role of MicroRNAs in *Penaeus monodon* Under Moderately Low Salinity Stress

Jianzhi Shi [1,2], Song Jiang [2], Yangyang Ding [2], Hongshan Diao [2], Wenzhe Li [2], Yundong Li [2], Jianhua Huang [2,3], Lishi Yang [2], Qibin Yang [1,*] and Falin Zhou [1,2,*]

1 Key Laboratory of Efficient Utilization and Processing of Marine Fishery Resources of Hainan Province, Sanya Tropical Fisheries Research Institute, Sanya 572018, China; shijianzhi1989@163.com

2 Key Laboratory of South China Sea Fishery Resources Exploitation and Utilization, Ministry of Agriculture and Rural Affairs, South China Sea Fisheries Research Institute, Chinese Academy of Fishery Sciences, Guangzhou 510300, China; tojiangsong@163.com (S.J.); dingyangyang93@163.com (Y.D.); 17875808801@163.com (H.D.); 15836677553@163.com (W.L.); liyd2019@163.com (Y.L.); huangjianhua@scsfri.ac.cn (J.H.); yangls2016@163.com (L.Y.)

3 Shenzhen Base of South China Sea Fisheries Research Institute, Chinese Academy of Fishery Sciences, Shenzhen 518108, China

* Correspondence: yangqibin1208@163.com (Q.Y.); zhoufalin@scsfri.ac.cn (F.Z.)

Simple Summary: This study investigates how microRNAs (miRNAs) help black tiger shrimp (*Penaeus monodon*) adapt to moderately low salinity stress (17 ppt), a common challenge in aquaculture due to climate change. Using high-throughput sequencing, researchers identified 118 miRNAs in shrimp hepatopancreas that showed significant changes in expression after salinity exposure. These miRNAs regulate genes involved in metabolism, immune response, and stress adaptation. Key findings revealed that low salinity disrupts growth-related processes (e.g., digestion, hormone secretion) and increases disease susceptibility, while miRNAs help coordinate stress responses by modulating pathways like immune signaling and nervous system function. The results provide new insights into shrimp stress tolerance mechanisms and highlight potential molecular targets for breeding more resilient shrimp varieties, supporting sustainable aquaculture practices.

Abstract: MicroRNAs (miRNAs) play crucial roles in regulating various biological processes in crustaceans, including stress responses. Under acute low salinity stress conditions, miRNAs exhibit dynamic expression patterns that significantly influence the physiological and molecular responses of the shrimp. However, research on miRNAs in *P. monodon* is very limited, and their functions under low salinity stress remain unclear. In this study, by using high-throughput sequencing technology, we identified miRNAs and investigated their regulatory mechanism in *P. monodon* under low salinity stress. A total of 118 miRNAs were differentially expressed after low salinity exposure. These miRNAs were found to target genes involved in metabolism, pathogen infection, immune response and stress signaling pathways. By modulating the expression of these target genes, miRNAs were able to fine-tune the stress response of *P. monodon*, thereby enhancing its tolerance to low salinity. This study provides new insights into the regulatory roles of miRNAs in the stress response of aquatic organisms and suggests potential targets for genetic improvement to enhance stress tolerance in *P. monodon* aquaculture.

Keywords: microRNA; *Peaneus monodon*; low salinity; regulatory roles

1. Introduction

The black tiger shrimp (*Penaeus monodon*) is a tropical shrimp commonly found in the Indo-Pacific region. With the rapid growth in global demand, *P. monodon* has become an economically important shrimp species [1]. In 2023, the production of *P. monodon* reached 128,000 tons, making it the second largest marine shrimp species cultured in China [2]. However, extreme climate changes, such as rising temperatures and increased occurrence of heavy rainfall, are currently impacting the culture environment. The rise of water temperature alters the growth and feed conversion of cultured shrimp, leading to high mortality rates [3]. Similarly, abnormal rainfall causes sudden shifts in salinity, which can affect the dynamics of microorganisms in water and the susceptibility of shrimp to pathogen infection [4]. Environmental stress has become a significant threat to the quantity and quality of shrimp aquaculture by affecting metabolic processes or increasing the risk of disease outbreaks.

As a euryhaline crustacean species, *P. monodon* can survive in a wide range of salinities [5]. This marine species has been cultured in low salinity groundwater in South China. However, it has been observed that low salinity significantly depresses the growth performance of shrimp [6]. This is primarily because, in order to maintain osmotic balance at low salinity, shrimp are forced to expend more energy for osmoregulation. Additionally, inland low salinity waters are deficient in K^+ and Mg^{2+}, which are crucial for the growth and survival of shrimp [7]. On the other hand, low salinity could increase the toxicity of nitrite (NO_2^-), nitrate (NO_3^-) and ammonia (NH_3), leading to environmental stress for shrimp [8]. Thus, further exploration is needed to understand the physiological adaptations of osmoregulation in shrimp, and a practical way to improve growth performance in *P. monodon* at low salinity needs to be identified.

The regulatory mechanisms of penaeid shrimp under low salinity conditions have garnered significant scientific interest. In *Litopenaeus vannamei*, previous studies on salinity adaptation mainly focused on the function of ion transporters and ion channels in osmoregulation [9,10]. Several studies have used omic approaches to reveal the genes and pathways that respond to salinity change in different tissues. Pathways related to signal transduction, oxidative, lipid and fatty acid metabolism were activated after low salinity stress [11–13]. Moreover, two vital genes—mitogen-activated protein kinase kinase and cold shock domain binding protein—involved in stress resistance and immunity were identified as regulators of low salinity in *P. monodon* [14,15]. Although our understanding of salinity adaptation in shrimp is increasing, information about the roles of microRNA (miRNA) in improving shrimp osmoregulation capacity remain limited.

MicroRNAs belong to a class of small non-coding RNAs, approximately 22 nucleotides in length, that regulate the post-transcription expression of messenger RNAs through sequence-specific base pairing and the consequent silencing of target mRNAs [16]. The miRNAs are involved in many biological processes, such as cellular differentiation, proliferation, apoptosis and immunity [17–20]. With the development of high-throughput sequencing technologies, there has been a rapid growth in the number of identified miRNAs in aquatic animals. Recently, several studies have demonstrated the role of miRNAs in the stress response of shrimps [21–23]. In a previous study, we also identified miRNAs in *P. monodon* under severe low-salinity stress (3 ppt). A total of 43 differentially expressed miRNAs were selected as putative key modulators, whose target genes formed a network to maintain homeostasis [24]. However, another dimension of salinity variation requires attention. Under the influence of climate change, frequent summer rainstorms in South China have led to a significant reduction in seawater salinity, decreasing from ~30 ppt to below 20 ppt, resulting in moderately low salinity stress on shrimp. It has been proven that the response of *P. monodon* varies with salinity levels. Extremely low salinity (<2.5 ppt) induced the lowest survival and growth rate, while moderately low salinity (10 ppt) had a relatively small impact [6]. Additionally, a difference in rearing salinity could also affect the microbiota and gene expressions of shrimp intestines [25]. Thus, to gain deeper insights into the regulatory role of microRNAs in *P. monodon*, further experiments related to moderately low salinity are needed.

In crustaceans, the hepatopancreas acts as a multifunctional organ, coordinating nutrient absorption and digestion, energy storage, lipid and carbohydrate metabolism, detoxification, and immunity [26]. Transcriptome analysis has been widely employed to investigate gene expression patterns in shrimp under low-salinity conditions [27,28]. However, the regulatory roles of miRNAs in the low-salinity adaptation of *P. monodon* remain poorly understood. In the present study, we constructed small RNA libraries from the hepatopancreas of black tiger shrimp at four time points after moderately low salinity stress (0, 6, 24, and 96 h). Based on differentially expressed miRNAs and their target genes, the molecular mechanisms under moderately low salinity stress were comprehensively analyzed. Our results will expand the understanding of shrimp miRNA regulation and provide a scientific basis for the breeding of improved varieties of *P. monodon*.

2. Materials and Methods

2.1. Animals Used for the Experiments

P. monodon for the experiments were acquired from the Shenzhen experimental base of the South China Sea Fisheries Research Institute, Chinese Academy of Fisheries Sciences, Guangzhou, China. Shrimp weighing 10 ± 2 g were cultivated in aerated and filtered seawater maintained at a temperature of 28–30 °C and a salinity level of 30–32 ppt, while being fed three times a day with commercial feed (Dongteng Feed, Zhanjiang, China). Approximately one third of the water in each tank was replaced every day.

2.2. Treatment of Low Salinity Stress

Following previous works, a salinity level of 17 ppt and a duration of 96 h were chosen to induce a moderately low salinity stress treatment [14,15]. Experimental shrimp were randomly selected and exposed to salinity 17. Three plastic tanks were set up, with each tank containing 30 shrimp. Hepatopancreas tissues of randomly selected individuals were collected at specified intervals of 0 (serving as the control), 6, 24, and 96 h post-stress exposure. All samples were placed in tubes and frozen in liquid nitrogen overnight.

2.3. Total RNA Extraction and Quality Analysis

Total RNA was extracted by using a Trizol reagent kit (Invitrogen, Carlsbad, CA, USA) according to the manufacturer's instructions. To eliminate residual DNA, the extracted total RNA was incubated with RNase-free DNase I (Takara Bio, Shiga, Japan) at 37 °C for 30 min. The concentration of RNA was assessed using a NanoDrop 2000 spectrophotometer (Thermo Scientific, Waltham, MA, USA), while agarose gel electrophoresis was utilized to determine the integrity of the RNA samples.

2.4. Small RNA Library Construction and Sequencing

Polyacrylamide gel electrophoresis (PAGE) was used to enrich RNA molecules falling within the 18–30 nucleotide (nt) size range. Following this, adapters were introduced and ligated to the RNA molecules. The ligated RNA products were then subjected to reverse transcription via PCR amplification. PCR products ranging from 140–160 bp in length were subsequently enriched to construct a cDNA library, which was sequenced on an Illumina Novaseq 6000 platform at Gene Denovo Biotechnology Co. (Guangzhou, China).

2.5. Bioinformatics Analysis of miRNA Transcriptome

Original data obtained from Illumina sequencing were preprocessed using fastp v.0.23.4 [29] to eliminate low-quality reads, adaptors and other contaminative sequences. Filtered reads were subjected to BLAST v.2.13.0 searches against the GeneBank and Rfam databases [30] to remove non-coding RNA sequences (including tRNAs, scRNAs, snoRNAs, snRNAs and rRNAs). Reads mapped to exons, introns and repeat sequences were also removed by aligning to the reference genome of *P. monodon*. The remaining reads were searched against the miRBase database [31] to identify known miRNAs. For unannotated sequences, a search was conducted using miRDeep2 v.0.1.3 [32] to predict novel miRNA candidates.

2.6. Expressed Analysis and Target Prediction of miRNA

Transcripts per million (TPM) were used to investigate the expression levels of miRNA. Differentially expressed miRNAs (DEMs) between the control group and experimental group were detected using the DESeq2 package [33] in R. In this study, the threshold for defining differentially expressed miRNAs was set as $|\log 2 \text{ fold change}| \geq 1$ and p value < 0.05. After obtaining the files of differentially expressed miRNAs, trend analysis was performed using the OmicShare tool [34] with default parameters.

The target genes of miRNA were predicted using a variety of software tools, including miRanda v.3.3a [35], RNAhybrid v.2.1.2 [36] and TargetScan v.7.0 [37]. We employed the default parameters of TargetScan to seek miRNA seed matches (2–8 nt sequences starting from the 5′ end of the miRNA) in 3′-UTR sequences. Additionally, miRanda and RNAhybrid were utilized to align entire miRNA sequences. The parameters of miRanda were set to a score >140 and free energy <−20 kcal/mol, respectively. The RNAhybrid software was run using the following parameters: free energy threshold <−20 kcal/mol and helix constraint from 2 to 8. The final set of predicted target genes for DEMs was determined by identifying the overlap between the results obtained from all the prediction tools.

2.7. Enrichment Analysis of Target Genes

To gain a better understanding of the biological activities regulated by DEMs enriched in trend modules, Gene Ontology (GO) [38] and KEGG pathway [39] enrichment analyses of target genes were conducted. Based on hypergeometric tests and FDR correction, terms or pathways exhibiting an FDR less than 0.05 were defined as significantly enriched. We employed the OmicShare tool [34] to select and visualize the top 20 results.

2.8. Validation of DEMs and Statistical Analysis

To assess the reliability of miRNA expression profiles from high-throughput sequencing, nine random DEMs were selected for the validation of their expression. Then, stem-loop qRT-PCR was conducted. Primers were designed and are listed in Supplementary Table S1. To normalize the miRNA expression levels, U6 snRNA was chosen to be an internal control. Further qPCR experiments were carried out in a LightCycler® 480 II Real-time PCR Instrument (Roche, Basel, Switzerland) using SYBR Green Master Mix (Qiagen, Hilden, Germany) following the manufacturer's recommendations. Each 25 μL PCR reaction volume was carried out with 12.5 μL PCR mastermix, 2 μL primers, 5.5 μL distilled water, and 5 μL of cDNA template. The amplification protocol consisted of an initial denaturation at 95 °C for 10 s, followed by 40 cycles at 95 °C for 15 s, 60 °C for 30 s, and 72 °C for 30 s. The relative expression of each DEM was calculated using the $2^{-\Delta\Delta Ct}$ method. Finally, the results were statistically analyzed using a one-way ANOVA with SPSS v.23.0 (IBM, Armonk, NY, USA).

3. Results

3.1. Identification and Classification of miRNAs

Twelve separate small RNA libraries were constructed from the hepatopancreas of *P. monodon*. Based on high-throughput sequencing, 159,919,915 raw reads were generated. After removing the low-quality and contaminated reads, 145,241,136 (90.82%) clean reads were acquired from all libraries. The sequence length distribution of small RNAs ranged from 18 nt to 29 nt, with 21 nt small RNAs being the most abundant, followed by 22 nt and 23 nt (Figure 1).

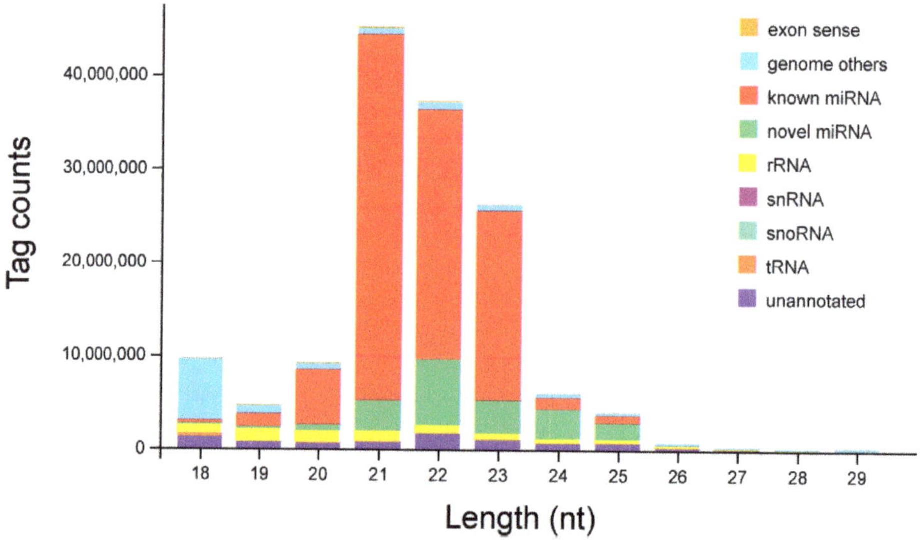

Figure 1. Length distribution of small RNA sequences.

For small RNA annotation, the clean reads were aligned to the genome and various databases. A total of 125,553,226 (86.44%) reads were mapped to the genome sequences. Then, small RNAs were annotated and classified into rRNAs, snRNAs, tRNAs, snoRNAs, miRNAs, and unannotated RNAs after being compared with the Rfam and GenBank databases (Figure 1). The number of miRNAs detected in each sample ranged from 479 to 605, with no significant differences (Figure S1). Finally, 684 miRNAs were identified from hepatopancreas tissues in total, including 488 known miRNAs and 196 novel miRNAs (Table S2).

3.2. Principal Component Analysis

In order to evaluate the relationships between all samples, a principal component analysis was conducted. Based on the two-dimensional results, the groups exhibit some overlap and are not completely separated from each other (Figure 2A). It seems that the 6 h group (6 h) is close to the control group (CK), while the 24 h and 96 h groups are similar. The same results can also be found in Figure 2B, which illustrates the relationships among samples in a three-dimensional representation. As the duration of low-salinity stress increases, the expression of miRNAs exhibits different patterns.

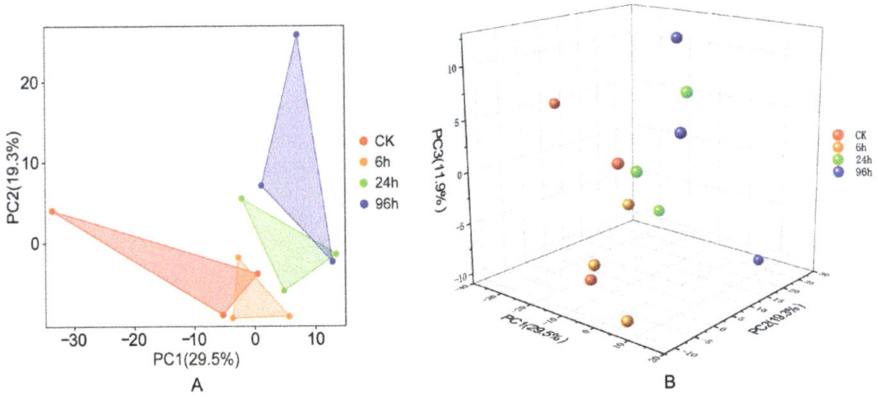

Figure 2. Relationships between samples according to principal component analysis. Groups subjected to different treatments are represented by different colors. (**A**) Visualization of 2D PCA results. (**B**) Visualization of 3D PCA results.

3.3. Expression Profiles of Differentially Expressed miRNA

Compared to the controls, miRNAs with a fold change greater than 2 and a p value less than 0.05, were identified as differentially expressed (DEM). Upon exposure to moderately low salinity treatments for 6 h, 24 h, and 96 h, there were 4, 14, and 44 miRNAs that showed up-regulated expression, respectively, while 22, 40, and 36 miRNAs demonstrated down-regulated expression (Figure 3, Table S3). An UpSet diagram analysis of the DEMs revealed that various miRNAs responded to different low-salinity conditions, with a total of 12 miRNAs being differentially expressed across all stages of treatment (Figure 3).

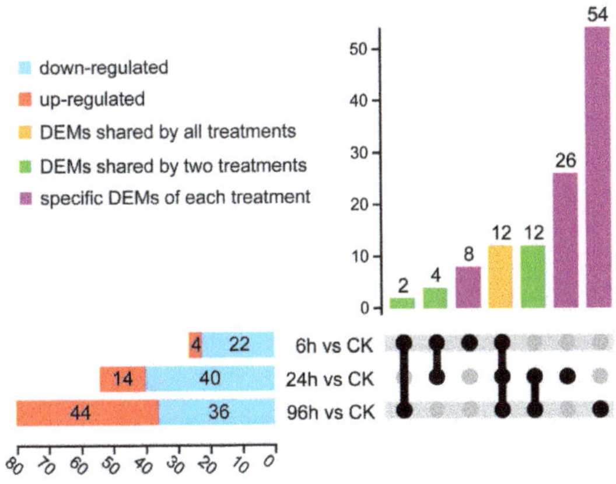

Figure 3. UpSet diagrams based on DEMs at different stages of moderately low salinity treatment.

3.4. Trend Analysis of Differentially Expressed miRNA

To explore the potential role of miRNAs in *P. monodon* under low-salinity stress, we performed an analysis of the expression trends of DEMs. After excluding the profiles that did not enrich any miRNAs, fifteen distinct trend profiles were obtained (Figure 4). Based on p values, the DEMs were most enriched in four profiles, including two descending and two rising trend modules. Distinct expression patterns were observed across four significant expression profiles. In profile 19, miRNAs were up-regulated consistently. In profiles 10 and 9, most miRNAs maintained their original expression levels within 24 h and were up-regulated or down-regulated from 24 h to 96 h. In contrast, expression levels of miRNAs in profile 1 were down-regulated over 0–24 h, while being up-regulated from 24 h to 96 h. It is commonly recognized that miRNAs play essential regulatory roles by targeting mRNA for cleavage or translational repression. Consequently, DEMs enriched in those four trend modules (profile 1, profile 9, profile 19, profile 10) were used for further analysis.

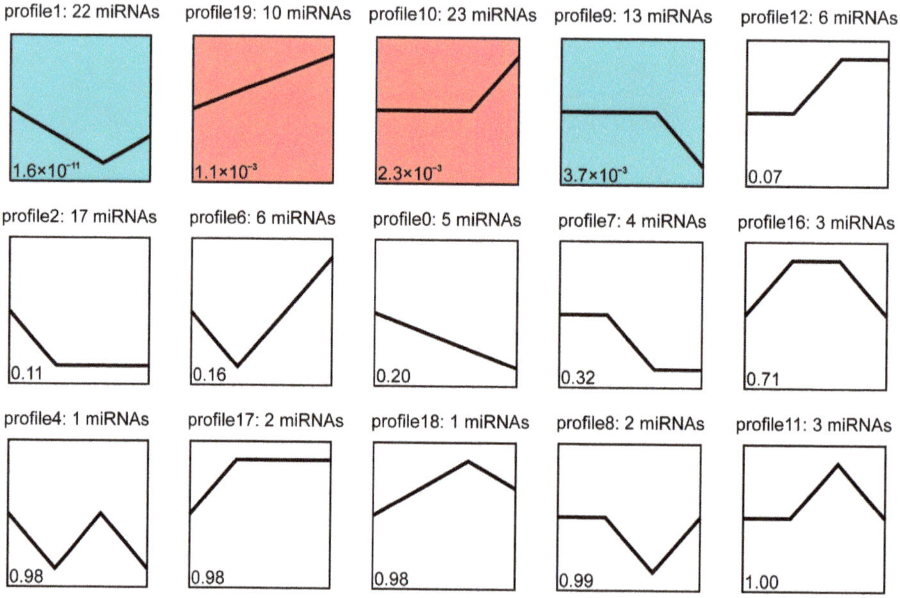

Figure 4. Schematic diagram of miRNA expression trends between treatments. The trend ID and the number of genes it contains are at the top. The colored trend blocks indicate a trend of significant enrichment, and different colors indicate different trends.

3.5. Prediction of miRNA Targets and Functional Enrichment Analysis

Based on the published genome of *P. monodon*, we predicted putative targets of the DEMs by using multiple tools. For all 118 DEMs, a total of 4907 transcripts were identified as targets, indicating that a single miRNA regulates multiple target genes. Subsequently, the target genes of DEMs enriched in four significant trend modules were subjected to GO enrichment analysis. The top 20 GO terms with the lowest Q values were selected as significantly enriched functional categories (Figure 5). Results of the GO analysis demonstrated that there are significant differences in the functions of target genes regulated by DEMs in rising trend modules (profile 10, profile 19) and descending modules (profile 1, profile 9). For up-regulated DEMs, pathways involved in development (GO:0055034, GO:0014043, GO:0031581, GO:1903430, GO:0007438), differentiation (GO:0001742, GO:0001779, GO:0060708, GO:0001746, GO:1904800), biosynthesis (GO:0006042, GO:0052699, GO:0052703) and metabolism (GO:0052698, GO:0052701) were enriched (Figure 5A). In terms of down-regulated DEMs, top terms were related to microtubule (GO:0031116, GO:0035371, GO:0051010, GO:0008017, GO:1990752), binding (GO:0005515, GO:0017075, GO:0005488, GO:0001067) and regulation processes (GO:0031116, GO:0048522, GO:0048518, GO:0031113, GO:0031112, GO:0045893, GO:0001067) (Figure 5B).

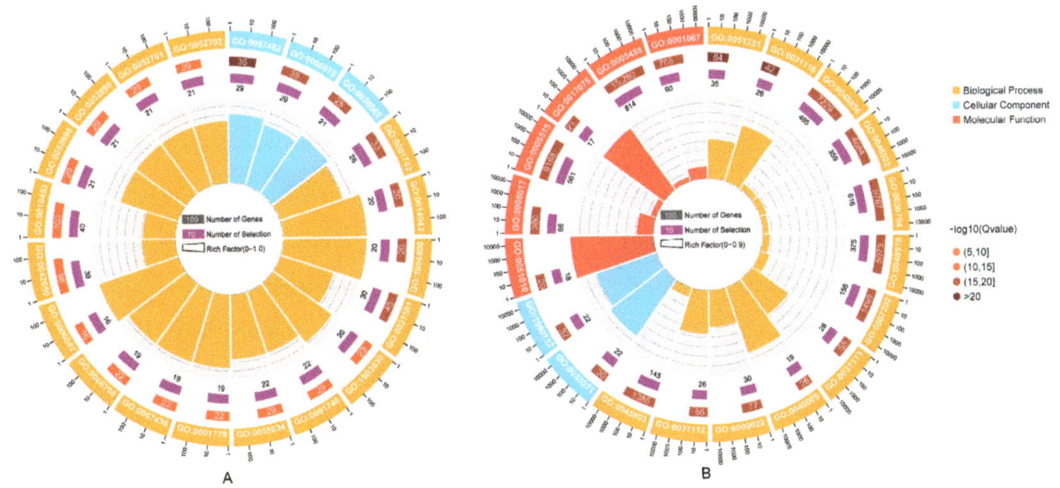

Figure 5. GO enrichment analysis of target genes. (**A**) Genes regulated by miRNA enriched in rising trend modules. (**B**) Genes regulated by miRNA enriched in descending trend modules.

Similar to the GO analysis, a KEGG pathway analysis was also performed on target genes regulated by DEMs in four significant trend modules. Similar to the GO enrichment analysis, the top 20 significantly enriched KEGG pathways were selected based on Q values. The results showed that target genes could be classified into several different categories, including organismal systems, cellular processes, environmental information processing and metabolism. However, significant differences were also found among pathways associated with different modules (Figure 6). Genes targeted by up-regulated DEMs were most concentrated on metabolism (alanine, aspartate and glutamate metabolism, neomycin, kanamycin and gentamicin biosynthesis, taurine and hypotaurine metabolism, amino sugar and nucleotide sugar metabolism, glutathione metabolism, arachidonic acid metabolism), the endocrine system (cortisol synthesis and secretion, parathyroid hormone synthesis, secretion and action, progesterone-mediated oocyte maturation, melanogenesis), the digestive system (gastric acid secretion, salivary secretion) and development (dorso-ventral axis formation). For down-regulated DEMs, target genes were associated with the nervous system (synaptic vesicle cycle, dopaminergic synapse, axon guidance, axon regeneration, neurotrophin signaling pathway, hedgehog signaling pathway), the immune system (Th1 and Th2 cell differentiation, IL-17 signaling pathway, Th17 cell differentiation), signal transduction (ras signaling pathway, MAPK signaling pathway, phosphatidylinositol signaling system) and environmental adaptation (thermogenesis, endocytosis).

Furthermore, the interactions between miRNAs and their target genes enriched in the top 20 KEGG pathways were predicted. The results showed that miRNAs with different expression trends also have different regulatory patterns (Figure 7). The down-regulated target genes are primarily controlled by several miRNAs, such as miR-8485, miR-3082, miR-1187, novel-m0129 and novel-m0071. On the contrary, genes targeted by down-regulated miRNAs are relatively dispersed. More down-regulated miRNAs participated in the top 20 KEGG pathways. It is noteworthy that some genes are regulated by three or more miRNAs, which are listed in Table 1. Functional analysis shows that these common target genes are primarily involved in processes such as metabolism, signal transduction, development and differentiation, and cytoskeleton regulation.

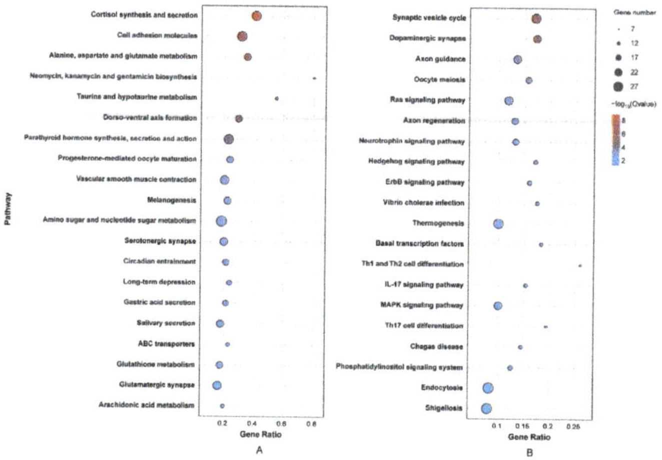

Figure 6. KEGG pathway analysis of target genes. (**A**) Genes regulated by miRNA enriched in rising trend modules. (**B**) Genes regulated by miRNA enriched in descending trend modules.

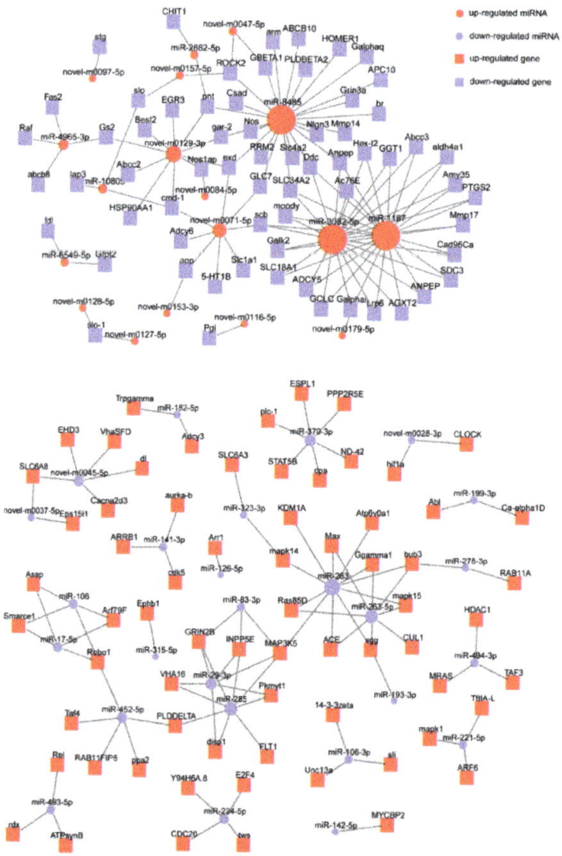

Figure 7. Interactions between DEMs and target genes enriched in KEGG pathway.

Table 1. Details of common target genes.

Gene Symbol	Description	Related miRNA	Gene Function
Anpep	aminopeptidase N-like	miR-1187, miR-3082-5p, miR-8485	Amino acid metabolism
cmd-1	calmodulin isoform X1	miR-10805, novel-m0071-5p, novel-m0129-3p	Signal transduction
Ddc	aromatic-L-amino-acid decarboxylase-like	miR-1187, miR-3082-5p, miR-8485	Amino acid metabolism
exd	homeobox protein extradenticle-like isoform X1	miR-8485, novel-m0071-5p, novel-m0129-3p	Development and differentiation
Galphai	guanine nucleotide-binding protein G(i) subunit alpha isoform X1	miR-1187, miR-3082-5p, novel-m0179-5p	Lipid metabolism
pnt	ETS-like protein pointed isoform X1	miR-2682-5p, miR-8485, novel-m0129-3p	Development and differentiation
ROCK	rho-associated protein kinase 1-like	miR-8485, novel-m0047-5p, novel-m0157-5p	Regulation of cytoskeleton
scb	integrin alpha-7-like	miR-1187, miR-3082-5p, novel-m0071-5p	Regulation of cytoskeleton
SLC34A2	sodium-dependent phosphate transport protein 2B-like	miR-1187, miR-3082-5p, miR-8485	Phosphate digestive absorption
MAP3K5	mitogen-activated protein kinase kinase kinase 15-like isoform X1	miR-263, miR-285, miR-29-3p, miR-83-3p	Signal transduction
bub3	mitotic checkpoint protein BUB3-like isoform X1	miR-263-5p, miR-263, miR-278-3p	Pathogen infection
GRIN2B	glutamate receptor ionotropic, NMDA 2B-like isoform X1	miR-285, miR-29-3p, miR-83-3p	Synapse maturation and plasticity
INPP5E	phosphatidylinositol polyphosphate 5-phosphatase type IV-like isoform X1	miR-285, miR-29-3p, miR-83-3p	Signal transduction
Robo1	protein sax-3-like isoform X1	miR-106, miR-17-5p, miR-452-5p	Axon guidance and regeneration

3.6. Stem-Loop qPCR Validation of Differentially Expressed miRNA

To verify the reliability of the high-throughput sequencing and bioinformatics analysis, six DEMs were randomly selected and subjected to stem-loop qPCR validation (Figure 8). The results revealed that the expression levels of these DEMs were consistent with the results of miRNA sequencing. The correlation coefficient of regression analysis between these two approaches was calculated to be 0.7634 ($p < 0.0001$), indicating that the miRNA expression profiling of sequencing data can reflect reliable expression levels in *P. monodon* hepatopancreas tissue under low-salinity treatment. Further analysis predicted 406 target genes regulated by the 6 DEMs, which are primarily associated with nervous system function, signal transduction, development, and regeneration, suggesting that miRNAs play a critical role in low salinity adaptation (Table S4).

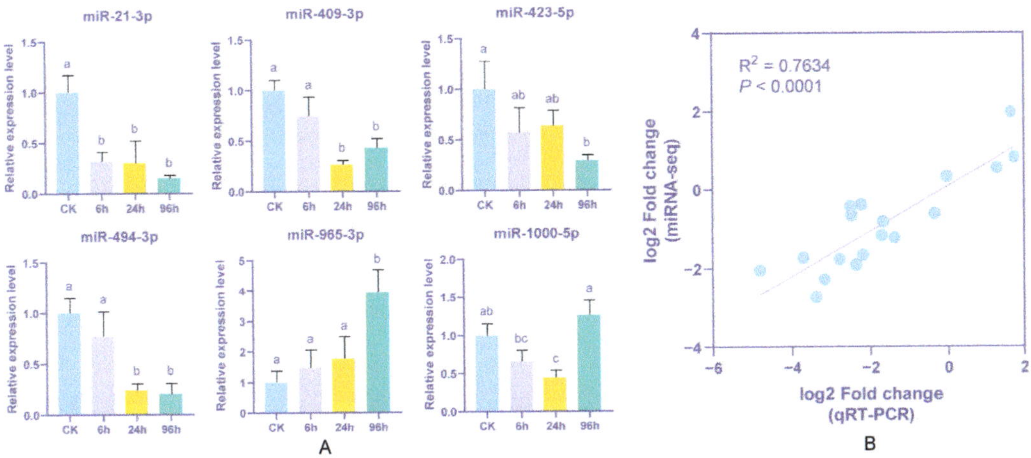

Figure 8. Expression levels of six miRNAs. Bars with different letters indicate significant differences (**A**). Comparison of relative fold change obtained from qPCR analysis and miRNA-seq data (**B**).

4. Discussion

As an environmental factor, salinity significantly impacted various parameters of the black tiger shrimp, including growth, survival and gene expression. In recent years, heavy rainfall caused by global climate change has frequently been observed in southern China, resulting in a decrease in water salinity from 30 ppt to below 20 ppt, which has severely affected the development of the *P. monodon* aquaculture industry [11]. Thus, there is an urgent need to understand how low salinity affects shrimp. Over the past few decades, a large number of miRNAs have been identified and proven to serve as key players in a robust adaptive response against stress in aquatic animals [21–23]. However, there are relatively few reports on the role miRNAs play in salinity response in crustaceans, especially *P. monodon*. In a previous study, we predicted the molecular mechanisms of *P. monodon* under severe low-salinity stress (3 ppt) [24]. Nevertheless, studies have indicated that varying low-salinity gradients differentially affect the physiological, biochemical, and gene expression profiles of *P. monodon* [6]. This suggests that distinct regulatory networks are activated in *P. monodon* to cope with different low-salinity conditions. Therefore, to better understand the molecular mechanisms underlying its low-salinity adaptation, research should focus not only on extreme low-salinity stress but also on moderately low salinity gradients. Based on small RNA transcriptome sequencing and bioinformatics analyses, a total of 684 microRNAs were identified from the hepatopancreas, including 481 known miRNAs, 196 novel miRNAs and 7 conserved miRNAs. The length of the miRNAs was mainly concentrated between 21 and 23 nt, which was consistent with other observations in *P. monodon* [40]. The miRNAs identified in this study lay the groundwork for future investigations into shrimp miRNAs and will help to fill the knowledge gap regarding their function in shrimp stress response.

To identify the miRNAs involved in the response to low-salinity exposure, miRNA expression levels at 6 h, 24 h and 96 h were compared with those in controls. A total of 118 microRNAs were differentially expressed, implying their potential roles in stress response. Compared with our previous work, the number of DEMs was lower. The reason for this might be that moderately low salinity has a minor impact on *P. monodon* and there is no need for a large number of DEMs to cope with stress. Similar results were also found in a previous study, which indicated that a salinity of 10 ppt had less effect on growth and gene expression compared to salinities of 5 ppt and 2.5 ppt [6]. Moreover, to investigate the expression patterns of candidate miRNAs in response to low salinity, DEMs were subjected to trend analysis. Finally, 68 DEMs with either rising or descending trends were significantly enriched and selected for subsequent analysis. As key endogenous mediators of RNA interference (RNAi), miRNAs are capable of modulating various biological processes in organisms, such as cell differentiation, development, metabolism, and apoptosis [41]. The expression patterns of DEMs suggested that a series of target genes were repressed or activated in *P. monodon* under low-salinity stress. Similarly, in a transcriptome study, it was found that low salinity induced the differential expression of genes, including 1140 up-regulated genes and 1531 down-regulated genes [42].

The target genes of DEMs were predicted using bioinformatic methods to characterize the function of miRNAs in shrimp stress response to low-salinity exposure. However, the prediction of target genes varies among software tools. Referring to published works in *L. vannamei* [20] and *Fenneropenaeus chinensis* [22], we employed multiple software tools to obtain target genes in the present study. A total of 4907 transcripts were identified as the targets of 118 DEMs, indicating that a single miRNA could regulate numerous target genes. Furthermore, enrichment analysis was conducted to elucidate the biological function of these predicated target genes. According to GO analysis, genes targeted by up-regulated DEMs were highly enriched in terms of development, differentiation, biosynthesis and metabolism. This might suggest that the growth of *P. monodon* was restricted by low salinity, which is consistent with previous studies on shrimp [6,11]. On the contrary, genes controlled by down-regulated DEMs were concentrated on microtubule, binding and regulation processes, indicating that a series of stress-related responses were activated to maintain cellular homeostasis [24]. Similarly, in the miRNA transcriptomes of *Apostichopus japonicus*, the salinity adaptation process was related to ion channels, transporters, and environmental signals, indicating the regulatory roles of miRNAs [43].

KEGG pathway analysis of potential target genes provides a powerful approach to investigate the molecular mechanisms of *P. monodon* under low-salinity stress conditions. In this study, target genes involved in amino acid metabolism, amino sugar metabolism and arachidonic acid metabolism were governed by up-regulated miRNAs. Amino acid metabolism is critical for physiological homeostasis [44]. Previous studies have demonstrated that stress significantly disrupts amino acid metabolic pathways in *L. vannamei*, resulting in a decrease in most amino acid concentrations [45]. Amino sugar metabolism, particularly chitin synthesis, is essential for crustacean exoskeletal integrity, which plays a potential role in growth [46]. As an important fatty acid, arachidonic acid could enhance the antioxidant and immune capacity of fish, resulting in improved growth [47]. Given that miRNAs function to silence mRNA expression, we can hypothesize that low-salinity stress suppresses metabolic processes in *P. monodon*, potentially further impacting its growth. Moreover, genes related to the digestive and endocrine systems are also managed by up-regulated miRNAs. Salivary and gastric acid are responsible for food digestion and mineral absorption. A decrease in salivary and gastric acid secretion means that low salinity affected the feeding behavior of shrimp, which could be a contributing factor to the slow growth observed in shrimp under low-salinity conditions [6]. Cortisol and parathyroid hormones, which belong to the endocrine system, play crucial roles in dealing with harsh environments [48,49]. In our study, changes in these two hormones suggested that energy metabolism and immune response were affected by stress. Based on the above results, it can be inferred that low salinity has impacts on a series of biological processes related to metabolism, digestion and hormone synthesis, ultimately causing retarded growth in *P. monodon*.

Low salinity can cause damage to the hepatopancreas and compromise the immune system, leading to outbreaks of disease and death in shrimp. When *L. vannamei* were exposed to low-salinity stress, the levels of *Vibrio* infection increased [4]. In this research, three pathways involved in pathogen infection were also significantly enriched, indicating the risk of disease outbreak. For crustaceans, the innate immune response is crucial for defense against microbial infections. Being closely associated with innate immune cells, endocytosis is important for maintaining normal cellular function by removing foreign substances and protecting the host from pathogen/virus attacks. A comparative proteome analysis of the hepatopancreas in *L. vannamei* revealed a significant up-regulation of the protein associated with endocytosis under long-term low-salinity stress [50]. Similar results were detected in this study, implying the potential role of endocytosis in shrimp innate immunity. It is commonly acknowledged that invertebrates, including crustaceans, do not possess an acquired immune system. Recently, increasing evidence has suggested that a specific acquired immunity may exist in invertebrates, which is called 'immune priming' [51]. After challenge with WSSV, pathways related to acquired immunity, such as Th1 and Th2 cell differentiation, the IL-17 signaling pathway and Th17 cell differentiation were detected in Pacific white shrimp [52]. Additionally, another study implied that thermogenesis may play a key role in the interactions between shrimp and pathogens across tissues [53]. Interestingly, these immune-related pathways were also found in this research, reflecting immune priming in *P. monodon*. These pathway changes indicate that low salinity increased disease susceptibility, triggering various immune responses in the hepatopancreas of *P. monodon*.

Although *P. monodon* has euryhalinity, its survival depends on its ability to sense and respond to salinity changes. As a key mediator of the stress response, the nervous system gathers information from sensory organs, processes it, and then coordinates actions. It was reported that the nervous system in crustaceans has the capability to regulate numerous physiological processes, thereby maintaining the organism's homeostasis under stressful conditions [54]. From the genome of *L. vannamei*, genes related to neural development are notably enriched for environment adaptation [55]. Our results showed that pathways related to synapse and axon were enriched under low salinity, which was consistent with a previous study [24]. It can be inferred that the nervous system is vital in adapting to salinity changes in *P. monodon*. Signal transduction serves as a bridge that connects sensing and responding. The renin-angiotensin system (RAS) is a hormone-based system that participates in osmoregulation and salinity adaptation by regulating the water balance and blood pressure of teleost fish [56]. A comparative analysis of Pacific white shrimp showed that the RAS signaling pathway was up-regulated under multiple environmental stressors [57]. As receptors for growth factors, the ErbB family mediate a variety of cellular responses. In aquatic animals, the ErbB signal is involved in development, homeostasis and pathologies [58,59]. Similar to ErbB, the mitogen-activated protein kinase (MAPK) pathway is a multifunctional signaling pathway which can modulate diverse cellular function, including growth, differentiation and responses to external stress. It has been demonstrated that genes linked to the MAPK signaling pathway are up-regulated in shrimp in response to low-salinity stress [14]. As a vital component of cell membranes, phosphatidylinositol is essential for signal transduction in response to osmotic stress. In *L. vannamei*, dietary myo-inositol can regulate osmotic pressure through the phosphatidylinositol signaling pathway, ultimately alleviating the damage induced by low-salinity stress [60]. Due to the down-regulation of relevant miRNAs, we can conclude that various signal pathways were activated to cope with acute low-salinity stress.

5. Conclusions

This study has demonstrated the regulatory role of miRNAs in black tiger shrimp under moderately low salinity stress (17 ppt). While low salinity impaired growth processes and increased disease vulnerability, miRNAs modulated compensatory pathways to alleviate stress effects, such as signal transduction and immune priming. These findings contrast with prior studies on severe salinity stress, highlighting distinct molecular mechanisms in moderate conditions. The identification of stress-responsive miRNAs provides potential biomarkers for breeding programs aimed at improving salinity tolerance in shrimp aquaculture. Further functional validation of these miRNAs could support the development of resilient shrimp strains, aiding sustainable aquaculture in fluctuating environments.

Supplementary Materials: The following supporting information can be downloaded at: https://www.mdpi.com/article/10.3390/biology14040440/s1, Figure S1: Number of miRNAs detected in each sample; Table S1: Primers used for miRNA expression validation; Table S2: Detected miRNAs and their expression levels; Table S3: Expression patterns of differentially expressed miRNAs across experimental groups; Table S4: Target genes regulated by six differentially expressed miRNAs.

Author Contributions: J.S.: Writing—original draft, Preparation, Writing—review & editing, Project administration. S.J.: Investigation, Formal analysis, Data curation. Y.D.: Methodology, Software, Investigation. H.D.: Investigation, Formal analysis. W.L.: Validation, Investigation. J.H.: Conceptualization, Methodology. Y.L.: Data curation. L.Y.: Visualization, Supervision. Q.Y.: Resources, Methodology, Writing—review & editing. F.Z.: Conceptualization, Methodology, Writing—review

& editing, Funding acquisition. All authors have read and agreed to the published version of the manuscript.

Funding: This work was supported by the National Key R & D Program of China (2022YFD2400104), the Rural Revitalization Strategy Special Fund Seed Industry Revitalization Project of Guangdong Province (2022SPY00006 and 2022SPY00002), the China Agriculture Research System of MOF and MARA (CARS-48), and the Central Public Interest Scientific Institution Basal Research Fund, South China Sea Fisheries Research Institute, CAFS (NO. 2023TS01, NO. 2024XT01 and NO. 2023TD34).

Institutional Review Board Statement: The animal study protocol was approved by the Animal Ethics Committee of South China Sea Fisheries Research Institute (protocol code No. nhdf2024-02).

Informed Consent Statement: Not applicable.

Data Availability Statement: Data will be made available on request.

Conflicts of Interest: The authors declare no conflicts of interest.

References

1. Angthong, P.; Uengwetwanit, T.; Arayamethakorn, S.; Rungrassamee, W. Transcriptomic analysis of the black tiger shrimp (*Penaeus monodon*) reveals insights into immune development in their early life stages. *Sci. Rep.* **2021**, *11*, 13881. [CrossRef] [PubMed]
2. Fisheries Administration Bureau, Ministry of Agriculture. *China Fishery Statistics Year Book*; China Agriculture Press: Beijing, China, 2024.
3. Ahmed, N.; Diana, J.S. Threatening "white gold": Impacts of climate change on shrimp farming in coastal Bangladesh. *Ocean Coast. Manag.* **2015**, *114*, 42–52. [CrossRef]
4. Chang, Y.T.; Huang, W.T.; Wu, P.L.; Kumar, R.; Wang, H.C.; Lu, H.P. Low salinity stress increases the risk of *Vibrio parahaemolyticus* infection and gut microbiota dysbiosis in Pacific white shrimp. *BMC Microbiol.* **2024**, *24*, 275. [CrossRef] [PubMed]
5. Ferraris, R.P.; Parado-Estepa, E.D.; De Jesus, E.G.; Ladja, J.M. Osmotic and chloride regulation in the hemolymph of the tiger prawn *Penaeus monodon* during molting in various salinities. *Mar. Biol.* **1987**, *95*, 377–385. [CrossRef]
6. Rahi, M.L.; Azad, K.N.; Tabassum, M.; Irin, H.H.; Hossain, K.S.; Aziz, D.; Moshtaghi, A.; Hurwood, D.A. Effects of Salinity on Physiological, Biochemical and Gene Expression Parameters of Black Tiger Shrimp (*Penaeus monodon*): Potential for Farming in Low-Salinity Environments. *Biology* **2021**, *10*, 1220. [CrossRef]
7. Roy, L.A.; Davis, D.A.; Saoud, I.P.; Henry, R.P. Effects of varying levels of aqueous potassium and magnesium on survival, growth, and respiration of the Pacific white shrimp, *Litopenaeus vannamei*, reared in low salinity waters. *Aquaculture* **2007**, *262*, 461–469. [CrossRef]
8. Guo, H.; Xian, J.A.; Zheng, P.H.; Lu, Y.P.; Wang, L.; Zhang, X.X.; Li, J.T.; Zhang, Z.L.; Wang, A.L. Dietary copper affects antioxidant status of shrimp (*Penaeus monodon*) reared in low salinity water. *Aquac. Rep.* **2022**, *22*, 100979. [CrossRef]
9. Pan, L.; Liu, H.; Zhao, Q. Effect of salinity on the biosynthesis of amines in *Litopenaeus vannamei* and the expression of gill related ion transporter genes. *J. Ocean Univ. China* **2014**, *13*, 453–459. [CrossRef]
10. Liu, M.; Liu, S.; Hu, Y.; Pan, L. Cloning and expression analysis of two carbonic anhydrase genes in white shrimp *Litopenaeus vannamei*, induced by pH and salinity stresses. *Aquaculture* **2015**, *448*, 391–400. [CrossRef]
11. Gao, W.; Tian, L.; Huang, T.; Yao, M.; Hu, W.; Xu, Q. Effect of salinity on the growth performance, osmolarity and metabolism-related gene expression in white shrimp *Litopenaeus vannamei*. *Aquac. Rep.* **2016**, *4*, 125–129. [CrossRef]
12. Li, E.; Wang, X.; Chen, K.; Xu, C.; Qin, J.G.; Chen, L. Physiological change and nutritional requirement of Pacific white shrimp *Litopenaeus vannamei* at low salinity. *Rev. Aquac.* **2017**, *9*, 57–75. [CrossRef]
13. Huang, M.; Dong, Y.; Zhang, Y.; Chen, Q.; Xie, J.; Xu, C.; Zhao, Q.; Li, E. Growth and Lipidomic Responses of Juvenile Pacific White Shrimp *Litopenaeus vannamei* to Low Salinity. *Front. Physiol.* **2019**, *10*, 1087. [CrossRef]
14. Fan, H.; Li, Y.; Yang, Q.; Jiang, S.; Yang, L.; Huang, J.; Jiang, S.; Zhou, F. Isolation and characterization of a MAPKK gene from *Penaeus monodon* in response to bacterial infection and low-salinity challenge. *Aquac. Rep.* **2021**, *20*, 100671. [CrossRef]
15. Si, M.; Li, Y.; Jiang, S.; Yang, Q.; Jiang, S.; Yang, L.; Huang, J.; Chen, X.; Zhou, F. A CSDE1/Unr gene from *Penaeus monodon*: Molecular characterization, expression and association with tolerance to low salt stress. *Aquaculture* **2022**, *561*, 738660. [CrossRef]
16. Bartel, D.P. MicroRNAs: Genomics, biogenesis, mechanism, and function. *Cell* **2004**, *116*, 281–297. [CrossRef]
17. Hao, R.; Zhu, X.; Tian, C.; Jiang, M.; Huang, Y.; Zhu, C. Integrated analysis of the role of miRNA-mRNA in determining different body colors of leopard coral grouper (*Plectropomus leopardus*). *Aquaculture* **2022**, *548*, 737575. [CrossRef]
18. Yue, C.; Li, Q.; Yu, H. Integrated analysis of miRNA and mRNA expression profiles identifies potential regulatory interactions during sexual development of Pacific oyster *Crassostrea gigas*. *Aquaculture* **2022**, *546*, 737294. [CrossRef]

19. Wang, H.; Wei, H.; Tang, L.; Lu, J.; Mu, C.; Wang, C. Identification and characterization of miRNAs in the gills of the mud crab (*Scylla paramamosain*) in response to a sudden drop in salinity. *BMC Genom.* **2018**, *19*, 609. [CrossRef]
20. Wu, W.; Dai, C.; Duan, X.; Wang, C.; Lin, X.; Ke, J.; Wang, Y.; Zhang, X.; Liu, H. miRNAs induced by white spot syndrome virus involve in immunity pathways in shrimp *Litopenaeus vannamei*. *Fish Shellfish Immun.* **2019**, *93*, 743–751. [CrossRef]
21. Guo, H.; Lu, Z.; Zhu, X.; Zhu, C.; Wang, C.; Shen, Y.; Wang, W. Differential expression of microRNAs in hemocytes from white shrimp *Litopenaeus vannamei* under copper stress. *Fish Shellfish Immun.* **2018**, *74*, 152–161. [CrossRef]
22. He, Y.; Li, Z.; Zhang, H.; Hu, S.; Wang, Q.; Li, J. Genome-wide identification of Chinese shrimp (*Fenneropenaeus chinensis*) microRNA responsive to low pH stress by deep sequencing. *Cell Stress Chaperon.* **2019**, *24*, 689–695. [CrossRef] [PubMed]
23. Wang, W.; Zhong, P.; Yi, J.; Xu, A.; Lin, W.; Guo, Z.; Wang, C.; Sun, C.; Chan, S. Potential role for microRNA in facilitating physiological adaptation to hypoxia in the Pacific whiteleg shrimp *Litopenaeus vannamei*. *Fish Shellfish Immun.* **2019**, *84*, 361–369. [CrossRef] [PubMed]
24. Shi, J.; Jiang, S.; Yang, Q.; Li, Y.; Yang, L.; Huang, J.; Jiang, S.; Zhou, F. Identification of microRNAs in black tiger shrimp (*Penaeus monodon*) under acute low-salinity stress. *Front. Mar. Sci.* **2024**, *11*, 1403559. [CrossRef]
25. Chaiyapechara, S.; Uengwetwanit, T.; Arayamethakorn, S.; Bunphimpapha, P.; Phromson, M.; Jangsutthivorawat, W.; Tala, S.; Karoonuthaisiri, N.; Rungrassamee, W. Understanding the host-microbe-environment interactions: Intestinal microbiota and transcriptomes of black tiger shrimp *Penaeus monodon* at different salinity levels. *Aquaculture* **2022**, *546*, 737371. [CrossRef]
26. Vogt, G. Functional cytology of the hepatopancreas of decapod crustaceans. *J. Morphol.* **2019**, *280*, 1405–1444. [CrossRef]
27. Chen, K.; Li, E.; Li, T.; Xu, C.; Wang, X.; Lin, H.; Qin, J.; Chen, L. Transcriptome and molecular pathway analysis of the hepatopancreas in the Pacific white shrimp *Litopenaeus vannamei* under chronic low-salinity stress. *PLoS ONE* **2015**, *10*, e0131503. [CrossRef]
28. Shekhar, M.S.; Kiruthika, J.; Ponniah, A.G. Identification and expression analysis of differentially expressed genes from shrimp (*Penaeus monodon*) in response to low salinity stress. *Fish Shellfish Immun.* **2013**, *35*, 1957–1968. [CrossRef]
29. Chen, S.; Zhou, Y.; Chen, Y.; Gu, J. fastp: An ultra-fast all-in-one FASTQ preprocessor. *Bioinformatics* **2018**, *34*, i884–i890. [CrossRef]
30. Kalvari, I.; Nawrocki, E.P.; Argasinska, J.; Quinones-Olvera, N.; Finn, R.D.; Bateman, A.; Petrov, A.I. Non-coding RNA analysis using the rfam database. *Curr. Protoc. Bioinf.* **2018**, *62*, e51. [CrossRef]
31. Kozomara, A.; Birgaoanu, M.; Griffiths-Jones, S. miRBase: From microRNA sequences to function. *Nucleic Acids Res.* **2019**, *47*, D155–D162. [CrossRef]
32. Mackowiak, S.D. Identification of Novel and Known miRNAs in Deep-Sequencing Data with miRDeep2. *Curr. Protoc. Bioinf.* **2011**, *36*, 12.10.1–12.10.15. [CrossRef] [PubMed]
33. Love, M.I.; Huber, W.; Anders, S. Moderated estimation of fold change and dispersion for RNA-seq data with DESeq2. *Genome Biol.* **2014**, *15*, 550. [CrossRef] [PubMed]
34. Mu, H.; Chen, J.; Huang, W.; Huang, G.; Deng, M.; Hong, S.; Ai, P.; Gao, C.; Zhou, H. OmicShare tools: A zero-code interactive online platform for biological data analysis and visualization. *iMeta* **2024**, *3*, e228. [CrossRef] [PubMed]
35. Betel, D.; Koppal, A.; Agius, P.; Sander, C.; Leslie, C. Comprehensive modeling of microRNA targets predicts functional non-conserved and non-canonical sites. *Genome. Biol.* **2010**, *11*, R90. [CrossRef]
36. Agarwal, V.; Bell, G.W.; Nam, J.-W.; Bartel, D.P. Predicting effective microRNA target sites in mammalian mRNAs. *eLife* **2015**, *4*, e05005. [CrossRef]
37. Kruger, J.; Rehmsmeier, M. RNAhybrid: MicroRNA target prediction easy, fast and flexible. *Nucleic Acids Res.* **2006**, *34*, W451–W454. [CrossRef]
38. Ashburner, M.; Ball, C.A.; Blake, J.A.; Botstein, D.; Butler, H.; Cherry, J.M.; Davis, A.P.; Dolinski, K.; Dwight, S.S.; Eppig, J.T.; et al. Gene ontology: Tool for the unification of biology. *Nat. Genet.* **2000**, *25*, 25–29. [CrossRef]
39. Kanehisa, M.; Goto, S. KEGG: Kyoto encyclopedia of genes and genomes. *Nucleic Acids Res.* **2000**, *28*, 27–30. [CrossRef]
40. Nantapojd, T.; Panyim, S.; Ongvarrasopone, C. Identification and expression of white spot syndrome virus-encoded microRNAs in infected *Penaeus monodon*. *Aquaculture* **2019**, *503*, 436–445. [CrossRef]
41. Li, Y.; Xu, Q.; Liu, H.; Dai, X. Differential expression of microRNAs in giant freshwater prawn (*Macrobrachium rosenbergii*) during the infection of *Vibrio parahaemolyticus*. *Fish Shellfish Immun.* **2024**, *153*, 109827. [CrossRef]
42. Li, Y.; Si, M.; Jiang, S.; Yang, Q.; Jiang, S.; Yang, L.; Huang, J.; Chen, X.; Zhou, F.; Li, E. Transcriptome and molecular regulatory mechanisms analysis of gills in the black tiger shrimp *Penaeus monodon* under chronic low-salinity stress. *Front. Physiol.* **2023**, *14*, 1118341. [CrossRef] [PubMed]
43. Tian, Y.; Shang, Y.; Guo, R.; Chang, Y.; Jiang, Y. Salinity stress-induced differentially expressed miRNAs and target genes in sea cucumbers *Apostichopus japonicus*. *Cell Stress Chaperon.* **2019**, *24*, 719–733. [CrossRef] [PubMed]
44. Huang, Z.; Aweya, J.; Zhu, C.; Tran, N.; Hong, Y.; Li, S.; Yao, D.; Zhang, Y. Modulation of Crustacean Innate Immune Response by Amino Acids and Their Metabolites: Inferences from Other Species. *Front. Immunol.* **2020**, *11*, 574721. [CrossRef] [PubMed]
45. Zhang, R.; Shi, X.; Guo, J.; Mao, X.; Fan, B. Acute stress response in gill of Pacific white shrimp *Litopenaeus vannamei* to high alkalinity. *Aquaculture* **2024**, *586*, 740766. [CrossRef]

46. Niu, M.; Gao, G.; Qin, K.; Chen, Y.; Wang, H.; Li, X.; Liang, G.; Wang, C.; Mu, C.; Su, Q. Multiple low salinity stress modes provided novel insight into the metabolic response of *Scylla paramamosain* adapting to inland saline-alkaline water. *Front. Mar. Sci.* **2022**, *9*, 977599. [CrossRef]
47. Nayak, S.; Koven, W.; Meiri, I.; Khozin-Goldberg, I.; Isakov, N.; Zibdeh, M.; Zilberg, D. Dietary arachidonic acid affects immune function and fatty acid composition in cultured rabbitfish *Siganus rivulatus*. *Fish Shellfish Immun.* **2017**, *68*, 46–53. [CrossRef]
48. Wang, Y.; Liu, Z.; Liu, C.; Liu, R.; Yang, C.; Wang, L.; Song, L. Cortisol modulates glucose metabolism and oxidative response after acute high temperature stress in Pacific oyster *Crassostrea gigas*. *Fish Shellfish Immun.* **2022**, *126*, 141–149. [CrossRef]
49. Zhang, L.; Zhang, S.; Dou, Y.; Shen, H.; Qiao, Y.; Cao, X.; Fan, X.; Hu, Y.; Qian, J. Hormonal regulation changes in shrimp infected with *Enterocytozoon hepatopenaei*: A transcriptomic analysis. *Aquac. Int.* **2024**, *32*, 841–2863. [CrossRef]
50. Xu, C.; Li, E.; Liu, Y.; Wang, X.; Qin, J.; Chen, L. Comparative proteome analysis of the hepatopancreas from the Pacific white shrimp *Litopenaeus vannamei* under long-term low salinity stress. *J. Proteom.* **2017**, *162*, 1–10. [CrossRef]
51. Medina Gomez, H.; Adame Rivas, G.; Hernández-Quintero, A.; González Hernández, A.; Torres Guzmán, J.C.; Mendoza, H.L.; Contreras-Garduño, J. The occurrence of immune priming can be species-specific in entomopathogens. *Microb. Pathog.* **2018**, *118*, 361–364. [CrossRef]
52. Zhai, Y.; Xu, R.; He, P.; Jia, R. A proteomics investigation of 'immune priming' in *Penaeus vannamei* as shown by isobaric tags for relative and absolute quantification. *Fish Shellfish Immun.* **2021**, *117*, 140–147. [CrossRef] [PubMed]
53. Wu, Z.; Chu, K.H.; Ma, K. Transcriptome Analysis of Multiple Tissues in the Shrimp *Penaeus vannamei* Reveals the Typical Physiological Response to Three Pathogens. *J. Mar. Sci. Eng.* **2023**, *11*, 389. [CrossRef]
54. Ren, X.; Wang, Q.; Shao, H.; Xu, Y.; Liu, P.; Li, J. Effects of Low Temperature on Shrimp and Crab Physiology, Behavior, and Growth: A Review. *Front. Mar. Sci.* **2021**, *8*, 746177. [CrossRef]
55. Zhang, X.; Yuan, J.; Sun, Y.; Li, S.; Gao, Y.; Yu, Y.; Liu, C.; Wang, Q.; Lv, X.; Zhang, X.; et al. Penaeid shrimp genome provides insights into benthic adaptation and frequent molting. *Nat. Commun.* **2019**, *10*, 356. [CrossRef]
56. Kumai, Y.; Bernier, N.J.; Perry, S.F. Angiotensin-II promotes Na^+ uptake in larval zebrafish, *Danio rerio*, in acidic and ion-poor water. *J. Endocrinol.* **2014**, *220*, 195–205. [CrossRef]
57. Farhadi, A.; Tang, S.; Huang, M.; Yu, Q.; Xu, C.; Li, E. Identification of key immune and stress related genes and pathways by comparative analysis of the gene expression profile under multiple environmental stressors in pacific white shrimp (*Litopenaeus vannamei*). *Fish Shellfish Immun.* **2023**, *135*, 108695. [CrossRef]
58. Chen, D.; Xin, Y.; Teng, J.; Zhao, X.; Lu, J.; Li, Y.; Wang, H. Comparative transcriptomic and molecular biology analyses to explore potential immune responses to *Vibrio parahaemolyticus* challenge in *Eriocheir sinensis*. *Front. Cell Infect. Microbiol.* **2024**, *14*, 1456130. [CrossRef]
59. Lai, X.; Zhang, C.; Wu, Y.; Yang, Y.; Zhang, M.; Qin, W.; Wang, R.; Shu, H. Comparative transcriptome analysis reveals physiological responses in liver tissues of *Epinephelus coioides* under acute hypoxia stress. *Comp. Biochem. Phys. D* **2022**, *43*, 101005. [CrossRef]
60. Li, Z.; Chang, T.; Han, F.; Fan, X.; Liu, W.; Wu, P.; Xu, C.; Li, E. Effects of myo-inositol on growth and biomarkers of environmental stress and metabolic regulation in Pacific white shrimp (*Litopenaeus vannamei*) reared at low salinity. *Comp. Biochem. Phys. D* **2024**, *50*, 101216. [CrossRef]

biology

MDPI

Article

Metabolic Response of Black Tiger Shrimp (*Penaeus monodon*) to Acute Ammonia Nitrogen Stress

Yangyang Ding [1,2], Shigui Jiang [1], Song Jiang [1], Yundong Li [1], Qibin Yang [1], Lishi Yang [1], Jianhua Huang [3], Jianzhi Shi [1], Pengying Li [1], Hongshan Diao [1] and Falin Zhou [1,2,3,*]

1 Key Laboratory of South China Sea Fishery Resources Exploitation and Utilization, Ministry of Agriculture, South China Sea Fisheries Research Institute, Chinese Academy of Fishery Sciences, Guangzhou 510300, China; dingyangyang93@163.com (Y.D.); jiangsg@21cn.com (S.J.); tojiangsong@163.com (S.J.); liyd2019@163.com (Y.L.); yangqibin1208@163.com (Q.Y.); yangls2016@163.com (L.Y.); shijianzhi1989@163.com (J.S.); lipengyingcafs@163.com (P.L.); 17875808801@163.com (H.D.)

2 State Key Laboratory of Mariculture Biobreeding and Sustainable Goods, Yellow Sea Fisheries Research Institute, Chinese Academy of Fishery Sciences, Qingdao 266071, China

3 Shenzhen Base of South China Sea Fisheries Research Institute, Chinese Academy of Fishery Sciences, Shenzhen 518121, China; huangjianhua@scsfri.ac.cn

* Correspondence: zhoufalin@scsfri.ac.cn

Simple Summary: Ammonia nitrogen is an extremely harmful pollutant in the aquatic environments, which could cause the death of the shrimp and thus hinder the promotion of new varieties of shrimp, resulting in a huge economic loss. However, the metabolic response of black tiger shrimp (*Penaeus monodon*) to acute ammonia nitrogen stress remains unclear. In this study, the histological change in the hepatopancreas, gills, and intestine, the metabolic product of ammonia nitrogen in the plasma, and the enzyme activity in the hepatopancreas were investigated. The results showed that acute ammonia nitrogen stress could result in severe tissue damage in black tiger shrimp. Meanwhile, black tiger shrimp could convert excessive ammonia nitrogen to urea for detoxification. In addition, black tiger shrimp might modulate antioxidants and apoptosis to reduce the damage caused by ammonia nitrogen.

Abstract: High concentrations of ammonia nitrogen could result in the death of aquatic animals and cause a huge economic loss in the aquaculture industry. However, the metabolic responses to acute ammonia nitrogen stress remain largely unknown in *Penaeus monodon*. In this study, we first investigated the histological change in tissues in *Penaeus monodon* under 96 h acute ammonia nitrogen stress. The result of the paraffin section showed that acute ammonia nitrogen stress induced severe epithelial detachment and lumen dilatation of the hepatopancreas, swollen and hemocyte infiltration of the gills, and mucosa exfoliation and shortened villi of the intestine in *Penaeus monodon*, suggesting the impairment of the normal physiological function in these tissues. We next examined the change in the metabolic product in the plasma and the enzyme activity in the hepatopancreas after ammonia nitrogen stress. Upon ammonia stress, both the concentration of ammonia and urea nitrogen significantly increased, while there was no significant increase in the concentration of uric acid, which is consistent with the results that the enzyme activity of glutamine synthetase (GS), glutamate dehydrogenase (GDH), and aspartate transaminase (GOT) became significantly elevated and the enzyme activity of adenosine deaminase (ADA) in the purine metabolism pathway significantly decreased after ammonia stress, suggesting that shrimp could convert excessive ammonia to urea for ammonia detoxification through the ammonia–nitrogen metabolism pathways. Interestingly, we also observed a significant increase in superoxide dismutase (SOD) activity, suggesting a potential role of this antioxidant enzyme

https://doi.org/10.3390/biology14050501

in the clearance of reactive oxygen species (ROS) induced via ammonia stress. Moreover, we found that acute ammonia nitrogen stress inhibited the enzyme activity of caspase 3 and caspase 8, suggesting an important role of apoptosis in protecting *Penaeus monodon* against acute ammonia stress. Overall, our findings revealed that *Penaeus monodon* may employ metabolic and purine pathways and undergo oxidative stress and apoptosis for ammonia detoxification under ammonia nitrogen stress, thus providing new insight into the metabolic response of shrimp to acute ammonia stress.

Keywords: black tiger shrimp (*Penaeus monodon*); ammonia nitrogen stress; metabolic response; antioxidant enzymes; caspase

1. Introduction

In aquatic environments, ammonia nitrogen (ammonia-N) is a common toxic and harmful substance, and its concentration is an important indicator of the water quality [1]. The origin of ammonia-N is mostly from the natural decomposition of organic matter, the metabolic byproducts of protein catabolism yielding, and the excessive effluent discharge [2]. Ammonia nitrogen refers to the total ammonia in the water, which primarily exists in two forms: ammonium (NH_4^+) and molecular ammonia (NH_3). These two forms are in a state of dynamic equilibrium, and the NH_4^+ to NH_3 ratios are affected by the salinity, temperature, and especially pH, which are also the determinants of ammonia-N toxicity [3]. An inverse relationship exists between salinity and ammonia-N toxicity; that is, a lower salinity increases the NH_4^+ to NH_3 ratio and thus makes it more toxic [4]. When the temperature increased, the ratio of NH_4^+ to NH_3 decreased so the ammonia-N became more toxic [2]. A higher pH will increase the ammonia-N toxicity due to the NH_3 form becoming more prevalent than NH_4^+ [5]. NH_3 is uncharged and highly lipophilic, allowing it to penetrate cell membranes and exert toxic effects on organisms, thus being more toxic than NH_4^+ [6]. Chronic ammonia-N toxicity can cause tissue damage, enhance gill permeability, impair ion exchange functions, and reduce the growth rate and reproductive capacity of crustaceans. It can also increase their susceptibility to pathogens [7]. Acute ammonia-N toxicity can cause aquatic organisms to experience convulsions, persistent hyperexcitability, loss of balance in water, hypoxia, and death. Additionally, it can inhibit the activity of various enzymes, cause severe oxidative damage, and trigger apoptosis [8,9].

In crustaceans, ammonia nitrogen is primarily generated through the transamination and deamination of amino acids via specific enzymes, and these reactions are usually reversible [10]. For example, glutamine synthetase (GS) can utilize NH_4^+ to synthesize glutamine, and glutamate dehydrogenase (GDH) can use NH_4^+ to synthesize glutamate. Thus, crustaceans detoxify ammonia nitrogen by synthesizing non-toxic amino acids such as glutamine and glutamate [11]. GS is an important enzyme in ammonia nitrogen metabolism in crustaceans and is widely found in plants and animals [12]. GDH is a mitochondrial enzyme widely present in plants, animals, and microorganisms. It is a key enzyme in amino acid catabolism, distributed in tissues such as the liver, kidney, and brain in animals, where it exhibits a high activity [13]. GDH is the only enzyme capable of using both NAD^+ and $NADP^+$ as redox cofactors [14]. Studies have shown that when GDH uses $NADP^+$ as a cofactor, it primarily catalyzes glutamate synthesis, whereas with NAD^+, it catalyzes the oxidative deamination of glutamate [15].

Under normal conditions, the main metabolite of amino acids in crustaceans is ammonia nitrogen, with only a small amount as urea and uric acid [16]. However, studies have shown that when the external environment (e.g., salinity, pH, or ammonia nitrogen) changed, urea and nitrite excretion significantly increased in shrimp [17]. Aspartate transaminase (GOT/AST) can catalyze the reaction between glutamate and oxaloacetate to produce aspartate, which contributes to ammonia production through transamination [16]. In the mammalian mitochondria, aspartate serves as a direct source of amino groups during urea formation [18]. Thus, GOT is closely related to ammonia production and excretion. Xanthine oxidoreductase (XOD) is a member of the molybdenum-containing enzyme family and is widely distributed in the various tissues of animals. It primarily participates in the physiological processes of hypoxanthine metabolism to xanthine and xanthine oxidation to uric acid in mammals [19]. XOD exists in two main forms: xanthine dehydrogenase (XDH) and xanthine oxidase (XO) [20]. These enzymes are the key components in the purine metabolism pathway, where uric acid is the final metabolic product, and reactive oxygen species (ROS) are metabolic byproducts [21]. Adenosine deaminase (ADA) can catalyze the irreversible deamination of adenosine to inosine, and the final metabolite is uric acid [22].

Ammonia stress can also increase the concentration of reactive oxygen species (ROS) in the organism and thus lead to oxidative stress in organisms [23]. The excessive production of ROS can lead to the severe impairment of cellular function. Therefore, the organism has evolved antioxidant systems to maintain the intracellular redox homeostasis as well as to maintain the normal cellular function [24]. Among them, superoxide dismutase (SOD) can catalyze the conversion of harmful superoxide radicals and oxygen radicals into less toxic hydrogen peroxide (H_2O_2), while catalase (CAT) can catalyze the conversion of toxic H_2O_2 into nontoxic water, and they play important roles in the protection of living organisms as important components of biological antioxidant systems [19]. In addition, programmed cell death plays important roles in embryonic development, organ maintenance, aging, and the coordination of immune responses and autoimmunity, and apoptosis is one of the most well-understood types [25]. Cells undergoing apoptosis not only encourage phagocytosis by phagocytes but also actively reduce inflammation and aid in the tissue repair within the surrounding area [25].

Black tiger shrimp (*Penaeus monodon*) is the largest species within the genus Penaeus, primarily distributed along the Indo-West Pacific coasts. As one of the three major cultured shrimp species globally, black tiger shrimp is a key marine aquaculture species in the South China Sea and Southeast Asia, holding significant economic value. In culture conditions, due to the intensification and high-density farming of black tiger shrimp, ammonia nitrogen has become one of the primary pollutants in the aquaculture water environment, which can directly damage the immune system of black tiger shrimp, indirectly increasing its susceptibility to pathogens. Acute ammonia nitrogen stress can induce oxidative stress, apoptosis, and purine metabolism detoxification pathways in black tiger shrimp [9]. However, the metabolic response of black tiger shrimp to acute ammonia nitrogen stress remains largely unknown. We hypothesized that black tiger shrimp can convert excessive ammonia to harmless substances for detoxification through the ammonia-metabolism pathway and purine-metabolism pathway. Meanwhile, shrimp can also modulate the antioxidants and apoptosis to reduce the damage caused by ammonia nitrogen. In this study, we investigated the products of ammonia nitrogen metabolism, the activity of enzymes related to ammonia nitrogen and the purine-metabolism pathway, as well as antioxidant and apoptosis-related enzymes after ammonia nitrogen stress, which will enhance the understanding of the metabolism response of black tiger shrimp to ammonia stress and facilitate the process of breeding new strains of ammonia-tolerant shrimp.

2. Materials and Methods

2.1. Experimental Animals

Healthy juvenile black tiger shrimp were obtained from the experimental base of the South China Sea Fisheries Research Institute (Shenzhen, China). Shrimp (average body length: 11 ± 0.5 cm, body weight: 21 ± 1 g) were selected and acclimatized for 2 weeks in black tanks containing 500 L of filtered seawater (30 ± 1‰, temperature 28 ± 1 °C, pH 7.6 ± 0.3, and dissolved oxygen ≥ 6.8 mg/L). Shrimp were fed daily with commercial pellets (Dongteng, Guangzhou, China) at a rate of 1% body weight per day. Half of the seawater was replaced daily to maintain the water quality. Shrimp were anesthetized with an overdose of eugenol before being sacrificed. All efforts were made to minimize the suffering of the animals.

2.2. 96 h Acute Ammonia Nitrogen Stress

The 96 h median lethal concentration (LC_{50}) of ammonia nitrogen for black tiger shrimp was determined through a preliminary experiment according to previous studies, with minor modification [26]. The LC_{50} was calculated using linear interpolation and determined to be 65.0 mg/L ammonium chloride (NH_4Cl) (Pubo, Guangzhou, China) (Figure S1). In the acute ammonia nitrogen stress experiment, shrimp were challenged with 65.0 mg/L ammonium chloride, with three replicates (50 shrimp per replicate). Shrimp without ammonium chloride treatment were set as the control group, with three replicates (50 shrimp per replicate). Samples were collected at 0, 6, 12, 24, 48, 72, and 96 h after ammonia nitrogen stress. Three shrimp were randomly selected from each replicate for mixed sampling, and a total of nine shrimp were used for analysis at each time point. The hemolymph was drawn using a 0.2 mL disposable syringe pre-washed with anticoagulant. Then, the hemolymph samples were centrifuged at 3000× rpm for 10 min at 4 °C to separate the plasma. The plasma samples were stored at −80 °C before their analysis. Mixed samples of hepatopancreas tissue were collected from three shrimp per time point and stored in liquid nitrogen immediately before enzyme activity analysis.

2.3. Histology Analysis

The hepatopancreas, gill, and intestine tissues of black tiger shrimp from the control group and the ammonia nitrogen stress group (96 h after stress) were dissected and fixed in 4% neutral buffered formalin overnight at 4 °C for the preparation of paraffin sections. To study the effect of ammonia stress on the tissue of black tiger shrimp, the embedded sections were stained with hematoxylin and eosin (HE) to visualize the morphological structure. Images were acquired at a 200× magnification and analyzed using the Nikon DS-U3 imaging system. The histopathological severity score was developed based on previous studies in the liver, gills, and intestine, with minor modification [27–29]. The following parameters were scored individually and added to generate the final pathology score: separation of the basement membrane from epithelial cells, lumen dilatation, filaments' level of swelling, hemocyte infiltration, mucosa exfoliation, shortened villi, and the percentage involvement of the tissue. The scoring method is presented in Table S1.

2.4. Measurement of Ammonia, Urea, and Uric Acid in the Plasma

Plasma samples from each time point were used to measure the metabolic products of black tiger shrimp under ammonia nitrogen stress. The Nanjing Jiancheng Bioengineering Institute (Nanjing, China) kits (Blood Ammonia assay kit (A086-1-1), Urea Assay Kit (C013-2-1), and Uric acid (UA) Test Kit (C012-2-1)) were employed to detect the concentration of ammonia, urea, and uric acid in the plasma samples, respectively, following the manufacturer's instructions. The concentration of ammonia in the plasma was measured through the Berthelot reaction. Ammonium was oxidized to chloramine and phenol, and then they were catalyzed to form blue indophenol [30]. The concentration of urea in the plasma can be hydrolyzed by urease to produce ammonium and carbon dioxide, and then, ammonium can be detected [31]. The uric acid in the plasma was oxidized to allantoin and hydrogen peroxide, and then the hydrogen peroxide can be detected [32].

2.5. Measurement of the Enzyme Activity in the Hepatopancreas

Hepatopancreas samples from each time point were used to measure the activity of ammonia nitrogen metabolism-related enzymes (including GS, GDH, GOT/AST, ADA, and XOD), the antioxidant enzymes, superoxide dismutase and catalase (SOD and CAT), and the apoptosis-related enzymes (caspase 3 and caspase 8). The same weight of hepatopancreas tissues was used to detect the protein concentration of each sample by using the Bradford method. Briefly, the hepatopancreas was homogenized before centrifugation, and then, the supernatant was mixed with Coomassie Brilliant Blue to detect the absorbance. The Nanjing Jiancheng Bioengineering Institute kits (glutamine synthetase assay kit; glutamate dehydrogenase assay kit; aspartate aminotransferase assay kit; xanthine oxidase assay kit; adenosine deaminase assay kit; superoxide dismutase assay kit; catalase assay kit; caspase 3 activity assay kit; caspase 8 activity assay kit) were employed to detect the enzymes' activity, respectively, according to the manufacturer's instructions.

2.6. Statistical Analysis

An independent t test was used to analyze the differences in the histopathological severity score between the groups. The data of the metabolic products and the enzyme activities were analyzed using one-way ANOVA followed by Tukey's multiple comparison test in SPSS 23.0 and presented as the mean $\pm$ the standard error of the mean (mean $\pm$ SEM). Figures were made using GraphPad Prism 8 software. It was considered statistically significant when $p < 0.05$.

3. Results

3.1. Effects of Ammonia Nitrogen Stress on Shrimp Tissues

To study the effect of ammonia stress on the tissues of black tiger shrimp, the paraffin sections of hepatopancreas, gill, and intestine tissues from the control group and the ammonia nitrogen stress group (96 h after stress) were stained with hematoxylin and eosin and analyzed. In the hepatopancreas, we observed the severe separation of the basement membrane from epithelial cells and lumen dilatation after acute ammonia nitrogen stress (Figure 1A, left). In the gills, the gill filaments were swollen and hemocytes infiltrated the gill tissue after ammonia nitrogen stress (Figure 1A, middle). In the intestine, mucosa showed exfoliation from the epithelial cells, and the villi became shortened after ammonia nitrogen stress (Figure 1A, right). We observed significant changes in the histopathological severity score in the hepatopancreas, gills, and intestine from the control shrimp and shrimp after 96 h of acute ammonia nitrogen stress (Figure 1B). These results indicated that ammonia nitrogen stress induced severe histological damage in the black tiger shrimp.

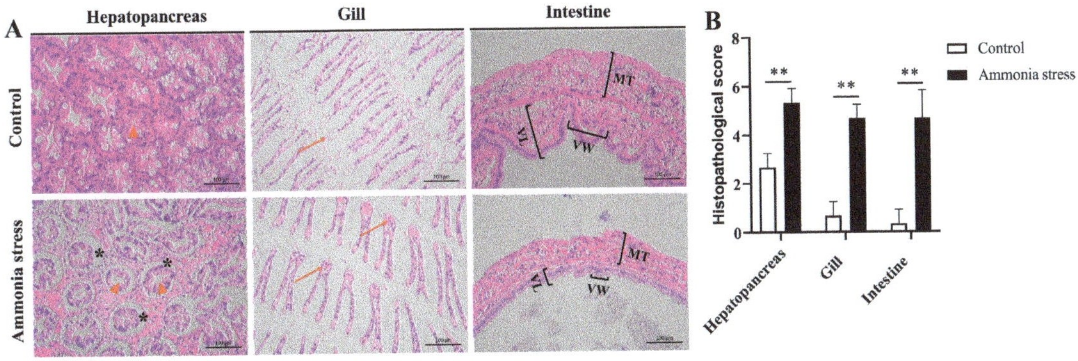

Figure 1. Histological changes in the hepatopancreas, gills, and intestine of black tiger shrimp under acute ammonia stress. (**A**) Images of hematoxylin and eosin staining on black tiger shrimp hepatopancreas (**left**), gill (**middle**), and intestine (**right**) paraffin sections from the control shrimp (**upper**) and ammonia nitrogen stress shrimp (**lower**) (n = 3 per group). Black asterisk indicates severe separation of the basement membrane from epithelial cells. The red triangle indicates lumen dilatation. Red arrows indicate infiltrated hemocytes in the gills. Mucosal thickness (MT), villi length (VL), and villi width (VW). Magnification, 200×. Scale bar, 100 μm. (**B**) Histopathological severity score of hepatopancreas, gills, and intestine in control shrimp and shrimp after 96 h of acute ammonia nitrogen stress (n = 3 per group). ** $p < 0.01$.

3.2. Effects of Ammonia Nitrogen Stress on the Metabolic Product in the Plasma

In order to study the metabolic responses of shrimp to ammonia stress, we detected the concentration of ammonia, urea nitrogen, and uric acid in the plasma of black tiger shrimp after acute ammonia nitrogen stress. The results showed that both the concentration of ammonia and urea nitrogen significantly increased in the plasma after ammonia stress, while the levels of uric acid did not change (Figure 2). In detail, under acute ammonia nitrogen stress, the concentration of ammonia in the plasma rose rapidly, with a significant increase at 6 h ($p < 0.05$). The concentration of ammonia continued to rise, peaking at 48 h (4.13 mmol/L), which was 3.58 times higher than the 0 h control (Figure 2A). Although it declined afterward, the concentration at 96 h remained significantly higher than the control ($p < 0.05$). The concentration of urea nitrogen in the plasma followed a trend similar to ammonia (Figure 2B), with a significant increase after stress. The concentration at 6 h was significantly higher than the control ($p < 0.05$) and peaked at 48 h (39 mmol/L), which was 24.88 times higher than the control level. It subsequently decreased but remained significantly higher than the control at 96 h ($p < 0.05$). In contrast, the uric acid concentration in the plasma showed no significant differences at all the examined time points after ammonia stress (Figure 2C). These results indicated that black tiger shrimp could induce the conversion of excessive ammonia to urea rather than uric acid for ammonia detoxification under acute ammonia nitrogen stress.

3.3. Effect of Ammonia Nitrogen Stress on the Activity of Metabolic-Related Enzymes

To further explore the function of ammonia nitrogen and purine metabolism-related enzymes in the metabolic response to acute ammonia stress, we conducted enzyme activity analysis to reveal the change in the activity of GS, GDH, GOT/AST, XOD, and ADA after ammonia stress. The protein concentration of the hepatopancreas after ammonia nitrogen stress is shown in Figure 3A. The activities of ammonia-related metabolic enzymes exhibited the following patterns: The activity of glutamine synthetase (GS) significantly decreased at 96 h ($p < 0.05$) (Figure 3B). Glutamate dehydrogenase (GDH) activity was significantly elevated at 24 h ($p < 0.01$), with no significant difference observed at 96 h

(Figure 3C). The activity of aspartate aminotransferase (GOT/AST) significantly increased ($p < 0.01$) (Figure 3D). The xanthine oxidase (XOD) activity was significantly higher than the control at 12 h ($p < 0.01$), then decreased, showing no significant difference from the control at 96 h (Figure 3E). The adenosine deaminase (ADA) activity was markedly reduced at 96 h ($p < 0.01$), showing a significant difference compared to the 0 h control (Figure 3F). The results of the increased activity of GOT supported the hypothesis that shrimp could convert ammonia to urea for detoxification. The change in GS and GDH activity also indicated that shrimp might convert ammonia to glutamine and glutamate for detoxification, warranting further research into their detoxification roles.

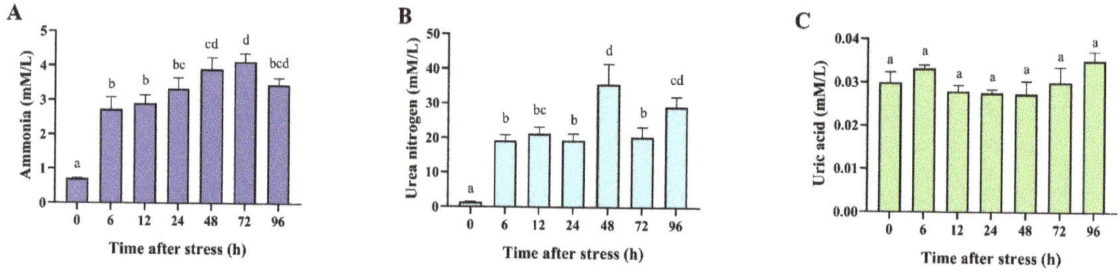

Figure 2. The metabolic products in the hemolymph of black tiger shrimp after 96 h of acute ammonia nitrogen stress. (**A**) The histogram represents the concentration of ammonia in the plasma of black tiger shrimp. (**B**) The histogram represents the concentration of urea nitrogen in the plasma of black tiger shrimp. (**C**) The histogram represents the concentration of uric acid in the plasma of black tiger shrimp. Each bar represents the mean ± standard error of the mean (mean ± SEM) ($n = 9$). Data with different letters are significantly different among groups ($p < 0.05$).

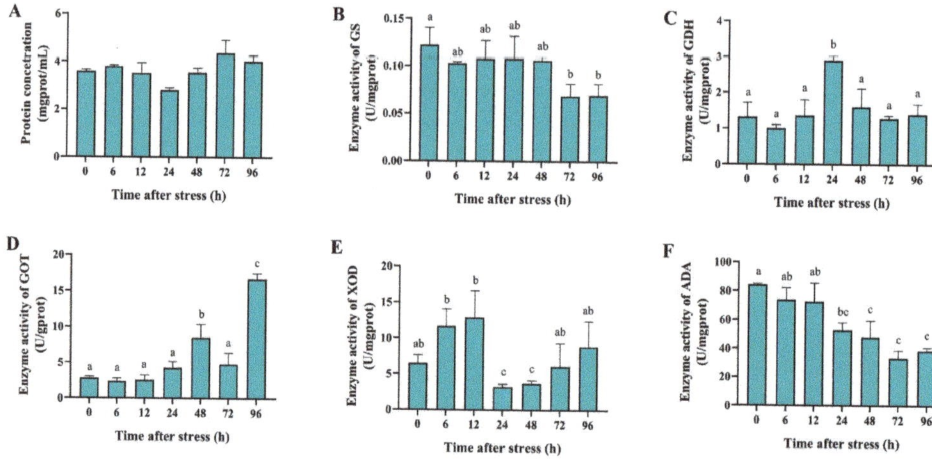

Figure 3. The enzyme activity of the ammonia nitrogen and purine metabolism-related enzymes in the hepatopancreas of black tiger shrimp after 96 h of acute ammonia nitrogen stress. (**A**) The histogram represents the protein concentration in the hepatopancreas of black tiger shrimp. (**B**) The histogram represents the enzyme activity of GS in the hepatopancreas of black tiger shrimp. (**C**) The histogram represents the enzyme activity of GDH in the hepatopancreas of black tiger shrimp. (**D**) The histogram represents the enzyme activity of GOT in the hepatopancreas of black tiger shrimp. (**E**) The histogram represents the enzyme activity of XOD in the hepatopancreas of black tiger shrimp. (**F**) The histogram represents the enzyme activity of ADA in the hepatopancreas of black tiger shrimp. Each bar represents the mean ± standard error of the mean (mean ± SEM) ($n = 9$). Data with different letters are significantly different among groups ($p < 0.05$).

3.4. Effect of Ammonia Nitrogen Stress on the Activity of Antioxidant and Apoptosis-Related Enzymes

To reveal whether the antioxidant enzyme plays a role in the metabolic response to acute ammonia nitrogen stress, we further conducted the enzyme activity analysis of SOD and CAT in the hepatopancreas of black tiger shrimp after ammonia stress. The activities of the antioxidant enzymes, superoxide dismutase (SOD) and catalase (CAT), showed the following changes: SOD activity significantly increased at 48 and 96 h ($p < 0.01$) (Figure 4A). The CAT activity increased at 24 h but was not significantly different compared with that of the control group (Figure 4B). The results indicated that shrimp could activate SOD to prevent cellular oxidative damage due to ROS induced by high concentrations of ammonia nitrogen.

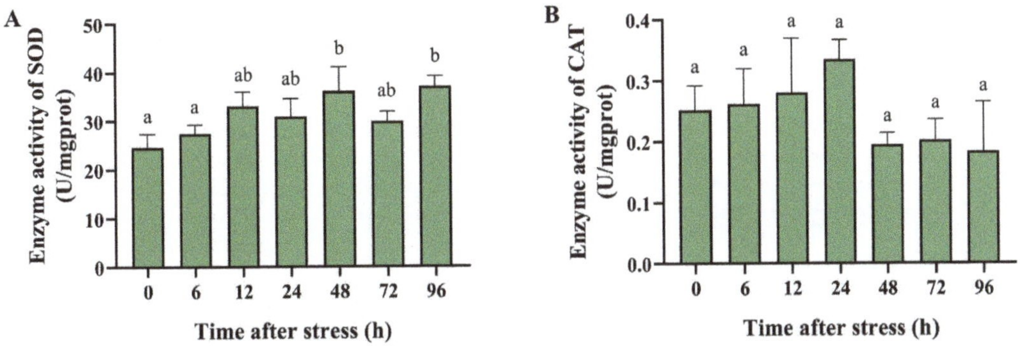

Figure 4. The enzyme activity of the antioxidant-related enzymes in the hepatopancreas of black tiger shrimp after 96 h of acute ammonia nitrogen stress. (**A**) The histogram represents the enzyme activity of SOD in the hepatopancreas of black tiger shrimp. (**B**) The histogram represents the enzyme activity of CAT in the hepatopancreas of black tiger shrimp. Each bar represents the mean ± standard error of the mean (mean ± SEM) ($n = 9$). Data with different letters are significantly different among groups ($p < 0.05$).

Previous studies found that ammonia nitrogen stress induced apoptosis in crustaceans [33]. We thus investigated the change in activity of the apoptosis-related enzymes caspase 3 and caspase 8 in the hepatopancreas of black tiger shrimp after ammonia stress. After high-concentration ammonia nitrogen stress, the activities of the apoptosis-related enzymes, caspase 3 and caspase 8, exhibited similar trends (Figure 5). The enzyme activities decreased immediately after the stress onset, increased at 12 h but not significantly differently from the control, and then showed a significant decline. At 96 h, their activities were significantly lower than those of the control ($p < 0.01$). The results indicated that ammonia stress suppressed the enzyme activity of the caspase family, and shrimp might modulate caspase-dependent apoptosis to reduce the damage caused by high concentrations of ammonia nitrogen.

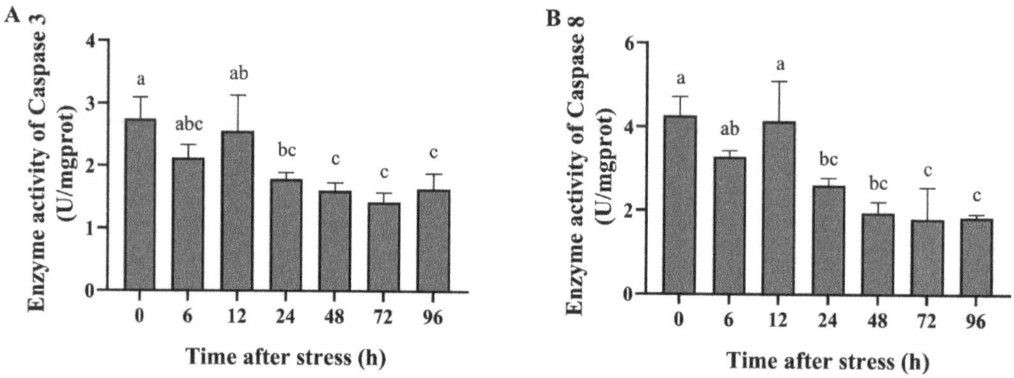

Figure 5. The enzyme activity of the apoptosis-related enzymes in the hepatopancreas of black tiger shrimp after 96 h of acute ammonia nitrogen stress. (**A**) The histogram represents the enzyme activity of caspase 3 in the hepatopancreas of black tiger shrimp. (**B**) The histogram represents the enzyme activity of caspase 8 in the hepatopancreas of black tiger shrimp. Each bar represents the mean ± standard error of the mean (mean ± SEM) ($n = 9$). Data with different letters are significantly different among groups ($p < 0.05$).

4. Discussion

In crustaceans, 60–70% of nitrogen (N) is excreted as ammonia, with smaller amounts excreted as urea and uric acid [16]. To investigate the detoxification mechanisms of black tiger shrimp, this study measured the metabolic product and the metabolic-related enzyme activity following high-concentration ammonia stress. The results showed that the concentrations of ammonia and urea nitrogen in the hemolymph significantly increased, while the uric acid levels showed no significant changes, suggesting that black tiger shrimp might convert excess ammonia into urea for detoxification, possibly through the urea–ornithine pathway. In the swimming crab (*Portunus trituberculatus*), significant increases in ammonia and urea were also observed in the hemolymph under ammonia nitrogen stress [11]. Importantly, the key enzymes in the ammonia nitrogen metabolic pathway, such as GS, GDH, and GOT/AST, play critical roles in ammonia detoxification in crustaceans [34–36]. In this study, the enzyme activity of GDH and GOT significantly increased, while the GS activity significantly decreased under ammonia stress. GDH can facilitate the conversion of NH_4^+ into non-toxic glutamate, while GOT transfers an amino group from glutamate to oxaloacetate, forming aspartate, a direct amino donor for urea synthesis [37]. The increased enzyme activity of GOT is in accordance with the observed increase in urea levels, indicating that black tiger shrimp may conduct ammonia detoxification through amino acid metabolic pathways. The change in GS and GDH activity suggested that black tiger shrimp may convert the excessive ammonia to glutamine and glutamate. Xanthine oxidase (XOD), a key enzyme in the purine metabolism pathway, catalyzes the conversion of hypoxanthine to xanthine and subsequently to uric acid [38]. In the land crab (*Gecarcoidea natalis*), excessive ammonia is converted to uric acid for detoxification [14]. Though the uric acid levels did not significantly change, the XOD activity significantly increased at 12 h after stress, suggesting a possible role for XOD in ammonia detoxification via purine metabolism. Conversely, ADA activity showed a significant decrease in this study. ADA can catalyze the conversion of adenosine into inosine; then, the inosine can produce hypoxanthine, which leads to the production of uric acid [39,40]. The results suggested that the unchanged levels of uric acid in the plasma after ammonia stress may be due to the suppression of the ADA activity of black tiger shrimp. In *Procambarus clarkii*, the ADA activity was significantly down-regulated to lower the purine metabolism after ammonia stress [41]. Moreover, uricase can oxidize uric acid, converting it into allantoin. Allantoin is subsequently hydrolyzed by allantoinase to form allantoic acid, which is ultimately converted into ammonia [42]. The suppression of ADA activity may also inhibit the excessive ammonia production in the body of black tiger shrimp.

The hepatopancreas is the main digestive and metabolic organ in shrimp, and thus plays a major role in the ammonia nitrogen stress response [43]. The gills and intestine are the external and internal borders of shrimp, which are in direct contact with the environment and thus the main target tissues of ammonia nitrogen [44]. In this study, acute ammonia nitrogen stress induced severe histopathological damage in the hepatopancreas, gills and intestine in black tiger shrimp, suggesting the impairment of the normal physiological function in these tissues. In *Litopenaeus vannamei*, ammonia exposure induced histopathological damage to the hepatopancreas and gills [45]. In addition, ammonia nitrogen stress could also induce tissue damage of the intestine mucosa and suppress the immune function of the intestine [46,47].

Ammonia nitrogen stress induces oxidative stress through increased reactive oxygen species (ROS) production [23]. For hosts, SOD and CAT play crucial roles in mitigating oxidative damage by converting ROS into less harmful molecules [24,48]. Thus, the activity of SOD and CAT is measured to indicate the cellular oxidant/antioxidant balance since they play important roles when protecting against cellular free radical damage [49]. The results showed that the enzyme activity of SOD significantly increased, suggesting that SOD may protect black tiger shrimp against oxidative damage caused by ammonia nitrogen stress through the clearance of ROS. Conversely, in *Litopenaeus vannamei* and *Marsupenaeus japonicus*, the activity of SOD and CAT was significantly decreased after ammonia exposure, while the malondialdehyde (MDA) levels significantly increased [45,50]. In *Ctenopharyngodon idellus*, the activity of SOD and CAT also significantly decreased when exposed to high environmental ammonia [51]. The increase in the activity of SOD but not CAT indicated that the oxidative stress produced by ROS was partially eliminated, but there may be residual and increased MDA levels. Interestingly, in *Takifugu obscurus*, the activity of SOD and CAT was significantly elevated after ammonia nitrogen stress [52]. So, we speculated that ammonia stress could induce oxidative stress in aquatic animals and influence the enzyme activity of antioxidants, but the responses of different animals may differ due to their different tolerance to ammonia and the different ways of treatment of ammonia in various studies.

Apoptosis plays a critical role in maintaining the homeostasis and defending against external and internal injuries in multicellular organisms. Apoptosis can be broadly categorized into caspase-independent apoptosis and caspase-dependent apoptosis [25]. Alterations in the activity of the caspase family of enzymes can be used as a marker for the initiation of apoptosis programs [52]. A recent study showed that the expression of caspase 2, which is involved in the initiation of apoptosis, and caspase 7, which is downstream of the apoptotic cascade, were significantly down-regulated in the black tiger shrimp under ammonia nitrogen stress [9]. Among the caspase family members, caspase 8 and caspase 3 are the most important initiator and executioner, respectively [25]. Therefore, in this study, we investigated the activity changes in caspase 8 and caspase 3 after ammonia nitrogen stress. The results showed that after a high concentration of ammonia nitrogen stress, the activities of both caspase 8 and caspase 3 significantly decreased. In *Litopenaeus vannamei*, ammonia exposure induces apoptosis in the hepatopancreas [53]. In *Ctenopharyngodon idellus*, the c-Myc-Bax-Caspase 9 apoptosis pathway was activated when exposed to high levels of ammonia [51]. The results suggested that ammonia stress suppressed the enzyme activity of the caspase family, and black tiger shrimp might modulate apoptosis to protect themselves under ammonia nitrogen stress.

5. Conclusions

In summary, acute ammonia nitrogen stress resulted in significant increases in ammonia and urea in the hemolymph, while uric acid remained unchanged in black tiger shrimp, suggesting that shrimp might convert excess ammonia into urea for ammonia detoxification. Importantly, the key metabolic and purine pathway enzymes, including GS, GDH, and GOT, showed notable activity changes, suggesting the vital roles of these enzymes in the metabolism of excessive ammonia. Moreover, the antioxidant enzyme SOD activity was significantly upregulated and the activities of the apoptosis-related proteins, caspase 8 and caspase 3, were significantly altered, indicating the responses involving anti-oxidative reaction and apoptosis. Overall, our findings revealed that black tiger shrimp undergo oxidative stress and apoptosis and employ metabolic and purine pathways for ammonia detoxification under ammonia nitrogen stress, thus providing new insights into the metabolic response of shrimp to acute ammonia stress.

Supplementary Materials: The following supporting information can be downloaded at: https://www.mdpi.com/article/10.3390/biology14050501/s1, Figure S1: The regression equation of the mortality and the concentration of NH_4Cl in the 96 h acute ammonia stress experiment; Table S1: Histopathological severity score.

Author Contributions: Conceptualization, Y.D.; methodology, S.J. (Shigui Jiang); software, S.J. (Song Jiang); validation, Y.L., Q.Y. and J.S.; formal analysis, L.Y.; investigation, Y.D.; resources, J.H.; data curation, P.L.; writing—original draft preparation, Y.D.; writing—review and editing, F.Z.; visualization, H.D.; supervision, F.Z.; project administration, F.Z.; funding acquisition, F.Z. All authors have read and agreed to the published version of the manuscript.

Funding: This work was supported by the National Key Research and Development Program of China (2024YFD2401200, 2024YFD2401800), the China Agriculture Research System of MOF and MARA (CARS-48), 2024 Provincial Rural Revitalization Strategy Special Fund Seed Industry Revitalization Project (2024-SPY-00-005), Central Public-interest Scientific Institution Basal Research Fund, CAFS (NO. 2023TD34), and the Central Public-interest Scientific Institution Basal Research Fund, South China Sea Fisheries Research Institute, CAFS (NO. 2024TS02).

Institutional Review Board Statement: All the experiments in this study were approved by the Animal Care and Use Committee of the South China Sea Fisheries Research Institute, Chinese Academy of Fishery Sciences (nhdf2023-18), and the collection and handling of the experimental animals were performed according to the regulations and guidelines established by this committee.

Informed Consent Statement: Not applicable.

Data Availability Statement: Data will be made available on request.

Conflicts of Interest: The authors declare no conflicts of interest.

References

1. Mengal, K.; Kor, G.; Kozák, P.; Niksirat, H. Effects of environmental factors on the cellular and molecular parameters of the immune system in decapods. *Comp. Biochem. Physiol. A Mol. Integr. Physiol.* **2023**, *276*, 111332. [CrossRef] [PubMed]
2. Romano, N.; Zeng, C. Toxic Effects of Ammonia, Nitrite, and Nitrate to Decapod Crustaceans: A Review on Factors Influencing their Toxicity, Physiological Consequences, and Coping Mechanisms. *Rev. Fish. Sci.* **2013**, *21*, 1–21. [CrossRef]
3. Weihrauch, D.; Morris, S.; Towle, D.W. Ammonia excretion in aquatic and terrestrial crabs. *J. Exp. Biol.* **2004**, *207*, 4491–4504. [CrossRef]
4. Chen, J.C.; Lin, C.Y. Lethal effects of ammonia on *Penaeus chinensis* Osbeck juveniles at different salinity levels. *J. Exp. Mar. Biol. Ecol.* **1992**, *156*, 139–148. [CrossRef]
5. Noor-Hamid, S.; Fortes, R.D.; Parado-Estepa, F. Effect of pH and ammonia on survival and growth of the early larval stages of *Penaeus monodon* Fabricius. *Aquaculture* **1994**, *125*, 67–72. [CrossRef]
6. Chen, J.-C.; Chen, C.-T. Changes of osmotic and electrolyte concentrations in the haemolymph of *Penaeus japonicus* exposed to ambient ammonia. *Comp. Biochem. Physiol. Part C Pharmacol. Toxicol. Endocrinol.* **1996**, *114*, 35–38. [CrossRef]

7. Chen, J.C.; Lin, C.Y. Oxygen consumption and ammonia-N excretion of *Penaeus chinensis* juveniles exposed to ambient ammonia at different salinity levels. *Comp. Biochem. Physiol. Part C Comp. Pharmacol. Toxicol.* **1992**, *102*, 287–291. [CrossRef]
8. Cheng, S.Y.; Chen, J.C. Hemocyanin oxygen affinity, and the fractionation of oxyhemocyanin and deoxyhemocyanin for *Penaeus monodon* exposed to elevated nitrite. *Aquat. Toxicol.* **1999**, *45*, 35–46. [CrossRef]
9. Li, Y.; Zhou, F.; Huang, J.; Yang, L.; Jiang, S.; Yang, Q.; He, J.; Jiang, S. Transcriptome reveals involvement of immune defense, oxidative imbalance, and apoptosis in ammonia-stress response of the black tiger shrimp (*Penaeus monodon*). *Fish Shellfish Immunol.* **2018**, *83*, 162–170. [CrossRef]
10. Liu, S.; Pan, L.; Liu, M.; Yang, L. Effects of ammonia exposure on nitrogen metabolism in gills and hemolymph of the swimming crab *Portunus trituberculatus*. *Aquaculture* **2014**, *432*, 351–359. [CrossRef]
11. Ren, Q.; Pan, L.; Zhao, Q.; Si, L. Ammonia and urea excretion in the swimming crab *Portunus trituberculatus* exposed to elevated ambient ammonia-N. *Comp. Biochem. Physiol. Part A Mol. Integr. Physiol.* **2015**, *187*, 48–54. [CrossRef] [PubMed]
12. Dong, X.; Liu, Q.; Kan, D.; Zhao, W.; Guo, H.; Lv, L. Effects of ammonia-N exposure on the growth, metabolizing enzymes, and metabolome of *Macrobrachium rosenbergii*. *Ecotoxicol. Environ. Saf.* **2020**, *189*, 110046. [CrossRef]
13. Li, C.; Zhao, T.; Ren, L.; Cui, D.; Zhan, Y.; Chang, Y. Characterization of a novel glutamate dehydrogenase gene and its response to heat stress in the sea urchin *Strongylocentrotus intermedius*. *Aquac. Rep.* **2023**, *28*, 101446. [CrossRef]
14. Linton, S.; Greenaway, P. Urate deposits in the gecarcinid land crab *Gecarcoidea natalis* are synthesised de novo from excess dietary nitrogen. *J. Exp. Biol.* **1997**, *200*, 2347–2354. [CrossRef] [PubMed]
15. Pahlich, E.; Joy, K.W. Glutamate Dehydrogenase from Pea Roots: Purification and Properties of the Enzyme. *Can. J. Biochem.* **1971**, *49*, 127. [CrossRef]
16. Michele, R. Nitrogen excretion in marine and fresh-water crustacea. *Biol. Rev.* **1987**, *62*, 1–24.
17. Chen, J.C.; Lin, C.Y. Responses of oxygen consumption, Ammonia-N excretion and Urea-N excretion of *Penaeus chinensis* exposed to ambient ammonia at different salinity and pH levels. *Aquaculture* **1995**, *136*, 243–255. [CrossRef]
18. Lee, W.C.; Chen, J.C. Nitrogenous excretion and arginase specific activity of kuruma shrimp *Marsupenaeus japonicus* exposed to elevated ambient nitrite. *J. Exp. Mar. Biol. Ecol.* **2004**, *308*, 103–111. [CrossRef]
19. Zhang, X.; Pan, L.; Wei, C.; Tong, R.; Li, Y.; Ding, M.; Wang, H. Crustacean hyperglycemic hormone (CHH) regulates the ammonia excretion and metabolism in white shrimp, *Litopenaeus vannamei* under ammonia-N stress. *Sci. Total Environ.* **2020**, *723*, 138128. [CrossRef]
20. Sekine, M.; Okamoto, K.; Pai, E.F.; Nagata, K.; Ichida, K.; Hille, R.; Nishino, T. Allopurinol and oxypurinol differ in their strength and mechanisms of inhibition of xanthine oxidoreductase. *J. Biol. Chem.* **2023**, *299*, 105189. [CrossRef]
21. Cancio, I.; Cajaraville, M.P. Seasonal variation of xanthine oxidoreductase activity in the digestive gland cells of the mussel *Mytilus galloprovincialis*: A biochemical, histochemical and immunochemical study. *Biol. Cell* **1999**, *91*, 605–615. [CrossRef] [PubMed]
22. Ding, Z.; Ge, W.; Xu, X.; Xu, X.; Sun, Q.; Xu, X.; Zhang, J. A crucial role of adenosine deaminase in regulating gluconeogenesis in mice. *J. Biol. Chem.* **2024**, *300*, 107425. [CrossRef]
23. Murthy, C.R.; Rama Rao, K.V.; Bai, G.; Norenberg, M.D. Ammonia-induced production of free radicals in primary cultures of rat astrocytes. *J. Neurosci. Res.* **2001**, *66*, 282–288. [CrossRef] [PubMed]
24. Halliwell, B. Antioxidant defence mechanisms: From the beginning to the end (of the beginning). *Free Radic. Res.* **1999**, *31*, 261–272. [CrossRef]
25. Bertheloot, D.; Latz, E.; Franklin, B.S. Necroptosis, pyroptosis and apoptosis: An intricate game of cell death. *Cell Mol. Immunol.* **2021**, *18*, 1106–1121. [CrossRef] [PubMed]
26. Shi, M.; Jiang, S.; Li, Y.; Yang, Q.; Jiang, S.; Yang, L.; Huang, J.; Zhou, F. Comprehensive expression analysis of the beta integrin from *Penaeus monodon* indicating its participation in innate immunity and ammonia nitrogen stress response. *Fish Shellfish Immunol.* **2020**, *98*, 887–898. [CrossRef]
27. Goodman, Z.D. Grading and staging systems for inflammation and fibrosis in chronic liver diseases. *J. Hepatol.* **2007**, *47*, 598–607. [CrossRef]
28. Zilberg, D.; Munday, B.L. Pathology of experimental amoebic gill disease in Atlantic salmon, *Salmo salar* L., and the effect of pre-maintenance of fish in sea water on the infection. *J. Fish Dis.* **2010**, *23*, 401–407. [CrossRef]
29. Ringø, E.; Salinas, I.; Olsen, R.E.; Nyhaug, A.; Myklebust, R.; Mayhew, T.M. Histological changes in intestine of Atlantic salmon (*Salmo salar* L.) following in vitro exposure to pathogenic and probiotic bacterial strains. *Cell Tissue Res.* **2007**, *328*, 109–116. [CrossRef]
30. Alwael, H.; Alharthi, A.S.; Dabi, M.M.; Oubaha, M.; El-Shahawi, M.S. A highly sensitive electrochemical sensing probe incorporating classical Berthelot's reaction and glassy carbon electrode for measuring ultra-trace levels of ammonia/ NH_4^+ in water. *Electrochem. Commun.* **2024**, *162*, 107686. [CrossRef]

31. Melara, F.; da Silva, L.K.; Mandelli, N.A.; Krein, D.D.C.; Chiomento, J.L.T.; Dettmer, A.; Piccin, J.S. Effect on N release by urea coating with chitosan, starch and urease inhibitor. *Int. J. Biol. Macromol.* **2025**, *303*, 140603. [CrossRef] [PubMed]
32. Sun, Q.; Miao, S.; Yu, W.; Jiang, E.-Y.; Gong, M.; Liu, G.; Luo, X.; Zhang, M.-Z. Visual detection of uric acid in serum through catalytic oxidation by a novel cellulose membrane biosensor with schiff base immobilized uricase. *Biosens. Bioelectron.* **2025**, *268*, 116912. [CrossRef] [PubMed]
33. Tong, R.; Li, Y.; Yu, X.; Zhang, N.; Liao, Q.; Pan, L. The mechanism of reactive oxygen species generation, DNA damage and apoptosis in hemocytes of *Litopenaeus vannamei* under ammonia nitrogen exposure. *Aquat. Toxicol.* **2024**, *272*, 106958. [CrossRef]
34. Tanguy, A.; Boutet, I.; Moraga, D. Molecular characterization of the glutamine synthetase gene in the Pacific oyster *Crassostrea gigas*: Expression study in response to xenobiotic exposure and developmental stage. *Biochim. Biophys. Acta (BBA)–Gene Struct. Expr.* **2005**, *1681*, 116–125. [CrossRef]
35. Wang, Y.; Li, E.; Yu, N.; Wang, X.; Cai, C.; Tang, B.; Chen, L.; Van Wormhoudt, A. Characterization and expression of glutamate dehydrogenase in response to acute salinity stress in the Chinese mitten crab, *Eriocheir sinensis*. *PLoS ONE* **2012**, *7*, e37316. [CrossRef]
36. Hlordzi, V.; Wang, J.; Kuebutornye, F.K.A.; Yang, X.; Tan, B.; Li, T.; Cui, Z.; Lv, S.; Lao, T.; Chi, S. Hydrolysed fish protein powder is better at the growth performance, hepatopancreas and intestinal development of Pacific white shrimp (*Litopenaeus vannamei*). *Aquac. Rep.* **2022**, *23*, 101025. [CrossRef]
37. Otto-Ślusarczyk, D.; Graboń, W.; Mielczarek-Puta, M. Aspartate aminotransferase--key enzyme in the human systemic metabolism. *Postep. Hig. Med. Dosw.* **2016**, *70*, 219–230. [CrossRef]
38. Fried, R.; Fried, L.W. Xanthine Oxidase (Xanthine Dehydrogenase). In *Methods of Enzymatic Analysis*; Academic Press: Cambridge, MA, USA, 1974; pp. 644–649.
39. Senger, M.R.; Rosemberg, D.B.; Seibt, K.J.; Dias, R.D.; Bogo, M.R.; Bonan, C.D. Influence of mercury chloride on adenosine deaminase activity and gene expression in zebrafish (*Danio rerio*) brain. *NeuroToxicology* **2010**, *31*, 291–296. [CrossRef]
40. Štefanić, Z.; Mikleušević, G.; Luić, M.; Bzowska, A.; Leščić Ašler, I. Structural characterization of purine nucleoside phosphorylase from human pathogen *Helicobacter pylori*. *Int. J. Biol. Macromol.* **2017**, *101*, 518–526. [CrossRef]
41. Wang, L.; Wang, H.; Shi, W.; Zhang, Y.; Chang, G.; Wu, N.; Xue, C.; Li, J. RNA-seq analysis uncovers effects of ammonia on metabolism, oxidant-antioxidant equilibrium and apoptosis in the red swamp crayfish (*Procambarus clarkii*). *Aquac. Rep.* **2020**, *18*, 100459. [CrossRef]
42. Rosemberg, D.B.; Rico, E.P.; Guidoti, M.R.; Dias, R.D.; Souza, D.O.; Bonan, C.D.; Bogo, M.R. Adenosine deaminase-related genes: Molecular identification, tissue expression pattern and truncated alternative splice isoform in adult zebrafish (*Danio rerio*). *Life Sci.* **2007**, *81*, 1526–1534. [CrossRef]
43. Liguo, Q.; Xiang, S.; Simeng, Y.; Qian, H.; Xiaoping, D.; Hailong, Z. Changes of Ammonia-Metabolizing Enzyme Activity and Gene Expression of Two Strains in Shrimp *Litopenaeus vannamei* Under Ammonia Stress. *Front. Physiol.* **2018**, *9*, 211.
44. Zhao, M.; Yao, D.; Li, S.; Zhang, Y.; Aweya, J.J. Effects of ammonia on shrimp physiology and immunity: A review. *Rev. Aquac.* **2020**, *12*, 2194–2211. [CrossRef]
45. Lin, L.; Zhuo, H.; Zhang, Y.; Li, J.; Zhou, X.; Wu, G.; Guo, C.; Liu, J. Effects of ammonia exposure and post-exposure recovery in pacific white shrimp, *Litopenaeus vannamei*: Histological, physiological and molecular responses. *Aquat. Toxicol.* **2024**, *277*, 107133. [CrossRef] [PubMed]
46. Duan, Y.; Liu, Q.; Wang, Y.; Zhang, J.; Xiong, D. Impairment of the intestine barrier function in *Litopenaeus vannamei* exposed to ammonia and nitrite stress. *Fish Shellfish Immunol.* **2018**, *78*, 279–288. [CrossRef]
47. Duan, Y.; Zhong, G.; Nan, Y.; Yang, Y.; Xiao, M.; Li, H. Effects of Nitrite Stress on the Antioxidant, Immunity, Energy Metabolism, and Microbial Community Status in the Intestine of *Litopenaeus vannamei*. *Antioxidants* **2024**, *13*, 1318. [CrossRef] [PubMed]
48. Rosa, A.C.; Corsi, D.; Cavi, N.; Bruni, N.; Dosio, F. Superoxide Dismutase Administration: A Review of Proposed Human Uses. *Molecules* **2021**, *26*, 1844. [CrossRef]
49. Batista, J.J.; Silva, F.C.O.; de Araújo, M.I.F.; de Almeida Moura Nunes, P.H.; de Oliveira Ferreira, S.A.; da Silva, L.A.; de Siqueira Patriota, L.L.; Napoleão, T.H.; Paiva, P.M.G.; de Carvalho, J.M.; et al. Parkia pendula polysaccharides have no acute toxicity and prevent ethanol-induced gastric ulcers via downregulation of TBARS, IL-6, and TNF-α and upregulation of SOD, CAT, and IL-10. *Int. J. Biol. Macromol.* **2025**, *309*, 142702. [CrossRef]
50. Ou, H.; Liang, J.; Liu, J. Effects of acute ammonia exposure on oxidative stress, endoplasmic reticulum stress and apoptosis in the kuruma shrimp (*Marsupenaeus japonicus*). *Aquac. Rep.* **2022**, *27*, 101383. [CrossRef]
51. Jin, J.; Wang, Y.; Wu, Z.; Hergazy, A.; Lan, J.; Zhao, L.; Liu, X.; Chen, N.; Lin, L. Transcriptomic analysis of liver from grass carp (*Ctenopharyngodon idellus*) exposed to high environmental ammonia reveals the activation of antioxidant and apoptosis pathways. *Fish Shellfish Immunol.* **2017**, *63*, 444–451. [CrossRef]

52. Cheng, C.H.; Yang, F.F.; Ling, R.Z.; Liao, S.A.; Miao, Y.T.; Ye, C.X.; Wang, A.L. Effects of ammonia exposure on apoptosis, oxidative stress and immune response in pufferfish (*Takifugu obscurus*). *Aquat. Toxicol.* **2015**, *164*, 61–71. [CrossRef] [PubMed]
53. Liang, Z.; Liu, R.; Zhao, D.; Wang, L.; Sun, M.; Wang, M.; Song, L. Ammonia exposure induces oxidative stress, endoplasmic reticulum stress and apoptosis in hepatopancreas of pacific white shrimp (*Litopenaeus vannamei*). *Fish Shellfish Immunol.* **2016**, *54*, 523–528. [CrossRef] [PubMed]

MDPI AG
Grosspeteranlage 5
4052 Basel
Switzerland
Tel.: +41 61 683 77 34

Biology Editorial Office
E-mail: biology@mdpi.com
www.mdpi.com/journal/biology

www.ingramcontent.com/pod-product-compliance
Lightning Source LLC
Chambersburg PA
CBHW061809210926
51850CB00006B/658
* 9 7 8 3 7 2 5 8 5 2 2 5 3 *